T0190176

Indian Institute of Metals Series

The study of metallurgy and materials science is vital for developing advanced materials for diverse applications. In the last decade, the progress in this field has been rapid and extensive, giving us a new array of materials, with a wide range of applications, and a variety of possibilities for processing and characterizing the materials. In order to make this growing volume of knowledge available, an initiative to publish a series of books in Metallurgy and Materials Science was taken during the Diamond Jubilee year of the Indian Institute of Metals (IIM) in the year 2006. Ten years later the series is now published in partnership with Springer.

This book series publishes different categories of publications: textbooks to satisfy the requirements of students and beginners in the field, monographs on select topics by experts in the field, professional books to cater to the needs of practicing engineers, and proceedings of select international conferences organized by IIM after mandatory peer review. The series publishes across all areas of materials sciences and metallurgy. An eminent panel of international and national experts acts as the advisory body in overseeing the selection of topics, important areas to be covered, and the selection of contributing authors.

More information about this series at http://www.springer.com/series/15453

Hem Shanker Ray · Saradindukumar Ray

Kinetics of Metallurgical Processes

Hem Shanker Ray
Formerly Director, Regional Research Laboratory
CSIR—Institute of Minerals and Metals Technology
Bhubaneswar, Odisha
India

Saradindukumar Ray
Formerly Head, Materials Technology Division
Indira Gandhi Centre for Atomic Research
Kalpakkam, Tamil Nadu
India

ISSN 2509-6400 ISSN 2509-6419 (electronic)
Indian Institute of Metals Series
ISBN 978-981-13-4479-4 ISBN 978-981-13-0686-0 (eBook)
https://doi.org/10.1007/978-981-13-0686-0

Printed on acid-free paper

This Springer imprint is published by the registered company Springer Nature Singapore Pte Ltd.
The registered company address is: 152 Beach Road, #21-01/04 Gateway East, Singapore 189721, Singapore

Believe nothing
Merely because you have been told it
Or because you yourself
have imagined it,
Or because it is traditional
Do not believe what your teachers
tell you
Out of respect for the teachers
But whatever, after due examination
and analysis,
You find to be conducive to the good,
to the benefit and
the welfare of all beings
That doctrine believe
and cling to
and take it as your guide.

—Lord Buddha

... Facts not yet accounted for
by available theories are of particular
value for sciences,
since it is on them that its
development primarily depends...

—A. Butlerov

Dedicated to
our students and colleagues but for whom
our ignorance would be much more
alarming.

Series Editors' Preface

The Indian Institute of Metals Series is an institutional partnership series focusing on metallurgy and materials sciences.

About the Indian Institute of Metals

The Indian Institute of Metals (IIM) is a premier professional body (since 1947) representing an eminent and dynamic group of metallurgists and materials scientists from R&D institutions, academia and industry mostly from India. It is a registered professional institute with the primary objective of promoting and advancing the study and practice of the science and technology of metals, alloys and novel materials. The institute is actively engaged in promoting academia–research and institute–industry interactions.

Genesis and History of the Series

The study of metallurgy and materials science is vital for developing advanced materials for diverse applications. In the last decade, the progress in this field has been rapid and extensive, giving us a new array of materials, with a wide range of applications and a variety of possibilities for processing and characterizing the materials. In order to make this growing volume of knowledge available, an initiative to publish a series of books in metallurgy and materials science was taken during the Diamond Jubilee year of the Indian Institute of Metals (IIM) in the year 2006. IIM entered into a partnership with Universities Press, Hyderabad, and as part of the IIM book series, 11 books were published, and a number of these have been co-published by CRC Press, USA. The books were authored by eminent professionals in academia, industry and R&D with outstanding background in their respective domains, thus generating unique resources of validated expertise of interest in metallurgy. The international character of the authors' and editors has enabled the books to command national and global readership. This book series includes different categories of publications: textbooks to satisfy the requirements of undergraduates and beginners in the field, monographs on select topics by experts in the field and proceedings of select international conferences organized by IIM after mandatory peer review. An eminent panel of international and national experts constitute the advisory body in overseeing the selection of topics, important areas to be covered, in the books and the selection of contributing authors.

Current Series Information

To increase the readership and to ensure wide dissemination among global readers, this new chapter of the series has been initiated with Springer. The goal is to continue publishing high-value content on metallurgy and materials science, focusing on current trends and applications. Readers interested in writing for the series may contact the undersigned series editor or the Springer publishing editor, Swati Meherishi.

About This Book

It is well known that while there are numerous books on metallurgical thermodynamics, the number of books on metallurgical kinetics is rather limited. This is because there is generally wide acceptance on the scope for the former, but not so for the latter for which useful information is scattered in the journals. The books that are available usually emphasize chemical kinetics.

The present book has been written based on the experience of the authors in teaching the subject in undergraduate and postgraduate levels. The book should also be useful to researchers in the area of materials science. The authors have tried to make a judicious selection of topics based on their teaching and research experience. The book's coverage is rather extensive, but teachers can make appropriate selection of topics for classroom teaching.

A major part of discussions in the book deals with heterogeneous processes because of their predominance in metallurgical transformations. The book covers the essential topics in kinetics in extractive metallurgy, mechanical metallurgy and physical metallurgy. A special feature of the book is explanations of the origins of the various kinetic equations—both isothermal and non-isothermal. There is also a chapter on thermal analysis techniques which are being increasingly used in kinetic studies.

Baldev Raj (Late)
Editor-in-chief, and
Director
National Institute of Advanced Studies, Bangalore

U. Kamachi Mudali
Co-editor-in-Chief
Outstanding Scientist and Associate Director
Indira Gandhi Centre for Atomic Research, Kalpakkam

Preface

Numerous books are available on the subject of chemical kinetics, and there are also some specialized monographs dealing with metallurgical processes. Some discussions of kinetics are also available in books dealing with the principles of extractive or other branches of metallurgy. However, although these books are certainly useful in their own ways, none of these is singly very useful for a full-fledged first course on metallurgical kinetics for teaching at undergraduate and postgraduate levels. Also, while the advisable scope and contents of an introductory course on metallurgical thermodynamics are generally widely accepted, the same is not true for metallurgical kinetics. For the latter case, much of the useful information remains scattered in books and research papers published in scholarly journals.

The present volume has been prepared keeping these two points in perspective. The approach employed in the present volume is entirely the authors' own. The basic scheme is as follows. The volume starts with a basic-level introduction to the key concept of thermal activation, followed by a discussion on empirical and semi-empirical approaches in kinetic studies. The next five chapters discuss kinetics of idealized systems and include a comprehensive discussion on nucleation and growth in the context of phase transformations in general. The subsequent four chapters take up application of the knowledge so gathered for analysing the kinetics of real systems. These include separate chapters devoted to non-isothermal kinetics and thermal analysis technique for non-isothermal conditions. Most of the concepts introduced are illustrated with examples from published research papers.

These chapters are biased towards process metallurgy, though Chap. 2 includes some examples from physical and mechanical metallurgy, and most of the discussion on nucleation and growth should be useful for physical metallurgy. In contrast, the final two chapters on kinetics of plastic deformation and creep fracture show how the concept of thermal activation can be integrated with additional inputs and intuitive reasoning to develop formulations for meaningful technological applications in these two areas. These two chapters, however, require a preliminary exposure to crystal defects as available in an introductory course on mechanical metallurgy for undergraduate or postgraduate students.

While the book is primarily meant for undergraduate and postgraduate students, there is much material that should be useful also for researchers. Portions of the book should also be useful in teaching of physical metallurgy, mechanical metallurgy, corrosion metallurgy, etc.

The authors will consider themselves rewarded if students and teachers find this volume useful.

Kolkata, India
May 2017

Hem Shanker Ray
Saradindukumar Ray

Acknowledgements

The authors gratefully acknowledge their debt to the published literature on the subject, books and journals, in preparing this volume. They also thank their numerous students and former colleagues for many useful discussions, and also help and cooperation. We should make particular mention of Profs. A. Ghosh, P. K. Sen, A. K. Lahiri, B. Sasmal, S. B. Sarkar, and (Ms.) A. Virkar. We have made use of many publications co-authored by many former colleagues, like S. Mukherjee, S. Prakash, P. Basu, P. S. Datta, U. Shyamaprasad, (Mrs.) G. Sasikala, K. G. Samuel, C. Phaniraj, M. Vasudevan, Sumantra Mondal, K. Sarveswara Rao, B. B. Agarwal, A. K. Jouhari, A. K. Tripathy, and Rezaul Haque, to name a few. We thank Dr. (Mrs.) G. Sasikala also for critically reading the first drafts of Chaps. 12 and 13 and suggesting several improvements.

The authors are thankful to Dr. Baldev Raj for his encouragements and for felicitating the initial processes that led to the selection of this volume for publication under the auspices of IIM. HSR wishes to thank Dr. U. Kamachi Mudali for his interest and help in this project.

Finally, we thank the publisher for helping us improve the text through numerous suggestions.

Hem Shanker Ray
Saradindukumar Ray

Acknowledgements

[illegible]

[illegible]

[illegible]

Contents

About the Authors

Prof. Hem Shanker Ray obtained his B.Tech. (Hons) in Metallurgical Engineering from the Indian Institute of Technology, Kharagpur, in 1962, and his Master of Applied Science (M.A.Sc.) and Ph.D. in Metallurgy from the University of Toronto, Canada, in 1963 and in 1966, respectively.

He was a lecturer and an assistant professor (1967–1980) at the Indian Institute of Technology, Kanpur, a manager/senior technologist (1977–1979) at Research and Development, Pilkington Research Centre, Lathom, England, and a professor of metallurgical engineering (1980–1990) and the dean of students' affairs (1988–1990) at the Indian Institute of Technology, Kharagpur. He also served as a research associate at the University of Toronto and the Indian Institute of Science, Bangalore. He was the director of the Regional Research Laboratory, Bhubaneswar, from 1990 to 2000, and an emeritus scientist at the Central Glass and Ceramic Research Institute, Kolkata.

His areas of interest include thermal analysis, kinetics, extractive metallurgy, energy and R&D administration, and he has published about 450 papers in reviewed journals and conference proceedings, seven textbooks, five popular science and other books, and some 60 popular science articles. He has also edited 14 conference proceedings and written 50 technical reports as well as numerous non-technical articles, editorials and fiction works in English and Bengali. He holds several patents including two in the USA.

He has received several awards in recognition of his work, including the National Metallurgists Day Award, G. D. Birla Award and two Kamani Gold Medals from the Indian Institute of Metals (IIM). Recently, he was conferred honorary membership of the IIM and has received the Distinguished Alumnus Award from IIT Kharagpur in 2014. He has worked for several companies and R&D laboratories as a consultant or member of boards/research councils and is associated with several professional bodies and was the president of the Indian Institute of Mineral Engineers and Indian Thermal Analysis Society. He is a vice president of the Indian Association for Productivity, Quality and Reliability,

Kolkata, the president of the Millennium Institute of Energy and Environment Management and the editor of several journals. He was the honourable director of Ramakrishna Vivekananda Mission Institute of Advanced Studies and the president of CSIR Pensioners Welfare Society (CPWSc), Kolkata.

Dr. Saradindukumar Ray obtained his B.E. in 1968 and Ph.D. in 1985. He served the Department of Atomic Energy from 1972 to 2007 and was the head of the Materials Technology Division, Indira Gandhi Centre for Atomic Research, from 1998. From September 2010 to May 2017, he served as the Ministry of Steel Chair Professor, Metallurgical and Materials Engineering Department, Jadavpur University. His areas of research interest are deformation and fracture, deformation processing and welding metallurgy. He served as the chief editor of the Transactions of the Indian Institute of Metals from 1996 to 2003. He is a recipient of the National Metallurgist's Day Best Metallurgist Award and MRSI Lecture Medal from the Materials Research Society of India. He is a fellow of the Indian National Academy of Engineering, fellow (life) of the Indian Institute of Metals and a member of several other professional societies connected with materials research and development. He has published more than 100 research papers in various journals and peer-reviewed conference proceedings and made numerous contributions in the form of keynote lectures, invited papers, technical presentations and reports.

Symbols

Symbols used in Chaps. 12 and 13 are listed separately. This is necessitated by the fact that many symbols commonly used in physical metallurgy and mechanical metallurgy literatures differ from those used in process metallurgy. Only symbols used globally are listed here. Other symbols are defined and used locally, and the same symbol may be used differently in different contexts.

Chapters 1–11

a	A constant
a_b	Area of a bubble
A	A constant, reactant, frequency factor, surface area
b	Constant
B	A constant, reactant, rate of temperature rise
c	A constant
C	A constant, reactant, concentration, specific heat
d	Density, diameter
D	A constant; reactant, diameter, diffusivity
D_0	Initial diameter
E	Potential energy, emf of a cell, activation energy
E_A	Energy of activation
$\dot{E}$	Stirring power density
E_D	Eddy diffusivity
f	Friction factor, shape factor
$f(\alpha)$	Function of α
F	Faraday constant, flow rate of gas
g	Acceleration due to gravity
$g(\alpha)$	Function of α
G	Free energy
G_A	Activation free energy

h_m	Mass transfer coefficient
H	Enthalpy
i	Current
i_d	Diffusion current, limiting current
J	Flux of reagent in diffusion
k	Rate constant
K	Degree Kelvin
K	Constant, equilibrium constant, thermal conductivity
K_B	Boltzmann constant
k_{eq}	Equilibrium constant
L	Length, latent heat
L_0	Length at zero time
m	A constant, mass
n	A constant, a number, order of reaction
N	Number of moles or atoms, Avagadro's number
N_0	Number of particles at zero time
N_{Gr}	Grashoff's number
N_{Nu}	Nusselt's number
N_{Re}	Reynolds number
N_{Sc}	Schmidt's number
N_{Sh}	Sherwood number
p	Probability, porosity, pressure, partial pressure
pH	$[-\log a_{H^+}]$
Q	Heat flux
r	Radius
R	Distance between atoms, gas constant, rate
S	Entropy
S_A	Activation entropy
t	Time
T	Temperature
T_e	Equilibrium temperature
U	Activation energy for addition of atom to embryo surface
v	Velocity, volume
v_p	Volume of pores
v_s	Volume of pores in sintered compact
v_x	Velocity in x direction
V	Velocity, volume
V_f	Velocity of forward reaction
V_b	Velocity of backward reaction
V_0	Original volume
V_s	Volume after sintering
V_{th}	Theoretical volume
w	Weight
$\dot{w}$	Rate of change of weight

W Weight
X Mole fraction
X_i Mole fraction of component i
y Thickness of product layer
τ Ratio of volume of products to that of reactants during gas–solid reaction
α Degree of reaction or conversion
δ Boundary layer thickness
γ A constant
υ Vibration frequency, kinematic viscosity
μ Viscosity
η Overpotential, viscosity
σ Stoichiometric factor, surface tension
ρ Density
ρ_{th} Theoretical density
ρ_0 Original density
τ Mixing time
τ_r Residence time
θ Wetting angle or contact angle

Chapters 12 and 13

Note:
For readers' convenience, Eqs. 12.23a–b and Fig. 12.11 (Sect. 12.5.3) generally retain the original notations of Hart, which differ from the rest of Chaps. 12 and 13. Interfacing, however, should not be a problem. Reader may please make a note of this.

Δa True activation area (see Eqs. 12.10b–c)
A Activation area (operational), Eq. 12.12
$\mathcal{A}$ Apparent activation area, Eq. 12.14b
$\mathcal{A}^{☆}$ Apparent activation area in the effective stress formalism
A_s A function of strain, Eq. 12.33
b Magnitude of Burgers vector of dislocations
C_v Concentration of vacancies
C_{MMG} A material-dependent constant (see Sect. 13.4)
D_{eff} Effective diffusivity for lattice plus dislocation pipe diffusion
D_{gb} Grain boundary diffusion coefficient
D_v The lattice vacancy diffusivity
E Young's modulus
F_C Strength contribution due to DSA, Eq. 12.48
F_{def} Deformation-dependent components of σ_D (MATMOD)
F_{sol} Deformation-independent components of σ_D (MATMOD)

F_1, F_2, F_3	Adjustable material parameters in Eq. 12.48
ΔF	Helmholtz free energy change for thermal activation
g	Scaled activation energy $\Delta G/(\mu b^3)$
g^0	$\Delta G^0/(\mu b^3)$, the maximum value of g
$\mathbf{G}(\cdot)$	Set of functions of the arguments defining $\dot{\Psi}$ (Eq. 12.4b)
ΔG	Gibbs free energy of thermal activation
ΔG^0	Value of ΔG for zero stress
ΔH	Activation enthalpy
k_B	Boltzmann constant
$\bar{l}$	Average dislocation link length
m	Reciprocal of strain rate sensitivity of flow stress (engineering) (Example 12.1)
m_d	Reciprocal of (engineering) strain rate sensitivity for the forest dislocations strengthening component alone
m_{sol}	Reciprocal of (engineering) strain rate sensitivities for the σ_{sol} component
m_t	Reciprocal of (engineering) strain rate sensitivities for the σ_t component
m^*	$-(\partial \ln T/\partial \ln \sigma)_{\Psi,\dot{\varepsilon}}$ (see Eq. 12.18c)
$\mathcal{M}$	Equivalent elastic modulus for specimen + machine
$\overline{M}$	Taylor orientation factor
n	Stress exponent in strain rate equation
n_f	Degree of freedom of least squares fit
n_{PL}	Stress exponent of steady-state creep in the power law regime
$\dot{p}_0$	Adjustable material parameters in Eq. 12.48
Q	Apparent activation enthalpy, Eq. 12.14a
Q_c	Activation energy for creep deformation
Q_r	Activation energy for creep rupture
Q_{SD}	Activation energy for self- (or in substitutional alloys, alloy-) diffusion
Q^*	Apparent activation enthalpy in effective stress formalism
$R_d(T)$	Temperature-dependent 'recovery parameter' (Sect. 12.7.4)
S	$(\partial \tau/\partial \ln \dot{\gamma})_{\Psi,T}$ (thermodynamic) strain rate sensitivity of flow stress (the definition $(\partial \sigma/\partial \ln \dot{\varepsilon})_{\Psi,T}$ is also used in the literature)
S_i	$\Delta\sigma_i/\Delta \ln \dot{\varepsilon}$, "instantaneous" strain rate sensitivity, without change in quasi-static ageing time t_a
S_f	$\Delta\sigma_n/\Delta \ln \dot{\varepsilon}$, "final" strain rate sensitivity, including change in quasi-static ageing time to that for new strain rate
s^*	Material-dependent parameters (Sect. 12.6.2)
ΔS	Activation entropy
t	Time
t_a	Quasi-steady-state ageing time for DSA
t_d	Duration of the yield transient
t_r	Rupture life

t_R	A characteristic time that defines the rate constant for primary creep
t_w	The period the mobile dislocations are held up at obstacles awaiting thermal activation
t_ω	Time for a given level of creep damage ω
t^*	Material-dependent parameter (Sect. 12.6.2)
T	Absolute temperature
T_M	Melting point in absolute scale
T/T_M	Homologous temperature
T_t	Material-dependent transition temperature in MATMOD
$\bar{v}$	Average speed of dislocations
ΔW	Activation work
Z	Zener–Hollomon parameter
Z_0	Zener–Hollomon parameter using $\dot{\varepsilon}_0$ (Sect. 13.6.2)
α	Factor in the stress–dislocation density relation (Eq. 12.1a)
χ	Standard error of fit
ε	Longitudinal plastic strain
ε_a	Longitudinal anelastic strain
ε_c	Critical strain for onset of serrations in tensile flow curve
ε_e	Longitudinal elastic strain
ε_P	Extent of primary creep (Fig. 13.9)
ε_r	True strain at fracture
ε_t	Longitudinal total strain
ε_T	Extent of tertiary creep (Fig. 13.9)
ε_0	Longitudinal strain when all the mobile dislocations accomplish one thermal activation (Eq. 12.11a)
$\varepsilon_{\theta=0}$	Strain for the saturation state ($\theta = A_s = 0$)
$\dot{\varepsilon}_0$	Factor in (longitudinal) strain rate equation (Chap. 12); $\dot{\varepsilon}_{t=0}$ (Chap. 13)
$\dot{\varepsilon}_m$	Minimum creep rate
$\dot{\varepsilon}_s$	Steady-state creep rate
γ	Plastic shear strain on glide plane
γ_s	Surface energy per unit area
γ_0	Crystallographic shear strain resulting when all the mobile dislocations accomplish one thermal activation (Eq. 12.11)
κ	Scaled error of least squares fit
λ	Damage tolerance parameter of Ashby and Dyson (Chap. 13). [Note: the same symbol has been used with different meanings in Chap. 12.]
Λ	Mean free path of dislocations
μ	Shear modulus
ν_G	Frequency of vibration of dislocations awaiting thermal activation
Γ	Stacking fault energy (SFE)
$\Gamma/\mu b$	Normalized SFE
θ	Uniaxial work hardening rate $(d\sigma/d\varepsilon)_{\dot{\varepsilon},T} = \overline{M}^2 \cdot \theta_\tau$
θ_0	Athermal uniaxial work hardening rate solely due to geometrical-statistical storage of dislocations

θ_y	Athermal work hardening rate in the absence of recovery in Alden's model of time-independent deformation (Sect. 12.5.4)
θ_τ	Strain hardening on slip plane $(d\tau/d\gamma)_{\dot{\gamma},T}$
θ'	Temperature correction factor in MATMOD
ψ	Scalar variable representing deformation substructure
$\boldsymbol{\Psi}$	Set of micro- and sub-structural parameters that characterize the material
ρ_m	Mobile dislocation density
σ	Uniaxial stress
σ_b	Athermal back stress
σ_d	Glide resistance from (forest) dislocation interactions
σ_D	"Drag stress" (corresponding to isotropic hardening)
σ_g	Glide resistances of grain boundaries
σ_p	Glide resistances of precipitates
σ_s	Stress for steady state
σ_{sol}	Glide resistances of solutes
σ_t	Glide resistance from other sources of strengthening which can be thermally activated
σ_V	Extrapolated steady-state stress (Voce stress)
σ_R	"Rest stress" (corresponding to kinematic hardening)
σ_0	Sum of all the strengthening contributions which do not vary with deformation level
$\widehat{\sigma}$	Uniaxial mechanical strength of the obstacle in polyslip (see Eq. 12.17)
$\Delta\sigma_i$	Maximum stress increase in affecting strain rate jump (Fig. 12.14)
$\Delta\sigma_n$	Stress increase in affecting strain rate jump after allowing for steady-state redistribution of solutes (Fig. 12.14)
$\Delta\sigma_a$	Strength of DSA (Fig. 12.14)
$\Delta\sigma_{WH}$	Strengthening due to work hardening (because of accumulation of plastic strain) during stress relaxation (Fig. 12.14)
$\Delta\sigma_h$	Permanent (non-transient) strengthening due to ageing under load during stress relaxation (Fig. 12.14)
$\left(\frac{\sigma_s}{\mu}\right)_{PLB}$	Modulus-reduced stress for the transition from PL to PLB regime of creep
τ_v	Extrapolated steady-state stress (Voce stress) for crystallographic shear
τ_s	Steady-state stress for $\theta_\tau = 0$
τ	Resolved shear stress on the glide plane of dislocation
$\tilde{\tau}$	Shear stress resisting dislocation glide on the slip plane
$\widehat{\tau}$	Maximum value of $\tilde{\tau}$
τ_μ	Athermal component of flow stress
τ^*	Effective stress or thermal component of flow stress
$\left(\frac{\widehat{\tau}_s}{\mu}\right)$	Scaled obstacle strength (Eq. 12.30)
ω	Creep damage parameter (initial and final values, respectively, ω_i and ω_f)

Abbreviations

DSA	Dynamic strain ageing
FE	Finite element
MATMOD	Material Modeling (an MEOS formulation)
MEOS	Mechanical equations of state
pdf	Probability distribution function
PL, PLB	Power law, Power paw breakdown (regimes of creep)
SFE	Stacking-fault energy
TASRA	Thermally activated strain rate analysis
TID	Time-independent deformation

Other Notations

One dot on top of a variable indicates its absolute differential with time t.
Δ before a variable indicates change in its magnitude
δ before a variable indicates its (infinitesimally) small increment

Chapter 1
Introduction

1.1 Preamble

1.1.1 Scope of Metallurgical Kinetics

The subject of kinetics deals with measurements and interpretation of reaction rates. It provides information which forms the quantitative basis underlying all theories of chemical reactivity. It serves to define the set of values for process variables that will best fulfil the requirements of practical operation. Thus, kinetics is essential in the research and development of new processes. Kinetics also forms a basis for process modelling, i.e. mathematical description of rate phenomena on the basis of suitable assumptions.

Both thermodynamics and kinetics are necessary for proper understanding of metallurgical processes. However, the subject of metallurgical kinetics is not as developed as that of thermodynamics. Accordingly, while there are a large number of books dealing with thermodynamics, there are neither many books on metallurgical kinetics nor the scope of the subject unambiguously defined.

Thermodynamics is interested only in the initial and final states of a transformation process; the reaction path, mechanism of conversion and the time required being of little importance. Time is not one of the thermodynamics variables while it is the most important parameter in kinetics. Thermodynamics deals with equilibrium which, in theory, can also be treated on the basis of that situation in which the rates of forward and backward reaction are equal. The converse is, however, not true. A reaction cannot be understood on the basis of thermodynamics alone.

The present book presents a brief text on the principles of metallurgical kinetics for those who have had some exposure to the subject of chemical kinetics. There is more emphasis on assumptions necessary for kinetic analysis and results predicted by theory for various situations rather than detailed derivations. Many actual experimental data are presented to illustrate the application of concepts. A number of preferences are included at the end of chapters to cover examples of applications

H. S. Ray and S. Ray, *Kinetics of Metallurgical Processes*,
Indian Institute of Metals Series, https://doi.org/10.1007/978-981-13-0686-0_1

of kinetic analysis. The write-up is aimed at students as well as researchers. A major part of the discussions centre around heterogeneous processes because of their predominance in metallurgical transformations.

1.1.2 Characteristic Features of Rate Processes

There are some characteristic features of every *rate process*. Some of the more important ones are as follows:

(a) For any transformation to occur, there must be a thermodynamic driving force, i.e. the change must involve a lowering of the free energy of the system.
(b) There is transformation to a more stable state only if sufficient energy is supplied to overcome a definite energy barrier. That is, a negative free energy change by itself does not guarantee onset of transformation.
(c) There is an underlying reaction mechanism, i.e. a logic which governs a transformation. This mechanism may change with time and extent of transformation.

A process is more easily understood if it is possible to assume some simplifications. These include the assumptions that the physical and chemical nature of the system undergoing change remain unaltered, that there is no reagent starvation, that the reaction is not heat transfer controlled, etc. The implications of these will be discussed subsequently.

Basic definitions
Before proceeding further, it will be appropriate to recapitulate some basic definitions.

Transformation is any rearrangement of atoms, ions or molecules of a system from one metastable state to another of lower free energy. It is also called reaction, transition or conversion. It can be a complete rearrangement of every particle (e.g. b.c.c. α—Fe changing to f.c.c. γ—Fe during heating), or only a fraction of particles (e.g. precipitation of excess solute during cooling of a solution or, conversely, dissolution of additional solute on heating of a saturated solution).

A *metastable state* is an intermediate state where the system is stable. However, it can be converted to a more stable state provided it is made to cross an energy barrier again. For example, quenching of carbon steel from a temperature above that of pearlite–austenite transformation produces a mixture of austenite and martensite which is metastable relative to the mixture of ferrite and cementite, and yet the quenched structure will remain unchanged essentially indefinitely at room temperature.

Reaction rate is the rate at which a reaction proceeds. It can be expressed either in terms of the rate of generation of any of the products or that of consumption of any of the reactants. For a wide range of reactions, there is no correlation between the driving force and the rate of reaction. This point is well illustrated by the

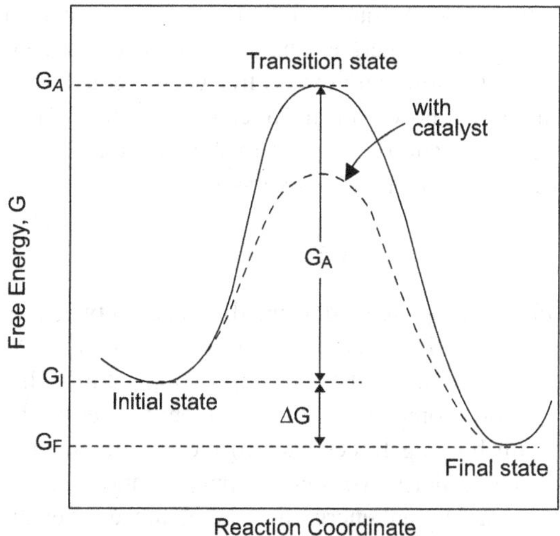

Fig. 1.1 Schematic diagram showing the energy barrier and effect of catalyst. G_A is the activation free energy

influence of catalysts on reactions. The balance of a chemical equilibrium is not changed by the catalyst, but the rate changes. This is due to lowering of the energy barrier (activation energy), Fig. 1.1. In Fig. 1.1, the reaction coordinate is any variable defining progress along the reaction path. This could be time, distance between reacting particles, etc.

1.2 Activation Energy

1.2.1 Energy Barrier for a Reaction

All the particles in a system do not undergo transformation at the same time. Reactants and products coexist throughout the transformation, and at any instant, only a small fraction of the available particles can be in the process of transformation for the simple reason that only a fraction of the particles has free energy in excess of the mean and, therefore, energetically suitable for transformations. It is always possible to find more than one set of configurational changes capable of providing a given transformation. Called "*reaction paths*", each of these is associated with a free energy curve similar to that shown in Fig. 1.1. However, because different transition configurations are involved, the activation energies are different.

The free energy of an atom or group of atoms during transformations first increases to a maximum and then decreases to the final value. The configuration associated with the maximum in the free energy curve is assumed to be the *transition* or *activation state*. As said earlier, at any instant the assembly of particles embraces a wide spectrum of energies and some particles have energies greatly in excess of the mean. Only those with excess free energy equal to or greater than G_A (Fig. 1.1) will transform. Those with insufficient free energy must wait until they receive the necessary activation energy from thermal fluctuations. The process is termed *thermal activation.* G_A may be expressed as

$$G_A = E_A - TS_A \tag{1.1}$$

where E_A is *activation energy* defined as the difference between the internal energy of the system in the transition state to that in the initial state, S_A is similarly an activation entropy change and T the (absolute) temperature. The internal energy, again, is divided into two components: (a) the potential energy of interaction of the atoms associated with binding forces and (b) the kinetic energy of the thermally induced motion of the particles. To a good approximation these two are independent, and changes in the internal energy may be obtained by evaluating changes in the two components separately (Burke 1965). The requirement that an atom must have the free energy of activation G_A before it can participate in a reaction is equivalent to the two conditions that

(a) an atom must have a thermal (kinetic) energy at least equal to E_A to enable it to overcome the potential barrier and
(b) simultaneously, the entropy requirements of the transition configuration are satisfied.

Either of these is a necessary but not a sufficient condition for reaction, whereas the possession of the free energy of activation is in itself the necessary and sufficient condition.

1.2.2 Potential Energy Surface

The concept of the barrier would become clearer if we consider interaction between three atoms *x*, *y*, *z* and the reaction

$$x + yz = xy + z$$

Consider potential energy surface for the interaction. There are three coordinates for each nucleus and, therefore, a total of nine, three of which may be taken as determining the centre of mass and three-determining orientation in space. Therefore, three coordinates remain to determine the relative positions of the nuclei. These three can be chosen to be the three internuclear distances between numbers of each possible pair of atoms. Let us assume that the atoms are always on a straight line. The two remaining coordinates may be chosen as x-y distance (R_{xy}) and y-z distance (R_{yz}) supposing that y is the intermediate atom. The potential energy,

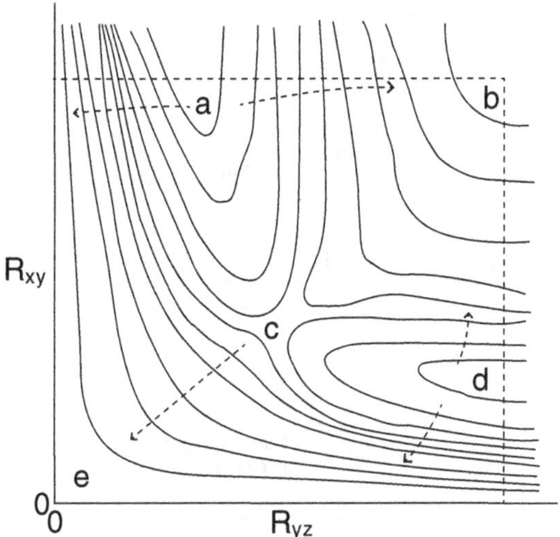

Fig. 1.2 Potential energy contour diagram for linear xyz system as a function of internuclear distance (*x*-*y*-*z* in straight line)

E, of the system can then be plotted in space as a function of two coordinates, and a contour map can be made as shown in Fig. 1.2. Here a and d represent potential energy minima. b and e represents maxima. c represents the minimum elevation called "*saddle point*". Regions around a and d represent valleys and that around d a plateau. The valleys join through a pass which go through the *saddle point.* We also note the following.

(i) a represents a configuration where *x* is far removed from yz molecule,
(ii) d corresponds to atom *z* being far removed from molecule *xy*,
(iii) b is a region where all three atoms are well separated,
(iv) c is where atoms are close together as in a collision of *x* with *yz* or *xy* with *z*.

The variation of *E* through the potential energy across sections through the energy surface at a and b would look like as shown in Fig. 1.3. The energy will not change rapidly with R_{xy} in the neighbourhood of a-b nor with R_yz in the neighbourhood of b-d since both such changes correspond to motion of an atom *z* or *x* already far removed from a molecule. As *x* approaches *yz* or *xy* approaches *z*, a van der Waals' repulsive force is expected to set in, thus raising energy. The variation of *E* as the system passes through the pass is also schematically shown in Fig. 1.3.

The van der Waals' repulsion is of the utmost importance because it is responsible for activation energies in general. Without it, all exothermic reactions would have zero or low activation energies and be immeasurably fast.

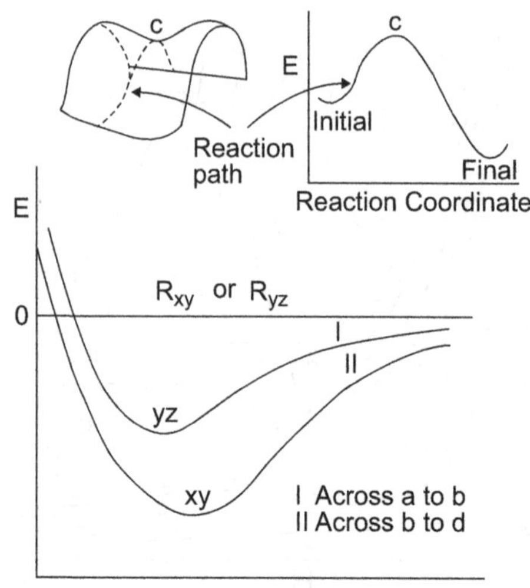

Fig. 1.3 Potential energy variations across sections

1.2.3 Rate of a Thermally Activated Process

The rate is given by the fraction of the total number of particles which reach the final configuration in unit time. Obviously this is proportional to the following, according to the collision theory.

(a) The frequency with which particles "attempt" to transform. In homogeneous reaction, it will equal collision frequency. For solids, one should consider the vibration frequency v.
(b) The fraction of the particles in the initial state having sufficient energy to surmount the potential barrier, given by $\exp(-E_A/K_B T)$ where K_B is Boltzmann's constant.
(c) The probability p that during the time a particle or particles have the requisite energy that satisfy the geometrical or other conditions necessary for the change.

Thus at constant temperature T, the rate equation should have the form

$$\text{Rate} = p \cdot \upsilon \cdot \exp\left[-\frac{E_A}{K_B T}\right] \tag{1.2}$$

υ is the vibration frequency. The pre-exponential factors change with the progress of a reaction and, therefore, one may write

$$\text{Rate} = \frac{d\alpha}{dt} = A\ \exp\left[-\frac{E}{K_B T}\right] \cdot f(\alpha) \tag{1.3}$$

where α is the degree of reaction. In Eq. 1.3, E_A is written as only E for simplification. A is a pre-exponential constant. At a given temperature and for fixed value of E, Eq. 1.3 is rewritten as

$$\frac{d\alpha}{dt} = k \cdot f(\alpha) \tag{1.4}$$

where k is the rate constant.

It should be noted that while the rate of a reaction may change with the progress of reaction, the rate constant remains unchanged provided E and T are unchanged.

1.2.4 Arrhenius Law

The reaction rate constant is expressed by the equation

$$k = A \cdot \exp\left(-\frac{E}{K_B T}\right) \tag{1.5}$$

Equation 1.5 is more commonly written using gas constant R ($R = NK_B$ where N is Avogadro number), and thus, we have a similar equation which is known as Arrhenius law.

$$k = A \cdot \exp\left(-\frac{E}{RT}\right) \tag{1.6}$$

A plot of ln k against reciprocal temperature should thus yield a straight line. The temperature effect on rate constant can be established from theory. Levenspiel (1969) has summarized the derivation of the equation on the basis of van't Hoff equation, collision theory and transition state theory. The predictions of the simpler version of the various theories for the temperature dependency of the rate constant may be summarized by the expression

$$k \propto T^m \exp\left(-\frac{E}{RT}\right) = k_0 T^m \exp\left(-\frac{E}{RT}\right) \quad 0 < m < 1 \tag{1.7}$$

For the complicated versions, m can be as high as 3 or 4. However, the exponential term is much more temperature sensitive than the T^m term. Therefore, the variation of k caused by the latter is effectively masked. The result is Eq. 1.6.

This can be shown in another way. Taking logarithms of Eq. 1.7 and differentiating with respect to T, one obtains

$$\frac{d(\ln k)}{dT} = \frac{m}{T} + \frac{E}{RT^2} = \frac{mRT + E}{RT^2}$$

As $mRT \ll E$ for most processes, we may ignore mRT and write

$$\frac{d(\ln k)}{dT} = \frac{E}{RT^2}, \quad \text{i.e. } k = A.\ \exp\left(-\frac{E}{RT}\right) \tag{1.6}$$

This expression has been found to fit experimental data well over wide temperature ranges and is strongly suggested from various standpoints as a reasonable first approximation in most processes. Equation 1.7 should be considered only if E value is very small.

Although various theories end up with a similar result, the assumptions and approximations made use of differ. Consider, for example, reaction between two reactants A and B to produce an intermediate complex which leads to the final product AB.

$$A + B \rightarrow AB^* \rightarrow AB \tag{1.8}$$

Collision theory views the rate as dependant on the number of energetic collision between A and B. The intermediate product AB^* is quite irrelevant, and it is assumed that its rapid breakdown into final product has no influence on the overall rate. On the other hand, transition state theory proposes that the reaction rate is governed by the rate of decomposition of the intermediate. The rate of formation of the intermediate is assumed to be so rapid that it is present in equilibrium concentrations at all times and how it is formed is of no concern. The collision theory thus views the first steps slow and rate controlling while the transition state theory views the second step, combined with the determination of the complex concentration to be rate controlling.

The proceeding discussion leads to a simple alternative derivation of Arrhenius law. If the number of particles in the transition state per unit volume is C_A and in the initial state C_I, then it is possible to write down an equilibrium constant k_{eq}

$$k_{eq} = \frac{C_A}{C_I} = \exp\left(-\frac{G_A}{RT}\right) \tag{1.9}$$

The reaction rate must be proportional to C_A and therefore

$$\frac{d\alpha}{dt} = \text{const} \cdot C_I \exp\left(-\frac{G_A}{RT}\right) = A\ \exp\left(-\frac{E}{RT}\right) f(\alpha) \tag{1.10}$$

In Eq. 1.10, the entropy component of G_A has been taken into the frequency factor A.

Levenspiel (1969) has given the following derivation of Arrhenius law. This is based on van't Hoff's equation. Consider the reaction

$$\mathrm{A} + \mathrm{B} = \mathrm{C} + \mathrm{D} \tag{1.11}$$

The rate of the forward reaction can be expressed in terms of the amount of A (or B) transformed in unit time or in terms of the amount of C (or D) produced in unit time. If C is the concentration, then the reaction rate is $\pm \frac{\mathrm{d}c}{\mathrm{d}t}$ the sign depending on the mode of definition used. The rate usually is a function of the concentrations and the temperature.

Effect of concentration
According to the law of mass action, the velocity V_f and V_b of the forward and backward reactions, respectively, are given by

$$V_f = k_f[\mathrm{A}]\,[\mathrm{B}] \tag{1.12}$$

$$V_b = k_b[\mathrm{C}]\,[\mathrm{D}] \tag{1.13}$$

Here [A], [B], [C], [D] represent concentration of A, B, C and D, and the k_f and k_b are rate constants for forward and backward reactions, respectively, when stable equilibrium is attained.

$$V_f = V_b \tag{1.14}$$

$$\frac{[\mathrm{C}] \cdot [\mathrm{D}]}{[\mathrm{A}] \cdot [\mathrm{B}]} = \frac{k_f}{k_b} = K \tag{1.15}$$

where K is the equilibrium constant of the reaction expressed in terms of concentration.

Effect of temperature on reaction rates
The temperature coefficient of the equilibrium constant k is given by van't Hoff equation

$$\frac{\mathrm{d}\ln k}{\mathrm{d}t} = \frac{\Delta H}{RT^2} \tag{1.16}$$

$$\frac{\mathrm{d}}{\mathrm{d}t}\ln k_f - \frac{\mathrm{d}}{\mathrm{d}t}\ln k_b = \frac{\Delta H}{RT^2} \tag{1.17}$$

$$\text{i.e.} \quad \frac{\mathrm{d}}{\mathrm{d}T}\ln k_f = \frac{E}{RT^2} + \mathrm{I} \tag{1.18}$$

$$\frac{\mathrm{d}}{\mathrm{d}T}\ln k_{\mathrm{b}} = \frac{E'}{RT^2} + \mathrm{I} \tag{1.19}$$

where $E' - E = \Delta H$ and I is a constant.

Assuming I to be zero and E to be temperature independent, on integration

$$\ln k_{\mathrm{f}} = -\frac{E}{RT} + \text{constant} \tag{1.20}$$

$$\ln k_{\mathrm{b}} = -\frac{E'}{RT} + \text{constant} \tag{1.21}$$

or, in general, the velocity constant or rate constant k is given by

$$k = A \cdot \exp\left(-\frac{E}{RT}\right) \tag{1.22}$$

where A, the constant of integration, is called the frequency factor and E the activation energy. Equation 1.22 is the well-known Arrhenius equation. Arrhenius found that a plot of $\ln k$ against reciprocal temperature was a straight line for a large number of reactions. It also means that the assumption $I = 0$ is generally valid.

1.2.5 Effect of Activation Energy on Reaction Rate

According to the Arrhenius equation, the magnitude of E dominates the reaction rate. For example, if E is 10 k.cal/mole, changing the temperature from 500 to 1000 K would increase the rate by a factor of about 150. Doubling E would decrease the rate at 500 K by a factor of 2×10^4 and at 1000 K by a factor of 150. The variation of rate is more rapid for higher values of E and lower values of T. This is easily understood by differentiating k with respect to E and $\ln k$ with respect to T. Figure 1.4 and Tables 1.1 and 1.2 summarize the results of some calculations that show the relative importance of activation energy and temperature on influencing reaction rate. We can draw the following important conclusions:

1. Reactions with high activation energies are very temperature sensitive. Conversely, low activation energies imply lower temperature sensitivity of rate.
2. Temperature sensitivity of rate is more pronounced at lower temperatures.
3. The pre-exponential factor is Arrhenius law has little effect on temperature sensitivity of reaction rate.

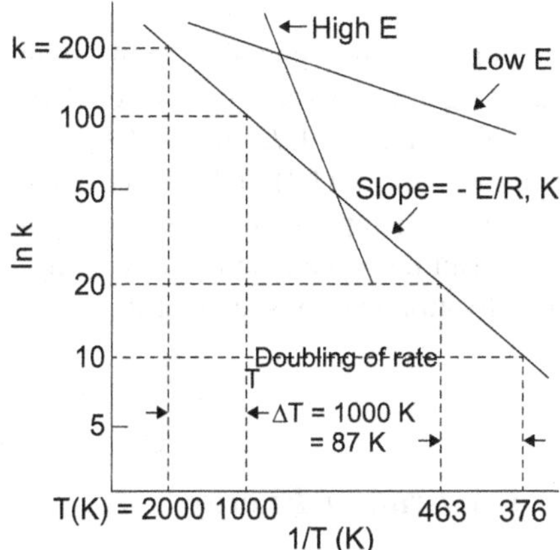

Fig. 1.4 Sketch showing temperature dependency of the reaction rate (Levenspiel 1969)

Table 1.1 Temperature rise needed to double the rate of reaction for activation energies and average temperature shown; hence shows temperature sensitivity of reaction (Levenspiel 1969)

Temperature (°C)	Activation energy E		
	10,000 cal	40,000 cal	70,000 cal
0	11	3	2
400	70	17	9
1000	273	62	37
2000	1037	197	107

Table 1.2 Relative rates of reaction as a function of activation energy and temperature (Levenspiel 1969)

Temperature (°C)	Activation energy E		
	10,000 cal	40,000 cal	70,000 cal
0	10^{48}	10^{24}	1
400	7×10^{52}	10^{43}	2×10^{33}
1000	2×10^{54}	10^{49}	10^{44}
2000	10^{55}	10^{52}	2×10^{49}

Example 1.1: At an average temperature of 400 °C, the temperature rise required to double rate constant is 70 °C. What is the activation energy?

Solution: $T_1 = 400 - 35 = 365$ °C = 638 K

$$k_1 = A\ \exp\left(-\frac{E}{RT_1}\right); \quad k_2 = 2k_1 = A\ \exp\left(-\frac{E}{RT_2}\right)$$

$$\ln\frac{k_2}{k_1} = \ln 2 = 0.693 = \frac{E}{RT}\left[\frac{70}{708 \times 638}\right]$$

which give E = 9000 cal/mole = 38 kJ/mole.

Limiting value of k
It should be noted k cannot go on increasing with temperature. In the limit $T \to \infty$, k should approach A according to the Arrhenius equation. However when T is large, the condition mentioned following Eq. 1.7, viz. $mRT \ll E$ is no longer valid, and thus, k may now be very large ($k = AT^m$) and yet dependent on temperature.

For most reactions, accessible temperature range corresponds to the lower, rising part of the k versus T plot. But there are reactions involving free atoms or radicals with very small or zero activation energy such that the upper part of k versus T is approached.

1.2.6 Dimension of Activation Energy

The numerical value of E does not depend on how we represent the reaction stoichiometry (i.e. number of moles of reactants). Therefore, many authors report E simply in terms of calories or Joules. However, the correct dimension is calories/mole so that E/RT term is dimensionless. The rate constant, therefore, has the same units as the pre-exponential term, the so-called collision frequency term, which is s^{-1}. The mole term in the unit for E pertains to the quantities associated with the molar representation of the rate-controlling step of the reaction. Numerically, E can be found without knowing what this is. Sometimes this may lead to ambiguity if too much importance is attached to the physical significance of the quantity. For example, E for densification of particulate systems should refer to per mole of empty space which is the "reactant" which gradually gates consumed in the process. To simplify matters, henceforth we will treat E merely as an experimentally determined parameter which is indicative of the reaction path.

1.2.7 Anti-Arrhenius Variation of Rate Constant

Not all reactions show, for the rate constant, the Arrhenius-type variation. We have already seen (Eq. 1.7) that if the activation energy is very small, then k varies in proportion to T^m. Many different types of variations have been seen in isolated cases. Some of these are schematically summarized in Fig. 1.5. The exceptional cases often known as anti-Arrhenius-type reactions generally involve changes in reaction mechanism with temperature, the change in E coming abruptly or gradually.

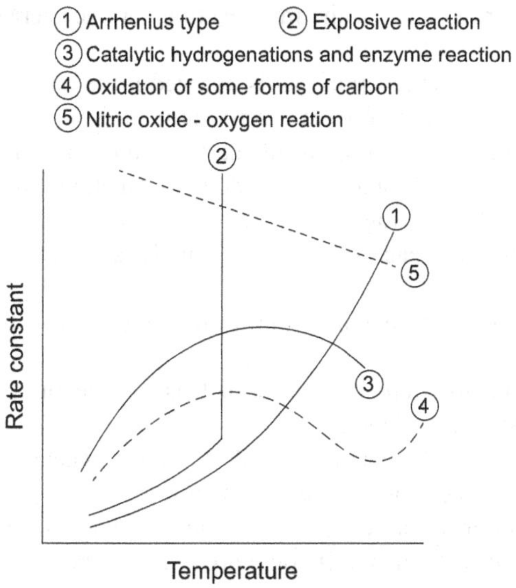

Fig. 1.5 Various types of variation of k with temperature

1.3 Review Questions

1. Discuss whether the following statements are correct or false.

    a. The rate of an exothermic reaction decreases with increasing temperature according to Le Chatelier's principle.
    b. Favourable kinetics can make a reaction that takes place even if it is thermodynamically not feasible.
    c. If a reaction is thermodynamically feasible, then it must take place.
    d. Kinetic barriers can stop a reaction even if there is a thermodynamic driving force for transformation.
    e. The rate of a reaction is more sensitive to temperature changes if the activation energy is smaller.
    f. The rate of a reaction depends on the free energy change for the reaction.
    g. The "saddle point" is the highest point on a potential energy surface.
    h. For a given activation energy, temperature changes have more effect on the rate constant at lower temperatures.
    i. A process is operating at a mean temperature; the temperature oscillating with a fixed frequency for a given time; the extent of reaction is more if the amplitude of temperature fluctuation is greater.
    j. A catalyst cannot change the driving force of a reaction.
    k. The rate constant of a reaction increases a hundredfold when the temperature rise is 200 °C. The rate constant will increasing more than fifty times if the temperature rise is 100 °C.

l. Both endothermic and exothermic reactions are accelerated by increase in temperature.
m. For any process, the reaction rate is doubled if the rate constant is doubled.
n. Activation energy cannot have a zero value.
o. If the temperature for a process fluctuated around an average value, then for the overall reaction for a given period, the fluctuations are not important and only the so-called average temperature matters.
p. Rate constant for a reaction must be infinitely large when the temperature is very high.
q. For a reaction to take place, the system must cross both enthalpy and entropy barriers.
r. All transformations are best described as change of a system from one metastable state to another.
s. Thermodynamic data cannot describe kinetics, whereas kinetic data may sometime give thermodynamic information.
t. Calcium carbonate is decomposing in a closed chamber. Its rate of decomposition will change if some argon is pumped into the chamber.

2. What are the differences in the basic postulates of the collision theory and the absolute reaction rate theory used for the derivation of the Arrhenius equation?
3. Why is reaction rate related to the activation energy barrier?
4. A given reaction can take place by one of two possible mechanisms—one characterized by a high value of activation energy and another by a low value of activation energy. Explain why the reaction mechanism may change as the temperature is increased.
5. Briefly discuss the basic states for any one method of derivation of the Arrhenius equation. Compare the underlying assumptions with those for other methods.
6. Examine Tables 1.1 and 1.2. Recalculate values and make two new tables by assuming the following values

   a. Temperatures (°C): 100, 500, 1000, 1500
   b. Activation energy $(E, \mathrm{kJ/mol})$: 20, 50, 100, 150

7. Discuss the influence of temperature and E on reaction rate constant (k) for $\ln k$ by plotting k against $T(k)$ for constant E values and k against E at constant temperatures.
8. The rate of a reaction is doubled by a certain rise in temperature just when the reaction begins, i.e. $\alpha = 0$. Will the rate be also doubled for $\alpha = 0.5$?
9. There is 60% reduction in the value of the rate constant of a reaction when the reaction temperature decreases from 60 to 40 °C. What is the activation energy value?
10. In a pyrometallurgical plant, the temperature variations during two shifts are different as shown below: although the average temperature is the same, the temperature fluctuation varies. In which shift the production will be more?

References

Burke, J.: The Kinetics of Phase Transformations in Metals, chap. 1. Pergamon Press Ltd, Oxford (1965)
Levenspiel, O.: Chemical Reaction Engineering, chap. 1. Wiley Eastern Pvt. Ltd, New York (1969)

Bibliography

Darken, L.S., Turkdogan, E.T.: Heterogeneous Kinetics at Elevated Temperatures. In: Belton, G.R., Worrell, W.L. Plenum Press, New York (1970)
Glasstone, S., Laidler, K.J., Eyring, H.: The Theory of Rate Processes. McGraw-Hill Book Co., New York (1941)
John, H.Y., Wadsworth, M.E. (eds.): Rate Processes in Extractive Metallurgy. Plenum Press, New York (1979)
Laidler, K.L.: Chemical Kinetics. McGraw-Hill Book Co., New York (1965)
Smith, J.M.: Chemical Engineering Kinetics. McGraw-Hill Book Co., New York (1970)
Szekeley, J., Themilis, N.J.: Rate Phenomena in Metallurgy. Wiley, New York (1971)

Chapter 2
Empirical and Semi-Empirical Kinetics

2.1 Introduction

2.1.1 General Rate Equation

Consider a chemical reaction

$$aA + bB + \cdots = cC + dD + \cdots \tag{2.1}$$

By stoichiometric considerations, one can also write

$$\frac{1}{a}\left(-\frac{dC_A}{dt}\right) = \frac{1}{b}\left(-\frac{dC_B}{dt}\right) = \frac{1}{c}\left(\frac{dC_C}{dt}\right) = \frac{1}{d}\left(\frac{dC_D}{dt}\right) \tag{2.2}$$

The term C_i indicates concentration, and each term within ( ) brackets indicates a rate of change of concentration. Each of the terms within ( ) brackets is an index of reaction rate, and these are interrelated.

Determination of concentration change affords a direct measure of reaction rate for solutions. However, in many cases, e.g. reaction of solids, this may not be useful. There could then be other parameters which can be measured to study the progress of a reaction, e.g. total weight of a reactants or product, volume of a product, etc. Alternately, there could be an "indirect" measure in terms of a property that is a function of changes occurring, e.g. variation in dimensions of a solid or that in mechanical, electrical or magnetic properties. However, if there is no direct correlation between such variations and the amounts of reactants consumed or products formed, then a kinetic analysis based on such "indirect" indices becomes empirical.

H. S. Ray and S. Ray, *Kinetics of Metallurgical Processes*,
Indian Institute of Metals Series, https://doi.org/10.1007/978-981-13-0686-0_2

Considering any particular reactant i one may write

$$-\frac{dC_i}{dt} = k \cdot C_A^a \cdot C_B^b \cdots \tag{2.3}$$

where $a+b+\cdots = n$ is the overall order of the reaction. By definition, the reaction is of order a w.r.t. A, or order b w.r.t. B and so on. The exponents a, b, c, etc., are usually simple integers but occasionally they may be fractions or even negative depending on the complexity of the reaction. For the nth-order reaction of a single component, one can write

$$-\frac{dC}{dt} = kC^n \tag{2.4}$$

where C is concentration and k is rate constant. For $n \neq 1$, one obtains by integration

$$\frac{1}{n-1}\left[\frac{1}{C^{n-1}} - \frac{1}{C_0^{n-1}}\right] = kt \tag{2.5}$$

Here C_0 is concentration at zero time and C is concentration at time t. For

$$n = 0, \quad C_0 - C = kt \tag{2.6}$$

$$n = 1, \quad \ln(C_0/C) = kt \tag{2.7}$$

It is interesting to note that while for a zero-order reaction, the time $t_{0.5}$ for half the reactant to get consumed depends upon the initial concentration, and for the first-order reaction this is not so. From Eq. 2.7, we have

$$t_{0.5} = \frac{0.69315}{k} \tag{2.8}$$

Thus *half-life* is independent of initial concentration. Radioactive decay and decomposition of several compounds (e.g. N_2O_5, acetone, etc.) follow first-order reactions, and the initial number of radioactive particles of concentration of compound decreases exponentially. There are many examples of first-order growth process; for example, population growth, bank deposit earning compound interests, etc.

For radioactive decay, say N_0 number of particles becomes N number of particles in time t. We have

$$\ln(N_0/N) = kt,$$

and, for $N = N_0/2$ at $t = t_{0.5}$,

$$\ln(2N_0/N_0) = kt_{0.5} \tag{2.9}$$

Hence,

$$\ln(N_0/N) = t \ln 2/t_{0.5} \tag{2.10}$$

Thermodynamic and kinetic orders of reactions

The order defined above refers to kinetics of a process and must be distinguished from thermodynamic order of a reaction. For a thermodynamically first-order reaction, there is a discontinuous change in the slope of the plot of free energy against temperature at the transition temperature: the first derivative of free energy with temperature is discontinuous at the transition temperature. This means that the transformation involves a finite latent heat of transformation. Most structural transformations in physical metallurgy, for example, are thermodynamically first order. In a thermodynamically second (or higher)-order transformation, discontinuity at the transition temperature is noted for the second (or higher) order of derivative of free energy with temperature. Examples of second-order transitions are some ordering transformations. For example, in β-brass, the order steadily falls to zero as the temperature rises. Thermodynamically second-order transition involves discontinuous change in specific heat at the transition temperature. Clearly, the thermodynamic order of a transformation has a clear-cut physical basis and must be an integer. In contrast, the kinetic order of a reaction is an empirical parameter and need not be an integer.

Example 2.1 A wooden antique has 25.6% as much C^{14} as a recently grown piece of wood. If the amount of C^{14} in the atmosphere was the same, when the old wood died as it is now what is the age of the antique? Half-life for C^{14} is 5600 yrs.

Solution

$$\ln(N_0/N) = t \ln 2/t_{0.5} \text{ and } N = 0.256N_0$$

$$\therefore\ t = \frac{t_{0.5}\ln(N_0/N)}{\ln 2} = \frac{5600}{0.693}\ln(1/0.256) \cong 11{,}000 \text{ yrs}$$

Example 2.2 Amount of a product from a reaction increases 20% per minute. What is the order of the reaction and what is the rate constant?

Solution

If the product amount is P, then $\frac{\mathrm{d}P}{P}/\mathrm{d}t = \frac{20}{100}$, i.e. $\frac{\mathrm{d}P}{\mathrm{d}t} = 0.2P$. Therefore, the reaction is first order and the rate constant is 0.2 min^{-1}.

Example 2.3 The energy consumption of a country during 1990 was twice that consumed during 1980. Assuming that the country consumed a total amount of energy up to 1980, equal to E units, how much did it consume during the period 1980–1990? Make suitable assumptions.

Solution

If we assume a natural growth in energy consumption, then we can assume a first-order model for the increase. That is

$$\frac{dE/E}{dt} = k$$

where k is the growth rate. Thus, $(dE/dt) = kE$, that is, the rate of energy consumption is proportional to energy consumption. Hence, the doubling time for both must be the same. Therefore, energy consumption during 1980–1990 must equal that consumed up to 1980, i.e. E units.

Example 2.4 During a reaction, the amount of product formed increases at a fixed rate of 10% per minute. If after 10 minutes the amount is 10 g, then what is the amount formed after 1 h?

Solution
Let the weight formed in time t be W. Then

$$\frac{dW/W}{dt} = 0.1, \text{ or } \quad \frac{dW}{W} = 0.1\,dt. \quad \text{i.e., } \ln W = 0.1\,\text{t}$$

We have $\frac{\ln 10}{\ln W} = \frac{\log 10}{\log W} = \frac{10}{60}$
Weight after an hour there is $W = 10^6$ g (if the reagent supply is unlimited).

2.1.2 *Degree of Reaction*

It is often more useful to express the kinetic equation in terms of a dimensionless parameter α, called the degree of reaction or conversion. At time t

$$\alpha = \frac{C_0 - C}{C_0 - C_\infty} \tag{2.11}$$

where C_∞ denotes concentration of this reactant after infinite time of reaction. If $C_\infty = 0$, then

$$\alpha = 1 - \frac{C}{C_0} \text{i.e., } C = C_0(1 - \alpha) \tag{2.12}$$

The differential form of the rate equation (Eq. 2.4) then becomes

$$\frac{d\alpha}{dt} = \frac{k \cdot C_0^n \cdot (1 - \alpha)^n}{C_0} = k' \cdot (1 - \alpha)^n \tag{2.13}$$

where k' is another constant—the rate constant when reaction rate is expressed in terms of α. We will write k for k'. The term $(1 - \alpha)$ denotes fraction of unreacted material which, together with the temperature-dependent rate constant, governs the reaction rate. In the simplest case of a zero-order reaction, n equals zero and the rate has a constant value k.

The integral form of rate equation is

$$\alpha = kt \tag{2.14}$$

For first-order reactions $(n = 1)$, one obtains

$$\frac{d\alpha}{dt} = k\,(1 - \alpha) \tag{2.15}$$

i.e.

$$-\ln(1 - \alpha) = kt \tag{2.16}$$

For higher orders, the integrated form of the rate equation is obtained by substituting Eq. 2.12 in Eq. 2.5. In general, the integral form of the rate equation is written as

$$g(\alpha) = kt \tag{2.17}$$

where $g(\alpha)$ is an appropriate function of α. Differentiating with respect to time,

$$\begin{aligned} g'(\alpha) \cdot \frac{d\alpha}{dt} &= k \\ \frac{d\alpha}{dt} = \frac{k}{g'(\alpha)} &= k \cdot f(\alpha) \end{aligned} \tag{2.18}$$

Therefore, $f(\alpha)$ in the differential form of rate equation equals $1/g'(\alpha)$.

One advantage of using dimensionless α is that the dimension of k is always t^{-1} irrespective of the order of reaction. Moreover, one can use a variety of parameters to express the progress of a reaction.

2.2 Evaluation of Kinetic Parameters *E* and *A*

The kinetic parameters E and A can be obtained using either a conventional integral approach or two differential approaches.

2.2.1 Integral Approach

If the integrated form of the kinetic law, i.e. Eq. 2.17, is known, then one can plot, for different isothermal experiments, values of $g(\alpha)$ against time. Each of these plots will be linear, the slope being the value of k at that temperature. Since

$$k = A\ \exp(-E/\mathrm{RT})$$

the plot of ln k versus reciprocal temperature would be a straight line. The slope of such a line gives $-E/R$ and intercepts ln A. Such an approach requires prior knowledge of the form of the function $g(\alpha)$. There is also an implicit assumption that the activation energy E does not change during the course of the reaction.

2.2.2 Differential Approaches

In the differential approach, it is not necessary to assume any kinetic law at all. One can evaluate the activation energy E without knowing the forms of $g(\alpha)$ or $f(\alpha)$.

We have, from Eq. 2.18,

$$\frac{\mathrm{d}\alpha}{\mathrm{d}t} = k \cdot f(\alpha) = A \cdot \exp\cdot(-E/\mathrm{RT}) \cdot f(\alpha) \tag{2.19}$$

Considering a fixed value of α, one obtains

$$\ln\left(\frac{\mathrm{d}\alpha}{\mathrm{d}t}\right)_{\alpha} = \ln A + \ln f(\alpha) - E/\mathrm{RT} \tag{2.20}$$

Since $f(\alpha)$ has a fixed value for a fixed α, it follows that a plot of the left-hand side against reciprocal temperature would be a straight line, the slope of which should yield the value of E/R.

In this approach, one has the advantage that it is not necessary to know the form of either $g(\alpha)$ or $f(\alpha)$. Also, one does not assume E to be independent of α, and, therefore, E can be calculated at different levels of α. However, the disadvantage is that one has to know $\mathrm{d}\alpha/\mathrm{d}t$ values. One can obtain these from $\alpha - t$ plots but calculation of slopes often involves uncertainties.

There is a second differential approach. Equation 2.19 can be rewritten, again for a fixed value of α, as

$$\frac{1}{A \cdot \exp\cdot(-E/\mathrm{RT})} \int_0^{\alpha} \frac{\mathrm{d}\alpha}{f(\alpha)} = \int_0^{t_{\alpha}} \mathrm{d}t \tag{2.21}$$

$$t_\alpha = \frac{\text{const.}}{A} \cdot \exp\left(\frac{E}{\text{RT}}\right) \tag{2.22}$$

This shows that, provided $f(\alpha)$ remains unchanged, the time required for a fixed value of α is directly proportional to $\exp(E/\text{RT})$. Thus, a plot of $\ln t_\alpha$ against reciprocal temperature should be a straight line with slope E/R. This approach has the same advantages as those of the first differential approach. In addition, it has the advantage of using t values which are more reliably determined than $\mathrm{d}\alpha/\mathrm{d}t$ values. However, the differential approaches cannot evaluate the pre-exponential constant A and, therefore, the rate constant k.

Ray and Kundu (1986) have reported a kinetic analysis of some data on gaseous reduction of Fe_2O_3 to wustite state. Isothermal data for three temperatures are shown in Fig. 2.1 as $\alpha - t$ and $g(\alpha) - t$ plots. Figure 2.2 shows calculation of E using different approaches.

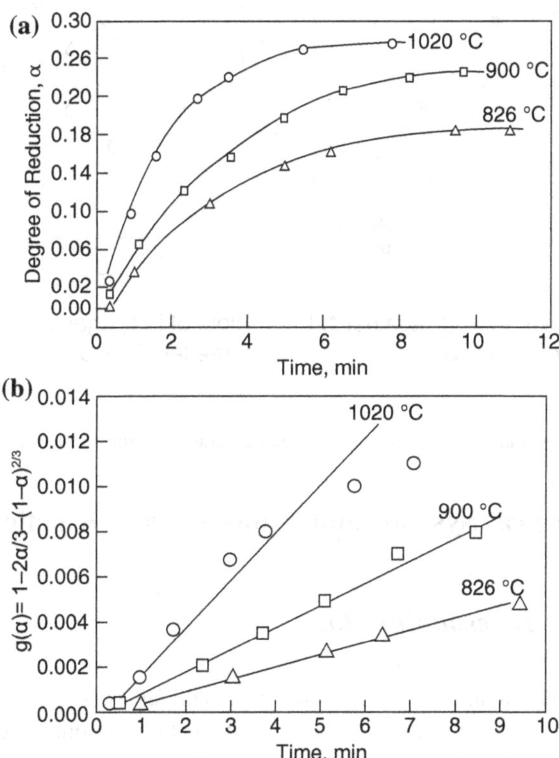

Fig. 2.1 **a** Kinetic data for isothermal reduction of iron ore (Ray and Kundu 1986). **b** Ginstling–Brounshtein plots to linearize kinetic data

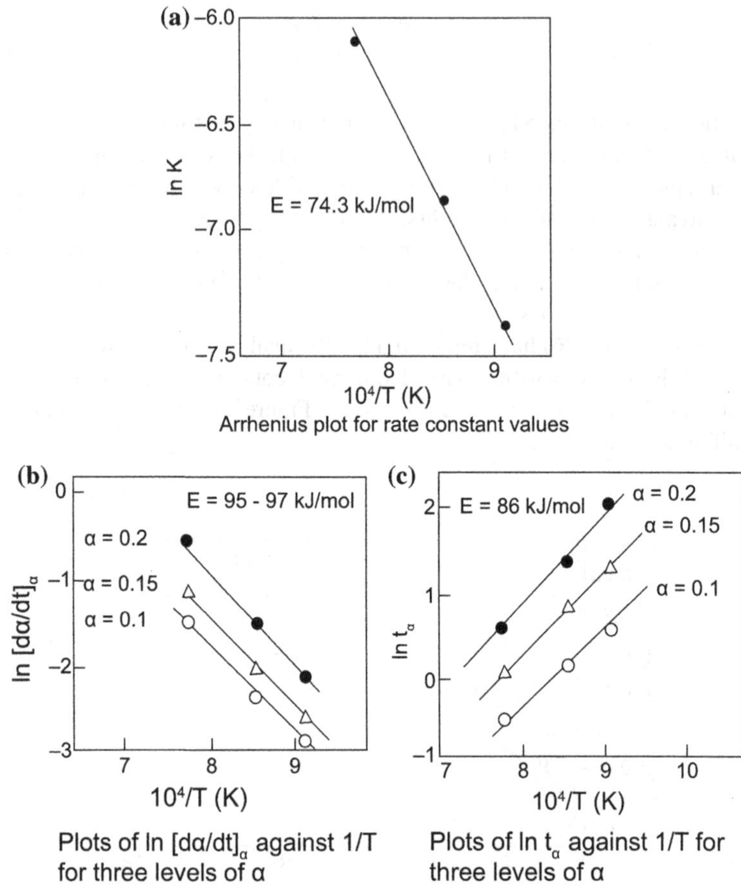

Fig. 2.2 Different approaches for evaluation of kinetic parameters for the data in Fig. 2.1

2.3 Metallurgical Systems and Approaches in Kinetic Analysis

2.3.1 *Factors Determining Rate*

Metallurgical transformations and reactions are mostly *heterogeneous* in nature, i.e. they involve more than one phase and there are distinct phase boundaries. Table 2.1 shows the different type of heterogeneous non-catalysed reactions of metallurgical importance as summarized by Habashi (1970).

In any given system, the rate of reaction at a given time depends basically on three factors, namely the nature of the system, the time of reaction and the temperature.

Table 2.1 Types of heterogeneous non-catalysed reactions of metallurgical importance (Habashi 1970)

Interface	Type	Examples
Solid–gas	$S_1 + G \rightarrow S_2$ $S_1 \rightarrow S_2 + G$ $S_1 + G_1 \rightarrow S_2 + G_2$	Physical Adsorption Chemical Oxidation of metals Decomposition of carbonates, or sulphates Oxidation of sulphide or gaseous reduction of oxides
Solid–liquid	$S \rightarrow L$ $S + L_1 \rightleftharpoons L_2$ $S + L_1 \rightarrow L_2$ $S + L_1 \rightarrow S_2 + L_2$	Physical Melting Dissolution–crystallization Chemical Leaching Cementation
Solid–solid	$S_1 \rightarrow S_2$ $S_1 + S_2 \rightarrow S_3 + G$ $S_1 + S_2 \rightarrow S_3 + S_4$	Physical Sintering, phase transformation Chemical Reduction of oxides by carbon Reduction of oxides or halides by metals
Liquid–gas	$L \rightleftharpoons G$ $L_1 + G_1 \rightarrow L_2 + G_2$ $L_1 + G \rightarrow L_2$	Physical Distillation–condensation Absorption Chemical Steelmaking by pneumatic process Absorption of gases in water
Liquid–liquid	$L_1 \rightleftharpoons L_2$	Solvent extraction Slag–metal reaction Liquid metal–liquid metal extraction

$$\text{Rate} = f(\text{nature of system}, t, T) \tag{2.23}$$

Rate is measured using a suitable parameter which changes with reaction. It has been mentioned previously that, at times, it may be difficult to have unambiguous index of rate. Uncertainties in kinetic studies can also arise out of poor incomplete characterization of the "nature" of the system. The word implies not only the system under test but also the environment. It encompasses chemical composition, presence of impurities which are inert and those which have a catalytic effect, distribution of impurities, physical factors such as size and shape of particles. Reproducible kinetic studies, therefore, require strict characterization (Ray et al. 1983).

Factors which lead to complications for real systems involving solid–gas reactions, for example, may be summarized as follows:

(a) Presence of impurities in the solid and the gas composition,
(b) Attempts at interpreting packed bed reaction rates using experimental data on singles particles of fixed geometry (sphere, cube, etc.),
(c) Multiparticle nature of systems with or without uniform particle size and shape,
(d) Reagent starvation and heat transfer problems,
(e) Variations in pore structure of solid during reaction,
(f) Non-isothermal condition, etc.

To simplify kinetic studies on a fixed system, it is often advantageous to fix the temperature and study the variation of rate with time. For studying the effect of temperature, such isothermal experiments are repeated at several temperatures. The initial section of this book deals with such isothermal approach. Non-isothermal experiments where time and temperature vary simultaneously are discussed in a latter chapter.

2.3.2 Approaches Towards Rate Laws

One of the main aims of a kinetic study is frequently to establish a rate law which can be used in prediction of reaction rates under a given set of circumstances. There can be basically three approaches in this, namely

(a) An empirical approach,
(b) A semi-empirical approach,
(c) A mechanistic approach.

When a system is rather complex and the transformation phenomena not well understood, it is not possible to establish a theoretical basis for mathematical formulations. In such cases, often it is advisable to employ an empirical approach with aims at establishing mathematical relationships to explain the trend in experimental results. Where some theoretical understanding is available, a semi-empirical approach may be possible. A rational mechanistic approach is possible only when the system is well-defined, and there is evidence to suggest a well-defined reaction mechanism. The preceding statements would be explained by the use of examples.

Samuel et al. (1987) studied the kinetics of degradation of impact toughness in type AISI 316 austenitic stainless steel due to thermal ageing at elevated temperatures. Isothermal ageing for various combinations of ageing durations t_α in the range 10–1000 h, and temperatures T in the range 823–1123 K, were used, and Charpy V impact energy C_v absorbed in half-thick standard specimens was used as the measure of toughness; Fig. 2.3a shows the test data. From these data, the $\ln t_\alpha - T$ combinations for various fixed levels of C_v can be determined by simple linear interpolations, Fig. 2.3b. Each of these constant C_v levels is taken to correspond to a fixed value for α in Eq. 2.22. Figure 2.3b suggests that most of the data can be described by Eq. 2.22, with a constant pre-exponential term. The value of E determined using a least square method so as to minimize the data spread was

Fig. 2.3 Thermal ageing-induced degradation of toughness in AISI-type 316 austenitic stainless steel (Samuel et al. 1987)

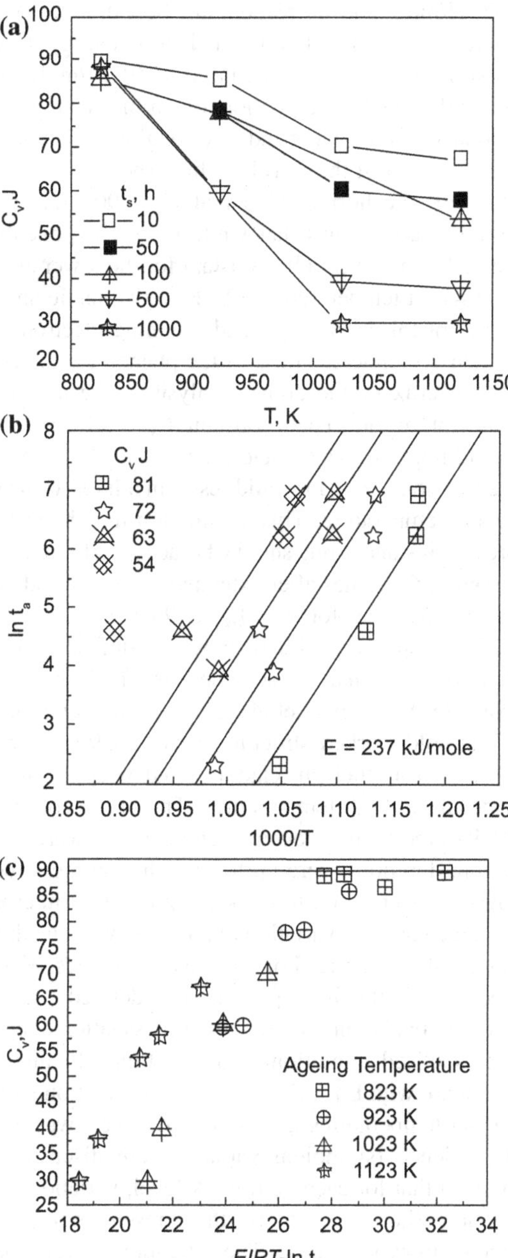

237 kJ/mol. The corresponding best-fit lines are also shown in Fig. 2.3b. It is interesting to know that the activation energy thus determined is similar to that for sensitization in this material ($\sim$240 kJ/mol). Thus, Samuel et al. (1987) had concluded that precipitation of carbides at grain boundaries is the mechanism responsible for ageing-induced toughness degradation in this material. Figure 2.3c plots C_v against the $E/\mathrm{RT}\ - \ln\, t_\alpha$ (negative of "temperature-compensated" ageing duration) the horizontal bar at $C_v = 90$ J corresponds to unaged material. This figure clearly shows that while the pre-exponential term in Eq. 2.22 can be considered to be essentially constant for the lower ageing temperatures, it is defined for the highest temperature, 1123 K. This kinetic analysis along with additional inputs from optical microscopy and scanning electron fractography led to a fairly comprehensive understanding of the phenomenon; see Samuel et al. (1987) for details.

The crux of the above analysis is the assumption that irrespective of ageing temperature, and each constant C_v level corresponds to fixed value for α. This would imply that a sufficiently prolonged ageing (corresponding to "completion of the reaction", $\alpha = 1$) would result in a fixed value of C_v that would be dependent of ageing temperature. This assumption may be questioned on a rigorous basis in the present instance, but should be acceptable as long as irrespective of ageing temperature, C_v values after extended ageing periods are sufficiently small compared to the initial value (for $\alpha = 0$), i.e. 90 J.

There are many situations where similar assumptions would be reasonable, and therefore the analysis scheme described above can be adopted or adapted. For example, Vasudevan et al. (1996) used similar analyses to examine the stability of initial cold-worked structure in alloy D9 (a Ti-modified version of AISI-type SS 316, used as fuel pin cladding and wrapper material in nuclear fast reactors) on exposure to high temperatures. The initial cold work levels were 15, 17.5, 20 and 22.5%; the thermal ageing temperatures were in the range 873–1123 K, and the ageing durations were up to 4810 h. Hardness measurements supported by metallography were used to track the process of microstructure evolution.

In general, the variation of hardness with the duration of ageing mimicked typical stages of annealing. For analysis of kinetics, Vasudevan and coworkers defined a fraction softening parameter, defined as $(H_0 - H_a)/(H_0 - H_r)$, where H_0 = initial hardness, H_a = hardness after ageing, and H_r = hardness in fully recrystallised condition. Thus, fraction softening corresponds to α as defined above (compare with Eq. 2.11). Variation of fraction softening with ageing times for fixed ageing temperatures and cold work levels was generally consistent with sigmoidal dependence (sigmoidal variations are discussed in Sect. 2.5). The authors also showed that for each cold work level, variations of $t_{0.5}$, the time for 50% recrystallization (i.e. $\alpha = 0.5$ in the present nomenclature) could be described by Arrhenius-type relation. The activation energies determined from these plots varied with the prior cold work level: $\sim$275 kJ/mol for the 15% cold-worked material, increasing to $\sim$420 kJ/mol for 20% cold work, and then sharply decreasing to $\sim$170 kJ/mol for 22.5% cold work.

2.4 The Empirical and Semi-empirical Approaches

This is best illustrated by the use of some illustrative examples which follow.

2.4.1 *Densification of Powder Compacts*

A good example of the empirical approach is provided by studies on densification kinetics. It should be noted that sintering and densification do not mean the same thing. Sintering is seemingly a simple process which transforms green powder agglomerates into dense products which are strong. The essential features are, therefore, densification and bonding of particles leading to product strength. The driving force for sintering is provided by the tendency of the agglomerate to reduce the total surface area of the powder particles and, therefore, total surface free energy. Densification merely implies reduction in porosity. It is often not well understood that sintering does not necessarily imply densification or vice versa. One can have one without the other.

In most cases, however, sintering leads to densification. The subject of sintering may be approached, for simplicity, from a point of view that any green compact, at best, is a two-phase material—porosity and the solid material, each with its own morphology, i.e. shape, size, distribution and amount (Bagchi and Sen 1982). Densification, a common feature of sintering, is the process of elimination of pores. For actual sintering, it is not possible to have a better, thoroughly rational approach unless one is restricted to study of bonding between two spherical particles or rods of cylindrical geometry. For an actual powder mass, we write following Eq. 2.19,

$$\frac{\mathrm{d}\alpha}{\mathrm{d}t} = k \cdot f(\alpha) \tag{2.24}$$

where α represents the fraction of originals porosity destroyed. α can be expressed as follows

$$\alpha = \frac{v_\mathrm{p} - v_\mathrm{s}}{v_\mathrm{p}} \tag{2.25}$$

where v_s and v_p denote, respectively, the volume of pores in the sintered compact and volume of pores in the original green compact. If

V_0 volume of the green compact,
V_s volume of the sintered compact,
V_th volume of the compact when it attains the theoretical density, ρ_th, and ρ porosity of the green compact expressed as fraction,

$$\alpha = \frac{v_p - v_s}{v_p} = \frac{V_0 - V_s}{V_0 - V_{th}} = \frac{1 - \frac{V_s}{V_0}}{1 - \frac{V_{th}}{V_0}} = \frac{1 - (V_s/V_0)}{1 - (\rho_0/\rho_{th})} \\ = \frac{1 - (V_s/V_0)}{p} = \frac{\Delta V/V_0}{p} \tag{2.26}$$

where ρ_0 is the density of the green compact and ΔV is the volume change.

Densification studies are usually carried out using a cylinder compact. For such a compact, there is change in both diameter and length as sintering progresses. While one can measure continuously changes in length easily using a dilatometer, continuous measurement of changes in diameter is not possible. It will, therefore, be useful if densification could be studied only in terms of length changes only. This is possible under some assumptions.

Let

- L_0 length of the green cylindrical compact
- D_0 diameter of the green cylindrical compact
- L length after densification for time t
- D diameter after densification for time t
- ΔL_∞ maximum change in length after densification for infinite time
- ΔD_∞ maximum change in diameter after densification for infinite time

We assume that at all values of t, $\frac{\Delta D}{\Delta L} \approx \frac{\Delta D_\infty}{\Delta L_\infty}$, i.e.

$$\Delta D = \frac{\Delta D_\infty}{\Delta L_\infty} \cdot \Delta L \tag{2.27}$$

We have $V_s = \frac{\pi}{4} \cdot LD^2$. Since $\Delta L = L_0 - L$ and $\Delta D = D_0 - D$,

$$V_s = \frac{\pi}{4}(L_0 - \Delta L)(D_0 - \Delta D)^2 = \frac{\pi}{4}(L_0 - \Delta L)\left(D_0 - \frac{\Delta D_\infty}{\Delta L_\infty} \cdot \Delta L\right)^2$$

But, $V_0 = \frac{\pi}{4} L_0 D_0^2$. Combining the last three equations

$$\frac{V_s}{V_0} = \frac{(L_0 - \Delta L)\left(D_0 - \frac{\Delta D_\infty}{\Delta L_\infty} \cdot \Delta L\right)^2}{L_0 D_0^2} = \left(1 - \frac{\Delta L}{L_0}\right)\left(1 - \frac{\Delta D_\infty}{\Delta L_\infty} \cdot \frac{\Delta L}{D_0}\right)^2 \tag{2.28}$$

If $\Delta D_\infty \cdot \Delta L \ll \Delta L_\infty \cdot D_0$, then

$$\frac{\Delta L}{L_0} \approx 1 - \frac{V_s}{V_0} \approx \frac{\Delta V}{V_0} \tag{2.29}$$

This will be specially true for long cylinders. Again if $\Delta D_\infty \approx L_\infty$ and $D_0 \approx L_0$, then

$$\frac{V_s}{V_0} \approx \left[1 - \frac{\Delta L}{L_0}\right]^3 \approx 1 - \frac{3\Delta L}{L0} \text{ if } \Delta L \ll L_0$$

i.e.

$$\frac{\Delta L}{L_0} \approx \frac{1}{3} \approx \frac{[V_0 - V_s]}{V_0} \approx \frac{1}{3}\frac{\Delta V}{V_0} \tag{2.30}$$

We thus see that $\Delta V/V_0$ can be used as a parameter indicative of α. Moreover in some cases, $(\Delta V/V_0)$ equals $(\Delta L/L_0)$ which can then be also used as a kinetic parameter. It is, generally, more common to measure $\Delta L/L_0$ for which dilatometers are convenient.

Densification kinetics, therefore, is commonly expressed in terms of an empirical equation of the type

$$\alpha = \frac{\Delta L}{L_0} = kt^n \text{ or } \alpha^{1/n} = k't \tag{2.31}$$

where the exponent is a constant roughly indicative of the reaction mechanism. The literature (Brophy et al. 1971) gives the following guidelines.

$n = 0.3$ indicates grain boundary diffusion,
$n = 0.4$ indicates volume diffusion from grain boundaries,
$n = 0.5$ indicates volume diffusion from grain boundaries and surface,
$n = 1.0$ indicates viscous flow.

Where such an equation holds, the volume of the exponent is easily obtained by plotting log $(\Delta L/L_0)$ against $\ln t$. The plot should be linear

$$\ln\left(\frac{\Delta L}{L_0}\right) = n \ \ln t + \ln k' \tag{2.32}$$

The slope gives the value of n and the intercept gives k. The differential form of the rate equation is obtained by differentiating Eq. (2.31). Thus,

$$\frac{d\alpha}{dt} = k' \ n \cdot \alpha^{(n-1)/n} \tag{2.33}$$

Such a simple approach may not be valid in an actual process of sintering, especially when the process is more complex such as in the case of liquid-phase sintering materials (e.g. W–Cu, W–Ag) and a variety of other materials including cermets and ceramic porcelains, where a liquid phase exists at the sintering temperatures. In such cases, densification is said to involve several sequential stages

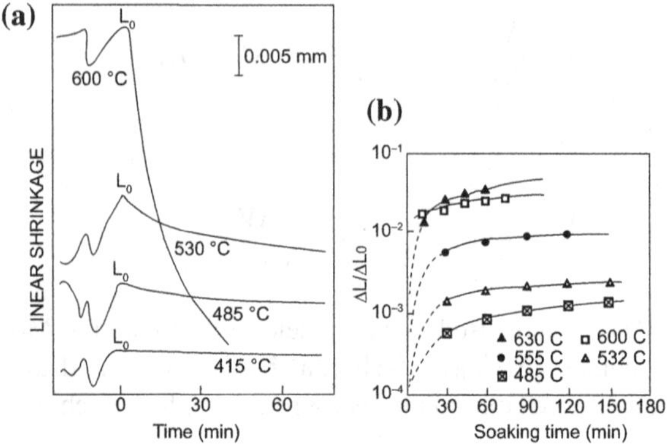

Fig. 2.4 **a** Isothermal linear shrinkage versus soaking time; **b** $\Delta L/L_0$ versus soaking time at different temperatures (Virkar et al. 1986)

with separate mechanism. Therefore, no single rate equation may be applicable. [1] Virkar et al. (1986) have reported an investigation on kinetics of sintering of monodisperse glass powder. Some of their data on variation of $\Delta L/L_0$ against time are shown in Fig. 2.4. Figure 2.5 shows a plot of $\log(\Delta L/L_0)$ values against log t values. If Eq. 2.31 is valid, then such a plot would be linear with slope n. Figure 2.5 shows linear plots whose slopes approximately equal 0.8. This figure also shows plots of $\log(\Delta L/L_0)^{1/0.8}$ against t. The slopes of such plots yield values of k. Figure 2.6 shows an Arrhenius plot with slopes so obtained. In the present case, n is near unity because sintering takes place mainly by viscous flow.

Example 2.5 A green powder compact of 30 pct porosity undergoes densification during sintering. If there is 20 pct increase in density, then what is the degree of densification?

Solution

According to Eq. 2.26, the degree of densification is

$$\alpha = \frac{1 - (V_s/V_0)}{p} = \frac{1 - \rho_0/\rho_s}{p}$$

we have, $\frac{\rho_s - \rho_0}{\rho_0} = \frac{20}{100}$ i.e., $\rho_s = 1.2\rho_0$

Hence, $\alpha = \frac{(\rho_s - \rho_0)}{\rho_s \cdot p} = \frac{\rho_s - \rho_0}{1.2\rho_0 \cdot p} = \frac{0.2}{1.2 \times 0.3} = 0.555$

[1]For review on sintering see Tendolkar (1983), Tendolkar and Sebastian (1980)

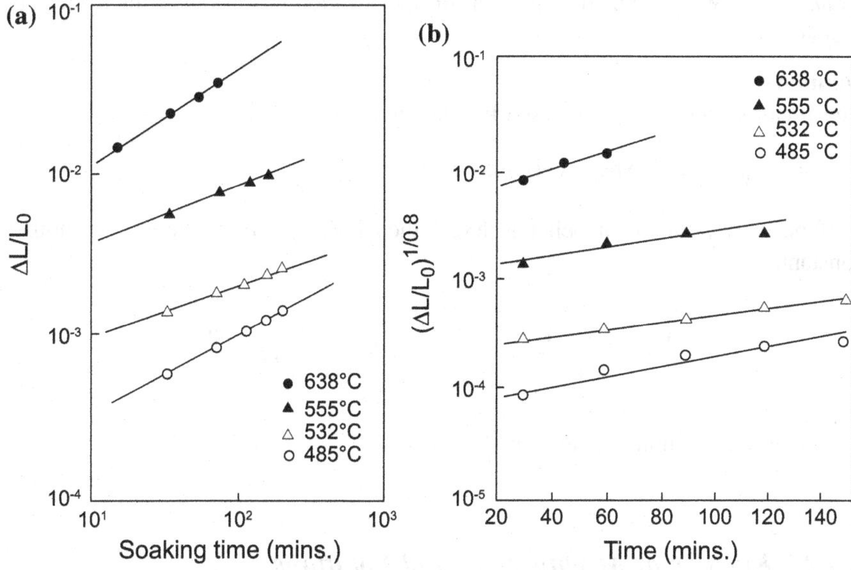

Fig. 2.5 **a** $\Delta L/L_0$ versus soaking time at different temperatures on log–log scale, **b** $\log(\Delta L/L_0)^{1/0.8}$ versus soaking time (Virkar et al. 1986)

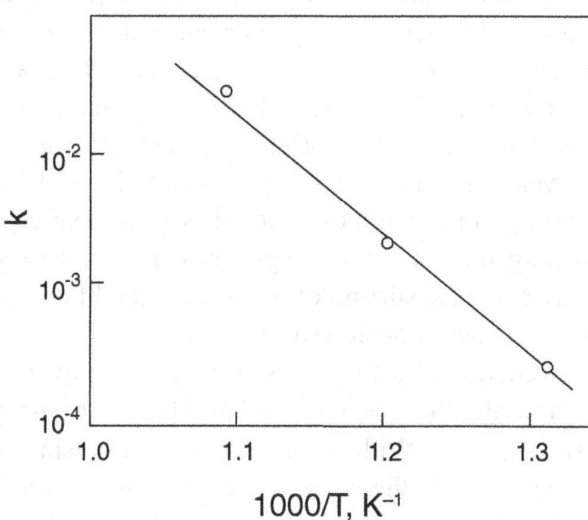

Fig. 2.6 Arrhenius plot of k values (Virkar et al. 1986)

Example 2.6 Write a kinetic equation for densification process in terms of density change.

Solution
Densification kinetics is expressed by the equation

$\alpha = \frac{1}{p} \cdot \frac{[\Delta V]}{[V_0]} = k' t^n$, where k' is a constant

If porosity p of green pellet is fixed, then $\Delta V / V_0 = kt^n$, where k is another constant.

$$\frac{\Delta V}{V_0} = \frac{V_0 - V}{V_0} = \frac{\frac{W}{\rho_0} - \frac{W}{\rho}}{\frac{W}{\rho_0}} = \frac{(\rho - \rho_0)}{\rho \rho_0} \cdot \rho_0 = \frac{\Delta \rho}{\rho}$$

Hence, the required equation is $\frac{\Delta \rho}{\rho} = kt^n$

2.4.2 *Kinetics of Mixing in Liquid Solutions*

As an example of semi-empirical approach let us consider another example—that of mixing of a solute in a solvent. If a solute is introduced in one region of a liquid, then it must spread because of concentration gradients. The solute is transferred to regions of lesser concentrations until the composition is uniform all throughout. The transfer involves the phenomena of diffusion, convection and turbulence; their relative importance depends on geometry of the system, liquid flow and stirring conditions and temperature gradients, among other conditions.

Many metallurgical processes make use of external agitation to accelerate mixing and a common method is to bubble a gas. Thus in bottom blowing steelmaking methods, the oxygen blown in not only refines the liquid metal but also causes intense stirring of bath. Such bubbling can have tremendous effects on mass and also heat transferred processes.

Agitation effect can be described in terms of the concept of effective diffusivity to describe the efficiency of mixing. In an unagitated liquid, it could be as low as 10^{-4} cm^2 s^{-1}. With a stream of bubbles rising in a tube with a superficial velocity of 0.08 m s^{-1}, the same would increase to a value of about 75 cm^2 s^{-1}.

Mixing process is generally studied by determining mixing time using a tracer technique. Figure 2.7 shows an apparatus used by Das et al. (1985) for studying effect of bubbling on mixing.

A small quantity of acid is added as tracer in one region of the liquid and a pH metre measures the variation of pH with time at a distant region. A typical curve has the shape shown in Fig. 2.8a, which is sigmoidal variation of pH with time. Obviously, it is difficult to determine the time of complete mixing. To make the latter assume a finite, easily determined value one considers time for 99% mixing or

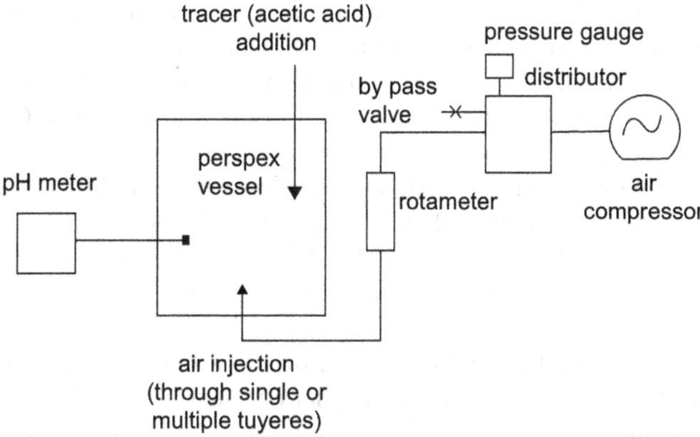

Fig. 2.7 Schematic representation of experimental set-up used in model study (Das et al. 1985)

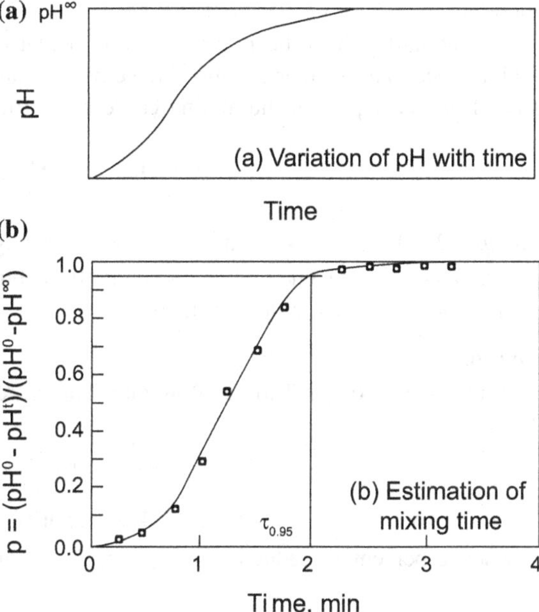

Fig. 2.8 Kinetic data for mixing of acid tracer in water **a** variation of pH with time; **b** estimation of mixing time

95% mixing, etc. To obtain a plot for mixing kinetics, one redraws Fig. 2.8a by expressing pH as a dimensionless parameter, degree of mixing α, as

$$\alpha = \frac{pH^0 - pH^t}{pH^0 - pH^\infty} \tag{2.34}$$

where superscripts 0, t and ∞ indicate, respectively, the initial condition, condition at time t, and that after infinite time. Figure 2.8b indicates how one determines time for 95% mixing ($\tau_{0.95}$). Mixing studies are perhaps better carried out using acid tracers and conductivity measurements instead of pH measurements. Conductivity directly relates to concentration.

In general, the mixing time τ is influenced by stirring according to the equation (Szekeley et al. 1979; Asai et al. 1983; Nakanishi et al. 1975)

$$\tau = K(\dot{E})^{-n} \tag{2.35}$$

where K is a constant (100–800) depending on the actual definition of τ (i.e. for what maximum degree of mixing), n is another constant (0.23–0.5) which depends on stirring mechanism and $\dot{E}$ is stirring power density. $\dot{E}$ is defined as the amount of kinetic energy dissipated by the rising bubbles per unit time per unit weight of the bath. It depends on bath weight, height and temperature volume of gas passed per unit time and densities of gas and liquid. Apparently, bubble diameter is not so important.

Mixing studies have been carried out in actual open hearths using radioactive gold as tracer, and similar equations have been obtained in this and other cases. In general, the later path of the mixing curve approximates the first-order equation

$$-\ln(1-\alpha) = kt \tag{2.36}$$

Example 2.7 During mixing in a liquid by bubble agitation, the value of exponent $n = 0.5$ and $\dot{E}$ has a value of 4 energy units. If the bubbling rate is quadrupled, then what is the percentage reduction in $\tau_{0.95}$?

Solution

$\dot{E}$ should be proportional to gas flow rate. Therefore,

$$(\tau_{0.95})_1 = K(4)^{\frac{1}{2}}, (\tau_{0.95})_2 = K(16)^{\frac{1}{2}}$$

$$\therefore \quad \tau_1/\tau_2 = 4/2 = 2 \text{ or } \tau_1 = 2\tau_2,$$

whence percentage reduction = $(\tau_1 - \tau_2)/\tau_1 = 50$.

2.5 Johnson–Mehl Equation

2.5.1 *Sigmoidal Variation of a Quantity*

Figure 2.8 for mixing kinetics shows a sigmoidal curve. Most phase transformations in physical metallurgy follow a similar kinetics. The distinctive feature is the initial increase in reaction rate followed by a gradual decrease. During phase transformations in metals and alloys, such initial rise in rate is caused by formation of increasing number of nuclei. The decreasing trend in rate is said to be caused by impingement of grains against one another and consequent impediment in growth. Accordingly, the kinetics is said to follow a nucleation and grain growth model.

2.5.2 *Kinetic Model for Nucleation and Grain Growth*

Kinetics of nucleation and grain growth is generally described by an empirical equation of the following form

$$\frac{\mathrm{d}\alpha}{\mathrm{d}t} = k^n\, t^{n-1}\,(1-\alpha) \tag{2.37}$$

where α denotes the degree of transformation, t, the time and k, a constant independent of α. The value of the exponent varies with reaction mechanism. In Eq. 2.37, the term t^{n-1} accounts for increasing rate in the initial stages, whereas the $(1-\alpha)$ term accounts for gradual decrease in rate as the reaction proceeds beyond a stage when rate is maximum. The increasing rate may be ascribed to nucleation, i.e. formation of new domains of the product. During later stages, the rate of growth of the new domains decreases because of mutual interference of neighbouring domains either through direct impingement or by long-range competition for solute atoms.

It should be noted that though k has the dimension t^{-1}, it is not a true rate constant because the right-hand side of Eq. 2.37 contains a t term. Integrating one obtains

$$\ln\left[\frac{1}{1-\alpha}\right] = \frac{k^n}{n}\, t^n = (k'\,t)^n \tag{2.38}$$

where k' equals $k/n^{1/n}$. k' is also, obviously, not a true rate constant. From Eq. 2.38, one also obtains

$$[-\ln(1-\alpha)]^{1/n} = (k'\,t) \tag{2.39}$$

This equation, or similar equations, has been proposed by Johnson and Mehl (1939), Fine (1978), Avrami (1939), Avrami (1940), Avrami (1941), Yerofeev and Mitekevich (1961), Young (1966) and Kolmogorov (1937). Henceforth, it will be referred to as the J-M Equation. The exponent n is said to be indicative of the growth mechanism. While $n = 2$ is for two-dimensional growth, $n = 3$ indicates three-dimensional growth involving initial random nucleation and then overlapping of growth nuclei. An n value equal to 4 has been said to result from two-stage nucleation. Equation 2.39 can be rewritten as

$$\log\left[\frac{1}{1-\alpha}\right]=\frac{1}{2.303}\,(\mathrm{k}'\,t)^n \tag{2.40}$$

$$\log\,\log\left[\frac{1}{1-\alpha}\right]=n\,\log\,t+n\,\log\,k-\log\,2.303 \tag{2.41}$$

Thus, a graph of log $\log[1/(1-\alpha)]$ versus log t is linear with slope n. k' is obtained from the intercept. This plot has been used by Sohn and Won (1985) and recommended by Burke (1965). k' can also be obtained directly from the $\alpha-t$ plots by noting that according to Eq. 2.39, k' equals the reciprocal time when α equals 0.6321. Thus, substituting $k'=1/t$ in Eq. (2.39), one obtain

$$\alpha=\frac{\mathrm{e}^{-1}}{\mathrm{e}}=0.6321 \tag{2.42}$$

Figure 2.9 shows some experimental data reported by Sarkar and Ray (1988), (1990) on reduction of hematite pellets to magnetite. Figure 2.10 shows linear plots according to Eq. 2.41 from which the value of n is obtained as equal to 3. The $g(\alpha)$ versus t plots (Eq. 2.39) are shown in Fig. 2.11.

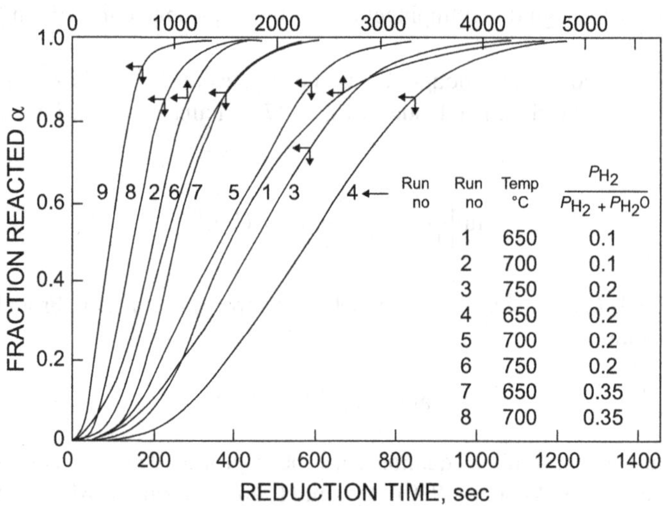

Fig. 2.9 Kinetic data for reduction of hematite particle to magnetite (Sarkar and Ray 1988, 1990)

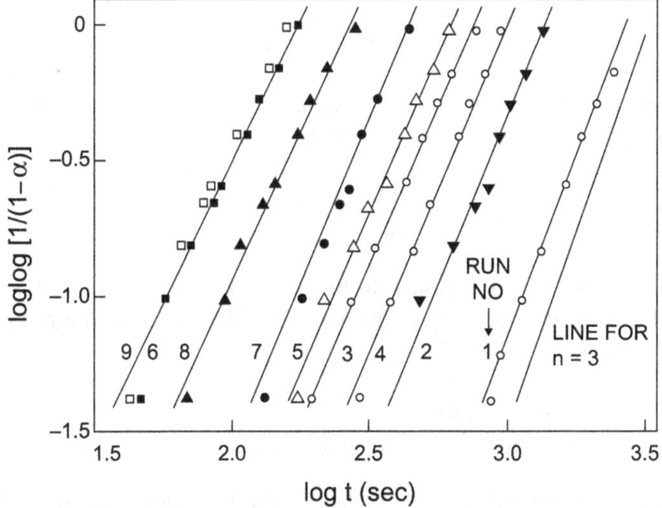

Fig. 2.10 Plots of kinetic data in Fig. 2.9 according to Eq. 2.41 and evaluation of exponent n in J-M equation (Sarkar and Ray 1988, 1990)

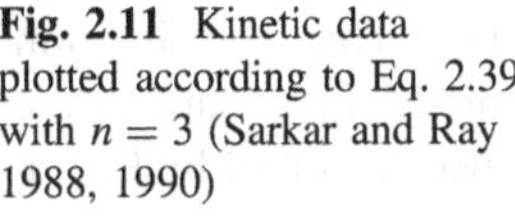

Fig. 2.11 Kinetic data plotted according to Eq. 2.39 with $n = 3$ (Sarkar and Ray 1988, 1990)

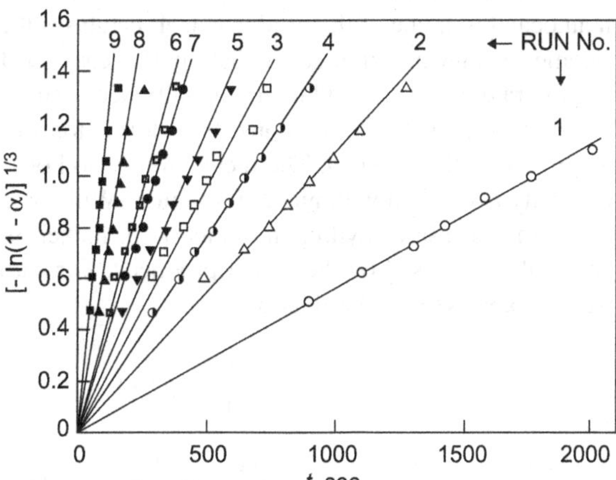

2.5.3 *Modification of Johnson–Mehl Equation*

For small values of α, i.e. for the initial part of the $\alpha - t$ plot $\mathrm{d}\alpha/\mathrm{d}t$ is proportional to t^{n-1}, i.e. α is proportional to t^n. Thus, $\mathrm{d}\alpha/\mathrm{d}t$ should be proportional to $\alpha^{(n-1)/n}$. It is thus plausible to assume that a relation of the following kind would be valid.

$$\frac{d\alpha}{dt} = k_m (1-\alpha)\,\alpha^{(m-1)/m} \tag{2.43}$$

where k_m is a true rate constant and exponent m is not necessarily equal to n. Equation 2.43 is analogues to the general expression

$$\frac{d\alpha}{dt} = k \cdot f(\alpha) \tag{2.44}$$

where $f(\alpha)$ denotes the functional form $(1-\alpha)\alpha^{(m-1)/m}$. Exponent m may be determined without the necessity of integration. Differentiating with respect to ln α Eq. 2.43 yields

$$\frac{d\ \ln(d\alpha/dt)}{d\ \ln\alpha} = \frac{d\ \ln k_m}{d\ \ln\alpha} - \frac{\alpha}{1-\alpha} + \frac{m-1}{m} \tag{2.45}$$

The procedure is to determine the gradient of the $\alpha - t$ plots of each of a series of α values. Values of ln $(d\alpha/dt)$ are then plotted against ln α. Since k_m is not a function of α, the first term in the right-hand side vanishes. The gradient of the plot thus gives $\alpha/(1-\alpha)$ and the intercept $(m-1)/m$. The latter yields the values of m for Eq. 2.43.

Autocatalytic reaction model

In autocatalysis, one of the products itself catalyses the reaction. Thus, the situation becomes somewhat similar to that in the case of Johnson–Mehl equation, the reaction rate first rising and then gradually decreasing. The mechanism of hydrogen reduction of nickel chloride consists of formation of nuclei of nickel followed by the growth of these nuclei. The incubation period observed at the start of reduction of usually associated with nuclei formation (Willams et al. 1981; Benton and Emett 1924). In the case of hydrogen reduction of nickel oxide, however, the reaction is autocatalytic, caused by the nickel produced. For autocatalytic reaction, the rate is usually expressed by the relation

$$\frac{d\alpha}{dt} = k\,\alpha\,(1-\alpha) \tag{2.46}$$

$$\frac{d\alpha}{\alpha} + \frac{d\alpha}{(1-\alpha)} = \mathrm{k}\,dt \tag{2.47}$$

Integrating, one obtains

$$\ln\alpha - \ln(1-\alpha) = kt + I \tag{2.48}$$

$$\ln[\alpha/(1-\alpha)] = \mathrm{k}t + I \tag{2.49}$$

where I is an integration constant. Equation 2.46 is not similar to modified Johnson–Mehl equation Eq. 2.43 because $(m-1)/m$ cannot be unity.

2.6 Reduced Time Plots

2.6.1 Preliminary Identification of Kinetic Equations

The identification of the kinetic model constitutes an important step in the analysis of high-temperature kinetic data for many metallurgical reactions. In many cases, a preliminary identification is possible using the so-called reduced time plots first introduced by Sharp et al. (1966) and Giess (1963). Although such plots have been widely used in the analysis of solid-state reactions, their application in metallurgy has been very limited. Ray (1983) has reviewed the subject.

In reduced time plots, the kinetic relationship is first expressed to the form:

$$g(\alpha) = kt \tag{2.50}$$

where α is the fraction reacted, k the rate constant and t the time. If $t_{0.5}$ be the time required to obtain 0.5 fraction reacted ($\alpha = 0.5$), then one obtains Eq. 2.50 in an altered form in terms of dimensionless time scale (reduced time)

$$g(\alpha) = A\ (t/t_{0.5}) \tag{2.51}$$

where A is a calculated constant dependent on the form of the function $g(\alpha)$. This expression is independent of the kinetic rate constants and is dimensionless. For a given reaction mechanism, a single equation represents all kinetic data irrespective of the nature of the system, temperature or any other factor that influence the rate.

Every mechanism has a unique dimensionless reduced time plot which is universally true for all reactions following this mechanism. These equations, therefore, offer a useful approach to the rapid selection of the appropriate rate equation.

The following is an example of how a kinetic equation is rewritten in terms of a reduced time plot. Suppose that a solid sphere is reacting with a fluid according to the rate expression

$$g(\alpha) = [1 - (1 - \alpha)^{1/3}] \tag{2.52}$$

where α is the fraction of solid reacted in time t. $g(\alpha)$ is a symbol to indicate this function. Then for $\alpha = 0.5,\ t = t_{0.5}$

$$[1 - (1 - 0.5)^{1/2}] = k \cdot t_{0.5} \tag{2.53}$$

Dividing Eq. 2.52 by Eq. 2.53, one gets

$$[1 - (1 - \alpha)^{1/3}] = 0.2063\,(t/t_{0.5}) \tag{2.54}$$

This reduced time equation can be used to obtain values of the left-hand side for various values of t provided the value of $t_{0.5}$ is known. This expression is independent of k and, therefore, valid for all systems and experimental conditions so long as the reaction mechanism follows Eq. 2.52.

2.6.2 Reduced Time Plots and Their Use

The appropriate reduced time expressions for a variety of other reaction mechanisms are available elsewhere (Ray 1983). In analysis of data, one first finds out $t_{0.5}$ in the $\alpha - t$ plot for a given temperature. The plot is then redrawn, for the same given temperature, by plotting α values of $t/t_{0.5}$. All isothermal $\alpha - t$ plots merge into a single α versus $t/t_{0.5}$ plot, if the kinetic data are truly isothermal and isokinetic. This α versus $t/t_{0.5}$ plot is matched against standard plots for different kinetic models. Table 2.2 lists the values of $t/t_{0.5}$ at different values of α for a number of kinetic models. Keatch and Dollimore (1975) have given graphical plots for several rate expressions in terms of reduced time.

The obvious drawback of the reduced time plots is that curves, and not straight lines, have to be matched. Secondly, if there are induction periods or initial time delays due to factors other than those associated with nucleation, then the time scale is faulty and so is the reduced time equation.

The first error may be countered by a double logarithmic analysis which gives straight lines of various slopes. However, not all equations have appropriate forms for such double logarithmic analysis. Moreover, the use of double logarithmic analysis desensitizes kinetic data and it does not get over the uncertainties in time scale due to induction periods, possible changes in reaction mechanisms, etc.

Table 2.2 Values of α and $t/t_{0.5}$ for some commonly used solid-state reaction equations

α	$D_1(\alpha)$	$D_2(\alpha)$	$D_3(\alpha)$	$D_4(\alpha)$	$F_1(\alpha)$	$R_2(\alpha)$	$R_3(\alpha)$	$A_2(\alpha)$	$A_3(\alpha)$
0.1	0.040	0.033	0.028	0.032	0.152	0.174	0.165	0.390	0.533
0.2	0.160	0.140	0.121	0.135	0.322	0.362	0.349	0.567	0.685
0.3	0.360	0.328	0.295	0.324	0.515	0.556	0.544	0.717	0.801
0.4	0.640	0.609	0.576	0.595	0.737	0.768	0.762	0.858	0.903
0.5	1.000	1.000	1.000	1.000	1.000	1.000	1.000	1.000	1.000
0.6	1.440	1.521	1.628	1.541	1.322	1.253	1.277	1.150	1.097
0.7	1.960	2.207	2.568	2.297	1.737	1.543	1.607	1.318	1.198
0.8	2.560	3.115	4.051	3.378	2.322	1.887	2.014	1.524	1.322

In Table 2.2

$D_1(\alpha) = \alpha^2 = (k/r^2)\, t$

$D_2(\alpha) = (1-\alpha)\ln(1-\alpha) + \alpha = (k/r^2)\, t$

$D_3(\alpha) = \left[1-(1-\alpha)^{1/3}\right]^2 = (k/r^2)\, t$

$D_4(\alpha) = (1-2\alpha/3) - (1-\alpha)^{2/3} = (k/r^2)\, t$

$F_1(\alpha) = \ln(1-\alpha) = -kt$

$R_2(\alpha) = \left[1-(1-\alpha)^{1/2}\right] = (k/r)\, t$

$R_2(\alpha) = \left[1-(1-\alpha)^{1/3}\right] = (k/r^2)\, t$

$A_2(\alpha) = [-\ln(1-\alpha)]^{-1/2} = kt$

$A_3(\alpha) = [-\ln(1-\alpha)]^{-1/3} = kt$

2.6.3 Examples of Application

Consider the data on reduction of Fe_3O_4 shown in Fig. 2.9. A reduced time plot for the same is shown in Fig. 2.12, which also shows the theoretical plots for the integrated forms of two J-M-type equations. This plot indicates that n is close to 3.

A better example is shown in Fig. 2.13a which shows kinetic plots for reduction of a central cylindrical column of iron ore fines by a surrounding layer of char fines, as reported by Mookherjee et al. (1986). Figure 2.13b shows data on kinetics of char gasification. Some of the kinetic data from Fig. 2.13a chosen at random are shown in Fig. 2.14 against standard reduced time plots for three kinetic models. The reduced time plot clearly identifies the correct equation. Figure 2.15a, b shows the linearized plots and Fig. 2.15c the Arrhenius plots. The latter have nearly identical slopes which indicate that reduction and gasification reactions are interrelated and there is a common activation energy. Additional information on reduced time plots is available elsewhere (Willams et al. 1981; Benton and Emett 1924; Keatch and Dollimore 1975; Ghosh and Ray 1984).

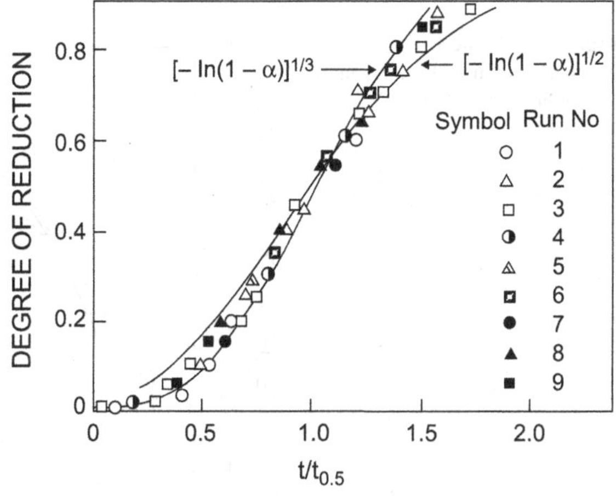

Fig. 2.12 Reduced time plot kinetic data shown in Fig. 2.9 for reduction of hematite pellets to magnetite (Sarkar and Ray 1988, 1990)

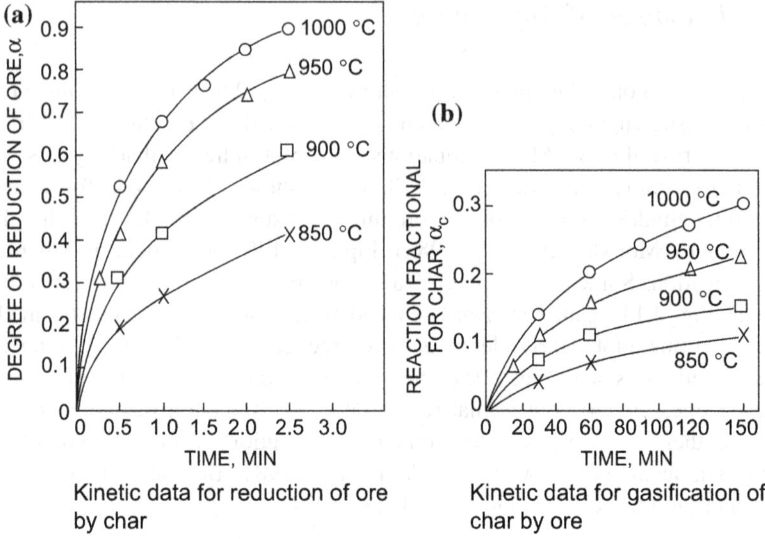

Fig. 2.13 Kinetic data for ore–char reaction: **a** kinetic data for reduction of ore by char; **b** kinetic data for gasification of char (Mookherjee et al. 1986)

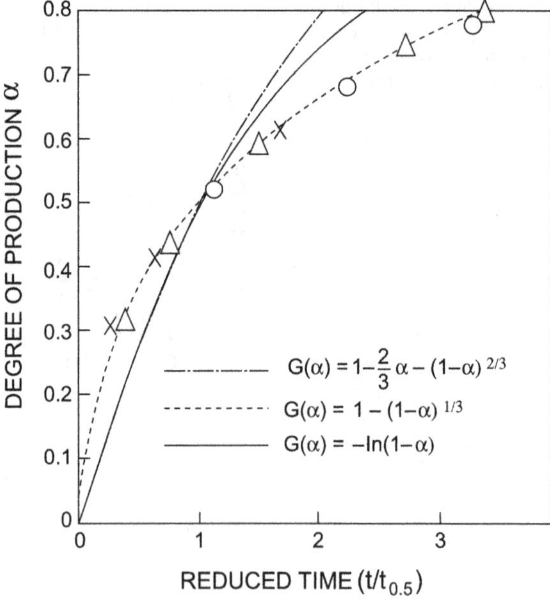

Fig. 2.14 Reduced time plot for data shown in Fig. 2.13a (Mookherjee et al. 1986)

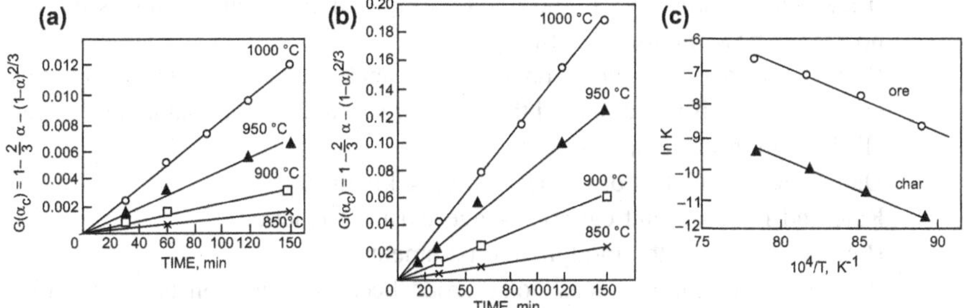

Fig. 2.15 **a** Analysis of kinetic data for ore–char reaction: Ginstling–Brounshtein plot for data shown in Fig. 2.13a. **b** Analysis of kinetic data for ore–char reaction: Ginstling–Brounshtein plot for kinetic data for char gasification in the presence of ore for data shown in Fig. 2.13b. **c** Analysis of kinetic data for ore–char reaction: Arrhenius plots for data shown in Fig. 2.15a, b

2.7 Review Questions

1. Discuss whether the following statements are correct or false.

   a. If the degree of a reaction is described in dimensionless terms, then reaction rate and rate constant have the same dimensions.
   b. The unit for rate constant depends on the order of reaction.
   c. The activation energy for a process cannot be determined unless one knows the rate equation that fits the kinetic data.
   d. The time for complete radioactive decay is a function of the half-life period.
   e. For any reaction, time required for 100% conversion is twice that required for 50% conversion.
   f. Rate constant can be evaluated from $\alpha - t$ plots only if the rate law is known.
   g. For any reaction, rate must gradually decrease.
   h. For a mixing process, if $\tau_{0.85}$ and $\tau_{0.90}$ are respectively 60 and 70 min, then $\tau_{0.95}$ equals 80 min.
   i. For the kinetic equation $1 - (1 - \alpha)^{1/3} = kt$, the order of reaction is 1/3.
   j. For study of densification kinetics $\Delta L/L_0$ is a better kinetic parameter if, for cylindrical compacts, L/D is smaller.
   k. During a phase transformation, the conductivity of the sample changes continuously. This change can be used to derive the kinetic law and evaluate kinetic parameters.
   l. The amount of product formed (P) during a reactions grows according to first-order kinetics. The doubling time for P will be equal to the doubling time of reaction rate.
   m. If the activation energy changes during the course of a reaction, then it is best evaluated using the integral method.

n. If a solute is added to a liquid, then the time of complete mixing is infinite no matter how intense the stirring is.
o. Primary isothermal kinetic data can be analysed and kinetic parameters evaluated only if it is assumed that the activation energy remains unchanged all through at all temperatures.
p. Densification kinetics describes sintering kinetics.
q. Rate and rate constant cannot change during a reaction.
r. Rate constant is a thermodynamic parameter.
s. A piece of aluminium does not corrode because oxidation is not thermodynamically feasible.
t. Leaching reactions are accelerated in autoclaves because of higher pressures.

2. In a solution reaction, a product is generated so that its concentration in solution increases 20% per hour. What is the kinetic law and what is the rate constant?
3. A ceramic ware manufacturer wants to produce dense products. What processes variables he needs to control and why?
4. Activation energy can be evaluated from $\alpha - t$ plots by employing either integral approaches or differential approaches. Discuss the relative merits and disadvantages of these two approaches.
5. In a reaction, a reagent is half consumed in one hour and three-fourth consumed in two hour. What is the order of the reaction?
6. Why do natural growth processes often follow a first-order kinetic law?
7. During a reaction the volume of a solid changes continuously. Can this be used to measure the reaction rate?
8. Why is it difficult to identify a true kinetic parameter in the study of kinetics of sintering of powder particles?
9. During a first-order reaction the amount of a reactant solid is reduced to one-fourth. Will the rate also decrease to one-fourth of the initial value?
10. For a reaction, 50% conversion is achieved at different times at different reaction temperatures as summarized.

    i. Temperature (K) 833 1000 1250
    ii. Time (min) 1^{05} 1^{03} 10

    Calculate the activation energy and the time required for 50% conversion at 909 K.
11. Show that if the differential form of the kinetic law is written as $\frac{d\alpha}{d_t} = k(1-\alpha)^n$, then the rate constant is given by the initial slope in $\alpha - t$ plots irrespective of the order of reaction.
12. Show that the $\alpha - t$ plots for autocatalysis cannot be analysed by Johnson–Mehl equation even though they are sigmoidal in nature.
13. Reduced time plots cannot be used to evaluate activation energy or rate constant. Why?

14. The $\alpha - t$ values for a reaction are as follows.

α	0.01	0.05	0.10	0.20	0.40
t, min	10	250	1000	4000	16000

Plot the data. Replot the data using reduced time $\theta = t/t_{0.2}$. Compare the second plot against the reduced time plots for the two following models and identify the correct kinetic law.

(a) $\alpha^2 = kt$
(b) $1 - (1 - \alpha)^{1/3} = kt$

15. Why is the Johnson–Mehl equation not a true kinetic equation? How can it be modified to become a true kinetic equation? Derived the modified form for n values 1.5, 2, 3 and 5.
16. Write the equation which relates mixing time to the stirring power density. Discuss the significance of the terms and the constants used.
17. Derived the reduced time equation in terms of $t/t_{0.2}$ for the following kinetic laws

    (a) $1 - (1 - \alpha)^{1/3} = kt,$
    (b) $1 - \frac{2}{3}\alpha - (1 - \alpha)^{2/3} = kt,$
    (c) $-\ln(1 - \alpha) = kt$

18. For a reaction, the values of α and t at various temperatures are as follows.

Reaction time t, h

α	1000 °C	950 °C	900 °C
0.3	0.2	0.25	0.50
0.4	0.3	0.45	0.90
0.5	0.5	0.70	1.50
0.7	1.15	1.60	3.60
0.9	2.5	2.50	–

Examine if these data fit any of the kinetic laws mentioned in the previous question.

19. Johnson–Mehl equation can be written as either $-\ln(1 - \alpha) = (k' t)^n$ or $-\ln(1 - \alpha) = k'' t^n$. Although k' and k'' are not true rate constants, one can obtain Arrhenius-type plots by plotting either $\ln k'$ or $\ln k''$ against $1/T$. Which plot should yield from the slope a more reliable value of E?
20. What is meant by an autocatalytic reaction? Derive the kinetic law for any particular mode of autocatalysis.
21. Differentiate between rate constant and specific rate constant.
22. Two different functions of α seem to explain a set of isothermal kinetic data for different temperatures. That is, one gets approximate straight lines using both

the equations; $g_1(\alpha) = k_1 t$ and $g_2(\alpha) = k_2 t$. How can one identify the correct model?

23. In a reaction, the amount of product formed increases at a fixed rate of 10% per minute. If after 10 min the amount is 10 g, then what is the amount formed in an hour?
24. The rate of a reaction is measured in terms of CO_2 evolved. The data for three temperatures and three reactions times are given in the following table. Find (a) the kinetic equation, (b) correct activation energy. Employ as many approaches as you can.

Total CO_2 evolved, cc

T °C/*t* min	5	15	35	55	65
650	1	1.5	6.5	10	10.2
700	1.2	3.9	9.8	12.8	13.1
750	1.8	6	15	22	23.0[a]

[a]This may be assumed to correspond to complete reaction

25. The kinetic data for isothermal reduction of iron ore at three temperatures summarized in the table.
Values of α

T °C/*t* min	1	2	3	4	5	6	7
826	0.03	0.976	0.10	0.13	0.14	0.16	0.178
900	0.05	0.15	0.14	0.17	0.21	0.21	0.230
1020	0.10	0.18	0.22	0.24	0.25	0.26	0.270

(a) Plot $\alpha - t$ plots for the three temperatures
(b) Plot integral kinetic plots $g(\alpha) = kt$ assuming

$$g(\alpha) = 1 - \frac{2}{3}\alpha - (1-\alpha)^{2/3}$$

(c) Plot Arrhenius plot and evaluate E
(d) Plot ln t_α versus $10^4/T$ at $\alpha = 0.2$, $\alpha = 0.15$ and $\alpha = 0.1$ and evaluate E
(e) Plot $\ln\left[\frac{d\alpha}{d_t}\right]_\alpha$ versus $10^4/T$ at $\alpha = 0.1, \alpha = 0.15, \alpha = 0.2$ and evaluate E. Comment on the different E values obtained.

26. Discuss the assumptions necessary to use $\Delta L/L_0$ as a kinetic parameter in the study of sintering.
27. In a mixing reaction, the mixing time is given by $\tau_{0.95} = 500(E)^{0.5}$. Assume the kinetic energy dissipated is directly proportional to volume of gas pass through

liquid in a given time. If $\tau_{0.95}$ is 2 min for a given gas flow rate of 50 litres per min, then find $\tau_{0.95}$ for flow rate of 500 l/min.

28. In a reaction, the α values at different values of t are as follows.

t, s	200	400	600	800	1200
α	0.025	0.22	0.53	0.80	0.92

Assuming Johnson–Mehl-type equation. Find the values of n and k.

29. Write a kinetic equation for densification process in terms of density change.

Appendix: Kinetics of Dynamic Recrystallization

The Johnson–Mehl equation, Eq. 2.38, has been widely used to describe the kinetics of static recrystallization process (Ye et al. 2002). For this application, $\alpha = X$, fraction recrystallized. However, during dynamic recrystallization (DRX) in rapid forming processes like hammer forging, the whole operation finishes within a fraction of second and it is not possible to measure time t to requisite precision, without very sophisticated experimental techniques. Therefore, for DRX studies on solution annealed (grain size ~ 200 μm) alloy D9 (a version of austenitic stainless steel, chosen for in-core application in fast nuclear reactors), (Mandal 2004) replaced t in Eq. 2.38 by true strain ε, to write a revised version of Johnson–Mehl equation in the form:

$$X = 1 - \exp(-k\varepsilon^n) \quad \text{i.e.,} \ln(-\ln(1 - X)) = \ln k + n \ \ln \ \varepsilon \tag{2.55}$$

This revised equation can be readily rationalized with Eq. 2.28, by considering an average constant true strain rate $\dot{\varepsilon}_0$ during the forging process so that $t = \varepsilon/\dot{\varepsilon}_0$. Compared to Eq. 2.28, the exponent n remains unchanged, but the t term now depends upon $\dot{\varepsilon}_0$ as well, and as such is expected vary for example from hammer forging to the slower process of hydraulic press forging. The results did support the expected linear variation of $\ln(-\ln(1 - X))$ with $\ln \varepsilon$, for both hammer forging and the slower press forging operations, Fig. 2.16. The temperatures mentioned in these figures are the starting temperatures for forging, and the lines correspond to least square fit of the data. The values for the exponent n are also indicated in these figures.

After hot working, the grain size in the product microstructure was smaller than the initial; grain size. This fact plus the small values for in the range of 1.07–1.41 led to the conclusion that alloy D-9 shows a growth control DRX. Large numbers of growing nuclei present mutually inhibit grain boundary mobility and restrict the grain growth. From the low n values, and also form metallographic studies, it was concluded that nucleation of new strain-free grain takes place in the interfacial area of grain and twin boundaries. For the hammer forging operation, n increased with increasing initial temperature, this was attributed to availability of higher thermal

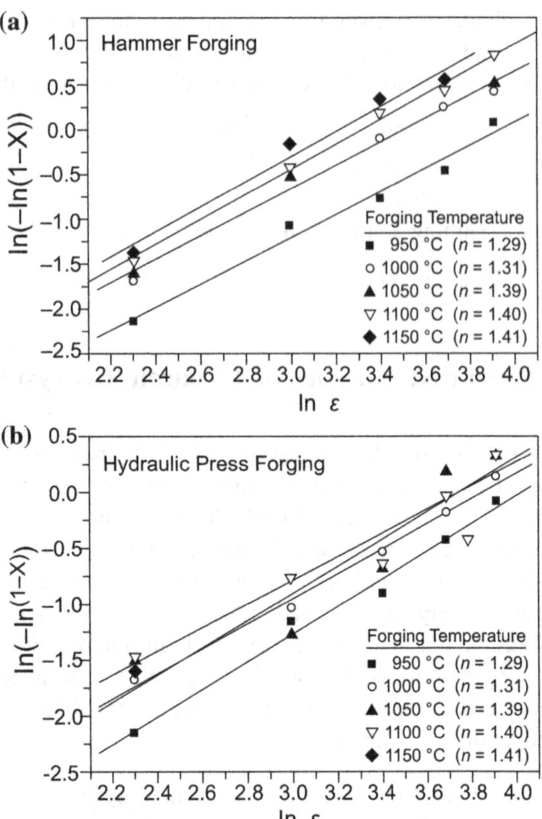

Fig. 2.16 Modified Johnson–Mehl plot of recrystallization kinetics at various temperatures in **a** forge hammer, **b** hydraulic press operation (Mandal 2004)

activation energy with increasing temperature. For the hydraulic press forging operation on the other hand, n values tended to decrease, particularly for the two high temperatures. This was attributed to the lower strain rate, which results in significant temperature drop during the forging operation.

Two-slope Behaviour with Johnson–Mehl Equation

It has been mentioned that of the many rate equations proposed for describing different reactions, most heterogeneous reactions of first order that are encountered in metallic alloys are best described by the Johnson–Mehl equation, Eq. 2.28. Occasionally, however, a two-slope behaviour in a plot of $\ln(-\ln(1-\alpha))$ against $\ln t$ is observed (Vandermeer and Gordon 1963; Bergmann et al. 1983), as shown schematically in Fig. 2.17.

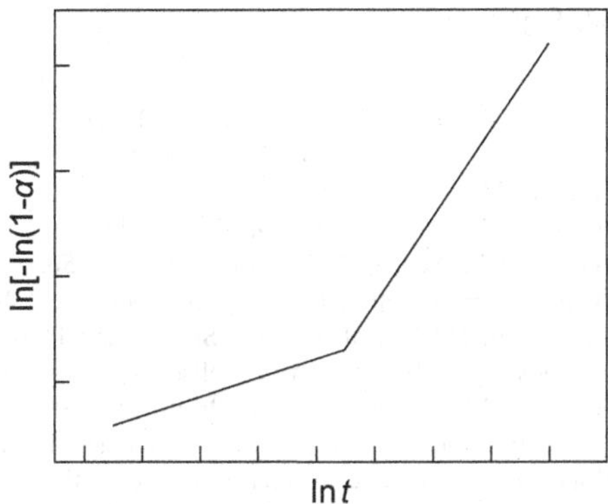

Fig. 2.17 Schematic of two-slope behaviour with Johnson–Mehl equation

One such example is the transformation of δ-ferrite phase in austenitic welds on exposure to high temperatures is considered here. During high-temperature exposure as encountered in service, the ferrite transforms to a variety of secondary phases such as $M_{23}C_6$ carbide, σ, χ, α' which degrade the mechanical and corrosion properties of the weld metal. The formation of these phases is influenced by a number of factors such as weld metal composition, ferrite content and its morphology, in addition to temperature and time of transformation. Johnson–Mehl equation has been employed by many investigators to describe the ferrite transformation in stainless steel weld metals. For example, Gill et al. (1992) determined the kinetics of transformation of δ-ferrite at 873, 923, 973 and 1023 K. In low carbon weld metal, the transformation could be described by the Johnson–Mehl equation; using microstructural evidences, it was concluded that the dissolution of ferrite takes place predominantly by its replacements byσ phase. In contrast, in the high carbon weld, transformation of δ-ferrite was represented by two slopes, indicating that δ-ferrite transforms rapidly during initial stages of ageing, which is followed by a region of sluggish transformation. The value of α where the change of slope occurs was found to increase with ageing temperature. Using microstructural evidences, Gill et al. (1992) attributed the two-slope behaviours to two competing reactions: replacement by carbides and austenite at initial ageing times which leaves the ferrite particles depleted of Cr and Mo, followed by transformation to σ and/or austenite at longer ageing times, which is sluggish. These authors systematically documented the variation in the value of the exponent(s) with weld chemistry and temperatures. Writing the Johnson–Mehl equation as $\ln(-\ln(1-\alpha)) = \ln b + n \ln t$, and they showed that the temperature dependence of b could be expressed in Arrhenius form, whence they determined values for apparent activation energy. The apparent activation energies are useful for data correlation, but as indicated above, bereft of fundamental significance.

References

Asai, S., Okamoto, T., He, J., Muchi, I.: Trans. ISIJ **23**, 43 (1983)

Avrami, M.: J. Chem. Phy. **7**, 1103 (1939)

Avrami, M.: J. Chem. Phy. **8**, 212 (1940)

Avrami, M.: J. Chem. Phy. **9**, 177 (1941)

Bagchi, T.P., Sen, P.K.: Thermochim. Acta **56**, 261 (1982)

Benton, A.F., Emett, P.H.: J. Amer. Chem. Soc. **46**, 2728 (1924)

Bergmann, H.W., Fritsh, H.U., Sprusil, B.: Phase transformations in crystalline and amorphus alloys, p. 199. Deutsche Gesellschaft Fur Metallkunde E. V. (1983)

Brophy, J.H., Rose, R.M., Wulff, J.: Structure and properties of materials, Vol. II, Thermodynamics properties. Wiley Eastern Pvt. Ltd., New Delhi (1971)

Burke, J.: The Kinetics of Phase Transformations in Metals. Pergamon Press, London (1965), Ch. 2

Das, A.K., Ray, H.S., Chatterjee, A.: Proc. Intl. Conf. on Progress in Metallurgical Research: Fundamental and Applied Aspects, Feb. (1985), IIT Kanpur, 299

Fine, M.E.: Phase transformations in condensed systems part II: growth of phases—A prototype model. The Pennsylvenia State University (1978)

Ghosh, A., Ray, H. S.: Principles of Extractive Metallurgy, Indian Institute of Technology, Calcutta (1984), Ch. 7

Giess, A.: J. Amer. Ceram Soc. **46**, 364 (1963)

Gill, T.P.S., Shankar, V., Vijayalaxmi, M., Rodriguez, P.: Scr. Metall. **27**, 313–318 (1992)

Habashi, F.: Principles of extractive metallurgy, vol. 1, p. 124. Gordon and Breach Science Publishers, New York, Pyrometallurgy (1970)

Johnson, W.A., Mehl, K.F.: Trans. TMS, AMIE **135**, 416 (1939)

Keatch, C.J., Dollimore, D.: An Introduction to Thermogravimetry, 2nd edn. Heyden, London (1975), Ch. 5

Mandal, S.: Hot working and modelling of the resulting microstuctures of Alloy D-9 Austenic stainless steel using artificial neural network, M Tech. Thesis, IIT, Kanpur (2004)

Mookherjee, S., Ray, H.S., Mukherjee, A.: Ironmaking Steelmaking **13**(5), 230 (1986)

Nakanishi, K., Fujii, T., Szekely, J.: Ironmaking and Steelmaking. **2**, 193 (1975)

Kolmogorov, A.N.: Izv. Akad, Nauk, USSR, Scv. Mat. **1**, 355 (1937)

Ray, H.S., Mukherjee, A., Mookherjee, S.: Reduction of iron ore by coal; Studies on real systems. In: Proceeding Silver Jubilee workshop on research needs on mineral processing and chemical metallurgy, p. C–34. Dept. of Met. Engg., IIT Bombay (1983)

Ray, H.S.: Trans. IIM **36**(1), 11 (1983)

Ray, H.S., Kundu, N.: Thermochim. Acta **101**, 107 (1986)

Samuel, K.G., Sreenivasan, P.R., Ray, S.K., Rodriguez, P.: J. Nucl. Mater. **150**, 78 (1987)

Sarkar, S.B., Ray, H.S.: Trans. ISIJ **28**, 1006–1013 (1988)

Sarkar, S.B., Ray, H.S.: J. Thermal Anal. **36**, 231–242 (1990)

Sharp, J.H., Brindley, G.W., Narahari Achar, B.N.: J. Amer. Ceram. Soc. **49**, 379 (1966)

Sohn, H.Y., Won, S.: Met. Trans. B **16B**, 831 (1985)

Szekeley, J., Lehner, T., Chang, C.W.: Ironmaking Steelmaking. **6**, 285 (1979)

Tendolkar, G.S.: *Trans. 11M*, **33**(4), 255 (1980)

Tendolkar, G.S.: *Trans. 11M*, **36**(2), 83 (1983)

Vandermeer, R.A., Gordon, P.: Recovery and recrystallization of metals, p. 211. Interscience Publisher, New York (1963)

Vasudevan, M., Venkateshan, S., Sivaprasad, P.V.: Mater. Sci. Technol. **12**, 338–344 (1996)

Virkar, A.N., Ray, H.S., Paul, A.: Kinetics of sintering of monodisperse glass powder. In: Procceeding XIV, Intl. Congr. on Glass, p. 161, New Delhi (1986)

Willams, D.T., El-Rahaiby, S.K., Rao, Y.K.: Met. Trans. B **1213**(161), 192 (1981)
Ye, W., Le Gall, R., Saindrenan, G.: Mater. Sci. Eng. A. **332**, 41–46 (2002)
Yerofeev, B.V., Mitekevich, N.I.: In: Reactivity of Solids, p. 273. Elsevier, Amsterdam (1961)
Young, D.A.: The international encyclopedia of physical chemistry and chemical physics. In: Tompking, F.C. (ed.) Topic 21, Pergamon Press, Oxford, vol. 1 (1966)

Chapter 3
Chemically Controlled Reactions

3.1 Mechanistic Approach in Kinetic Analysis

3.1.1 Introduction

It has been mentioned earlier that a reaction seldom occurs as simply as represented by chemical equation. For example, consider the reactions

$$Fe_2O_3(s) + WO_3(s) \xrightarrow{800^\circ C} Fe_2WO_6(s) \quad (3.1)$$

$$2(CaO \cdot MgO)(s) + Fe{-}Si(s) \longrightarrow 1200^\circ C2CaO \cdot SiO_2(s) + 2Mg(g) + Fe \quad (3.2)$$

There is evidence to suggest that the first reaction involves WO_3 vapours (Thomas and Ropital 1985) and the second a liquid Ca–Fe–Si phase (Ray et al. 1985). That is, neither of them is a truly solid–solid reaction.[1]

A reaction generally involves several steps in series and/or parallel. In the case of heterogeneous reactions, the reactions occur at the interface between different phases and may also require heat transfer in addition to mass transfer steps. The interplay of these, namely mass and heat transfer, affects the overall reaction rate. In the mechanistic approach, kinetic models and postulate on the basis of some idealized schemes of reaction mechanism and actual kinetic data are matched against them.

---

[1]The rate of reaction represented by Eq. 3.1 is sensitive to total pressure in reaction chamber.

H. S. Ray and S. Ray, *Kinetics of Metallurgical Processes*,
Indian Institute of Metals Series, https://doi.org/10.1007/978-981-13-0686-0_3

3.1.2 Reaction Steps and Rate-Controlling Step

Consider a simple reaction

$$A + B = AB \tag{3.3}$$

Let us assume that A and AB dissolve only in phase 1 and B dissolves in phase 2 only. Then the only possible meeting place is the interface dividing the two phases. Atoms of A and B must diffuse on this interface for the reaction to take place and, then, molecules of product AB must diffuse into phase 1 so that reaction can continue. The steps in the reaction then are as follows:

(a) Diffusion of A from the bulk of phase 1 to the interface,
(b) Diffusion of B from the bulk of phase 2 to the interface,
(c) Chemical reaction at the interface through.

   (i) adsorption of reactant species,
   (ii) chemical reaction,
   (iii) desorption of reaction products species.

(d) Diffusion of AB away from the interface into the bulk of phase 1.

Some of these steps are in series because if any of them is prevented, then the overall reaction will not occur. It will be like introducing a very high resistance in an electrical circuit and thus blocking current flow. The overall reaction may be considered analogous to a series circuit. If any of the resistances is very much greater compared to the rest, then it becomes the dominant factor in deciding the current. Similarly, when a reaction occurs as a result of a series of steps, then the slowest step—the step with the largest resistance—becomes the rate controlling. If, however, the resistances of two or more steps become comparable, then the situation becomes complicated and a "mixed control" is said to prevail.

Consider a simple analogy of a reaction involving two consecutive steps, shown Fig. 3.1. Water is maintained at a constant level H_A in reservoir A. Water passes into a second reservoir B through an orifice O_A. Tank B discharges water through orifice O_B. Water level in B, H_B, depends upon the relative resistances offered by the two orifices. If $O_B \ll O_A$, then, obviously $H_B \approx H_A$ and the flow through O_B is independent of O_A. In such a case, orifice O_B controls the water flow and is characterized by material accumulation just prior to it. If, however, $O_B \gg O_A$, then flow is controlled by O_A and O_B has little role, all water coming through O_A escaping as fast as it can come. In this case, $H_B \approx 0$, and the flow ignores O_B. When the resistances offered by the orifices are comparable, then both of them influence the flow (mixed control).

When reaction steps are in series, then the slowest step becomes rate controlling if other steps are comparatively fast. The situation is reversed when reaction steps are in parallel and are mutually independent. In this case, the fastest step tends to become rate controlling. There can also be series–parallel combinations. Consider,

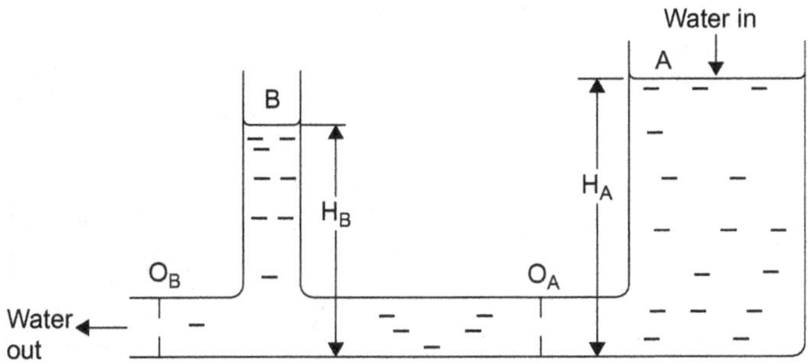

Fig. 3.1 Control of water flow in a pipe with two orifices in series

for example, the desulphurization reaction in iron blast furnace. The basic reaction is

$$\underline{S} + (O^{2-}) = (S^{2-}) + \underline{O} \tag{3.4}$$

where _ and () denote, respectively, metal and slag phases. In order to allow the reaction to proceed, oxygen should not be allowed to accumulate in liquid iron. Oxidation of other solutes ensures this. The reactions are

$$\underline{Mn} + \underline{O} = (MnO) \tag{3.5}$$

$$\underline{Si} + 2\underline{O} = (SiO_2) \tag{3.6}$$

$$C + O = CO(g) \tag{3.7}$$

The three preceding equations represent reactions in parallel. However, the individual steps in them, as well as the desulphurization reaction, are in series.

Assumption of a single rate-controlling step considerably simplifies the mathematical treatment of kinetics. As mentioned earlier, however, this may be an approximation of a real situation where the influences of other steps are also present. When a reaction step is assumed to be rate controlling, the rates of other steps are, by definition, much greater. Therefore, the other steps can reach thermodynamic equilibrium. The rate-controlling step, however, cannot be in equilibrium because the product of this step is quickly consumed by the subsequent step and, therefore, has no time to come to equilibrium with the reactant. Slowness of the step also implies an accumulation of material prior to the step. The rate of reaction calculated using such a concept of rate-controlling step is termed as "virtual maximum rate". The "actual" rate, in real situation, would perhaps be less because of the resistance offered by the other steps in series.

Some of the preceding concepts can be used to identify the rate-controlling step by mere physical examination of a partially reacted mass. For example, if a dense spherical pellet or a spherical particle of hematite is made to react with a reducing

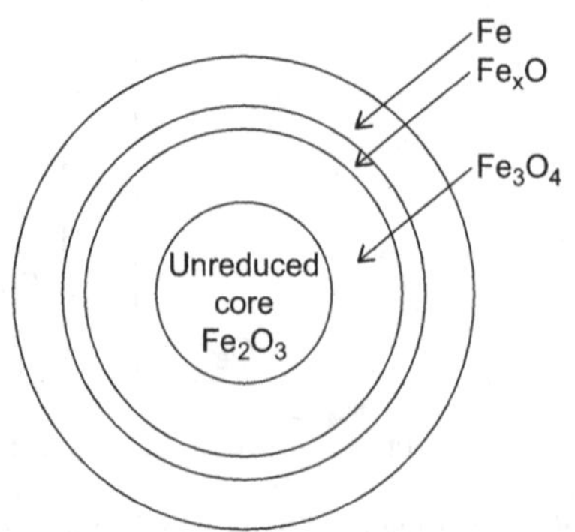

Fig. 3.2 Cross-sectional view of a partially reduced hematite sphere (lump or pellet)

gas such as H_2 or CO, then the reduction proceeds from the external layers to the interior. If a partially reduced particle is cooled and sectioned, then one would see layers of different oxides as shown schematically in Fig. 3.2. The very fact the internal layers exist indicate the rate-controlling step is reduction of wustite to iron which occurs at the Fe_xO/Fe interface. If this were not so, then Fe_xO would be reduced as fast as it was produced from Fe_3O_4.

3.2 Reaction Between a Solid and a Fluid

3.2.1 Reaction Steps in Solid/Fluid Reaction

In a previous example (Eq. 3.3), we considered a reaction which involved a few clearly identifiable steps. If the product forms a separate phase, then there can be more steps. For example, creation of a new phase would involve nucleation and growth steps also. Consider Fig. 3.3 which shows a partially reacted solid during its reaction with a surrounding fluid. We assume that the solid remains at a constant temperature throughout, and heat of reaction is negligible. We also assume that there is no reagent starvation, i.e. there is excess fluid at all times and the reaction is not hindered on this count. One can envisage the following sub-steps in the reaction:

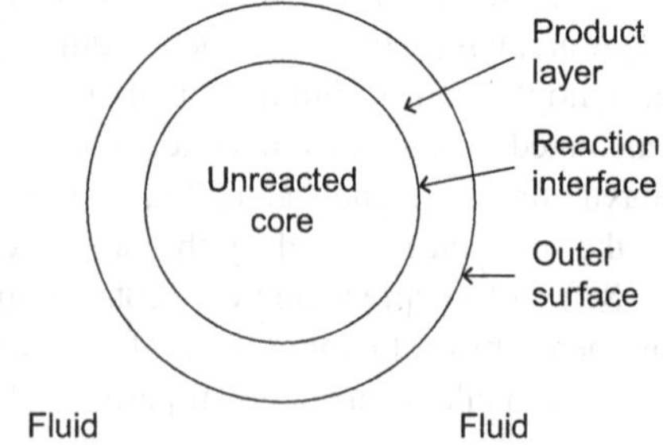

Fig. 3.3 Unreacted core and product layer in a partially reacted solid surrounded by a fluid

(a) Diffusion of reactant in bulk fluid towards the outer surface,
(b) Diffusion of reactant through product or residue layer,
(c) Adsorption of reactant at the reaction interface (in case of a gas),
(d) Reaction at the reaction interface,
(e) Desorption of product gases,
(f) Nucleation and growth of new phases formed,
(g) Diffusion of product fluids outward through product or residue layer,
(h) Diffusion of product fluids outward from the outer surface.

3.2.2 Concentration Profiles of Reactant Fluid

One should note that the reaction interface need not always be sharp. It may be diffused when the solid is porous and the fluid is able to penetrate somewhat before being consumed completely. In an extreme case, the solid may be so porous that the fluid penetrates it entirely with reaction occurring simultaneously everywhere so that there is no unreacted core as such at any time. The concentration profile of the diffusing species will depend on the rate-controlling step and the porosity of the solid. For example, if the chemical reaction at the interface is rate controlling, then the diffusion processes through the product layer is rapid. Therefore, there will be no concentration profile for the defusing fluid. Again, if the solid is porous so that the fluid penetrates while reacting, then the concentration will drop gradually. Figures 3.4, 3.5, 3.6, 3.7, 3.8, 3.9 and 3.10 depict some typical situations and concentration profiles.[2]

Consider another example, that of the reaction expressed by Eq. 3.4. For this reaction, sulphur in metal and oxygen in slag diffuse towards the interface of slag and metal. These react, and the products diffuse away, sulphur in slag and oxygen in metal. The concentration profile for a particular solute, again, depends on several factors. Figure 3.11 schematically shows the various situations for metal concentration profiles in metal and slag when chemical reaction at the interface is rate controlling, then diffusion steps are fast and, therefore, there are hardly any concentration profiles. For metal phase diffusion control, there is concentration only in the metal phases and not in the slag phase and vice versa. Presence of concentration profiles in both phases indicates a mixed control.

It should be remembered that existence of a rate-controlling process is not a must. It is possible to have a situation when all the sub-steps are quite fast and the overall reaction rate is controlled by the supply of the reactants only. This is not truly a kinetic problem. One can assure thermodynamic equilibrium at all steps and analyse the process on the basis of material balance calculations. Consider, for example, dehydrogenation of liquid steel by purging with argon. If the removal of

---

[2]From O. Levenspiel, *Chemical Reaction Engineering*. 2nd. Ed., Wiley Eastern Ltd., New Delhi (1972), Chap. 12.

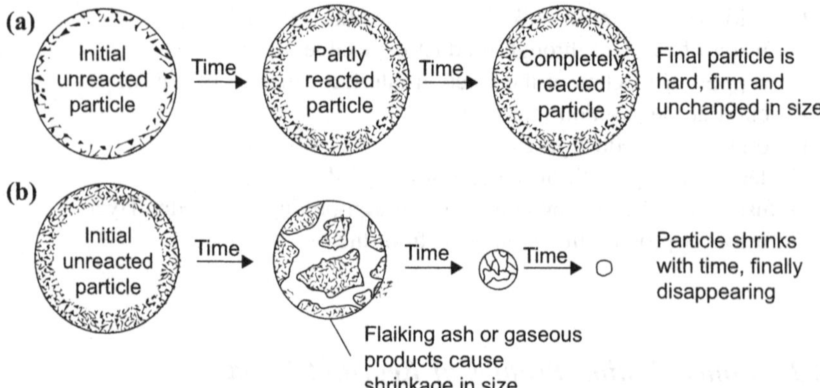

Fig. 3.4 Schematic representation of solid participating in gas–solid reaction: **a** solid product is formed; **b** no solid product is formed

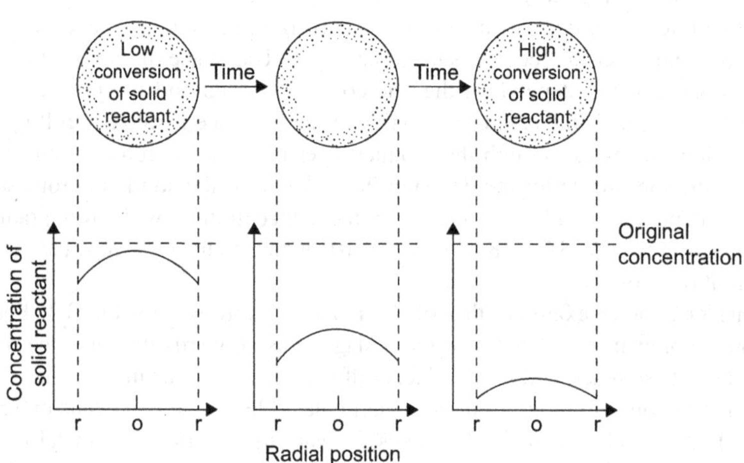

Fig. 3.5 Schematic representation of the unreacted core model

dissolved hydrogen gas from the metal phase to the argon bubbles is rapid, then one can assume equilibrium for the reaction

$$2\underline{H} = H_2(\text{argon}) \tag{3.8}$$

The rate of removal hydrogen can be calculated on the basis of argon flow rate and p_{H_2} (in argon according to Sievert's law for Eq. 3.8).

We will now try to derive from theory, and the kinetic models for some assumed rate-controlling steps under some simplifying assumptions.

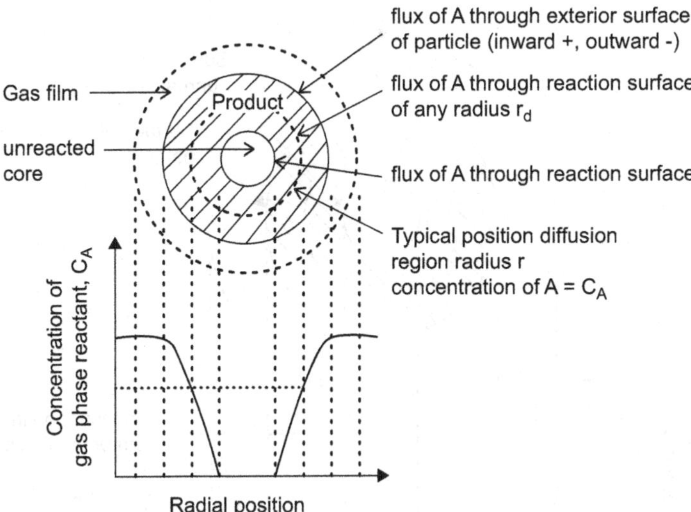

Fig. 3.6 Representation of a reacting particle when the diffusion through the product layer is rate controlling

Fig. 3.7 Representation of a reacting particle when chemical reaction is rate controlling

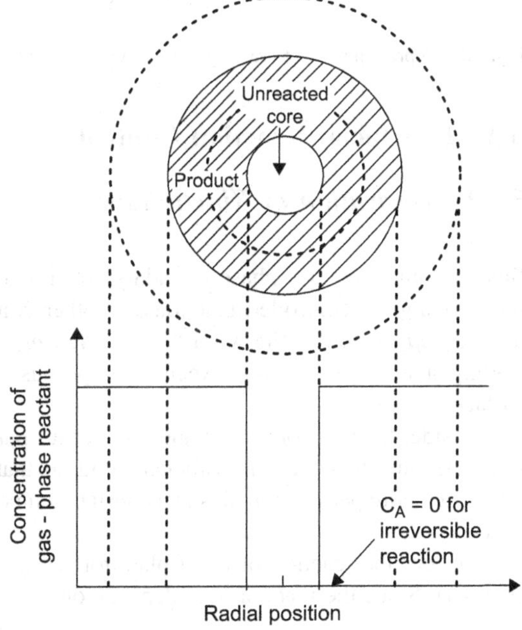

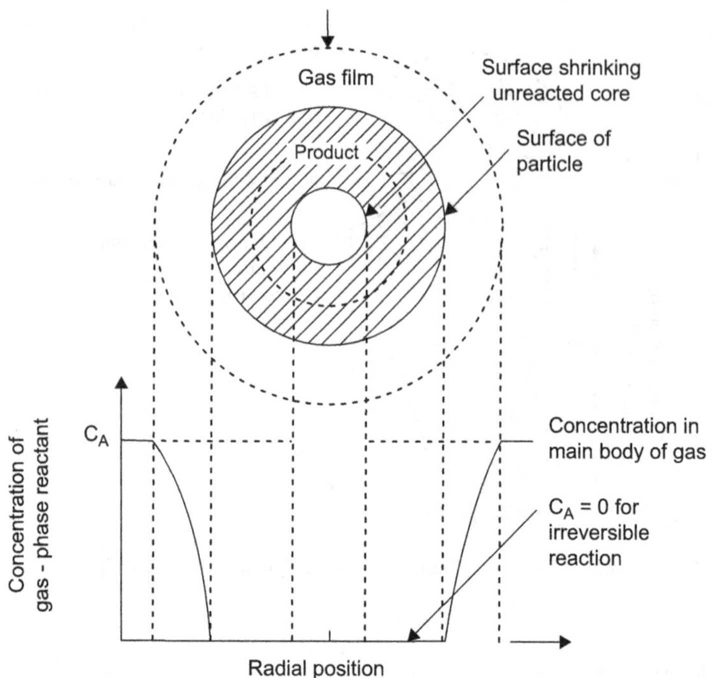

Fig. 3.8 Representation of reacting particle when gas film diffusion is the controlling resistance

3.3 Interfacial Reaction Control

3.3.1 Derivation of Kinetic Law

Since an interface is a phase boundary, such reactions are also sometimes called phase boundary controlled reactions. Another common nomenclature in the literature is *topochemical*, the word "topo" referring to surface. However, this is misleading since all reactions between a non-porous solid and a fluid has to occur at a surface anyway.

Consider the reaction of a single non-porous particle of solid with a surrounding fluid. Assume that there is sufficient fluid so that there is no reagent starvation. Also, the entire particle remains at a constant temperature and reaction heat does not influence it.

Consider the reaction of a flat plate for which the reaction surface area is constant (A). Since the reaction rate depends on area, we have

$$-\frac{\mathrm{d}W}{\mathrm{d}t} = \mathrm{k}^{'} AC \tag{3.9}$$

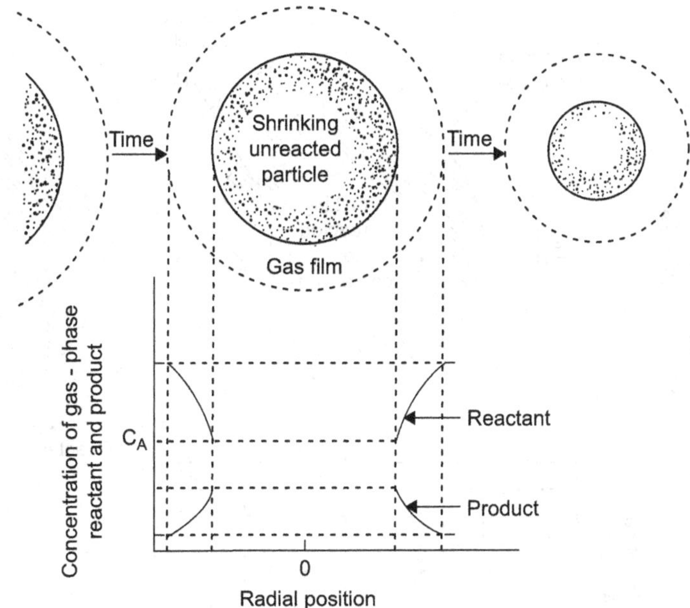

Fig. 3.9 Representation of concentration of reactants and products between a shrinking solid particle and gas

where

W = weight of solid plate
t = time of reaction
C = concentration of reagent in fluid
k' = a constant

At $t = t_0$, $W = W_0$. Hence, integration of Eq. 3.9 gives

$$W_0 - W = kt \tag{3.10}$$

where $k = k'C$. Such an equation is followed in many cases, e.g. dissolution of platelets of Mg,Zn and Fe in 0.1 N H_2SO_4.

The kinetic equation becomes less simple when the solid particle is a sphere for which the reaction interface must gradually decrease. If the reaction is phase boundary controlled, then the presence or absence of a product layer is immaterial. We have, again,

$$-\frac{dW}{dt} = k'AC \tag{3.11}$$

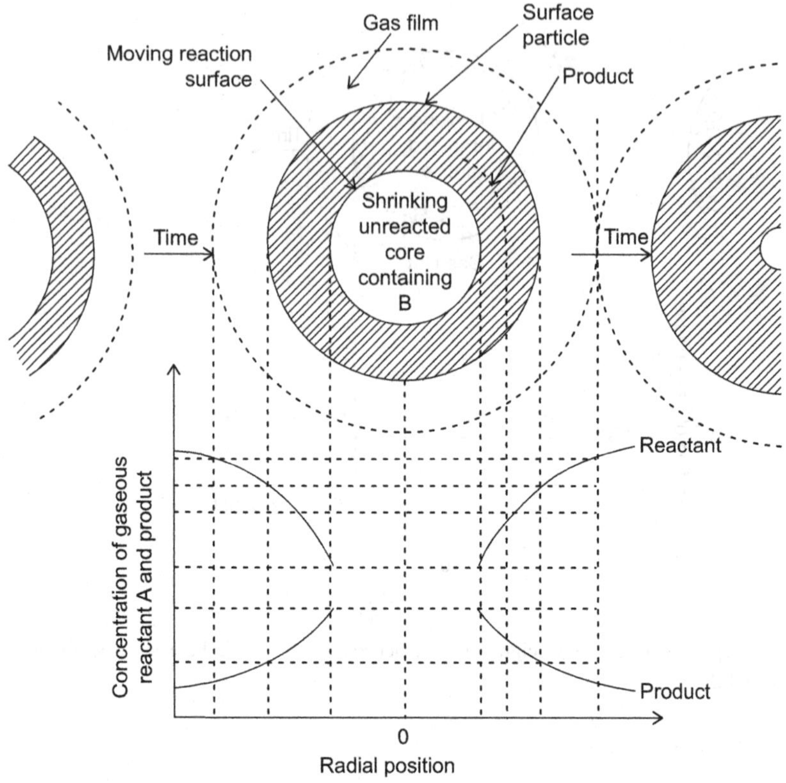

Fig. 3.10 Representation of concentrations of reactions and products for a particle of unchanging size

where W refers to the weight of the sphere and A the decreasing surface area. If r is the radius of the unreacted spherical core at a given time t, then one obtains

$$-\frac{\mathrm{d}}{\mathrm{d}t}\left[\frac{4}{3}\pi\pi r^3\rho\right] = k \cdot 4\pi r^2\rho, \quad \text{i.e.,} \; -\frac{\mathrm{d}r}{\mathrm{d}t} = \frac{k}{\rho} \tag{3.12}$$

where ρ is the density of the solid.

Equation 3.12 implies that the radius decreases at a constant rate. It is easy to show that this should also be true for a cylindrical rod-shaped particle for which length is taken as almost invariant.

If $r = r_0$ at $t = t_0$, integration of Eq. 3.12 gives

$$r_0 - r = \frac{kt}{\rho} \tag{3.13}$$

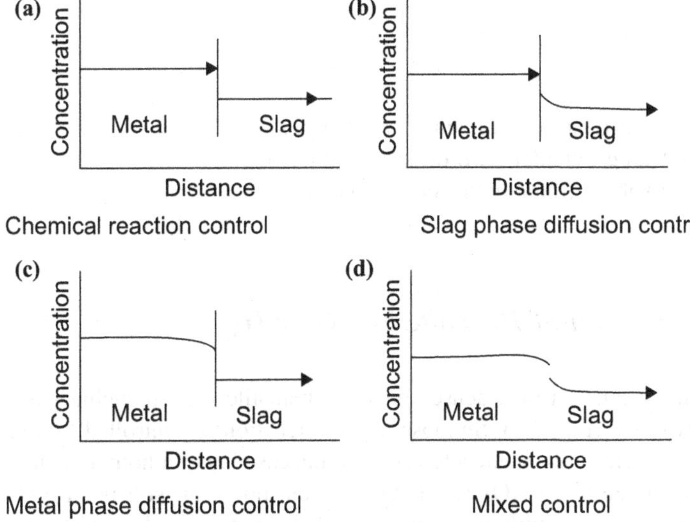

Fig. 3.11 Schematic representation of concentration profile of sulphur in metal and slag phases under different rate-controlling steps

If the initial weight is W_0, then

$$\alpha = \frac{W_0 - W}{W_0}, \frac{W}{W_0} = \left[\frac{r}{r_0}\right]^3 = (1 - \alpha) \quad \text{or} \quad \frac{r}{r_0} = (1 - \alpha)^{1/3} \tag{3.14}$$

Combining Eqs. 3.13 and 3.14, one obtains

$$1 - (1 - \alpha)^{1/3} = \frac{k}{\rho\, r_0} t \tag{3.15}$$

This equation has been derived among others, by McKewan (1958).

The same equation is obtained if the solid particle is assumed to be cubicle or cylindrical with $l = r$. For a plate, the change is in one dimension; for a long cylindrical particle ($l \gg r$), the change can be assumed to occur in two dimensions. In this case, one can derived the kinetic equation as

$$1 - (1 - \alpha)^{1/2} = \frac{k}{\rho\, r_0} t \tag{3.16}$$

We, therefore, see that although the reaction mechanism is the same in all cases, the form of the kinetic law changes with particle shape. The general equation is

$$1 - (1 - \alpha)^{1/n} = kt \tag{3.17}$$

where

$n = 1$ for a flat plate (one-dimensional change),
$n = 2$ for a long cylinder (two-dimensional change),
$n = 3$ for a sphere (three-dimensional change).

3.3.2 *Reduction of Hematite by CO or H_2*

Consider the reduction of a dense sphere of hematite by a reducing gas as mentioned previously (Fig. 3.2). Suppose that the reaction is controlled by reaction at the FeO/Fe interface. The unreacted core in this case is not a homogenous material but consists of Fe_2O_3, Fe_3O_4 and FeO. We can, however, still use Eq. 3.15 provided we make the simplifying assumption that the oxygen density is uniform all throughout in the unreacted core and α is expressed not in terms of weight of particle but the fraction of oxygen removed. In Eq. 3.15, ρ therefore will now mean oxygen density.

In essence, then, it is taken to be a reaction of a uniform oxygen-containing sphere with reducing gasses, the oxygen being consumed gradually.

The kinetic equation will remain valid provided the reduced iron layer does not hinder the access of the reducing gas to the interface. If this product layer has any role to play, then the reaction is no longer interface controlled.

3.3.3 *Reaction of a Monodisperse Ensemble*

A monodisperse ensemble is a collection of uniformly sized particles. If a multiparticle system consisting of spherical particles reacts with a fluid, then provided there is no reagent starvation, the equation expressing kinetics of reaction of a single particle will still hold. To prove this, we note that every single sphere must follow an identical equation.

$$(1 - \alpha)^{1/3} = 1 - \frac{k}{\rho r_0} t, \quad \text{i.e.,} \quad \alpha = 1 - \left(1 - \frac{kt}{\rho r_0}\right)^3 \tag{3.18}$$

The degree of reaction for the whole ensemble is

$$\alpha^* = \frac{m_1\alpha_1 + m_2\alpha_2 + \cdots}{m_1 + m_2 + \cdots} = \alpha \tag{3.19}$$

where $m_1,\ m_2,\ \ldots$, etc., are masses of particles $1,\ 2,\ \ldots$, etc., and $\alpha_1,\ \alpha_2,\ \ldots$, etc., are degrees of reaction ($\alpha_1 = \alpha_2 = \cdots = \alpha$). Thus, we also have

$$1 - (1 - \alpha^*)^{1/3} = \frac{k}{\rho\, r_0}\, t \tag{3.20}$$

The equation, however, will not be valid when particles are of varying size. This subject will be discussed in a latter section.

3.4 Influence of Solid Geometry

Combustion of solid propellants:

Burning of rocket propellants is discussed here as an example of interface-controlled reaction. The main job of a propellant is to impart motion to a projectile. For space exploration and military applications, both solid and liquid propellants are used. However, from the point of view of safety, reliability, simplicity and long storage life, solids propellants are often preferred.

For proper functional requirements of rockets/missiles, the ballistician desires that the propellant burn in a definite pattern. This is achieved by applying polymeric materials, which do not readily burn, over some portions of the propellants surface. The coating materials are known as inhibitors.

Suppose the solid mass is in the shape of a hollow cylinder, and the outside surfaces are coated. The interior surface will progressively increase with reaction and so will be the reaction rate and rocket thrust. On the other hand, if the outside surfaces are left free and interiors inhibited, then reaction rate will diminish gradually. If reaction occurs both on the outer and the inner surfaces, then the rate will be nearly uniform although not entirely so. Figure 3.12 schematically illustrates this. Actual shapes can be more complicated. Figure 3.13 shows some typical shapes of solid propellant grains.

The phase boundary control kinetic equation has been found to be valid in many cases.

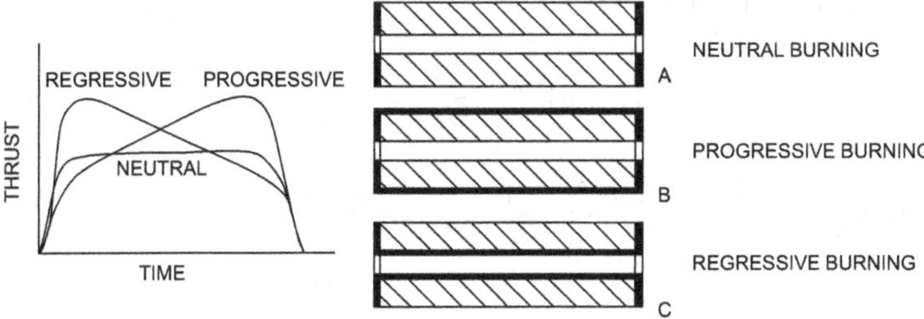

Fig. 3.12 Simple modes of inhibition and most of burning of solid fuel (Mckewan 1958)

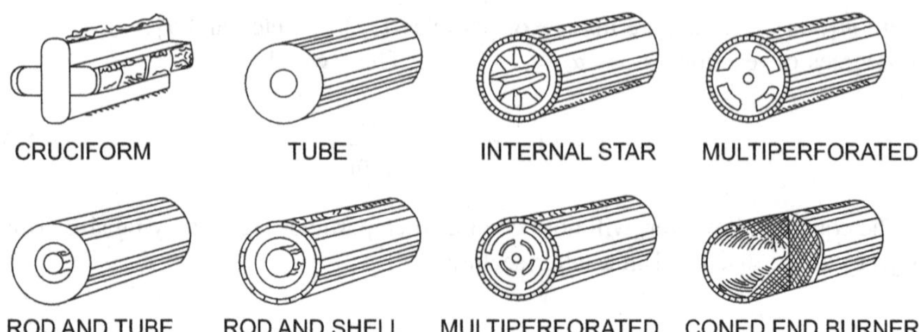

Fig. 3.13 Complicated shapes of solid propellants

Example 3.1 A solid sphere and a cube of the same material weigh the same. Each has also the same density. They react separately with a fluid and the reaction is phase boundary controlled. Which one will take longer to undergo fifty per cent reaction?

Solution

For a sphere of original radius r_0, $1-(1-\alpha)^{1/3}=\frac{k}{\rho r_0}\cdot t$.

Let the equivalent cube has the side l_0. Then weight $W=l_0^3\rho=\frac{4}{3}\pi r_0^3\rho$, so that

$$l_0=\left(\frac{4\pi}{3}\right)^{1/3}\cdot r_0$$

For cube, $-\frac{\mathrm{d}w}{\mathrm{d}t}=k\cdot A=k\cdot 6l^2$

$$\therefore\quad -\frac{\mathrm{d}l^3\rho}{\mathrm{d}t}=6kl^2,\ \text{which gives}\ -\frac{\mathrm{d}l}{\mathrm{d}t}=\frac{2k}{\rho},\quad \text{i.e. } l_0-l=\frac{2k}{\rho}\cdot t$$

Now, $\alpha=1-\left(\frac{l}{l_0}\right)^3$, i.e. $1-\frac{l}{l_0}=1-(1-\alpha)^{1/3}$.

Substituting from the previous equation, $1-(1-\alpha)^{1/3}=\frac{2k}{\rho l_0}\cdot t$

For a fixed value of α, let the reaction time for cube and sphere be, respectively, t_c and t_s. We have, for a fixed α

$$1-(1-\alpha)^{1/3}=\frac{2k}{\rho l_0}\cdot t_c=\frac{k}{\rho r_0}\cdot t_s$$

which gives $\frac{t_c}{t_s}=\frac{l_0}{2r_0}=\frac{1}{2}\left[\frac{4\pi}{3}\right]^{1/3}$, or $\left(\frac{t_c}{t_s}\right)^3=\frac{4\pi}{8\times 3}=\frac{12.57}{24}$.

Hence, $t_s>t_c$.

3.5 Examples of Leaching Reactions

3.5.1 Ammonia Leaching of Chalcopyrite Concentrate

Consider another example of phase boundary control reaction model, namely the leaching of chalcopyrite by ammonia, as described by Reilly and Scott (1977). When particles are spherical, then for a non-porous particle, of radius r, undergoing dissolution in ammonia solution and we have the rate equation

$$-\frac{\mathrm{d}\,[\mathrm{CuFe\ S_2}]}{\mathrm{d}t} = \frac{\mathrm{d}\,[\mathrm{Cu^{2+}}]}{\mathrm{d}t} = -\rho_s 4\pi r^2 \frac{\mathrm{d}r}{\mathrm{d}t}\left[\frac{W_0}{(4/3)\pi r_0^3 \rho_s}\right] \tag{3.21}$$

where ρ_s is the density of the solid, W_0 the solids concentration and, r_0 the original particle size. The expression within the brackets merely gives the total number of solid particles, n. If it is now assumed that, at constant reaction conditions, the rate of dissolution per unit of external surface, k ,is a constant independent of particle size, then Eq. 3.21 can also be written as

$$k = -\frac{-\rho_s 4\pi r^2 \frac{\mathrm{d}_r}{\mathrm{d}_t}\cdot n}{4\pi r^2 \cdot n}, \quad \text{i.e.} \quad -\frac{\mathrm{d}r}{\mathrm{d}t} = \frac{k}{\rho_s} \tag{3.22}$$

Assuming that the number of particles per unit volume of liquid remains constant and we have

$$\frac{r_0 - r}{r_0} = \frac{kt}{r_0 \rho_s} = 1 - (1-\alpha)^{1/3} \tag{3.23}$$

where r_0, r are average particle sizes.

Equation 3.22 does not necessarily assume that surface chemical reaction is rate controlling. As long as reaction conditions at the solid surface remain reasonably constant, the rate-controlling mechanism may be transport limited or chemically limited. Basically, the equation allows for the variation of surface area with time at constant reaction rate. This is discussed again subsequently.

3.5.2 Acid Leaching of Chalcopyrite Concentrate

Leaching of chalcopyrite in an autoclave has also been found to follow the chemical control equation, Eq. 3.23. The reaction is

$$\mathrm{CuFeS_2} + \frac{17}{4}\mathrm{O_2} + \frac{1}{2}\mathrm{H_2SO_4} = \mathrm{CuSO_4} + \frac{1}{2}\mathrm{Fe_2(SO_4)_3} + \frac{1}{2}\mathrm{H_2O} \tag{3.24}$$

The rate constant increases with decrease in particle size and increase in p_{O_2}. The reaction rate can be expressed in terms of decrease in moles of $CuFeS_2(n_c)$, consumption of moles of $O_2(n_o)$, or increase in copper ion concentration (Cu^{2+}). We have

$$-\frac{dn_c}{dt} = -\frac{4}{17}\cdot\frac{dn_O}{dt} = \frac{1}{V_s}\frac{d(Cu^{2+})}{dt} \tag{3.25}$$

where V_s = solution volume.

It should be noted that the phase boundary control equation will also be observed for some situations other than chemical reaction control (Reilly and Scott 1977). For example, if the process is controlled by diffusion to the liquid boundary layer of solid particles (boundary layer thickness δ at the receding mineral surface), then we have diffusion equation[3]

$$-\frac{dn_c}{dt} = \frac{4\pi r^2 D(C - C_s)}{\sigma\,\delta} \tag{3.26}$$

where C and C_s are, respectively, bulk and surface concentration of copper ions, D is diffusion coefficient and σ the stoichiometric factor which represent the number of moles of the diffusing species required for each mole of metal value released by the reaction.

If C_s is negligible compared to C and C remains essentially constant, then the reaction rate becomes proportional to surface area $4\pi r^2$—a requirement same as that for chemical control. Therefore, it is not always possible to deduce reaction mechanism from kinetic equations only. One has to consider the magnitude of activation energy also. This will be discussed again later.

3.6 Chemisorption

3.6.1 Role of the Interfacial Region

An interface is the meeting place of two phases. However, it is not just a two-dimensional area. Most interfaces are considered to have a thickness of a few atomic dimensions, say, 10 Å. The volume of a unit area of interface would, therefore, be about 10^7 in angstrom units. An interfacial reaction is a reaction that strictly takes place not at a surface but within a special region (three-dimensional) which connects two dissimilar phases.

The interfacial region is structurally different, and the reactants undergo some change when incorporated into this. The process of incorporation is known as

[3]See Chap. 5.

adsorption when solute is a gas. Consider the phase boundary reduction of an iron ore particle by a reducing gas the molecules of which first impinge on solid. Only those molecules can react which are not immediately reflected back into gas space, because reaction requires a certain time to take place. Some of the molecules do remain, however, because they become attached by physical or chemical forces, the binding process being known as adsorption. Physical adsorption is characterized by short duration of stay ($\sim 10^{-7}$ s). Chemical adsorption (chemisorption), however, results in stronger and, therefore, less transitory attachment. Chemisorption predominates at high temperature and is irreversible. Thus, oxygen chemisorbed on charcoal surface cannot be released as elemental gas; it can be desorbed only as carbon monoxide. Whenever one encounters rate control by interfacial chemical reaction, adsorption/desorption phenomena have to be taken into account.

Chemisorption can be a rate-controlling step. For example, the rate of dissolution of nitrogen in liquid iron decreases significantly if oxygen or sulphur content of iron increases. This is attributed to the fact that oxygen and sulphur, being surface active, occupy surface sites obstructing chemisorption of nitrogen.

3.6.2 *Gasification of Carbon*

For a detailed analysis, let us consider gasification of carbon by CO_2.

$$CO_2 + C = 2CO \tag{3.27}$$

To determine the rate equation, assume the following reaction mechanism

$$CO_2\ (g) + C_f \xrightarrow{k_1} CO\ (g) + C_O \tag{3.28}$$

$$C_O + CO\ (g) \xrightarrow{k_2} CO_2\ (g) + C_f \tag{3.29}$$

$$C_O \xrightarrow{k_3} CO\ (g) + n\,C_f \tag{3.30}$$

Here C_f represents a free, active carbon site and C_O a carbon site to which one oxygen is absorbed. k_1 and k_2 are similar in magnitude but k_3 is much larger than the rate constant for the reverse reaction. If n_C is the number of moles of carbon gasified, then the rate of gasification is

$$-\frac{dn_C}{dt} = k_3 C_{C_O}$$

where C_{C_O} is the concentration of occupied sites on the carbon surface.

Under steady-state conditions, the net rate of formation of occupied site must be zero. Hence,

$$\frac{dC_{C_O}}{dt} = k_1 p_{CO_2} C_{C_f} - k_2 p_{CO} C_{C_O} - k_3 C_{C_O} = 0 \tag{3.31}$$

$$C_{C_O} + C_{C_f} = C_{C_t}(\text{total number of sites}) \tag{3.32}$$

Solving for C_{C_O}, one obtains

$$C_{C_O} = \frac{k_1 C_{C_t} \cdot p_{CO_2}}{k_1 p_{CO_2} + k_2 p_{CO} + k_3} \tag{3.33}$$

Hence, we have, for rate of gasification,

$$-\frac{dn_c}{dt} = \frac{k_3 k_1 C_{C_t} \cdot p_{CO_2}}{k_1 \cdot p_{CO_2} + k_2 \cdot p_{CO} + k_3} \tag{3.34}$$

If $k_1 C_{C_t} = k, \frac{k_1}{k_3} = l$, and $\frac{k_2}{k_3} = m$, then we get

$$-\frac{dn_c}{dt} = \frac{k \cdot p_{CO_2}}{l \cdot p_{CO_2} + m \cdot p_{CO} + 1} \tag{3.35}$$

Under most experimental conditions $k_1 p_{CO} > k_3$ and $k_2 p_{CO} > k_3$, i.e. the last step is rate controlling. This gives the simplified equation

$$-\frac{dn_c}{dt} = \frac{k_3 k_1 \cdot C_{C_t} \cdot p_{CO_2}}{k_1 \cdot p_{CO_2} + k_2 \cdot p_{CO}} = \frac{k_3 k_1 \cdot C_{C_t}}{k_1 + p_{CO_2}/p_{CO}} \tag{3.36}$$

where $k_1 = k_1/k_2$ is the equilibrium constant for reaction of CO_2 with free carbon sites.

3.7 Total Internal Reaction

3.7.1 *Reaction of a Porus Particle*

A solid–gas reaction can have different visible features. The size of the solid sphere can increase, decrease or remain unchanged. It can disappear gradually by being consumed without producing any solid product. The reaction is understood better by examining the cross-section of a partially reacted solid. The product layer may be non-porous or porous, coherent or flaky with cracks. The core–product interface may be sharp or diffuse.

In the total internal reaction model, there is no unreduced core as such because reaction proceeds uniformly at every point. Such a situation prevails when the solid is so porous that the fluid can permeate freely within it.

When there is no resistance to diffusion and, in addition, reaction is not restricted to a core–product interface, the reaction follows a first-order kinetics, namely the equation.

$$-\ln(1-\alpha) = kt \tag{3.37}$$

This model implies the basic assumption that reaction occurs randomly, the rate being proportional to the fraction of unreacted materials, i.e.

$$\frac{d\alpha}{dt} = k(1-\alpha) \tag{3.38}$$

3.7.2 Reduction of Ore–Coal Pellets

Several workers have demonstrated the superior reduction behaviour of iron oxide pellets containing carbonaceous materials. It has been shown that internal generation of carbon monoxide accelerates the reduction kinetics manyfold. Moreover, the carbon gasification reaction

$$CO_2 + C = 2CO \tag{3.39}$$

implies a higher pressure of the reducing gasses ensuring effective permeation of CO all throughout. Basu (1987) reported some kinetic date on reduction in Fe_2O_3—coal pellets of different sizes (10–20 mm) and oxide/coal ratio at different temperatures. In all cases, the extent of reduction was studied in terms of a degree of reduction, f, defined as the ratio of weight loss, measured at given intervals, to the maximum possible loss of weight.[4] The latter is obtained from the weight of oxygen in iron oxide and the volatiles and fixed carbon in coal. Isothermal kinetic data fitted well (Basu 1987) into Eq. 3.37 as is shown in Fig. 3.14. The pseudo-rate constant so defined, i.e. the k values, fits into the Arrhenius-type equation which yield, for the so-called activation energy, a value of about 97 kJ/mole. A similar value was obtained by employing differential approaches to the primary kinetic data.

As expected, reduction rates increased with temperature and decreased with time. Increase in coal/iron oxide ratio increased rate constant almost linearly, the effect being more pronounced at higher temperatures. However, in the range 10–20 mm, the pellets size had practically no effect on rate.

Use of pellets provides an attractive approach as regards utilization of ore fines. However, reduction of ordinary pellets suffers from several disadvantages. Firstly, reduction of pellets requires reaction times comparable to those required for lump ore. Secondly, they are often likely to swell and decrepitate. Lastly, pellets made

[4] f is related to α but the relationship is complicated; see Sect. 3.8.3.

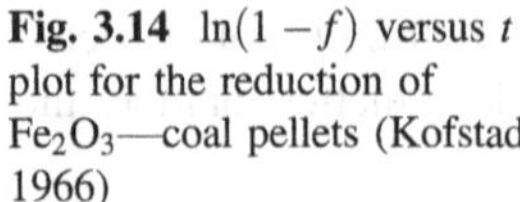

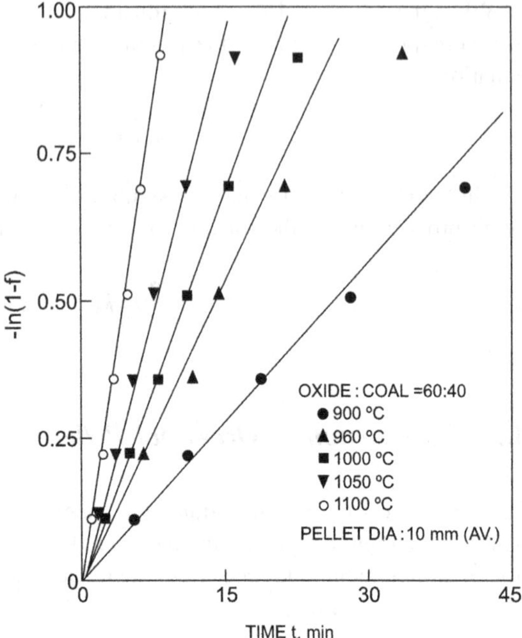

Fig. 3.14 $\ln(1-f)$ versus t plot for the reduction of Fe_2O_3—coal pellets (Kofstad 1966)

from iron ore fines. Moreover, it will accelerate reduction kinetics by 5–10 times and eliminate swelling. Pellets may also be expected to have better thermal properties.

3.8 Relationship Between the Degree of Reduction (α) and the Fraction of Reaction (f)

Thermogravimetric studies on the reduction of iron oxide by solid carbonaceous reductants are complicated because both the reduction of oxide and the oxidation of carbon contribute to the weight loss. It is not possible to delineate the two unless the released gases are analysed and their volume measured. Accordingly, such reactions have been studied with the help of gas chromatograph attachment with the reduction chamber. Even this method would run into trouble when actual coal is used in place of carbon.

Rao (1971) first defined a degree of reaction at a given instant as the ratio of weight loss of the mixture at that instant to the maximum possible weight loss for complete reduction. He reported different values for the maximum possible weight loss corresponding to the complete reduction for different C/Fe_2O_3 ratios. This concept can be used in the study of carbothermic reduction of ilmenite (Rao 1971). One can define the degree of reaction as the ratio of weight loss of ilmenite at a given time to the maximum possible weight loss obtained from stoichiometry of the following reaction

$$FeTiO_3 = TiO_2 + Fe + \frac{1}{2}O_2 \tag{3.40}$$

The weight loss for ilmenite at any given time can be calculated from the total weight loss of the mixture with the assumption that the product gas was composed of equilibrium CO/CO_2 ratio over carbon and ilmenite at the given temperature.

Consider reduction of iron ore by coal which contains carbon, moisture and volatile matter. The fraction f at any time t is calculated using the relation.

$$f = 100 \frac{\Delta W_t}{\Delta W_{max}} = \frac{100}{W_0} \frac{\Delta W_t}{M} \tag{3.41}$$

where

M = weight loss per cent corresponding to maximum possible removable oxygen from ore plus fixed carbon, volatile and moisture from coal
ΔW_t = weight loss per cent at time t
W_0 = original weight.

Use of parameter f:

Although the parameter f has been widely used in many kinetic studies, one can justifiably raise the following questions:

1. f is a pseudo-kinetic parameter, i.e. it is not proportional to the degree of oxygen removal α. Therefore, studies using f as a kinetic parameter can at best be accepted as semi-empirical.
2. While, at least in the initial stages, removal of carbon may necessarily imply simultaneous removal of oxygen from the ore, the same may not be valid at a later stage, i.e. beyond a certain degree of reaction, further removal of oxygen becomes more and more difficult, and the carbonaceous material may be gasified by the oxygen in the ore. In such an eventuality, f will continue to increase due to removal of carbon alone while α will remain practically unchanged. This is likely to be the case particularly when the coal percentage is more.

Mookherjee et al. (1985) have shown that for their system of ore fines surrounded by coal fines, f could be used as a useful parameter up to a value of 0.6, beyond which use of f led to ambiguities. They also showed that excellent results were, however, obtained when experiments were carried out under inert gas atmosphere which precluded the possibility of carbon gasification by aerial oxidation. It should be noted, however, that reduction rate is considerably decelerated under inert gas atmosphere.

Therefore, while f is convenient and useful parameter, there is a need for re-examination of its relation with α unambiguously. Agarwal et al. (1990) carried out such an analysis, as detailed in the following.

3.8.1 Theory

To derive a relationship between f and α, Agarwal et al. (1990) proceed with their basic definitions. For their particular system of ore–coal mixture, the degree of reduction f equals the ratio of weight of oxygen removed in time t to the weight of total removable oxygen (i.e. oxygen in iron oxides), i.e.

$$f = \frac{\text{Oxygen removed}}{K_1 \cdot W_t} \tag{3.42}$$

where

K_1 = constant for a particular type of iron ore depending upon its chemical composition. It is defined as per cent of oxygen (attached with iron only) in the iron ore. It is also assumed that other oxides present in the iron ore are not reducible.
W_1 = total weight of ore only.

Again, by definitions, the degree or fraction of reaction f = (weight loss in time t)/(maximum possible weight loss). Or,

$$f = \frac{\begin{matrix}\text{Oxygen}\\ \text{removed}\\ \text{from ore}\end{matrix} + \begin{matrix}\text{Moisture, volatile}\\ \text{matter and fixed carbon}\\ \text{removed from coal}\end{matrix}}{\begin{matrix}\text{Total}\\ \text{oxygen}\\ \text{in ore}\end{matrix} + \begin{matrix}\text{Total moisture, volatile}\\ \text{matter and fixed}\\ \text{carbon in coal}\end{matrix}} = \frac{\text{Oxygen removed from ore} + \text{Weight loss from coal}}{K_1 W_1 + K_2 W_2} \tag{3.43}$$

where

K_2 = constant (in %) for a particular type of coal and is defined as the percentage of moisture, volatile matter and fixed carbon in coal.
W_2 = total weight of coal in the mixture.

From Eqs. 3.42 and 3.43

$$\alpha = f\left(1 + \frac{K_2 W_2}{K_1 W_1}\right) - \frac{\text{weight loss from coal}}{K_1 W_1} \tag{3.44}$$

The above equation shows that, in general, the relation between α and f is not linear unless the second term in the right-hand side is negligible, such as, say, for low content of coal in ore–coal mixture. Specifically, if there is no coal in the mixture, then $\alpha = f$. The general relationship is curvilinear.

3.8.2 Experimental

Agarwal et al. (1990) prepared composite pellets of ore and coal for carrying out reduction tests. Iron ore from Baitarini (Orissa) and washed coal from Dugdha (Bihar) in the ratio 90 : 10 were ground to $-200+300$ mesh and were pelletized using 3.9% dextrine as binder. The pellets (8–9 g in each run) in the size range 10–12 mm dried in air for 36 h. Cold crushing strength of 30–35 kg/pellet was achieved. The chemical analysis of iron ore and proximate analysis of coal are given in Table 3.1.

From the chemical analysis of ore and coal, it is possible to calculate the values of K_1 and K_2 for a mixture of ore: coal in the ratio 90:10 in the following manner.

The oxygen attached to iron in iron oxides is 29.1%. Hence, $K_1 = 29.1\%$. The percentage of fixed carbon, volatile matter and moisture in coal is 73%. Hence, $K_2 = 73\%$. These values of K_1 and K_2 have been calculated without considering the binder which was used in pellet making (iron ore = 86.5%, coal = 9.6%, dextrin (binder) = 3.9%). Considering the fact that all the dextrine added is the value of K_2 becomes 0.81 (say K_2'). These values will, of course, change with the nature of the ore, coal and ore–coal ratio. The exact form of Eq. 3.44 for the present case becomes

Table 3.1 Analysis of ore and coal

Iron ore (Chemical composition, (mass %) Baitarini (Orissa))	Cloaking coal (Proximate analysis, (mass %) Dugdha coal (Bihar))
Fe (T) = 67.78	(Dry basis)
Fe_2O_3 = 91.37	Moisture = 0.4
FeO = 4.9	Volatile matter = 16.84
SiO_2 = 1.41	Ash = 27.11
Al_2O_3 = 2.13	Fixed carbon = 56.02
CaO = traces	
MgO = traces	
P = 0.07	
LOI = 0.85	

$$\begin{aligned}\alpha &= f\left(1 + \frac{K_2' \cdot \text{Weight of coal and dextrine}}{K_1 \cdot \text{Weight of ore}}\right) - \frac{\text{Weight loss from coal and dextrine}}{K_1 \cdot \text{Weight of ore}} \\ &= 1.45f - \frac{\text{Weight loss from coal and dextrine}}{0.29 \times 0.865 \times \cdot \text{Weight of pellet sample}} \\ &= 1.45f - 3.97\,\frac{\text{Weight loss from coal and dextrine}}{\text{Weight of pellet sample}} \\ &= 1.45f - 3.97Q\end{aligned} \tag{3.45}$$

where

$$Q = \frac{\text{Weight loss from coal} + \text{Weight loss from dextrine}}{\text{Weight of pellet sample}}$$

It is found that Q increases with time, i.e. with α. The minimum and maximum values of Q are respectively 0 and 0.109.

Reduction tests were carried out using silica crucibles with lids. The pellets in the crucible were transferred to a muffle furnace maintained at the desired test temperature. Inside the crucible, the environment was that generated by the reaction of pellets. After thermal exposure for the predetermined period, the crucible was quickly shifted to a desiccator to prevent oxidation of reduced pellets. Experiments were performed at 950, 1000, 1050 or 1100 °C, and the duration of exposure was 15, 30 or 45 min. The reduced pellets were analysed for α values. f values were determined by dividing measured weight loss by 36.57, the maximum possible weight loss as experimentally determined. This value was in close agreement with the calculated value of 36.08%.

3.8.3 *Results*

Values of α and f obtained from the reduction tests are summarized in Table 3.2. Figure 3.15 shows the relation between experimentally determined α and f. The figure also shows two theoretical lines, drawn on the basis of Eq. 3.44. The line placed higher is derived by assuming $Q = 0$ which should be valid only at the beginning of reduction. The line placed lower is based on the maximum possible value of this term ($Q = 0.109$) which is valid towards the end of reduction. As expected, the experimental data lie approximately along a diagonal, the terminal values being somewhat lower.

It is interesting to note that Rao (1971) expressed the kinetics of reduction of ore–coal mixtures by an equation of the form

Table 3.2 Values of α and f for the composite pellet

Expt. No.	Temperature (°C)	Time (min)	Degree of reduction (α)	Degree of reaction (f)
1	950	15	0.293	0.428
2	950	30	0.318	0.494
3	950	45	0.163	0.436
4	1000	15	0.403	0.460
5	1000	30	0.733	0.781
6	1000	45	0.709	0.769
7	1050	15	0.477	0.525
8	1050	30	0.657	0.754
9	1050	45	0.749	0.804
10	1100	15	0.753	0.830
11	1100	30	0.815	0.954
12	1100	45	0.889	0.995

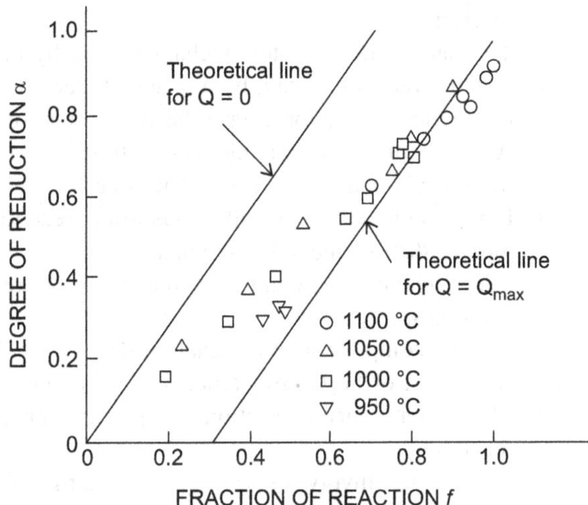

Fig. 3.15 Relationship between the degree of reduction α and fraction of reaction on f (Rao 1971)

$$-\ln(1 - mf) = kt \tag{3.46}$$

where m is a constant with value close to unity. As per Eq. 3.45, α is nearly proportional to f when Q is very small. Thus, if in Eq. 3.46, mf is replaced by α, then Eq. 3.46 represents first-order kinetics.

Equation 3.45 has been derived on the basis of a simplistic assumption that there is no aerial oxidation of carbon. This assumption appears valid in the initial stages of reduction, because α continues to rise with increase in f. However, it appears from Eq. 3.45 that there may be some aerial oxidation towards the end.

In the theoretical derivation, it has been implicitly assumed that oxygen removal would be complete when f approaches unity. This, however, may not be so when the amount of carbon is inadequate from the stoichiometric point of view. This may also explain why for the data analysed, and α remains less that unity as f approaches unity.

3.9 Review Questions

1. Discuss whether the following statements are correct or false.

   a. When a reaction takes place to a series of steps, then there is always a rate-controlling step.
   b. During a steady-state multistep process, all steps have the same rate except the rate-controlling step.
   c. Two-phase reactions are more likely than three-phase reactions.
   d. A maximum of three phases can react simultaneously in a heterogeneous reaction.
   e. The rate-controlling step is characterized by thermodynamic equilibrium.
   f. When a reaction occurs by a series of steps, then the step characterized by the highest activation energy becomes rate controlling at all temperatures.
   g. A solid sphere is reacting with a fluid, the reaction being controlled by interfacial chemical reaction. The time for complete reaction is infinity.
   h. For a chemically controlled gas–solid reaction, the order of reaction is independent of the solid geometry.
   i. For a chemically controlled reaction, the reaction rate must gradually decreases.
   j. Rate of complete internal reaction depends on both pellet size and shape.
   k. Reduction of Fe_2O_3 by carbon is a solid-state direct reduction.
   l. Most heterogeneous reactions are likely to proceed to a series of two-phase reactions.
   m. If a reaction involves several parallel steps, then the fastest step defines the rate.
   n. There must be material accumulation somewhere after a rate-controlling step.
   o. The virtual maximum rate of reaction is the rate of the fastest step when a reaction occurs by a series of steps.
   p. The mechanism of a reaction is reflected in the chemical equation representing the equation.
   q. Total internal reaction during a gas–solid reaction requires that the solid be totally porous.

2. A solid sphere and a cube of the same material weigh the same and both have the same density. They react separately with a fluid, and the reaction is phase boundary controlled. Which one will take longer to undergo 50% reaction? How much longer?
3. A long cylinder ($l \gg r$) is dissolving in a liquid according to phase boundary reaction mechanism. If the original radius r_0 is reduced to $r_0/2$ in one hour, how long will it take for the radius to be reduced to $r_0/3$?
4. Derive the kinetic law for reaction of a cube and a long cylinder ($l \gg r$) with a gas assuming the reaction to be chemically controlled.
5. A solid fuel-fired rocket is to have a fixed acceleration in space. Show that a hollow solid cylinder (fuel) of concentric geometry cannot give constant acceleration.
6. Show that, for phase boundary-controlled reaction, the general kinetic equation may be written as $1 - (1 - \alpha)^{1/n} = kt$ where n denotes the number of dimensions along which there is movement of the phase boundary.
7. Kinetics of reduction of Fe_2O_3 by a reducing gas can be studied by continuous of weight loss. What assumptions are necessary to define weight as a kinetic parameter?
8. A sphere, a cube and a long cylinder ($l \gg r$), each made of the same dense solid, react separately with a liquid and are completely consumed after identical reaction time. Find the ratio of their weights if the reaction is controlled by chemical reaction at the interface.
9. Refer to Fig. 3.2 which shows concentric layers of Fe_2O_3, Fe_3O_4, Fe_xO and Fe. How will the picture change if rate was controlled by $Fe_2O_3 \rightarrow Fe_xO$ step in reduction?
10. The diameter of a spherical solid reduces at a constant rate when the solid reacts with a liquid. Does this necessarily imply interfacial chemical reaction control?
11. Reduction of an ore–coal pellet proceeds much faster than that of a dense pellet of ore embedded in coal. Why?

References

Agarwal, J.P.: Science Reporter, July 1983, p. 401; March 1986, p. 148

Agarwal, B.B., Prasad, K.K., Ray, H.S.: ISIJ Intl. **30**, 997 (1990)

Basu, P.: Reduction of Fe_2O_3—coal pellets. M. Tech. Thesis, Indian Institute of Technology, Kharagpur (1987)

Basu, P., Sarkar, S.B., Ray, H.S.: Trans. Ind. Inst. Metals, **42**, 165 (1989)

Dutrizac, J.R., MacDonald, R., Ingraham, T.R.: Metall. Trans. **1**(225), 3083 (1970)

Kofstad, P.: High Temperature Oxidation of Metals, p. 14 (1966)

McKewan, W.M.: Trans. TMS-AIME **212**, 791 (1958)

Mookherjee, S., Ray, H.S., Mukherjee, A.: Thermochim Acta. **95**, 235 (1985)

Rao, Y.K.: Metall. Trans. **2**, 1459 (1971)

Ray, H.S., Sridhar, R., Abraham, K.P.: Extraction of Nonferrous Metals (Ch. 6). Affiliated East-West Press Pvt. Ltd., New Delhi (1985)

Reilly, I.G., Scott, D.S.: Can. J. Chem. Eng. **55**, 527 (1977)
Rosenquist, T.: Principles of Extractive Metalurgy (Ch. 5). McGraw-Hill, New Delhi (1974)
Thomas, G., Ropital, F.: J. Thermal Anal. **30**, 121 (1985)
Wodsworth, M.E.: In: Sohn, H.Y., Wodsworth, M.E. (eds.) Rate Processes in Extractive Metallurgy. Plenum Press, New York (1979). Ch. 2

Chapter 4
Diffusion Through Product Layer

4.1 Introduction

4.1.1 Role of Reaction Product

If a solid reaction product is formed on the reacting solid, the kinetics of reaction may be governed by the character of the coating and the diffusion process. If the coating is very porous, then there will be no resistance to reagents reaching the reaction interface and the coating will have no effect. In case the porosity is low or non-existent, then the reagent has to diffuse through the protective film (by pore diffusion and lattice diffusion, respectively) before it reaches the interface. In this case, the reaction rate is likely to decrease progressively because of the gradual thickening of the product layer. Figure 4.1 shows schematically the two different situations that are normally encountered. Table 4.1 gives a list of some heterogeneous reactions where a solid reaction product or residue is formed.

Through a product or residue layer becomes rate controlling then the reaction rate must gradually decrease as the layer thickness increases and, therefore, the diffusion path increases. The decrease in rate would be enhanced if the area of the core/product interface also decreases as would be the case for the reaction of a spherical solid. The kinetic equation must be derived in terms of the diffusion equation taking into account the geometrical factors.

4.1.2 Reaction of a Flat Plate

Consider a flat plate of area A and a small thickness d. We assume that only one face of the plate reacts. We also assume that during reaction of the solid plate, A remains unchanged and only d diminishes. Reaction produces a product of thickness y which increases gradually. If diffusion of reactant through this product

H. S. Ray and S. Ray, *Kinetics of Metallurgical Processes*,
Indian Institute of Metals Series, https://doi.org/10.1007/978-981-13-0686-0_4

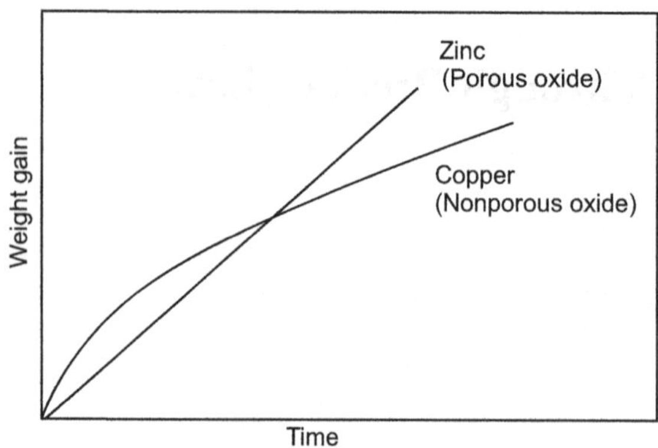

Fig. 4.1 Schematic diagram indicating influence of nature of reaction product on oxidation behaviour of metals

Table 4.1 Formation of a solid reaction product in heterogeneous reactions

	Process	Example
Solid–gas	Oxidation of metals	Metal + O_2 → Oxide
	Oxidation of sulphides	Sulphide + O_2 → Oxide + SO_2
	Reduction of oxides by gases	Oxide + CO → Metal + CO_2
	Thermal decomposition of carbonates or sulphates	Carbonate → Oxide + CO_2 Sulphate → Oxide + SO_3
	Oxidation of coal	Coal → Ash + H_2O + V.M. + CO_2
Solid–liquid	Leaching of sulphides with liberation of elemental sulphur	$CuS + 2Fe^{3+} \rightarrow Cu^{2+} + S + 2Fe^{2+}$
	Double decomposition	$NiS + Cu^{2+} \rightarrow CuS + Ni^{2+}$
	Cementation	$Cu^{2+} + Fe \rightarrow Cu + Fe^{2+}$

layer is rate controlling, then the rate of reaction is given by the rate of diffusion of reagent. Hence,

$$\text{Rate} = \frac{\mathrm{d}W}{\mathrm{d}t} = \frac{DA \cdot C}{y} \tag{4.1}$$

where W is the weight of product at time t, D the diffusion coefficient and C the reagent concentration (fixed). Since $W = Ay\rho$, we obtain

$$\frac{\mathrm{d}y}{\mathrm{d}t} = \frac{k'}{y} \tag{4.2}$$

where $k' = \frac{DC}{\rho}$ = constant. Here, ρ is density. Since $y = 0$ at $t = 0$, we obtain, by integration,

$$y^2 = kt \tag{4.3}$$

where k is a constant. Again, if the weight of the oxide on complete reaction is W_o, then

$$\alpha = \frac{W}{W_o} \tag{4.4}$$

Since W is proportional to y and α is proportional to W, we also have for kinetic equations, the forms

$$W^2 = kt \tag{4.5}$$

$$\alpha^2 = kt \tag{4.6}$$

Equations 4.3, 4.5 and 4.6 are forms of parabolic law. Figure 4.2 shows how oxidation of iron follows the parabolic law.

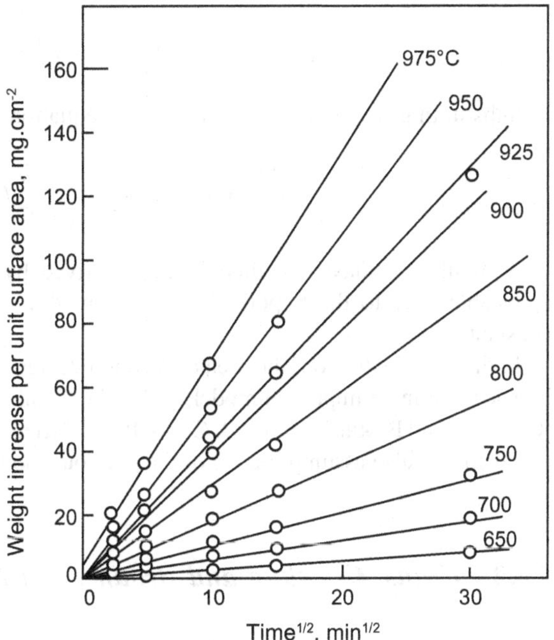

Fig. 4.2 Oxidation of iron—the parabolic law (Habashi 1970)

4.2 Reaction of a Spherical Solid

4.2.1 Jander's Equation

The parabolic law will not hold good for a spherical particle. When the reaction interface is a spherical core then, with progressive reaction, the sphere contracts and the area of reaction, A, decreases. To derive the simplest diffusion control model for a spherical geometry, we make the following assumptions:

(a) The degree of reaction is so small that the product layer thickness y is small.
(b) The density of the product layer is the same as that of the reactant so that y, the product layer thickness, also is the thickness of reactant consumed when the radius decreases from an initial value r_0 to r at time t.

We have

$$\alpha = \frac{r_0^3 - r^3}{r_0^3} = 1 - (r/r_0)^3 \tag{4.7}$$

$$\frac{r_0 - r}{r_0} = 1 - (1 - \alpha)^{1/3} \tag{4.8}$$

Since $(r_0 - r) = y$, we get

$$y = r_0[1 - (1 - \alpha)^{1/3}] \tag{4.9}$$

Substituting in Eq. 4.3, we get Jander's equation

$$[1 - (1 - \alpha)^{1/3}]^2 = \frac{k}{r_0^2} t \tag{4.10}$$

It should be noted that this equation derived by Jander and Anorg (1927) is applicable only to the hypothetical situation of y nearly zero where A is nearly constant.

In the actual situation, the area of reaction decreases progressively. This is taken into account in an improved model, called the Crank–Ginstling–Brounshtein model (Ginstling and Brounshtein 1959; Crank 1957; Carter 1961; Valensi 1935). However, in this model also assumption (b) of the previous model remains unchanged.

4.2.2 Crank–Ginstling and Brounshtein Model

If J is the number of molecules of the reagent diffusing in time t through the product layer (Fig. 4.3), then according to Fick's law

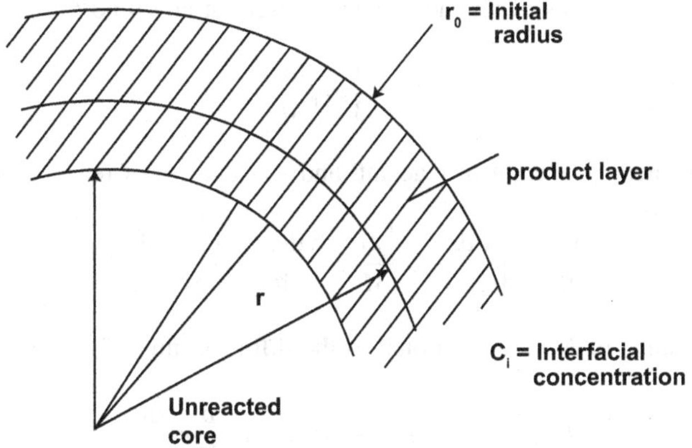

Fig. 4.3 Schematic diagram showing formation of product layer and meanings of symbols

$$J = -AD\frac{\mathrm{d}C}{\mathrm{d}r} = -4\pi r^2 \cdot D\frac{\mathrm{d}C}{\mathrm{d}r} \quad (4.11)$$

$$\int\limits_{Ci}^{C} \mathrm{d}C = -\frac{J}{4\pi D}\int\limits_{r_1}^{r_0} \frac{\mathrm{d}r}{r^2} \quad (4.12)$$

$$C - C_\mathrm{i} = -\frac{J}{4\pi D}\frac{(r_0 - r_1)}{r_0 r_1} \quad (4.13)$$

where the symbols are explained in Fig. 4.3.

For a diffusion-controlled process C_i, the concentration of diffusing species at interface equals zero. Hence,

$$J = -4\pi D\left(\frac{r_0\, r_1}{r_0 - r_1}\right)C \quad (4.14)$$

C may be taken as constant. We have,

$$\alpha = \frac{\frac{4}{3}\pi r_0^3 - \frac{4}{3}\pi r_1^3}{\frac{4}{3}\pi r_0^3} = 1 - \left(\frac{r_1}{r_0}\right)^3 \quad (4.15)$$

i.e.

$$r_1 = r_0(1-\alpha)^{1/3} \quad (4.16)$$

The number of moles of unreacted solid present at any time t is

$$N = \frac{4}{3}\pi r_1^3 \frac{\rho}{M} \tag{4.17}$$

where M = molecular weight of reactant, and ρ = density of reactant. Therefore,

$$\frac{\mathrm{d}N}{\mathrm{d}t} = \frac{\mathrm{d}N}{\mathrm{d}r_1}\frac{\mathrm{d}r_1}{\mathrm{d}t} = \frac{\frac{4}{3}\pi\rho}{M}3r_1^2\frac{\mathrm{d}r_1}{\mathrm{d}t} = \frac{4\pi\rho r_1^2}{M}\frac{\mathrm{d}r_1}{\mathrm{d}t} \tag{4.18}$$

Rate of change of N is proportional to the diffusion flux. Therefore,

$$J = -4\pi D\left(\frac{r_0\, r_1}{r_0 - r_1}\right)\mathrm{C} = \beta 4\pi \frac{\rho}{M} r_1^2 \frac{\mathrm{d}r_1}{\mathrm{d}t}$$

where β is the stoichiometry factor. Therefore,

$$-\frac{MDC}{\beta\rho}\mathrm{d}t = \frac{r_1(r_0 - r_1)\mathrm{d}r_1}{r_0} = \left(r_1 - \frac{r_1^2}{r_0}\right)\mathrm{d}r_1 \tag{4.19}$$

$$-\frac{MDC}{\beta\rho}\int_0^t \mathrm{d}t = \int_{r_0}^{r_1}\left(r_1 - \frac{r_1^2}{r_0}\right)\mathrm{d}r_1 \tag{4.20}$$

$$\begin{aligned} \left[-\frac{MDC}{\beta\rho}t\right]_0^t &= \left[\frac{r_1^2}{2}\right]_{r_0}^{r_1} - \left[\frac{r_1^3}{3r_0}\right]_{r_0}^{r_1} \\ -\frac{MDC}{\beta\rho}t &= \frac{r_1^2}{2} - \frac{r_0^2}{2} - \frac{r_1^3}{3r_0} + \frac{r_0^3}{3r_0} = \frac{1}{2}r_1^2 - \frac{1}{6}r_0^2 - \frac{1}{3}\frac{r_1^3}{r_0} \end{aligned} \tag{4.21}$$

Substitute for r_1 in terms of α

$$\begin{aligned} -\tfrac{MDC}{\beta\rho}t &= \tfrac{1}{2}r_0^2(1-\alpha)^{2/3} - \tfrac{1}{6}r_0^2 - \tfrac{1}{3}\tfrac{r_0^3}{r_0}(1-\alpha) \\ -\tfrac{MDC}{\beta\rho r_0^2}t &= \tfrac{1}{2}(1-\alpha)^{2/3} - \tfrac{1}{6} - \tfrac{1}{3}(1-\alpha) \end{aligned} \tag{4.22}$$

$$\frac{2MDC}{\beta\rho r_0^2}t = 1 - \frac{2}{3}\alpha - (1-\alpha)^{2/3} \tag{4.23}$$

This means that the plot of $\left[1 - \frac{2}{3}\alpha - (1-\alpha)^{2/3}\right]$ against t should yield a straight line.

4.2.3 Crank–Carter–Valensi Model

Let

r_0	original radius,
r_1	radius of unreacted core,
$r_2 - r_1$	thickness of product layer,
J	number of moles of reagent diffusing in time t,
C	concentration of reagent, assumed constant,
C_i	concentration of reagent at the interface,
W_O	original weight,
W	weight of unreacted core,
ρ_R	density of reactant,
ρ_P	density of product,
M_R	molecular weight of reactant,
M_P	molecular weight of product,
N	number of moles of reactant,
β	stoichiometry factor,
V_R	molecular volume of reactant,
V_P	molecular volume of product.

We have, for fraction reacted (Eq. 4.15), $\alpha = \frac{W_o - W}{W_o} = 1 - \left(\frac{r}{r_0}\right)^3$.

Referring to Fig. 4.3 (Eq. 4.11), $J = -AD\frac{dc}{dr} = -4\pi r^2 D\frac{dc}{dr}$.

Therefore,

$$\int_{C_i}^{C} dC = -\frac{J}{4\pi D}\int_{r_1}^{r_2}\frac{dr}{r^2}$$

$$J = -4\pi D\left(\frac{r_1 r_2}{r_2 - r_1}\right)(C - C_i) \tag{4.24}$$

Since $C_i = 0$, we get

$$J = -4\pi D\left(\frac{r_1 r_2}{r_2 - r_1}\right)C \tag{4.25}$$

The rate of change of W is proportional to J, and the flux of reagent diffusing through the spherical shell of thickness $r_2 - r_1$:

$$N = \frac{W}{M_R} = \frac{4}{3}\pi r_1^3\frac{\rho_R}{M_R} \tag{4.26}$$

$$\frac{\mathrm{d}N}{\mathrm{d}t} = \frac{4\pi r_1^2 \rho_\mathrm{R}}{M_\mathrm{R}} \frac{\mathrm{d}r_1}{\mathrm{d}t} \tag{4.27}$$

We have,

$$J = \beta \frac{\mathrm{d}N}{\mathrm{d}t} \tag{4.28}$$

Hence,

$$\begin{aligned} -4\pi D\left(\frac{r_1 r_2}{r_2 - r_1}\right) C &= \beta \frac{4\pi r_1^2 \rho_\mathrm{R}}{M_\mathrm{R}} \frac{\mathrm{d}r_1}{\mathrm{d}t} \\ -\frac{M_\mathrm{R} D C}{\beta \rho_\mathrm{R}} \mathrm{d}t &= \left(r_1 - \frac{r_1^2}{r_0}\right) \mathrm{d}r_1 \end{aligned} \tag{4.29}$$

The number of molecules consumed equals number of molecules formed times stoichiometry factor:

$$\frac{\left(\frac{4}{3}\pi r_0^3 - \frac{4}{3}\pi r_1^3\right)\rho_\mathrm{R}}{M_\mathrm{R}} = \frac{\left(\frac{4}{3}\pi r_2^3 - \frac{4}{3}\pi r_1^3\right)\rho_\mathrm{P}}{M_\mathrm{P}} \beta$$

$$\frac{1}{\beta} \frac{\rho_\mathrm{R} M_\mathrm{P}}{\rho_\mathrm{P} M_\mathrm{R}} = \frac{r_2^3 - r_1^3}{r_0^3 - r_1^3}$$

$$\frac{1}{\beta} \frac{M_\mathrm{P}/\rho_\mathrm{P}}{M_\mathrm{R}/\rho_\mathrm{R}} = \frac{1}{\beta} \frac{V_\mathrm{P}}{V_\mathrm{R}} = Z, \quad \text{say}$$

Therefore,

$$Z = \frac{r_2^3 - r_1^3}{r_0^3 - r_1^3}$$

$$r_2^3 - r_1^3 = Z(r_0^3 - r_1^3)$$

$$r_2^3 = [Z r_0^3 + r_1^3(1 - Z)]$$

$$r_2 = [Z r_0^3 + r_1^3(1 - Z)]^{1/3} \tag{4.30}$$

It should be noted that if $Z = 1$ then $r_2 = r_0$. Substituting the value of r_2 from Eq. 4.30 into Eq. 4.29

$$-\frac{M_\mathrm{R} D C}{\beta \rho_\mathrm{R}} \mathrm{d}t = \left(r_1 - \frac{r_1^2}{[Z r_0^3 + r_1^3(1 - Z)]^{1/3}}\right) \mathrm{d}r_1 \tag{4.31}$$

Integrating from the limits r_0 to r_1 for $t = 0$ and $t = t$

$$-\frac{M_R DC}{\beta\rho_R}t = \frac{1}{2}r_1^2 + \frac{Zr_0^2 - [Zr_0^3 + r_1^3(1-Z)]^{2/3}}{2(1-Z)}$$

$$-2(1-Z)\frac{M_R DC}{\beta\rho_R}t = (1-Z)r_1^2 + Zr_0^2 - [Zr_0^3 + r_1^3(1-Z)]^{2/3}$$

$$Zr_0^2 + 2(1-Z)\frac{M_R DC}{\alpha\rho_R}t = [Zr_0^3 + r_1^3(1-Z)]^{2/3} - (1-Z)r_1^2 \quad (4.32)$$

where $\alpha = 1 - \left(\frac{r_1}{r_0}\right)^3$ (Eq. 4.15) and $r_1 = r_0(1-\alpha)^{1/3}$ (Eq. 4.16). Substituting the value of r_1 in terms of R into Eq. 4.32, we get

$$\begin{aligned} Zr_0^2 + 2(1-Z)\frac{M_R DC}{\beta\rho_R}t &= [Zr_0^3 + (1-Z)r_0^3(1-\alpha)]^{2/3} - (1-Z)r_0^2(1-\alpha)^{2/3} \\ &= r_0^2[Z + (1-Z)(1-\alpha)]^{2/3} - (1-Z)r_0^2(1-\alpha)^{2/3} \end{aligned} \quad (4.33)$$

Dividing by r_0^2, we get

$$\begin{aligned} Z + 2(1-Z)\frac{M_R DC}{\beta\rho_R r_0^2}t &= [Z + (1-Z)(1-\alpha)]^{2/3} - (1-Z)(1-\alpha)^{2/3} \\ &= [Z + (1-Z)(1-\alpha)]^{2/3} - (1-Z)(1-\alpha)^{2/3} \end{aligned} \quad (4.34)$$

Equation 4.34 can be rewritten to express the kinetic equation as follows

$$\frac{[1 + (Z-1)\alpha]^{2/3} + (Z-1)(1-\alpha)^{2/3} - Z}{2(1-Z)} = \frac{M_R DC}{\beta\rho_R r_0^2}\cdot t \quad (4.35)$$

4.2.4 *Comparison of the Three Diffusion Control Models*

It has been mentioned that while the Carter–Valensi equation gives an exact kinetic equation, the Crank–Ginstling–Brounshtein equation applies if it is assumed that there is no volume change in solid during reaction; that is, the density of product is equal to that of the reactant, and Jander's equation is even more approximate, being worth considering only when $\alpha \to 0$.

Equivalence of the last two models can be proved using La Hospital's rule. Consider the ratio

$$\frac{1-\frac{2}{3}\alpha-(1-\alpha)^{2/3}}{[1-(1-\alpha)^{1/3}]^2}$$

At $\alpha = 0$, the ratio equals $0/0$. Differentiating with respect to α

$$\frac{0-\frac{2}{3}-\frac{2}{3}(1-\alpha)^{-1/3}(-1)}{2[1-(1-\alpha)^{1/3}]\left[0-\frac{1}{3}(1-\alpha)^{-2/3}(-1)\right]}$$

This ratio is also $0/0$ at $\alpha = 0$. Differentiating once again with respect to α, we obtain

$$\frac{\frac{2}{9}(1-\alpha)^{-4/3}}{\frac{4}{9}(1-\alpha)^{-5/3}-\frac{2}{9}(1-\alpha)^{-4/3}}$$

which equals 1 at $\alpha = 0$.

To prove equivalence of Crank–Ginstling–Brounshtein equation with Carter–Valensi equation, consider the form of $g(\alpha)$ for the later, namely

$$\frac{[1+(Z-1)\alpha]^{2/3}+(Z-1)(1-\alpha)^{2/3}-Z}{(1-Z)}$$

Here, Z represents the ratio of the density of the product to that of the reactant solid. The ratio equals $0/0$ at $Z = 1$. Differentiating with respect to α, one obtains

$$\frac{\frac{2}{3}[1+(Z-1)\alpha]^{-1/3}\cdot\alpha+(1-\alpha)^{2/3}-1}{-1}$$

which equals $1-\frac{2}{3}\alpha-(1-\alpha)^{2/3}$ at $Z = 1$. Thus, if we consider the ratio of the expressions for the two equations, repeated differentiation would eventually give a ratio of unity.

Example 4.1 Show, without using La Hospital's rule, that the Ginstling–Brounshtein equation reduces to the Jander's equation when $\alpha \to 0$.

Solution

From the binomial expansion

$$(a+b)^n = a^n + na^{n-1}b + \frac{n(n-1)}{2!}a^{n-2}b^2 + \cdots + b^n$$

$$(1-\alpha)^{2/3} = 1 + \frac{2}{3}\cdot 1\cdot(\alpha) + \frac{2}{3}\frac{\left(\frac{2}{3}-1\right)\cdot 1}{2!}\alpha^2 + \cdots$$
$$+(-\alpha)^{2/3} = 1 - \frac{2}{3}\alpha - \frac{1}{9}\alpha^2 \cdots$$

Hence for small values of α

$$1 - \frac{2}{3}\alpha - (1-\alpha)^{2/3} = 1 - \frac{2}{3}\alpha - 1 + \frac{2}{3}\alpha + \frac{\alpha^2}{9} + \cdots = \frac{\alpha^2}{9}$$
$$[1-(1-\alpha)^{1/3}]^2 = \left[1 - 1 + \frac{\alpha}{3} + \frac{1}{9}\alpha^2 + \cdots\right]^2 = \frac{\alpha^2}{9}$$

Hence for $\alpha \to 0$, $1 - \frac{2}{3}\alpha - (1-\alpha)^{2/3} = [1-(1-\alpha)^{1/3}]^2$, i.e. the Ginstling–Brounshtein equation reduces to the Jander's equation.

Role of Particle Size

In all the three models for reactions controlled by diffusion through product layer, the rate constant is proportional to $1/r_0^2$. For chemical reaction control, i.e. phase boundary control, the rate constant is proportional to (Eq. 3.14) $1/r_0$. This means that, for a given mass of solid reductant, if the particle size is reduced then the rate constant is increased more for the diffusion-controlled processes.

Diffusion control is more sensitive to particle size because diffusion not only depends on the available surface area through which species move but also the length of the diffusion path. The chemical control depends only on the interfacial area. The dependence of k on r_0 can be a test for identification of reaction mechanism. It is convenient to plot $\log k$ versus $\log r_0$. A slope of 2 identifies the diffusion model.

Example 4.2 A mass of spherical solid particles of 2 mm size is leached by a liquid with no reagent starvation. Fifty per cent reaction is complete within two hours. Determine the leaching time for the same extent of reduction if the particle size is reduced to 1 mm assuming that the rate controlling step is

(a) Chemical reaction at the phase boundary,
(b) Diffusion through a product layer.

Solution
For (a), the kinetic equation applicable is Eq. 3.14,

$$g(\alpha) = 1 - (1 - \alpha)^{1/3} = \frac{k}{r_0}t$$

For a fixed value of α, this reduces to $r_0 =$ const. t. Hence when the particle size is reduced from 2 to 1 mm, the reaction time is reduced from 2 to 1 h.

For (b), we can assume the Ginstling–Brounshtein equation, i.e. Eq. 4.23.

$$g(\alpha) = 1 - \frac{2\alpha}{3} - (1 - \alpha)^{2/3} = \frac{k}{r_0^2}t$$

For a fixed value of α, this reduces to $r_0^2 =$ const. t. Hence, when the size is reduced from 2 to 1 mm, the reaction time will be reduced from 2 to 30 min.

Example 4.3 A reaction is 48.8% complete in 1 h. What will be the time required for complete reaction if the rate control follows:

(a) Parabolic law?
(b) Ginstling–Brounshtein equation?
(c) Phase boundary control equation?

Solution For (a), $\alpha^2 = kt$, $\therefore$ $(0.4888)^2 k \cdot 1$. For $\alpha = 1, 1^2 = kt$,

$$\therefore \quad t = \frac{1}{(0.488)^2} = 4.201 \text{ h}.$$

For (b), $1 - \frac{2}{3}\alpha - (1 - \alpha)^{2/3} = kt$. The LHS equals 0.0347 for $\alpha = 0.488$ and $1/3$ for $\alpha = 1$. Hence,

$$\frac{1}{3 \times 0.0347} = \frac{t}{1} \quad \therefore\ t = 8.76 \text{ h}.$$

For (c), $1 - (1 - \alpha)^{1/3} = kt$. The LHS equals 0.2 for $\alpha = 0.488$ and 1 for $\alpha = 1$. Hence,

$$\frac{1}{0.2} = \frac{t}{1} \quad \therefore\ t = 5 \text{ h}.$$

It should be noted that for complete internal reaction, the rate constant is independent of particle radius (Eq. 3.37). When a particle is completely porous and the fluid has free access to all regions, then there is no interface at and the outer gross dimension of the solid has no relevance. However, in the microscale, the individual grains may behave as dense regions and thus the grain size may influence the rate constant.

4.3 Diffusion Through Porous Solid

4.3.1 Diffusion and Flow of Gases Through Pores

Transport of gas through a porous solid is of considerable theoretical and practical importance. Accordingly, the subject has received considerable attention. Some fundamental concepts concerning diffusion and flow of gases through pores have been discussed elsewhere (Ray 1976; Abraham et al. 1973; Scott and Dullien 1962; Present 1958; Youngquist 1970).

Diffusion and flow through porous solids involve several uncertainties regarding the pore structure, e.g. roughness, shape and orientation of the pores, their mutual interconnections and lengths. To account for these uncertainties, an adjustable empirical constant called the "tortuosity factor" (τ) is generally used. Thus, if one is concerned with a mass flux J through a porous sample of overall surface area A and bulk porosity ε, then the flux can be considered to be equivalent to one across a pore surface area $A\varepsilon/\tau$ if all the pores constitute parallel paths. Here, $A\varepsilon$ gives the effective pore area on the surface and τ is incorporated to account for uncertainties in the pore structure. The experiments to be described measure ε/τ and τ.

Theories on diffusion and flow of gases through porous media are based on relatively well-defined equations where pores are considered analogous to capillaries. When the total pressure on either side of a capillary is the same and gaseous species migrate because of partial pressure gradient, then the mass flux is diffusive in nature. When the capillary radius,r, is small compared to the mean free path, λ (say $r/\lambda < 0.1$), then the flux is said to involve Knudsen diffusion. Ordinary molecular diffusion takes place when the capillary dimension becomes comparable to the mean free path. When the total pressures on either side of a capillary are unequal, then the gas migration is governed also by the pressure gradient. At low gradients, the flow may be simple streamline flow, called Poiseulle's flow. At high gradients with high Reynold's number, however, the flow turns turbulent. In general, in the presence of a pressure gradient both diffusion and flow contribute to the net flux through the capillary.

4.3.2 Two Simple Experiments to Study Flow and Diffusion (Ray 1976; Abraham et al. 1973)

Experiment 1 Flow under pressure gradients

The experimental arrangement for investigating flow under considerable pressure gradients is shown in Fig. 4.4. A porous cylinder is sealed between two glass tubes and the curved surface coated with a sealant as shown. The open flat surfaces are exposed to a vacuum/gas line on one side and a manometer on the other side. The manometer is connected to a large flask which approximates the total volume.

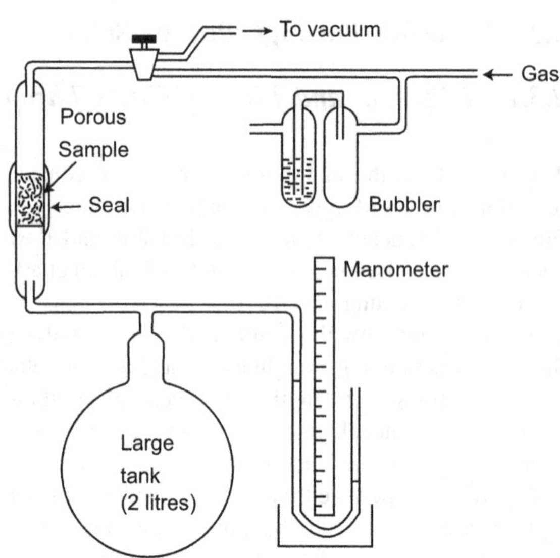

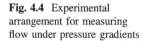

Fig. 4.4 Experimental arrangement for measuring flow under pressure gradients

The entire system is first evacuated and checked for leaks. It is then suddenly connected to the gas line to establish immediately a pressure by using a simple bubbler arrangement. The gas flows slowly through the sample pores, and the manometer reading gradually falls indicating a gradual build-up of pressure. The pressure readings are recorded over an extended period of time. The pressure–time plot allows calculation of ε/τ as shown subsequently.

Experiment 2 Diffusion of gas due to difference in partial pressures
In the diffusion experiments, equal pressures are maintained on either side of the porous sample through which diffusion takes place unidirectionally. An experimental setup for hydrogen diffusion is shown in Fig. 4.5. The hydrogen concentration in the gases is measured using a hydrogen concentration cell. The porous cylindrical sample is located across two parallel gas streams as shown. Purified argon is passed through S_1S_7 while purified hydrogen is passed through S_5S_6. This arrangement allows diffusion of hydrogen through the pores against a stream of argon. As a result of the diffusion, the argon stream is enriched by hydrogen and vice versa. The hydrogen intake in the argon stream is measured accurately by the concentration cell.

The gas passing out through S_7 is made to bubble at a hydrogen indicator electrode of a hydrogen concentration cell. The reference electrode of the cell is constituted by platinum foil at which hydrogen is continuously bubbled at 1 atm pressure. Knowing the flow rate of argon and the emf of the cell, the concentration of hydrogen in argon strain may be calculated. From this, one can calculate the effective diffusion coefficient of hydrogen in the pores, as well as the "tortuosity" factor. These values are independent of argon flow rate as will be shown later. The theoretical basis of the experiments is discussed briefly.

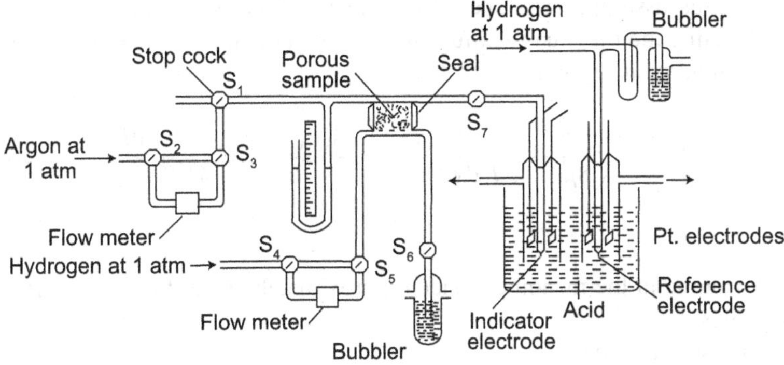

Fig. 4.5 Experimental arrangement for measuring diffusion

4.3.3 *Theory of Experiments*

Theory of Experiment 1 Involving Flow

When a finite pressure gradient exists across a capillary, both diffusion and flow contribute to the net flux. It has been shown (Abraham et al. 1973) that, for a uniform capillary, forced flow flux

$$J = -\left(\frac{r^2P}{8\mu} + \frac{4rR/T}{3M\bar{v}}\right)\frac{1}{RT}\frac{\mathrm{d}P}{\mathrm{d}x} \tag{4.36}$$

where

r radius of capillary,
P pressure at a distance,
x along the length of a capillary,
$\mathrm{d}P/\mathrm{d}x$ pressure gradient,
T absolute temperature,
M molecular weight of the gas,
$\bar{v}$ mean molecular speed,
R universal gas constant and
μ viscosity of the gas.

For the flow equation for the actual sample, one has to consider the effective pore area as well as the "tortuosity" factor. Thus, flux through the actual sample is

$$J' = -\left(\frac{r^2P}{8\mu} + \frac{4rRT}{3M\bar{v}}\right)\frac{1}{RT}\frac{\mathrm{d}P}{\mathrm{d}x}\cdot\frac{\varepsilon}{\tau} \tag{4.37}$$

The second term within the bracket, which accounts for molecular diffusion, is negligible compared to the first, which is the flow term and may, therefore, be neglected. This assumption is specially valid when pressure gradients are large. Thus, neglecting the flux due to diffusion and remembering that flux also equals the rate of accumulation in the bulb:

$$J' = -\left(\frac{r^2 P}{8\mu} + \frac{1}{RT}\frac{dP}{dx}\right)\frac{\varepsilon}{\tau} = \frac{1}{A}\cdot\frac{d(C_t \cdot V)}{dt} = \frac{V}{ART}\cdot\frac{dP_t}{dt} \tag{4.38}$$

where

V total volume in the manometer side (approximately equal to the volume of the large flask),
P_t gas pressure at time,
t as shown by the manometer and
C_t moles per unit volume at time t.

Since there is no gas accumulation inside the capillary pores:

$$\frac{dJ'}{dx} = 0 \tag{4.39}$$

The equation

$$J' = -\frac{r^2 P}{8\mu}\cdot\frac{1}{RT}\cdot\frac{dP}{dx}\cdot\frac{\varepsilon}{\tau}$$

may now be integrated using the boundary conditions: at $x = 0$, $P = P_0$ and at $x = L$, $P = P_t$ (variable with time) where L is the thickness of the sample. The result is

$$J' = \frac{r^2}{8\mu RT}\frac{\varepsilon}{2L\tau}\left(P_0^2 - P_t^2\right) \tag{4.40}$$

Equating Eq. 4.40 with the right-hand side of Eq. 4.38,

$$\frac{dP_t}{dt} = \frac{ART}{V}\cdot\frac{r^2}{8\mu RT}\frac{\varepsilon}{2L\tau}\left(P_0^2 - P_t^2\right) \tag{4.41}$$

$$\frac{dP_t}{\left(P_0^2 - P_t^2\right)} = \frac{Ar^2}{16\mu VL}\frac{\varepsilon}{\tau}dt \tag{4.42}$$

Integration of the above equation yields

$$\log\left(\frac{P_0 + P_t}{P_0 - P_t}\right) - \log\left(\frac{P_0 + P_t^0}{P_0 - P_t^0}\right) = \left(\frac{Ar^2}{8\mu VL}\frac{\varepsilon}{\tau}\cdot\frac{P_0}{2.303}\right)t \tag{4.43}$$

where P_t^0 denotes the pressure reading of the manometer at zero time. The second term in the left-hand side vanishes if P_t^0 is zero, i.e. if manometer readings are recorded right from the beginning. Equation 4.43 indicates that:

(a) a plot of $\log[(P_0 + P_t)/(P_0 - P_t)]$ against time should be a straight line with slope

$$\frac{1}{2.303}\frac{Ar^2P_0}{8\mu VL}\left(\frac{\varepsilon}{\tau}\right)$$

(b) the slopes plotted against $1/\mu$ should also give a linear plot with a slope

$$\frac{1}{2.303}\frac{Ar^2P_0}{8\mu VL}\left(\frac{\varepsilon}{\tau}\right)$$

From these plots, ε/τ may be found, provided values of A,r, P_0, μ, V and L are all known. Moreover, from statistical considerations ε equals the bulk porosity of the sample (Hillard and Cahn 1961). Knowledge of porosity thus yields τ.

The derivation of Eq. 4.43 is based on a number of assumptions at different steps. The validity of these assumptions taken together may be checked by analysing the equation itself. Thus for three or four different gases, linear plots should be obtained for each gas, each plot having a different slope. These slopes plotted against the reciprocal viscosity should also give linear plots (Ray 1976).

It should be remembered, however, that the equations have hitherto considered a single average pore radius r. The treatment, therefore, is valid only when the pore size in the sample is reasonably uniform. Various uncertainties may be introduced if there is an appreciable pore size distribution.

Theory of Experiment 2 Involving Diffusion

The emf of the hydrogen concentration cell is given by

$$E = \frac{1.987\,T}{2F}\ln\left(\frac{1}{p_{H_2}}\right) \tag{4.44}$$

where T is the absolute temperature, F the Faraday constant (cal) and p_{H_2} the unknown partial pressure of hydrogen in the argon stream (atm). E is in volts. Measurement of the cell emf E thus yields p_{H_2}.

Again, p_{H_2} may be expressed in terms of molar fluxes:

$$p_{H_2} = \frac{\dot{n}H_2}{\dot{n}H_2 + \dot{n}Ar} \tag{4.45}$$

where $\dot{n}_i$ is the flux of species in the gas (g mol s^{-1}). According to Avogadro's law, the molar flux terms may be replaced by corresponding volumetric rates at STP.

For "steady-state" flow:

$$\begin{array}{ll}\text{rate of hydrogen flow} = & \text{rate of hydrogen input} \\ \text{through solid} & \text{in argon stream}\end{array} \tag{4.46}$$

Hence,

$$\frac{p_{H_2} \cdot \dot{n}_{Ar}}{(1 - p_{H_2})} = \frac{A}{L}(1 - p_{H_2}) \cdot \frac{D^{eff}}{RT} \tag{4.47}$$

where $\dot{n}_{Ar}$ is the flow rate of argon, p_{H_2} the steady-state hydrogen partial pressure in argon, and D^{eff} is the effective diffusion coefficient given by the equation:

$$D^{eff} = D \cdot \frac{\varepsilon}{\tau} \tag{4.48}$$

where D is the diffusion coefficient.

Equation 4.47 shows that when $\dot{n}_{Ar}$, A, L, R and T are known and p_{H_2} is measured by the emf cell, then the effective diffusion coefficient can be calculated. It is interesting to note that D^{eff} should be independent of different $\dot{n}$Ar values, each of which would set p_{H_2} at a given value. Equation 4.47, however, neglects argon flow in counter diffusion. The fluxes of the two gases argon and hydrogen are given by the following expressions:

$$J_{H_2} = -(D_{H_2})^{eff} \frac{\partial C_{H_2}}{\partial x} + X_{H_2} \delta_{H_2} J \tag{4.49}$$

$$J_{Ar} = -(D_{Ar})^{eff} \frac{\partial C_{Ar}}{\partial x} + X_{Ar} \cdot \delta_{Ar} J \tag{4.50}$$

where J_i stands for flux, $(D_i)^{eff}$ for effective diffusion coefficient, C_i for concentration and X_i for mole fraction, all for species i; x is the distance along the axis of the specimen. J is the total flux given by

$$J = J_{H_2} + J_{Ar}$$

δ_{H_2} and δ_{Ar} equal, by definition, $\frac{D_{H_2}}{D_{H_2}/\mathrm{Ar}}$ and $\frac{D_{Ar}}{D_{H_2}/\mathrm{Ar}}$, respectively. The conservation equations of hydrogen and argon are given by the following relationships:

$$\frac{\partial}{\partial x}\left(r^2 J_{H_2}\right) = 0 \tag{4.51}$$

$$\frac{\partial}{\partial x}\left(r^2 J_{Ar}\right) = 0 \tag{4.52}$$

$$\text{Also,}\quad \text{at } x = 0, C_{H_2} = C_{H_2} \quad \text{and at} x = L, C_{H_2} = C_{H_2} \tag{4.53}$$

Equations 4.49 and 4.50, after rearrangement, can be integrated using relationships 4.51–4.53. The result is the following:

$$(D_{H_2})^{eff} = \frac{K_2 L J_{H_2}}{\ln[(1 - C_{H_2}K_2)/(1 - C^0_{H_2}K_2)]} \tag{4.54}$$

where

$$K_2 = \frac{\beta_1 \delta_2}{C_{H_2O}} \quad \text{and} \quad \beta_1 = 1 - \left(\frac{M_{H_2}}{M_{Ar}}\right)^{1/2} \tag{4.55}$$

M_i being the molecular weight of gas i.

Using Eq. 4.48, one obtains from Eq. 4.54

$$\frac{\varepsilon}{\tau} = \frac{K_2 L J_{H_2}}{D_{H_2} \cdot \ln[(1 - C_{H_2}K_2)/(1 - C^0_{H_2}K_2)]} \tag{4.56}$$

For Eq. 4.56, the value of C_{H_2} is known from a knowledge of D_{H_2}, calculated from the measured emf values. Knowing p_{H_2} and n_{Ar}, one can obtain the values of $\dot{n}_{H_2}$ and J_{H_2} using Eq. (4.45). The value of D_{H_2} is known from the following relationship:

$$\frac{1}{D_{H_2}} = \frac{1}{D^k_{H_2}} + \frac{1}{D_{H_2}/\text{Ar}} \tag{4.57}$$

where $D^k_{H_2}$ is the Knudsen diffusion coefficient and D_{H_2}/Ar is the binary diffusion coefficient for the diffusion of hydrogen in argon. This can be calculated using Gilliland's formula (1934). The Knudsen diffusion coefficient $D^k_{H_2}$ is given by

$$D^k_{H_2} = \frac{2}{3} r_p \left[\frac{8 \times 8.314 \times 10^7 T}{M_{H_2}}\right]^{1/2} \tag{4.58}$$

where r_p is the pore radius. The value of r_p is calculated using the following equation

$$r_p = \frac{2V_p}{S} \tag{4.59}$$

Table 4.2 Calculated values of $D_{H_2}^{eff}$ and (ε/τ) at different flow rates

Argon flow rate (ml/min)	p_{H_2} (atm)	$D_{H_2}^{eff}$ (cm^2/s)	ε/τ
71	0.0295	0.0530	0.075
77	0.0273	0.0531	0.075
87	0.0243	0.0532	0.075
99	0.0231	0.0572	0.081

where V_p is the total pore volume and S the total surface area. V_p is obtainable from density measurements and S from porosimetry using nitrogen adsorption.

Thus using sample dimension L and evaluation C_{H_2}, it is possible to determine the value of (ε/τ) using Eq. 4.56. One can then also obtain a value for τ.

For gas–solid reactions involving porous oxide pellets, τ is usually assumed to be around 2 (Spitzer et al. 1966).

The results of the diffusion experiment carried out by Abraham et al. (1973) using hydrogen and argon and a porous nickel pellet are shown in Table 4.2. The effective diffusion coefficient was calculated using Eq. 4.54. The results show that the effective diffusion coefficient and (ε/τ) remain almost unchanged at different flow rates.

It may be noted that while the experiment is described for measuring hydrogen diffusion only, the basic principle may be employed for many other gasses, provided there is a suitable arrangement for analysis of the argon stream.

4.3.4 Reaction of Porous Solid

It is possible that diffusion of gases through porous solid becomes the rate controlling step in gas–solid reaction. When this is so, the reaction rate is expected to increase when porosity is increased by changing the pore size. The increase would be caused by faster diffusion of gases through pores, as well as availability of larger pore surface area for chemical reaction. If the rate of interfacial reaction is very slow compared to rate of diffusion through pores, then the gas composition inside the pellet is substantially the same as that outside the pellet. If diffusion is slow, then the reactant gets depleted by chemical reaction as it diffuses inwards. Here, the entire internal surface area is not effective as a reaction site and the rate does not increase proportionately with increase in the pore surface area.

While in the case of non-porous solid reactant, a sharp boundary exists between the unreacted core and the completely reacted zone; in the case of porous solid, there is a gradual change in the degree of conversion throughout the solid. If the porosity is small, then this will be limited over a small distance. The product will also be porous. The situation where diffusion through a porous product layer is rate controlling has been discussed by Szekely and Evans (1970, 1971a, b). They have shown the kinetics in this case to be identical to that of the diffusion-controlled shrinking core model developed for non-porous solids.

4.4 Some Examples

4.4.1 Single Particle Studies

Figure 4.2 shows results of parabolic oxidation of platelets of iron. Similar parabolic law has been reported for solid lead, zinc, nickel and copper (Bagshaw et al. 1969). Some liquid metals also oxidize according to parabolic law (Drousy and Mascro' 1969).

While there are reports of many kinetic data which follow Ginstling–Brounshtein equation, there are not many which make use of the exact equation given by Carter and Valensi.

Majumdar et al. (1975, 1976a, b) have reported an investigation on the kinetics of reaction of $ZrCl_4$ or $HfCl_4$ vapours and their mixtures with solid sodium chloride vapours, as a function of pressure, temperature and sphere size. The reactions are of the type

$$2NaCl(s) + ZrCl_4(g) = 2Na_2ZrCl_6(s) \tag{4.60}$$

The product, the hexachloro zirconate, has a lower density and, therefore, while it completely envelops the unreacted NaCl core, the overall dimension of the solid sphere increases continuously. Majumdar et al. studied kinetics by measuring gain in weight of the solid and analysed the kinetic data using Carter–Valensi equation. Figure 4.6a, b shows some kinetic data. Figure 4.7 shows how the data fit into Carter–Valensi equation. The influence of pressure, particle size and temperature is summarized in Fig. 4.8 (a, b and c, respectively). Figure 4.8a shows that the rate constant K_{cv} is proportional to square root of the partial pressure of gas. Figure 4.8b shows that, as expected, the rate constant varies with, approximately, the square of

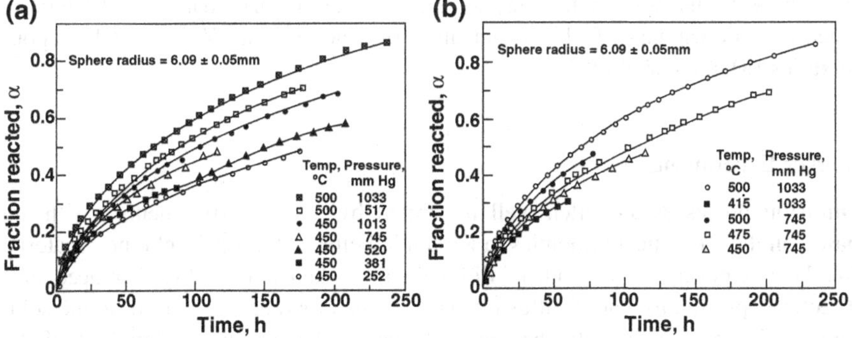

Fig. 4.6 Effect of pressure and temperature on reaction of NaCl spheres with $ZrCl_4$ vapour: **a** effect of $ZrCl_4$ pressure on the reaction rate (Mazumdar et al. 1975, 1976a, b) **b** effect of temperature on the reaction rate (Mazumdar et al. 1975, 1976a, b)

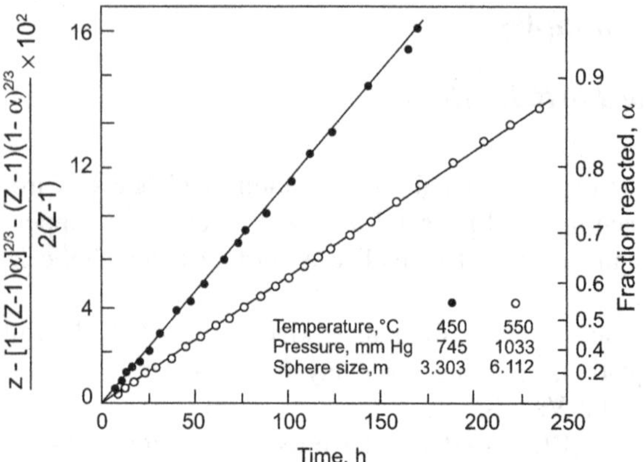

Fig. 4.7 Plot of experimental data according to Carter–Valensi rate expression (Eq. 4.35) (Mazumdar et al. 1975, 1976a, b)

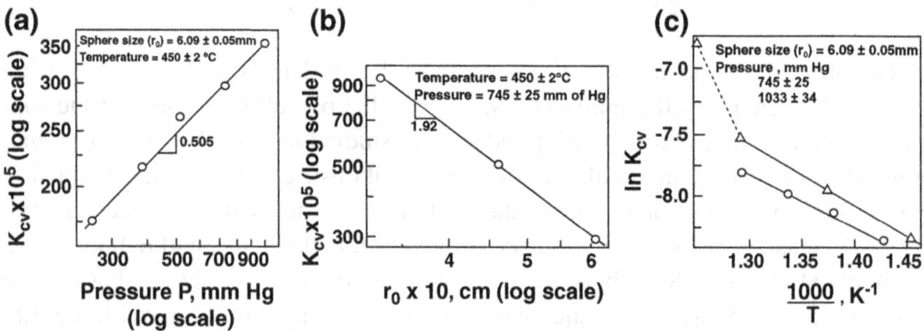

Fig. 4.8 Effect of various parameters on rate constant: **a** effect of pressure on rate constant K_{cv} **b** effect of sphere size on rate constant K_{cv} (Mazumdar et al. 1975, 1976a, b) **c** effect of temperature on rate constant K_{cv} (Mazumdar et al. 1975, 1976a, b)

the sphere radius. Similar findings were established (Mazumdar et al. 1976a) for reaction with vapours of $HfCl_4$ and also for reaction with $ZrCl_4$–$HfCl_4$ vapour mixtures (Majumdar et al. 1976b).

Marker Experiments

Diffusion processes are often well illustrated by marker experiments which are based on the theoretical principles in Kirkendall effect. If a small light inert material (marker) is placed in the path of diffusion flux, then it is bodily displaced in a direction opposite to that of diffusion. Thus if such markers are placed on the solid unreacted sphere, then with the onset of gas–solid reaction and build-up of the product layer, they will be displaced outward or inward. If diffusion involves an inward flux, then in the extreme case, they will gradually move outward remaining on the outer surface of the product layer. Conversely when there is outward diffusion, the markers gradually move inward remaining in the extreme case at the core/product interface. There can be intermediate situations where markers move but remain embedded in the product layer.

Majumdar et al. (1975, 1976a, b) carried out marker experiments using fine platinum wires laid on cylindrical samples of NaCl. Detailed marker studies showed that the reaction involved ionic diffusion, ions moving outward from unreacted core, through the product layer.

4.4.2 Studies on Packed Beds

Gas–solid reactions in many packed beds have been found to follow the Ginstling–Brounshtein (GB) equation. Figure 4.9 shows some data reported by Prakash and Ray (1987) on isothermal reduction of an iron ore by a non-coking coal, the ore–coal mixture being a packed bed. The figure shows the linear GB plot.

The integral and differential approaches in the evaluation of activation energy, discussed previously in Chap. 2, are shown in Fig. 4.10a–c. The general agreement of *E* values determined separately by integral and differential approaches further confirms the kinetic law shown. Reduction of a packed bed or ore particles by reducing gas also follows the same law. Figures 4.11 and 4.12 show the data.

Mookherjea et al. (1986) have reported a kinetic study on isothermal reduction (850–1050 °C) of a cylindrical column of iron ore fines surrounded by coal or char fines. They showed that the overall degree of reduction of ore as well as the degree

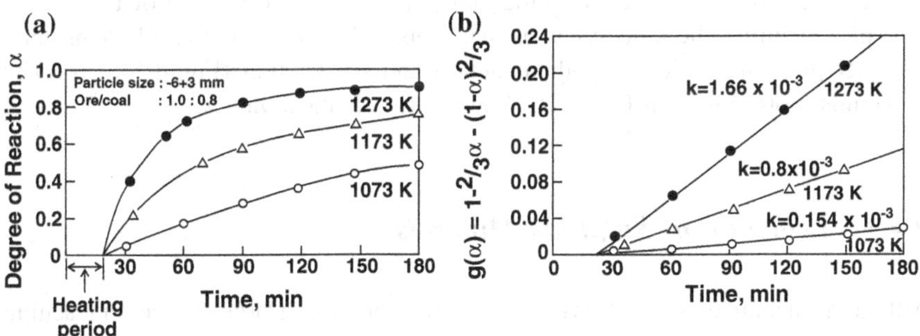

Fig. 4.9 Kinetic plots for the isothermal reduction of Barajamda iron ore by Raniganj non-coking coal in terms of degree of reaction (Prakash and Ray 1987)

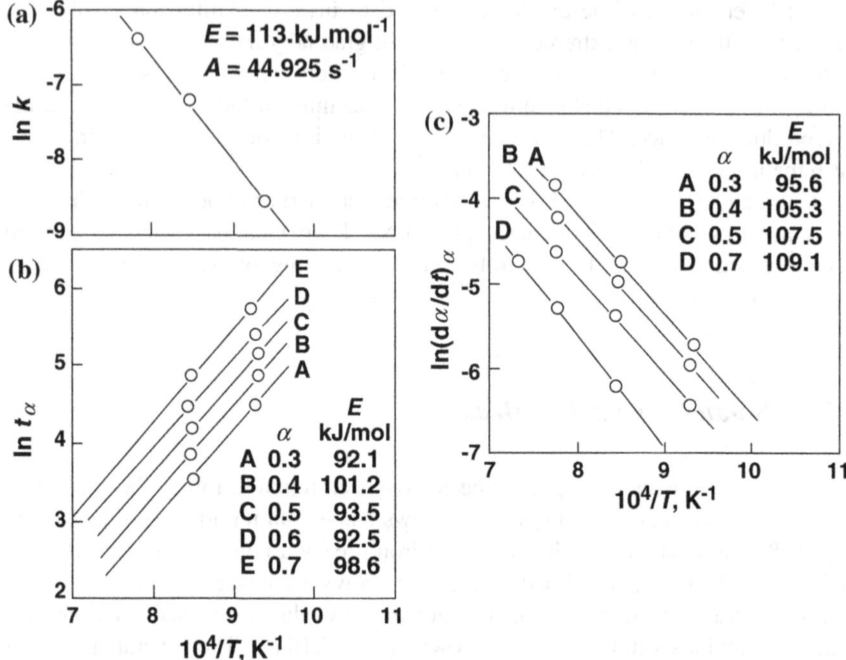

Fig. 4.10 Evaluation of kinetic parameters, using different approaches in the case of reduction for the data shown in Fig. 4.9 (Prakash and Ray 1987)

of gasification of char both follows the GB equation. The apparent activation energy values of the two processes are similar showing that these reactions are coupled at all times. Figure 4.13a, b shows the data on reduction and gasification. Figure 4.14 shows the reduction time plot (Chap. 2) for some data of Fig. 4.13a. This plot identifies the kinetic law as that proposed by Ginstling and Brounshtein. The GB plots are valid for both reduction and gasification (Fig. 4.15a, b). The Arrhenius plots shown in Fig. 4.16 show nearly identical values.

4.4.3 Leaching of Sulphide Minerals

Although leaching of minerals is often a complex process, a large number of leaching reactions conform to simple kinetic laws. There are many examples where product layer diffusion control equations apply. Leaching of finely ground chalcopyrite by ferric sulphate is controlled by diffusion mechanism. The reaction is written as

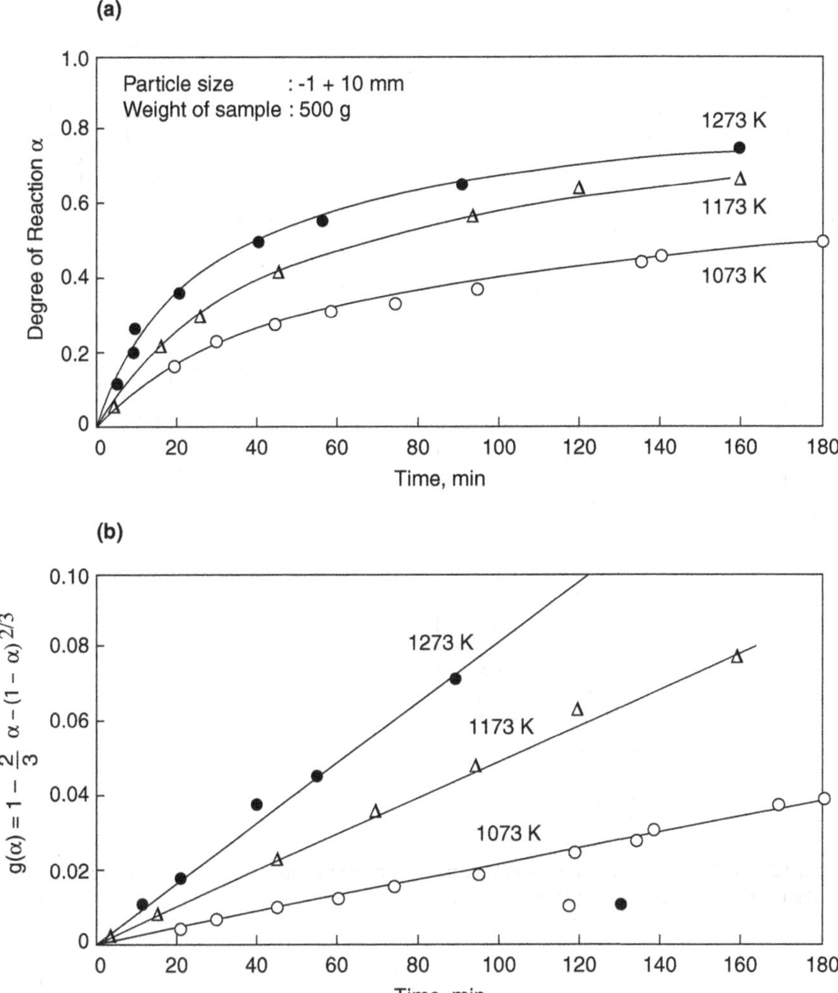

Fig. 4.11 Kinetic plots for the isothermal reduction of Kiriburu iron ore by reducing gas ($CO{:}N_2$ = 30:70) (Prakash and Ray 1987)

$$CuFeS_2 + 4Fe^{3+} = Cu^{2+} + 5Fe^{2+} + 2S^0 \tag{4.61}$$

The kinetics follows the GB equation because the sulphur produced in the reaction adheres to the mineral surface forming a tightly bonded diffusion layer. The reaction rate is extremely low. Dutrizac et al. (1970) have reported kinetic studies on

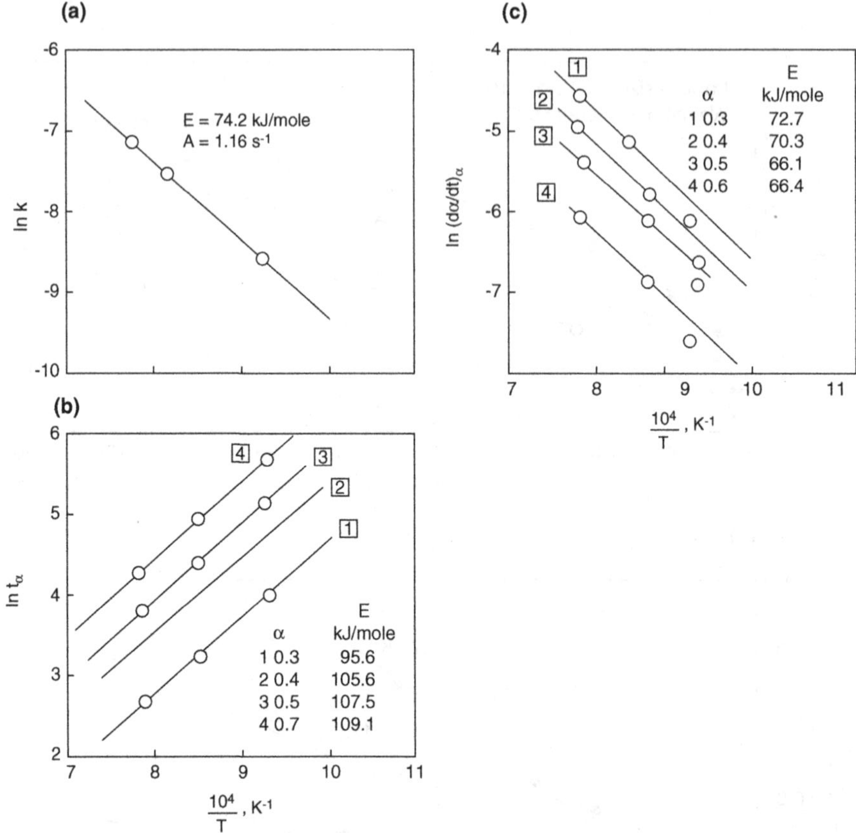

Fig. 4.12 Evaluation of kinetic parameters, using different approaches, for gaseous reduction for the data shown in Fig. 4.11 (Prakash and Ray 1987)

leaching of several synthetic minerals. They have shown that the leaching of bornite ($CuFeS_4$) below 35 °C follows a parabolic law, the reaction being controlled by diffusion through a non-stoichiometric product layer ($Cu_{5-x}FeS_4$).

Simple equations, however, can only apply when particles are of uniform size. If there is random distribution of sizes, then no simple equation will fit the kinetic data. In general, leaching of sulphide minerals is a complex process and the subject will be discussed in greater detail in a later chapter.

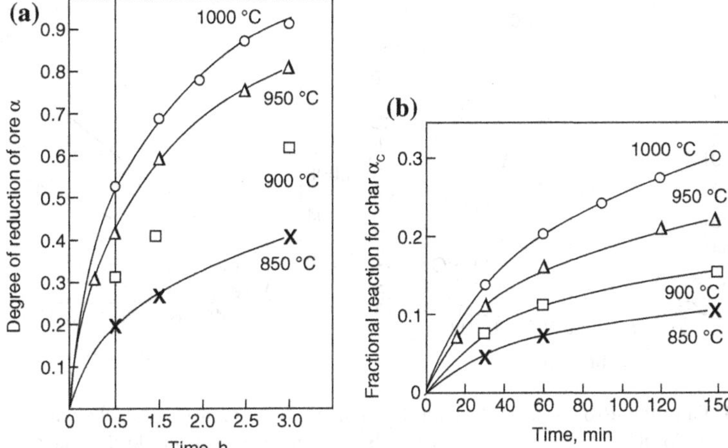

Fig. 4.13 Kinetic data for reaction of iron ore with coal char **a** kinetics of reduction of ore by char in quartz crucibles (argon atmosphere); **b** kinetics of gasification of char in the presence of ore in argon atmosphere (Mookherjea et al. 1986)

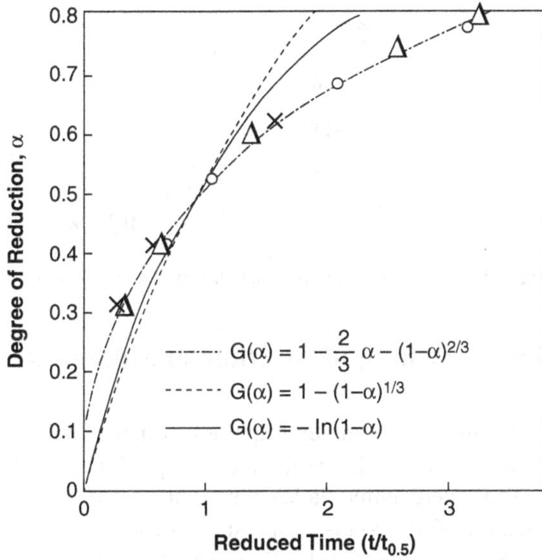

Fig. 4.14 Reduced time plot for data shown in Fig. 4.13a (Mookherjea et al. 1986)

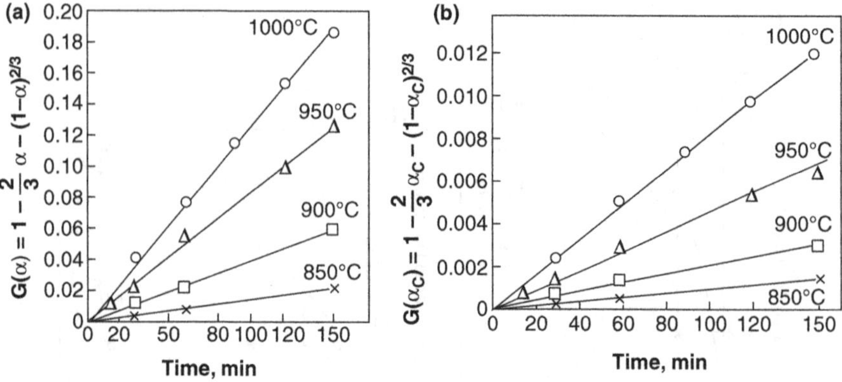

Fig. 4.15 Ginstling–Brounshtein plots for reaction of iron ore with coal char: **a** Ginstling–Brounshtein plot for data shown in Fig. 4.13a. **b** Ginstling–Brounshtein plot for kinetic data for char gasification in the presence of ore for data shown in Fig. 4.13b (Mookherjea et al. 1986)

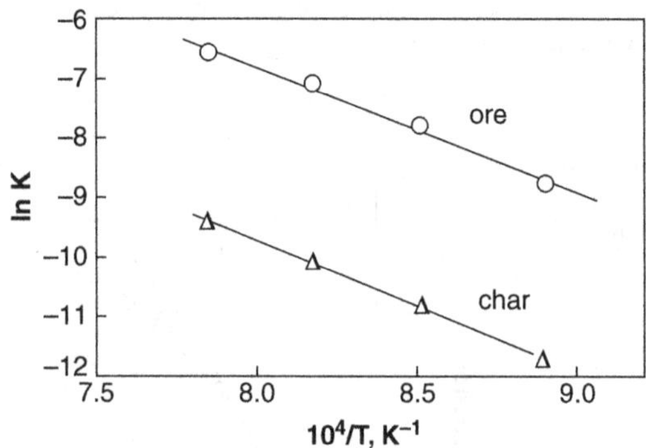

Fig. 4.16 Arrhenius plots for data shown in Fig. 4.15a, b (Mookherjea et al. 1986)

4.5 Leaching of Multimetal Sulphides

There are some complex sulphides, which contain both copper and zinc and, sometimes, lead also. These multimetal sulphides can be leached by various solvents. We consider here ammonia leaching. Sarveswara Rao et al. (1992) have discussed dissolution of 1% solids in the slurry of a particular ore (~200 + 300 mesh) in the temperature range 70–100%. It was shown that dissolution kinetics of Cu and Zn is different. The dissolution reactions are written as follows:

$$CuFeS_2 + \frac{17}{4}O_2 + 4NH_3 + 2OH^- \rightarrow Cu(NH_3)_4^{2+} + \frac{1}{2}Fe_2O_3 + SO_4^{2-} + H_2O \quad (4.62)$$

$$ZnS + 4NH_3 + 2O_2 \rightarrow Zn(NH_3)_4SO_4 \quad (4.63)$$

Table 4.3 Chemical analysis and mineral phases identified

S. no.	Element (%)	−200 + 3000 (mesh)	−140 + 500 (mesh)	Phases identified
1.	Cu	4.55	3.60	$CuFeS_2$
2.	Zn	22.20	19.95	$ZnS \cdot ZnO(SO_4)_2$
3.	Pb	9.38	10.00	$PbS \cdot PbSO_4$

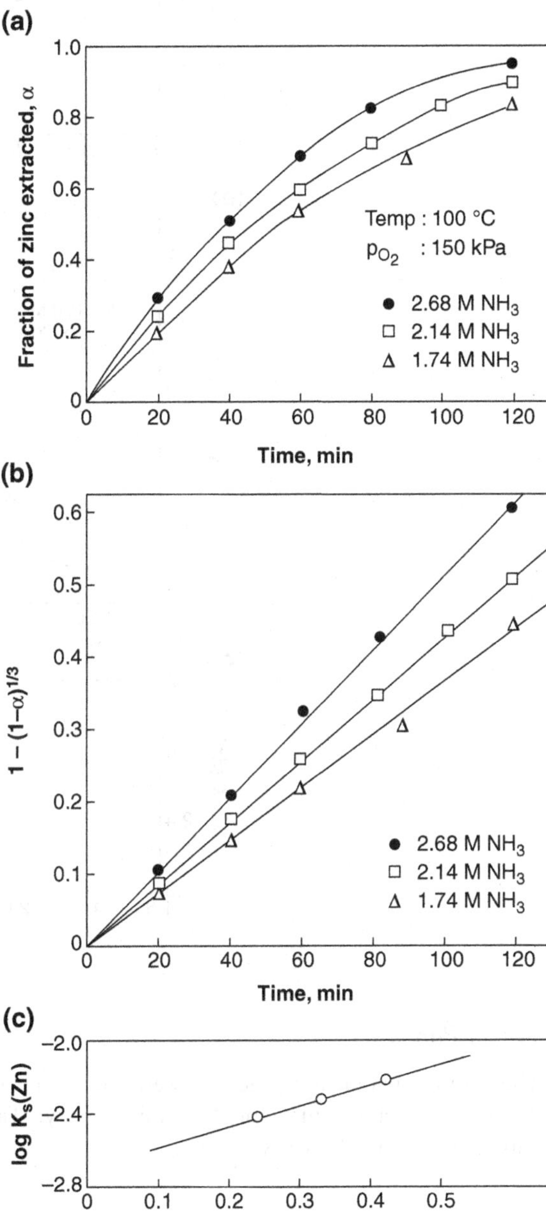

Fig. 4.17 Kinetic plots showing effect of ammonia concentration (Sarveswara Rao et al. 1992)

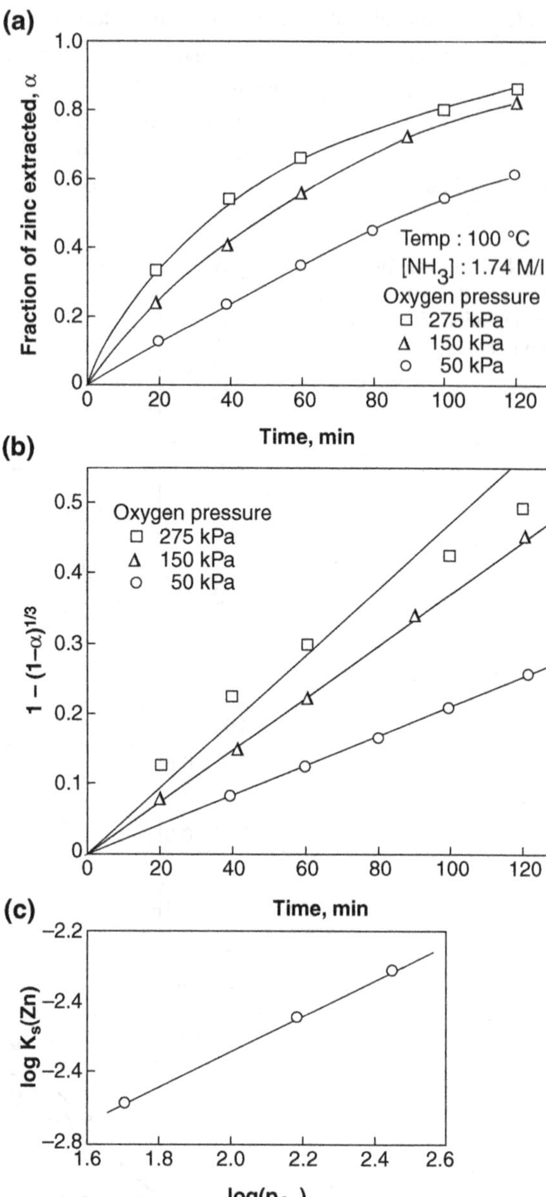

Fig. 4.18 Kinetic plots showing effect of oxygen partial pressure (Sarveswara Rao et al. 1992)

Effect of Parameters

The various parameters selected were: ammonia concentration, oxygen partial pressure and temperature for 1% solid, slurry. Table 4.3 shows the chemical analysis and mineral phases present.

Figures 4.17 and 4.18 indicate, respectively, the kinetic plots showing effect of ammonia concentration and that of oxygen partial pressure. Accordingly, the combined effect of the two parameters can be expressed as follows:

$$1-(1-\alpha)^{1/3}=k_{Zn}[NH_3][p_{O_2}]^{1/2}t \tag{4.64}$$

Figure 4.19 shows the effect of temperature and calculation of activation energy by the differential and integral methods.

The kinetics of copper dissolution, however, follows a different kinetic law, viz. the GB equation. Figures 4.20, 4.21 and 4.22 show the effects of ammonia concentration, oxygen pressure and temperature. In this case, the kinetic equation is as follows:

$$1-2/3(\alpha)-(1-\alpha)^{2/3}=k_{Cu}[NH_3]^2[p_{O_2}]^{1/2} \tag{4.65}$$

Thus, it is possible to have two different kinetic equations for dissolution of two different metal sulphides coming from the same source. That the kinetic model is valid is substantiated in each case by close agreement of E values calculated by integral and differential methods.

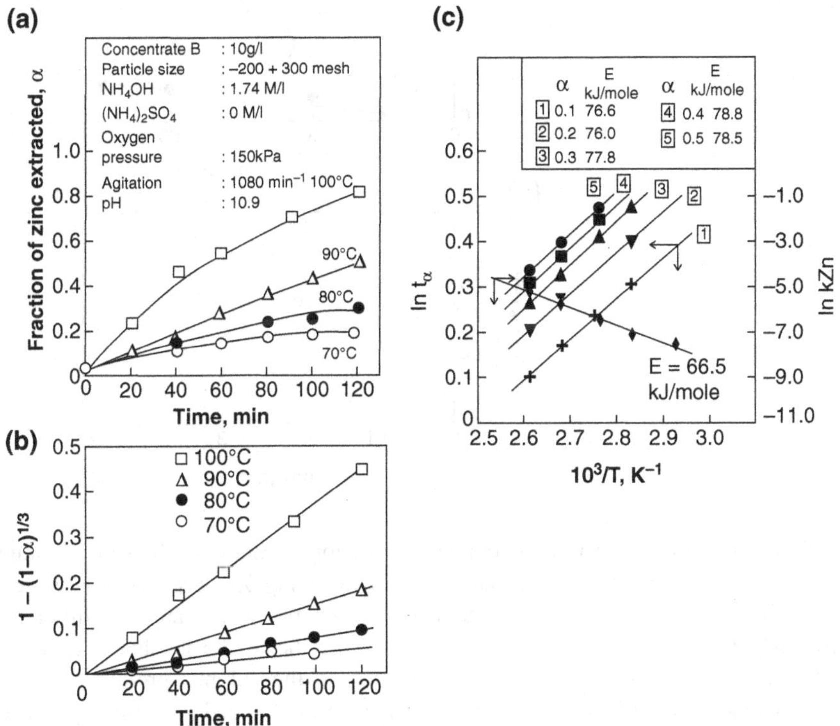

Fig. 4.19 Kinetic plots showing effect of temperature on Zn dissolution and evaluation of activation energy (Sarveswara Rao et al. 1992)

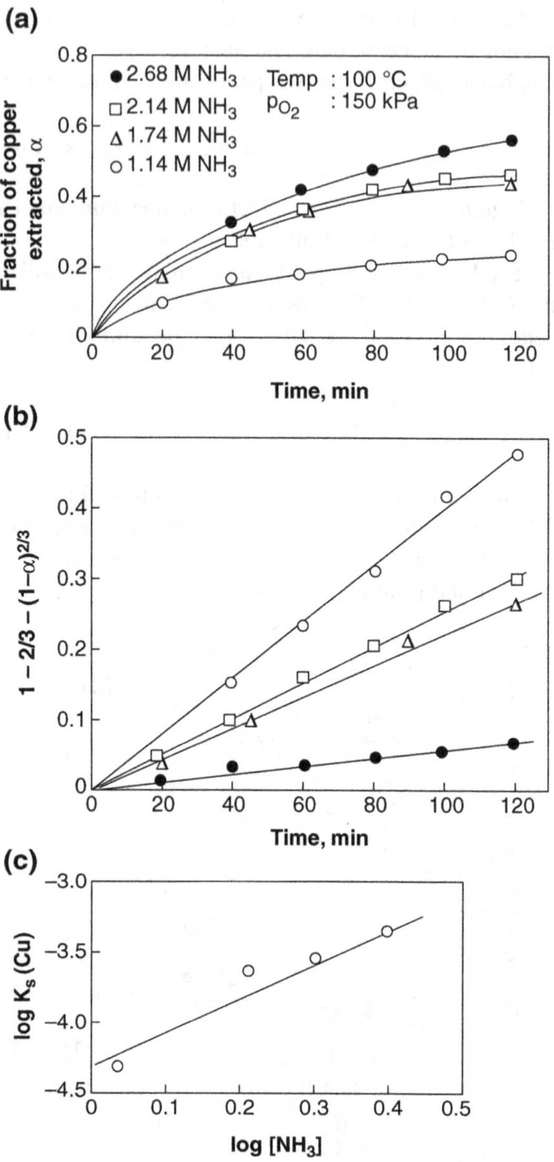

Fig. 4.20 Kinetic plot showing effect of ammonia concentration on Cu dissolution (Sarveswara Rao et al. 1992)

Obviously, in the present case, copper dissolution is controlled by some residue layer, such as goethite, which is produced during leaching. Zn dissolution, however, is not diffusion-controlled through the activation energy values are very different. Obviously, not much physical significance can be ascribed to the absolute value of E.

The kinetics follows the GB equation because the sulphur produced in the reaction adheres to the mineral surface forming a tightly bonded diffusion layer. The reaction rate is extremely low; Dutrizac et al. (1970) have reported kinetic studies on

Fig. 4.21 Kinetic plots for effect of oxygen partial pressure on Cu dissolution (Sarveswara Rao et al. 1992)

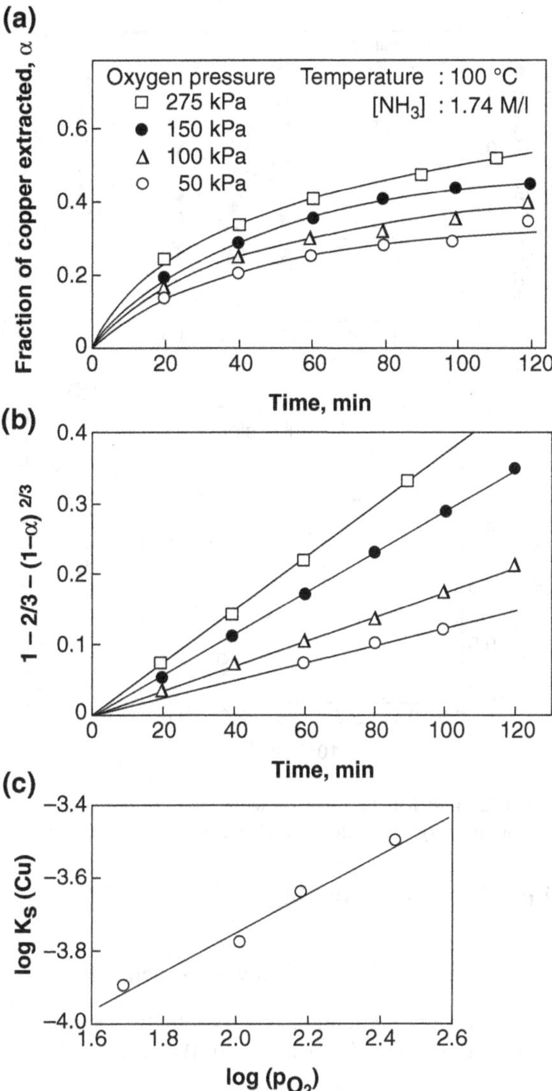

leaching of several synthetic minerals. They have shown that the leaching of bornite ($CuFeS_4$) below 35 °C follows a parabolic law, the reaction being controlled by diffusion through a non-stoichiometric product layer ($Cu_{5-x}FeS_4$).

Simple equations, however, can only apply when particles are of uniform size. If there is random distribution of sizes, then no simple equation will fit the kinetic data. In general, leaching of sulphide minerals is a complex process and the subject will be discussed in greater detail in a later chapter.

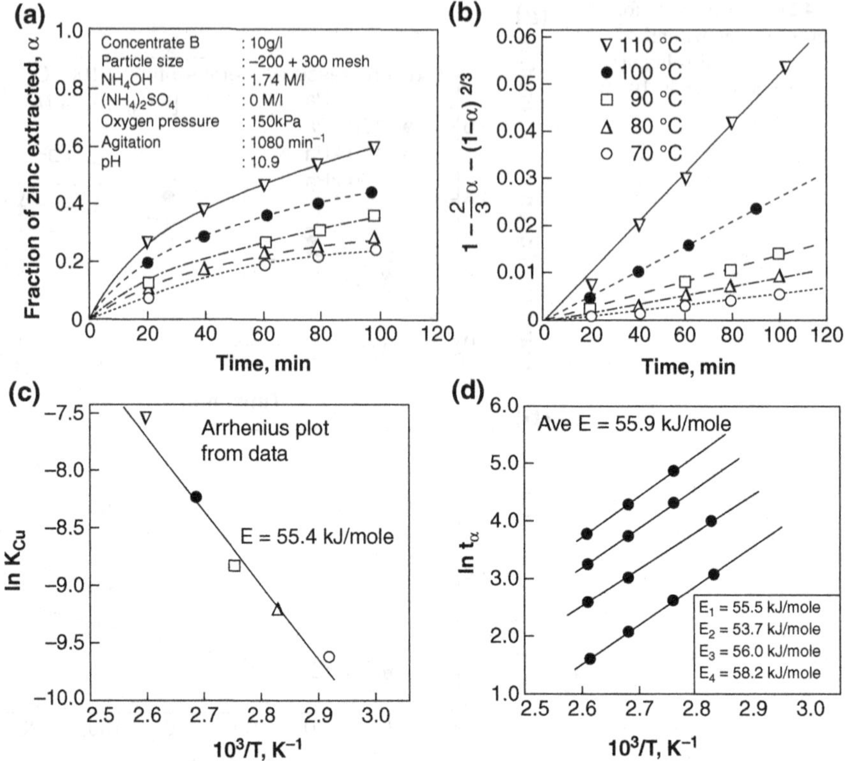

Fig. 4.22 a–d Kinetic plots showing effect of temperature on Cu dissolution and evaluation of activation energy by different and integral methods (Sarveswara Rao et al. 1992)

4.6 Review Questions

1. Discuss whether the following statements are correct or false.

   (a) Jander's equation is a hypothetical equation.
   (b) Ginstling–Brounshtein (GB) equation reduces to Jander's equation for large values of α.
   (c) Parabolic law is a special case of GB equation.
   (d) There can be no single reduced time plot for the Carter–Valensi (CV) equation.
   (e) For a reaction controlled by product layer diffusion, the rate must gradually decrease.
   (f) For a gas–solid reaction, the particle size influences reaction time more strongly when reaction is controlled chemically than by product layer diffusion.
   (g) At high temperatures, mass transfer is more likely to be rate controlling than chemical reaction.

(h) If, during oxidation of a metal, diffusion through product layer is rate controlling, then the density of the product layer must be greater than that of the reactant solid.
(i) For porous solids, “tortuosity” factor must always be less than one.
(j) “Tortuosity” factor for the solid (porous) depends on the nature of the gas flowing through.

2. A solid sphere reacts with a liquid such that it requires 4 h for 50% reaction. If the sphere is split into two equal spheres, then what will be the time required for 75% reaction provided that

   (a) Reaction is chemically controlled?
   (b) Reaction is controlled by product layer diffusion?

3. Derive Ginstling–Brounshtein (GB) equation for reaction of a solid cube with a fluid.
4. What is the significance of the parameter Z in the Carter–Valensi equation?
5. A spherical particle reacts with a gas such that the product layer reaches a limiting thickness beyond which the excess layers formed fall off. If diffusion through such a layer is rate controlling, then derive the kinetic law.
6. Draw the reduced time plots for the three equations for product layer diffusion control. Show that for proper delineation the α values must be sufficiently high.
7. The rate constant for a chemically controlled reaction for leaching of some solid particles is trebled when the solid is ground to a particular size. By what factor will the rate constant increase for the same degree of combination if the reaction kinetics followed GB equation?
8. What is the form of Jander’s equation for a cube and a long cylinder ($l \gg r$)?
9. Why does the parabolic law not hold for solids with curved surfaces even if the reaction be controlled by diffusion through product layer?
10. What role is assigned to “tortuosity factor” in diffusion of gas through porous solid?
11. During reduction of iron oxide by solid carbon, the reduction reaction and the oxidation reaction are sometimes found to have the same activation energy. What does this imply?
12. A spherical particle of Fe_2O_3 is being reduced by hydrogen at 900 °C. The reaction is controlled by diffusion of H_2–H_2O through the outer shell of porous metallic iron. Find the rate of reaction when α is 0, 0.5 and 0.9. Make necessary assumptions.
    Given:

    Diameter of particle = 2 cm
    Diffusion coefficient of H_2–H_2O = 3 cm^2/s
    p_{H_2} in equilibrium with Fe_xO–Fe = 0.62

13. Ammonia leaching of a sulphide concentrate (particle size—100 mesh) follows Ginstling–Brounshtein kinetic equation. If 50% reaction at a given temperature

takes 2 h, then calculate the time for 90% reaction when the material is ground to 300 mesh.

14. The kinetics of reaction of a spherical particle with a gas follows the Carter–Valensi equation. The values of α at various times are as follows.

t(h)	1.25	7.75	23.50	86.00
α	0.056	0.17	0.298	0.56

Calculate the value of Z.
If the original particle diameter is 2 cm, find the diameter of the reacted particle after 100 h of reaction.

References

Abraham, K.P., Deb Roy, T., Ray, H.S.: Proceedings of the 2nd National Heat and Mass Transfer Conference on IIT, vol. 436. Kanpur, India (1973)
Bagshaw, N.E., Foonoy, J.R., Harris, M.R.: Brit. Corr. J. **4**, 301 (1969)
Basu, S.N., Ghosh, A.: J. ISI. **765** (1970)
Carter, R.E.: J. Chem. Phys. **34**, 2010 (1961)
Crank, J.: Trans. Far. Soc. **53**, 1083 (1957)
Drousy, M., Mascro', C.: Met. Rav. **3**(2), 25 (1969)
Dutrizac, J.E., MacDonald, R., Ingraham, T.R.: Met. Trans. B **1B**(225), 3083 (1970)
Ghosh, A.: Kinetics of Extractive Metallurgy. Lecture Notes, IIT, Kanpur (1976)
Gilliland, E.R.: Ind. Eng. Chem. **26**, 681 (1934)
Ginstling, A.M., Brounshtein, B.I.: J. Appl. Chem. USSR. **23**, 1249 (1959)
Habashi, F.: Principles of Extractive Metallurgy, vol. 1, Chapter 7. Gordon and Breach Science Pub., New York (1970)
Hillard, J.H., Cahn, J.W.: Trans. TMS Amer. Inst. Met. Eng. **221**, 349 (1961)
Jander, W., Anorg, Z.: Allg. Chem. **163**, 1 (1927)
Majumdar, S., Ray, H.S., Kapur, P.C.: Met. Trans B **7B**, 217 (1976)
Mazumdar, S., Luthra, K.L., Ray, H.S., Kapur, P.C.: Met. Trans. B **6B**, 607 (1975)
Mazumdar, S., Ray, H.S., Kapur, P.C.: J. Appl. Chem. Biotechnol. **26**(5), 259 (1976)
Mookherjea, S., Ray, H.S., Mukherjee, A.: lronmaking Steelmaking **13**(5), 229 (1986)
Prakash, S., Ray, H.S.: Thermochim. Acta **3**, 143 (1987)
Present, R.D.: Kinetic Theory of Gases. McGraw-Hill Book. Co. Inc., New York (1958)
Ray, H.S.: J. Appl. Chem. Biotechnol. **26**, 436 (1976)
Sarveswara Rao, K., Anand, S., Dasand, R.P.H.S.: Ray Min. Proc. Extr. Metall. Rev. **10**, 11 (1992)
Scott, D.S., Dullien, F.A.L.: Amer. Inst. Chem. Eng. J. **8**(113), 293 (1962)
Spitzer, R.E., Manning, F.S., Philbrook, W.O.: Trans. TMS Am. Inst. Met. Eng. **226**, 726 (1966)
Szekely, J., Evans, J.W.: Chem. Eng. Sci. **25**, 1091 (1970)
Szekely, J., Evans, J.W.: Met. Trans. **28**, 1691 (1971a)
Szekely, J., Evans, J.W.: Chem. Eng. Sci. **26**, 1901 (1971b)
Trushenski, S.P., Li, K., Philbrook, W.O.: Met. Trans. **5**, 1149 (1974)
Valensi, G.: Bull. Soc. France. **5**, 668 (1935)
Wen, C.Y.: Ind. Eng. Chem. (1968)
Youngquist, G.R.: Ind. Eng. Chem. **62**, 52 (1970)

Chapter 5
Fluid Phase Mass Transfer

5.1 Introduction

5.1.1 *Role of Mass Transfer Across a Boundary Layer*

We have previously considered kinetics of solid–fluid heterogeneous reactions controlled either by reaction at an interface or by diffusion across a product layer. In the present chapter, we will discuss reactions which are controlled by a different rate-controlling step, namely fluid phase mass transfer across a boundary layer. A variety of reactions come under this category. We will consider examples of plain dissolution of a solid in liquid, complex leaching reactions, electrochemical phenomena controlled by ionic diffusion in liquids, etc. However, it is first necessary to introduce some fundamental concepts regarding mass transfer in fluids.

5.1.2 *Mechanism of Mass Transfer in Fluids*

There are three mechanisms of mass transfer in liquids and gases, namely diffusion, convection and eddy diffusion (turbulence). In diffusion, an element of solute is transferred in the fluid under the driving force of a concentration gradient. In convection, mass transfer is caused by the laminar flow of the bulk fluid itself, i.e. the solute is carried along. Eddy diffusion, which operates only in turbulent flow, causes sudden mutual exchange of volume elements in two isolated regions of the fluid. When these regions have different concentrations, such exchange amounts to a net transfer. Turbulence sets in at high fluid velocities when the velocities of volume elements in fluid are not coordinated with those of the neighbours. As a consequence, therefore, many elements are torn away from their surroundings continuously and transferred elsewhere. Since this is equivalent to random motion of the eddy pockets, the mathematics is analogous to that of molecular diffusion. The total flux J of species i across a fixed plane normal to the x-direction is given by the equation

H. S. Ray and S. Ray, *Kinetics of Metallurgical Processes*,
Indian Institute of Metals Series, https://doi.org/10.1007/978-981-13-0686-0_5

$$J_{ix} = C_i v_{ix} = -D_i \frac{\partial C_i}{\partial x} + C_i \bar{v}_x - E_D \frac{\partial C_i}{\partial x} \tag{5.1}$$

where

C_i Concentration of species i.
v_{ix} Overall velocity of the species along the x-direction w.r.t. stationary coordinates.
D_i Diffusion coefficient of i.
$\bar{v}_x$ Fluid flow (convection) velocity along the x-direction.
E_D Eddy diffusivity.

In the turbulent flow region

$$C_i \bar{v}_x \gg -D_i \frac{\partial C_i}{\partial x} \quad \text{and } E_D \gg D,$$

and therefore, the role of molecular diffusion is minimal so long as the bulk liquid is concerned. The situation, however, is different when one considers the situation at a solid–fluid interface.

5.1.3 Mass Transfer at Solid–Fluid Interface

We shall be concerned here with the fluid layer in contact with the solid. The layer adjacent to the solid surface cannot slip past the solid and is, therefore, stagnant.

There is a concentration profile across this boundary layer when a solute moves from the solid to the bulk liquid or vice versa.

Consider, for example, dissolution of a solid in a liquid. The liquid film adjacent to the solid may be taken to be saturated, provided the release of atoms or molecules from the bulk solid is sufficiently rapid. The solute diffuses into the bulk liquid from the saturated layer. Most of the concentration drop takes place within a relatively finite thin layer known as the boundary layer. The effective boundary layer thickness δ is arbitrarily defined as the distance by which 99% (or a similar figure) drop is achieved. Alternatively, it is also obtained by drawing a tangent (Fig. 5.1). A concentration profile exists in the boundary layer only when liquid phase mass transfer is rate controlling.

Since the fluid layer adjacent to the solid surface cannot slip past the solid, there must also be a velocity profile so that flow near boundary layer is laminar. We have

$$(\bar{v}_x)_{x=0} = 0 \tag{5.2}$$

$$(E_D)_{x=0} = 0 \tag{5.3}$$

Therefore, one may conclude that mass transfer across the boundary layer is by diffusion only. This does not mean that convection and turbulence in the bulk liquid

Fig. 5.1 Concentration profile of solute near solid–fluid interface Schematic of Mn concentration profile near the interface

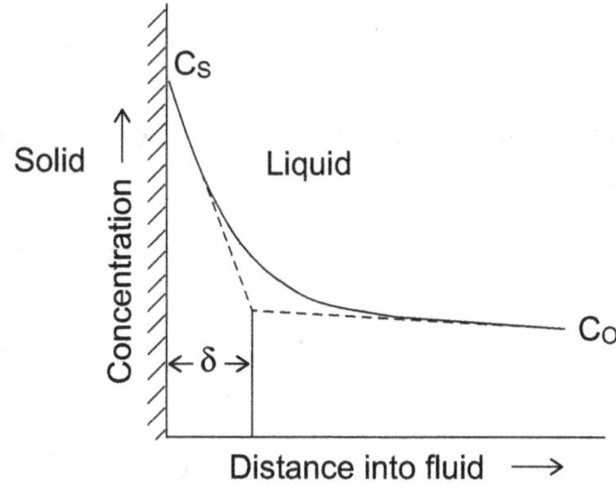

have no effect on the mass transfer at the interface. Increased turbulence, for example, will have an indirect effect through effect on the boundary layer thickness.

For diffusion across the boundary layer, we have Fick's law.

$$J = D\frac{C_s - C_o}{\delta} \tag{5.4}$$

where C_s and C_o denote, respectively, concentrations at the interface and in the bulk fluid. The concentration profile and the symbols are indicated schematically in Fig. 5.1. When there is enhanced convective and turbulent flow in the bulk liquid, the value of δ decreases and, therefore, the diffusion flux J increases.

5.2 Dissolution of a Solid in a Liquid

5.2.1 Kinetics of Dissolution Reaction

When a solid dissolves in a liquid, the dissolution is often accelerated by the stirring of the liquid. Stirring reduces the effective boundary layer thickness and increases the mass transfer coefficient h_m which is defined as D/δ. The appropriate equation for dissolution is Eq. 5.4. Suppose dn moles of solid is dissolved in time dt to change the concentration by an amount dC. If V is the total volume of the solution and A the area of dissolution (i.e. mass transfer interface), then

$$\frac{\mathrm{d}n}{\mathrm{d}t} = \frac{V}{A}\cdot\frac{\mathrm{d}C}{\mathrm{d}t} = \frac{D}{\delta}(C_s - C_o) \tag{5.5}$$

$$\text{i.e.} \quad \frac{\mathrm{d}C}{C_s - C_o} = \frac{D}{\delta}\cdot\frac{A}{V}\cdot\mathrm{d}t$$

$$\text{which gives,}\ \ln(C_s - C_o) = -\frac{D}{\delta}\cdot\frac{A}{V}\cdot t + \mathrm{I} \tag{5.6}$$

where I is the integration constant. At $t = 0$, with C_i as the initial concentration in the bulk,

$$\mathrm{I} = \ln(C_s - C_i) \tag{5.7}$$

$$\therefore\quad \frac{C_s - C_o}{C_s - C_i} = \exp\cdot\left(-\frac{D}{\delta}\cdot\frac{A}{V}\cdot t\right) \tag{5.8}$$

$$\therefore\quad C_o = C_s - (C_s - C_i)\exp\cdot\left(-\frac{D}{\delta}\cdot\frac{A}{V}\cdot t\right)$$

$$= C_s - (C_s - C_i)\left[1 - \frac{D}{\delta}\cdot\frac{A}{V}\cdot t + \frac{D^2\cdot A^2}{\delta^2\cdot V^2}\cdot t^2 + \cdots\right]$$

Therefore, for small intervals of time t,

$$C_o = C_s - (C_s - C_i)\left[1 - \frac{D}{\delta}\cdot\frac{A}{V}\cdot t\right] \tag{5.9}$$

To express Eq. 5.7 in terms of the degree of reaction α, we note that

$$\alpha = \frac{C_i - C_o}{C_i - C_s} \tag{5.10}$$

$$\text{i.e.}\quad 1 - \alpha = \frac{C_s - C_o}{C_s - C_i} \tag{5.11}$$

Hence, Eq. 5.8 may be written as

$$-\ln(1 - \alpha) = kt \tag{5.12}$$

where the rate constant k equals $\frac{DA}{\delta V}$. Equation 5.12 shows that the reaction is first order.

Ray et al. (1972). have reported an investigation on the dissolution of carbon from a clay–graphite crucible. Riebling (1969) has described dissolution of silica in glass. Both the reactions are diffusion controlled.

5.2.2 *Effect of Stirring*

It has been mentioned earlier that stirring of bulk liquid diminishes the boundary layer thickness, δ. However, increased stirring can have effect only up to a limit beyond which further stirring ceases to have more effect. At this limiting value, the liquid mass transfer ceases to be rate controlling because the enhanced rate becomes comparable to the first step which involves release of the solute from the solid surface. Dissolution rate can now be enhanced further only by an increase in temperature which will increase diffusion coefficient *D*.

Example 5.1 A solid plate is dissolving in a liquid and the dissolution process is controlled by mass transfer across a boundary layer. If it takes 2 h to achieve 50% reaction, then what is the time required for 75% dissolution?

Solution

We have $\frac{C_s - C_o}{C_s - C_i} = 1 - \alpha = \exp(-kt)$ which gives Eq. 5.12, i.e. $-\ln(1-\alpha) = kt$

As shown in Sect. 2.1, the "half-life" during a first-order reaction is independent of the initial concentration. Since the dissolution process follows a first-order reaction, it will take equal time intervals for the first 50% reaction and the subsequent 25% reaction.

Hence, total reaction time for 75% reaction = 4 h. Ans.

Example 5.2 Manganese is going out from liquid iron to slag phase. Draw schematically the concentration profile at the interface and calculate per cent manganese in metal after 10 min.

Data supplied	Initial Mn content of steel = 0.5%
	Initial Mn content of slag = 0%
	D_{Mn} (liquid iron) = 5×10^{-5} cm²/s
	δ_{Mn} (liquid iron) = 2×10^{-3} cm
	Density of liquid iron = 7.16 g/cc
	Conc. of Mn at interface = 0.0013%
	Volume of liq. metal/area of interface = 2.5 cm

Assume that there is concentration profile only in the metal layer and not in the slag layer.

Solution

The concentration profile may be represented schematically as shown below.

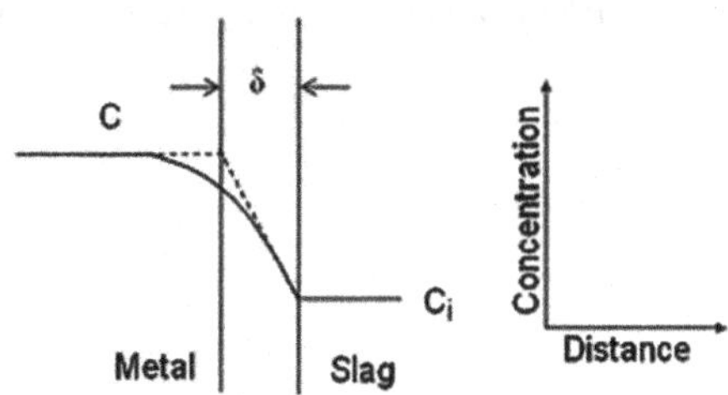

Schematic of concentration profile near the interface (see Fig. 5.1).

C is the instantaneous Mn concentration in metal and it decreases progressively.

C_i is equilibrium concentration at the interface, and it remains unchanged. The initial Mn concentration in metal is C_o.

$$\frac{dn}{dt} = \frac{V}{A} \cdot \frac{dC}{dt} = -\frac{D}{\delta}(C - C_i)$$

$$\ln\left[\frac{C - C_i}{C_o - C_i}\right] = -\frac{D}{\delta} \cdot \frac{A}{V} t$$

$$\ln\left[\frac{C - 0.0013}{0.5 - 0.0013}\right] = -\frac{5 \times 10^{-5}}{2 \times 10^{-3}} \times \frac{1}{2.5} t = -\frac{1}{100}$$

where t is in sec. In the problem, $t = 10$ min $= 600$ s.

$$\therefore \ \ln\left[\frac{0.5 - 0.0013}{C - 0.0013}\right] = 6 \quad \text{Ans.}$$

$$\therefore \ C = \frac{0.4897}{e^6} + 0.0013 = \frac{0.4897}{403.4} + 0.0013 = 0.002$$

The density of liquid iron does not come into the picture.

Example 5.3 A boundary layer mass transfer-controlled process is complete up to 50% in two hours. What is the time for complete reaction?

Solution

Such a reaction follows first-order kinetics

$$-\ln(1 - \alpha) = kt \quad \text{i.e.} \quad \alpha = 1 - e^{-kt}$$

For $\alpha = 1$ (i.e. complete reaction)

$$e^{-kt} = 0 \quad \text{i.e.} \quad \frac{1}{e^{kt}} = 0 \quad \text{i.e.} \quad t = \infty$$

The time of complete reaction is always infinite, irrespective of the value of k or time for 50% reaction. Ans.

5.3 Electrochemical Reactions

5.3.1 Electrochemical Mechanism in Leaching

Some leaching reactions are electrochemical in nature (Wadsworth and Wadia 1955). An electrochemical reaction differs from a chemical reaction in that in the former the electrons travel over distances much larger than merely atomic dimensions. During the dissolution or oxidation of a metal (or metallic compounds), the electrons produced are not restricted to say a particular location but are free to move through the solid. Thus, a specific number of electrons may be released to the oxidizing medium anywhere on the surface. When this happens, the element would oxidize at certain specific areas called anodic areas. At other areas, the electrons would be transferred to the oxidizing agent in a manner similar to the galvanic corrosion of metals. For example, the corrosion of iron is represented by the reactions

$$Fe = Fe^{2+} + 2e \quad \text{(anodic)} \tag{5.13}$$

$$\frac{1}{2}O_2 + H_2O + 2e = 2OH^- \quad \text{(cathodic)} \tag{5.14}$$

Examples of leaching by an electrochemical mechanism are the dissolution of gold and silver in a cyanide solution, copper in ammonia solution, and some oxide and sulphide minerals in acids.

Cyanidation of gold and silver

Cyanidation is a process in which gold and silver are leached by a cyanide solution. The commonly used cyanides are sodium cyanide, potassium cyanide and calcium cyanide. The dissolution of gold by NaCN is represented by the overall reaction

$$2Au + 4NaCN + 2H_2O + O_2 = 2NaAu(CN)_2 + 2NaOH + H_2O_2 \tag{5.15}$$

The cyanidation of silver can also be expressed by reactions similar to reaction 5.15.

A cyanidation reaction can be considered to be a combination of two steps—an oxidation step and a reduction step. The oxidation step is

$$Au + 2CN^- = Au(CN)_2 + 2e, \tag{5.16}$$

while the reduction step may be written as

$$O_2 + 2H_2O + 4e = 4OH^- \tag{5.17}$$

corresponding to reaction 5.15. The surface being leached is, therefore, divided into anode and cathode areas, represented, respectively, by A_1 and A_2. Further, A_1 and A_2 give, respectively, the sum of all the elemental anode and cathode areas. We assume the boundary layer thickness at the anode and cathode areas to be δ, as shown in Fig. 5.2.

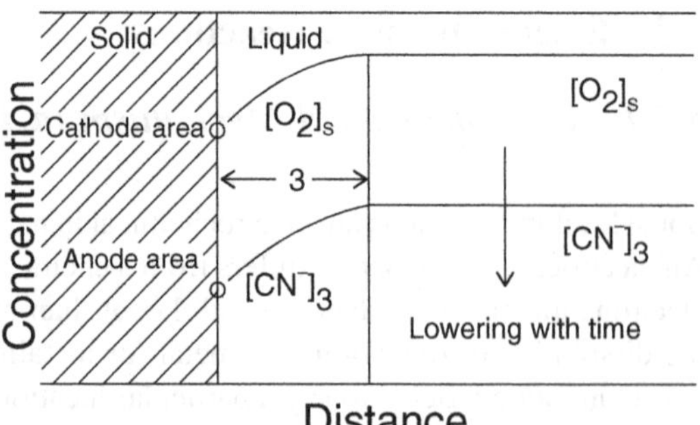

Fig. 5.2 Concentration profile of O_2 and CN^- near metal surface

It is assumed that the kinetics of the reactions are basically determined by the diffusion of both the dissolved O_2 and the CN^- ion and not by the chemical reactions at the surface. The appropriate equation for the diffusion of oxygen from the bulk to the cathode area is

$$\frac{d[O_2]}{dt} = D_{O_2}A_1\{[O_2]_B - [O_2]_S\}/\delta \tag{5.18}$$

where $[O_2]_B$ and $[O_2]_S$ are the concentrations of the dissolved oxygen at the bulk and surface, respectively, D_{O_2} the diffusion coefficient of oxygen and t the time. Similarly, for the diffusion of a CN^- ion, we have

$$\frac{d[CN^-]}{dt} = D_{\overline{CN}}A_2[CN^-]_B - [CN^-]_S/\delta \tag{5.19}$$

Assuming that the surface reactions proceed very fast, $[O_2]_S$ and $[CN^-]_S$ can be taken as the equilibrium values. As these values must necessarily be very low, $[O_2]_S$ and $[CN^-]_S$ can be eliminated from the previous two equations.

We have molar rate of dissolution of the metal(R) = 2 × molar rate of dissolution of oxygen = $\frac{1}{2}$ × molar rate of consumption of cyanide.

Hence,

$$R = 2D_{O_2}A_1[O_2]_B/\delta \tag{5.20}$$

that is,

$$A_1 = \frac{-R\delta}{2D_{O_2}[O_2]_B} \tag{5.21}$$

Similarly,

$$A_2 = \frac{2R\delta}{D_{CN} - [CN^-]_B} \tag{5.22}$$

The total area A is given by

$$A = (A_1 + A_2) = \frac{R\delta\{D_{CN} - [CN^-]_B + 4D_{O_2}[O_2]_B\}}{2D_{O_2}[O_2]_B D_{CN} - [CN^-]_B} \quad (5.23)$$

From Eq. 5.23, we make the following observations:

(1) If $[CN^-]_B$ is small as compared with $[O_2]_B$, then Eq. 5.23 reduces to

$$R = \frac{AD_{CN} - [CN^-]_B}{2\delta} \quad (5.24)$$

The absence of an $[O_2]_B$ term in Eq. 5.24 implies that oxygen should have no role to play when $[CN^-]_B$ is too low compared with $[O_2]_B$ and when the rate of dissolution depends only on the cyanide concentration.

(2) If $[CN^-]_B$ is high and $[O_2]_B$ is negligible as compared with $[CN^-]_B$, then Eq. 5.23 reduces to

$$R = 2AD_{O_2}[O_2]_B/\delta \quad (5.25)$$

Equation 5.25 shows that when the cyanide concentration is high, oxygen plays an important role in cyanidation kinetics. In this case, the diffusion of CN^- is already fast because of the high concentration gradient of CN^-. Therefore, the rate-determining factor should be the rate of oxygen diffusion. Thus, the rate of dissolution may be accelerated using higher pressures to ensure a higher degree of oxygen solubility in the leaching medium and, consequently, a higher concentration gradient of oxygen.

Experimental verification of the foregoing observations has proved that the diffusion phenomenon did indeed control cyanidation kinetics.

(3) In intermediate ranges of $[CN^-]_B$ concentration, both $[O_2]_B$ and $[CN^-]_B$ influence cyanidation kinetics, as expressed by Eq. 5.23. When

$$D_{CN} - [CN^-]_B = 4D_{O_2}[O_2]_B$$

Equation 5.23 reduces to

$$R = D_{O_2}D_{CN^-}A[O_2]_B^{1/2}[CN^-]_B^{1/2}(2\delta)\cdot \quad (5.26)$$

In Eq. 5.26, the value of D_{CN^-} is 1.83×10^{-5} $cm^2\ s^{-1}$ and that of D_{O_2} is 2.76×10^{-5} $cm^2\ s^{-1}$. The thickness of the boundary layer varies between 2×10^{-3} and 9×10^{-3} cm, depending on the speed and the method of agitation.

By comparing, we conclude that the rate of dissolution changes its greater dependence on $[CN^-]_B$ to that on $[O_2]_B$ at conditions given by Eq. 5.26. If the average ratio of D_{O_2}/D_{CN^-} is taken as 1.5, then the $[CN^-]_B/[O_2]_B$ ratio works out to be 6; actual experimental values range from 4.6 to 7.4 (Wadsworth and Wadia 1955).

It should be noted that if a surface chemical reaction were rate controlling, then, too, the equations for the rate would be somewhat similar. For an anode reaction, we have

$$R = K_1 A_1 [CN^-]_B^a, \tag{5.27}$$

and for a cathode reaction, we have

$$R = K_2 A_2 [O_2]_B^b, \tag{5.28}$$

The values of a and b would be determined by the actual order of the anode and cathode reactions, respectively. Their values could be unity, depending on the reaction mechanism. In such a situation, therefore, we would obtain an equation indistinguishable from the diffusion equation. However, an experimentally observed change of dependence, as already mentioned, would be valid only for the diffusion mechanism. As already discussed, a more categorical method of deciding the control mechanism is obtained by considering the dependence of the reaction rate on temperature.

Dissolution of copper in ammonia solution
We now discuss the dissolution of copper in an ammonia solution which is electrochemical in nature. In this case, too, some elemental areas on the metal surface are visualized as anodic and some as cathodic. Let A_1 and A_2 represent, respectively, the sum of all the elemental anode and cathode areas. The anode reaction is

$$\mathrm{Cu} + 4\mathrm{NH_3} = \mathrm{Cu\,(NH_3)_4^{2+}} + 2\mathrm{e} \tag{5.29}$$

and the cathode reaction is

$$\frac{1}{2}\mathrm{O_2} + \mathrm{H_2O} + 2\mathrm{e} = 2\mathrm{OH^-} \tag{5.30}$$

We know that the solubility of oxygen in ammonia solution is low. The rate of reaction 5.30 is, therefore, controlled by the diffusion of dissolved oxygen from the bulk to the cathode areas, unless the ammonia concentration is very low. The rationale behind this conclusion is as follows. If it is assumed that the oxygen reacts, on the surface as soon as it reaches it, and that the ammonia solution is sufficiently strong, then the reaction would depend only on oxygen diffusion. The corresponding equation is

$$R = 2k_1 A_1 [O_2]_B, \tag{5.31}$$

where R is the rate of dissolution and k_1 a constant $(= D_{O_2}/\delta)$. The factor 2 accounts for the fact that two atoms of copper are oxidized by one molecule of oxygen.

By the same rationale, the rate of reaction at the cathode is given by $k_2A_2[NH_3]_B$, where k_2 represents either the rate constant for the surface reaction or the mass transfer coefficient for the diffusion of ammonia. Under steady-state conditions, the rates of the anode and cathode reactions must be the same, and since $(A_1 + A_2)$ is equal to the total area A, we get

$$2k_1A_1[O_2]_B = k_2A_2[NH_3]_B = R, \tag{5.32}$$

$$\frac{R}{A} = \frac{2k_1k_2[O_2][NH_3]_B}{2k_1[O_2]_B + k_2[NH_3]_B} \tag{5.33}$$

Equation 5.33 shows that the rate of reaction depends on $[O_2]_B$ when $[NH_3]_B$ is large compared with $[O_2]_B$. On the other hand if $[NH_3]_B$ is very small compared with $[O_2]_B$, then the rate would depend only on $[NH_3]_B$. In practice, however, ammonia solutions are seldom sufficiently dilute to conform to the latter observation.

The electrochemical mechanism of dissolution is not restricted to metals; it can also be applied to semiconductors such as sulphides or oxides.

Electrochemical dissolution of sulphide minerals

The dissolution of various sulphide minerals in a leaching medium is interpreted as an electrochemical reaction similar to the corrosion of metals. Oxygen is reduced at the cathode areas, and at the anode areas, sulphides are dissolved, liberating electrons to complete the couple. The reduction of oxygen may occur through a series of steps involving peroxide intermediates.

As an example, consider the leaching of chalcopyrite ($CuFeS_2$) in an acid solution under oxygen pressure. The overall reaction is

$$4CuFeS_2 + 17O_2 + 4H^+ = 4Cu^{2+} + 4Fe^{3+} + 8SO_4^{2-} + 2H_2O \tag{5.34}$$

The cathode reactions are

$$O_2 + 2H^+ + 2e = H_2O_2, \tag{5.35}$$

$$H_2O_2 + 2H^+ + 2e = 2H_2O, \tag{5.36}$$

and the anode reactions are

$$CuFeS_2 + 8H_2O = Cu^{2+} + Fe^{2+} + 2SO_4^{2-} + 16e, \tag{5.37}$$

$$Fe^{2+} = Fe^{3+} + e \tag{5.38}$$

It should be noted that the electrochemical dissolution mechanism involves the diffusion of various ions in the aqueous medium as well as the transfer of the electrical charge across the solid–electrolyte interface. The kinetics of a charge - transfer reaction is influenced by the potential gradient at the electrode surface.

Thus, in the areas where the charge-transfer reactions are slow, the leaching rate can be influenced by applying an external voltage. Accordingly, attempts have been made to accelerate the leaching rate of some sulphide minerals by applying suitable potentials externally.

Role of oxygen

Oxygen plays an important role in many leaching operations. In most reactions, it is directly involved as a reactant, for example, in the cyanidation process for the dissolution of gold and silver. The reaction in this instance cannot proceed unless oxygen is made available. The equilibrium of this reaction would depend on the partial pressure of oxygen. The kinetics of leaching also would depend on the oxygen pressure if the rate-controlling mechanism were to involve the diffusion of oxygen either in the gas phase or in the aqueous solution. Oxygen pressure would also influence the diffusion processes in an aqueous solution by controlling the concentration of the dissolved oxygen. In some leaching processes, oxygen may have an indirect role; for example, in bacterial leaching, microorganisms serve as catalysts. Since these organisms necessarily need oxygen for survival, all catalytic activity would cease in the absence of oxygen. We have discussed the cyanidation of gold and silver earlier and shown that the partial pressure of oxygen governs the kinetics of dissolution when the cyanide concentration is high.

The dissolution of copper in ammonia, too, has been discussed earlier, and, in this case also, it has been shown that the partial pressure of oxygen determines the rate when the concentration of the dissolved oxygen is low and that of ammonia is high.

Oxygen is also involved in carbonate leaching of uranium ores. The reaction here is

$$U_3O_8 + \frac{1}{2}O_2 + 6HCO_3^- = 3UO_2(CO_3)_2^{4-} + 3H_2O \tag{5.39}$$

This reaction is believed to proceed according to the steps

$$U_3O_8 + \frac{1}{2}O_2 = 3UO_3 \tag{5.40}$$

$$UO_3 + 2HCO_3^- = UO_2(CO_3)_2^{4-} + H_2O \tag{5.41}$$

It is also believed that reaction 5.40 is a slow one, and, therefore, rate determining. Accordingly, the rate of dissolution of the oxide should be proportional to $[O_2]_B^{1/2}$.

5.3.2 Electrodeposition and Limiting Current

The liquid phase diffusion gives rise to an industrially very important phenomenon, that of limiting current. In an electrolytic cell, the metal output rate is proportional to the current (Faraday's law). Since current that passes between the electrodes depends directly on the voltage applied, one can increase the rate of metal deposition by increasing the voltage applied. However, it is always found that the current can only be increased up to a limit. Beyond a limiting value, further increases in voltage only cause heating, the current having reached a maximum. This phenomenon can be understood thus.

The deposition of metal on the cathode may be looked upon as a two-stage process. The cations diffuse to the cathode, and then at the cathode surface, they are discharged to be incorporated into the solid. The second step is fast and, therefore, the deposition reaction is diffusion controlled. This diffusion requires a concentration gradient, and therefore, near the electrode surface, there is a concentration profile, the local concentration near the electrode being less than that in the bulk liquid. Lowering of the local concentration brings about a deviation of the electrode potential, the deviation being known as concentration overpotential.

The overpotential would increase if the current density i (current per unit area of electrode) increases, because the latter is related to the rate of the electrochemical reaction per unit area per unit time. Obviously, the slow step in the electrochemical reaction will contribute most towards the overpotential. If mass transfer is rate controlling, the overpotential goes down if either the temperature or stirring is increased or if concentration of solute is higher. On the other hand, if the electrochemical reaction is controlled by the slow surface reaction step, then temperature and concentration have similar effect but stirring would have no influence. The nature of the surface and the presence of catalyst or inhibitor would also be important.

Mass transfer-controlled reaction–concentration overpotential
Let us write an electrode reaction in the generalized fashion as follows:

$$\mathrm{O\ (oxidized\ species)} + \mathrm{ze}^- \rightarrow \mathrm{R\ (reduced\ species)} \tag{5.42}$$

We may write

$$E = E^{\mathrm{O}} - \frac{\mathrm{R}T}{z\mathrm{F}} \ln \frac{a_{\mathrm{R}}^{\mathrm{O}}}{a_{\mathrm{O}}^{\mathrm{O}}} = E^{\mathrm{O}} + \frac{\mathrm{R}T}{z\mathrm{F}} \ln \frac{a_{\mathrm{O}}^{\mathrm{O}}}{a_{\mathrm{R}}^{\mathrm{O}}} \tag{5.43}$$

where

E Single electrode potential
$a_{\mathrm{O}}^{\mathrm{O}}$ Activity of the oxidized species in the bulk
$a_{\mathrm{R}}^{\mathrm{O}}$ Activity of the reduced species in the bulk

Equation 5.43 is the well-known Nernst equation for electrode potential. It is applicable strictly only when interfacial equilibrium prevails and is true for mass transfer-controlled reactions.

Let us consider a cathode reaction and assume that the transport of the oxidized species (e.g. Cd^{2-}) towards the electrode is rate controlling. When a current is passed, the concentration or activity of the oxidized species at the interface, $a_{\mathrm{O}}^{\mathrm{S}}$, would be lower than the activity in the bulk, $a_{\mathrm{O}}^{\mathrm{O}}$. Then we have

$$E + \eta_{\mathrm{o}} = E^{\circ} + \frac{\mathrm{R}T}{z\mathrm{F}} \ln \frac{a_{\mathrm{O}}^{\mathrm{S}}}{a_{\mathrm{R}}^{\mathrm{O}}} \tag{5.44}$$

Combining Eqs. 5.43 and 5.44, we obtain for cathodic overpotent

$$\eta_{\mathrm{c}} = \frac{\mathrm{R}T}{z\mathrm{F}} \ln \frac{a_{\mathrm{O}}^{\mathrm{S}}}{a_{\mathrm{O}}^{\mathrm{O}}} \tag{5.45}$$

Now, for dilute aqueous solutions, as well as fused salts, activity may be taken as proportional to concentration. Therefore, one obtains

$$\eta_{\mathrm{c}} = \frac{\mathrm{R}T}{z\mathrm{F}} \ln \frac{C_{\mathrm{O}}^{\mathrm{S}}}{C_{\mathrm{O}}^{\mathrm{O}}} \tag{5.46}$$

Similar equations would be valid for the anode also. Figure 5.3 shows schematically the concentration changes near an electrode surface due to passage of a current when mass transfer of reactant (i.e. O in case of cathode and R in case of anode) is rate controlling.

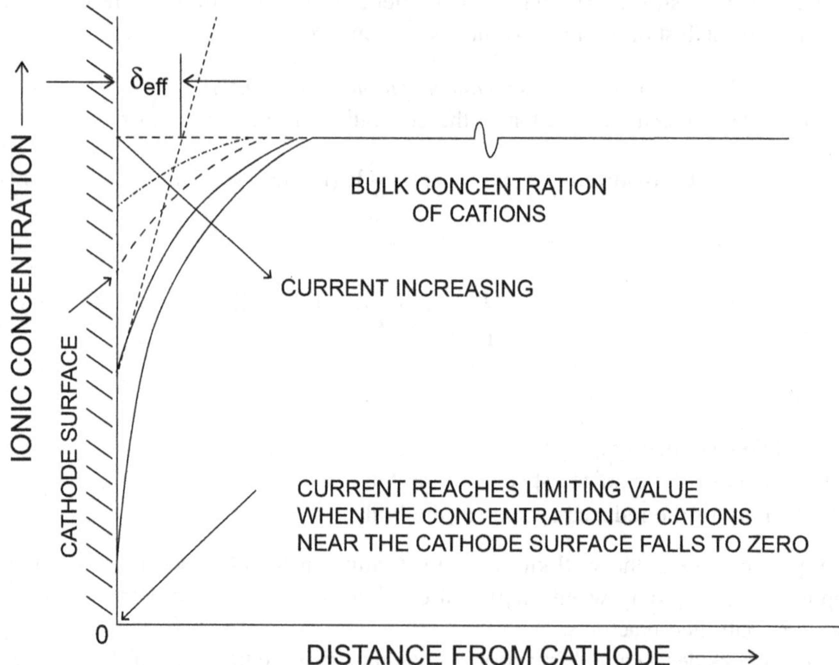

Fig. 5.3 Concentration profile near the cathode surface

The diffusion of ions is given by the equation

$$\frac{1}{A}\frac{\mathrm{d}n}{\mathrm{d}t} = h_m(C_\mathrm{O}^\mathrm{O} - C_\mathrm{O}^\mathrm{S}) \tag{5.47}$$

where $h_m = \frac{D}{\delta}$.

Again, from Faraday's laws of electrolysis,

$$\frac{1}{A}\frac{\mathrm{d}n}{\mathrm{d}t} = \frac{i}{z\mathrm{F}} \tag{5.48}$$

where the LHS is molar flux of oxidized species per unit area.

Combining Eqs. 5.47 and 5.48, we obtain

$$i = 2\mathrm{F}h_m(C_\mathrm{O}^\mathrm{O} - C_\mathrm{O}^\mathrm{S}) \tag{5.49}$$

h_m depends on fluid flow, which is roughly independent of i. Therefore with increase in i, C_O^S will decrease and gradually approach the limiting value of zero. When this happens, then i cannot be increased any further and it reaches a limiting maximum (i_d). For this limiting case and assuming δ to be independent of i, we obtain

$$i_\mathrm{d} = z\mathrm{F}h_m C_\mathrm{O}^\mathrm{O} \tag{5.50}$$

$$\text{Now,}\quad \eta_\mathrm{c} = \frac{\mathrm{R}T}{z\mathrm{F}}\ln\frac{C_\mathrm{O}^\mathrm{S}}{C_\mathrm{O}^\mathrm{O}} = \frac{\mathrm{R}T}{z\mathrm{F}}\ln\left[1 - \frac{C_\mathrm{O}^\mathrm{O} - C_\mathrm{O}^\mathrm{S}}{C_\mathrm{O}^\mathrm{O}}\right] = \frac{\mathrm{R}T}{z\mathrm{F}}\ln 1 - (i/i_\mathrm{d}) \tag{5.51}$$

This is the relationship between η_c and i for this situation.

For aqueous solutions $D \approx 10^{-9}$ m^2 s^{-1}, $\delta_\mathrm{eff} \approx 10^{-7}$ m, and C is generally around 10^3 mol m^{-3}. Substituting these values in Eq. 5.51, one obtains for i_d a value of approximately 10^5 A m^{-2}. Hence, η_c is hardly a few millivolts. This is true for fused salt electrolysis also. Therefore, we conclude that overpotential due to slow mass transport, i.e. concentration overpotential, is negligible in industrial electrolysis. However, the concept of the limiting current is of utmost importance. It limits the rate of production for a given cell. The output can only be increased by increasing the total electrode surface area either by using larger electrodes or by using multiple electrodes.

The phenomenon of limiting current is used in voltammetry for identifying the metal ions being discharged the exact nature of discharge reactions and also quantitative examination of ions in solution (Ray 1977).

5.4 Dimensionless Numbers and Their Use

Definitions of dimensionless groups
The mass transfer coefficient h_m (which equals D/δ) is a function of several variables which can be classified into three categories.

a. Variables connected with fluid flow,
b. Variables connected with mass transfer properties of the fluid,
c. Geometry and size of the system.

Dimensionless groups formed out of various combinations of these variables are used to simplify correlations. The dimensionless numbers can be used in estimating reaction rates when flow in liquids is rate controlling (Von Bogdandy and Engell 1971). Only some commonly used dimensionless numbers are described.

Sherwood number, N_{Sh}: By definition,

$$N_{Sh} = h_m \cdot \frac{d}{D} = \frac{d}{\delta} \tag{5.52}$$

where d is a characteristic length. For a spherical solid, d is the diameter.

Reynolds number, N_{Re}: The Reynolds number, which characterizes liquid flow is defined by the equation

$$N_{Re} = \frac{vd}{\upsilon} \tag{5.53}$$

where v is liquid velocity and υ the kinematic viscosity (μ/ρ where μ is viscosity and ρ the density of liquid).

Schmidt number, N_{Sc}: This is defined by the relationship

$$N_{Sc} = \frac{\upsilon}{D} = \frac{\mu}{\rho D} \tag{5.54}$$

The three above-mentioned numbers are interrelated by the following relationship.

$$N_{Sh} = C + C' N_{Re}^m N_{Sc}^n \tag{5.55}$$

where C and C' are constants.

For a sphere in a flowing fluid $C = 2$. If the velocity v is zero, then N_{Sh} equals 0.2. C' has a value of about 0.6 m and n are also constants where values depend on different facts. For turbulent flow, the exponent m is between 0.5 and 0.8 while $n = 1/3$. For laminar flows, the material transport coefficient h_m is roughly proportional to the square root of flow velocity v.

The Nusselt's number for mass, N_{Nu}, is the same as N_{Sh}.

Grasheff's number, N_{Gr}: This is used for fluid flow under natural conditions. N_{Gr} is defined by the relation

$$N_{Gr} = \frac{g d^3 \rho^2}{\mu^2} \left| \frac{\Delta\rho}{\rho} \right| \tag{5.56}$$

where $|\Delta\rho|$ is the absolute value of the difference in density between surface and bulk.

Numerous other dimensionless numbers have been defined in the literature for better understanding and analysis of transfer of heat and mass. A detailed discussion of the subject, however, is beyond the scope of the present book.

In general,

$$N_{u_m} = \text{function of } (N_{Re} \cdot N_{Sc}, \text{ geometry}) \text{ for forced convection} \tag{5.57}$$

and

$$N_{u_m} = \text{function of } (N_{Gr} \cdot N_{Sc}, \text{ geometry}) \text{ for natural convection} \tag{5.58}$$

When geometry is fixed then one can often write,

$$\text{For forced convection} \quad N_{Nu} = BN_{Re}^{m} \cdot N_{Sc}^{n} \tag{5.59}$$

$$\text{For natural convection} \quad N_{Nu} = B' N_{Gr}^{m'} \cdot N_{Sc}^{n'} \tag{5.60}$$

where B, B', m, n, m' and n' are all constants. The values of the last four lie in the range 0 to 1.

One example is given to show the use of dimensionless groups. This is based on problems discussed by Ghosh (1972).

Example 5.4 A plate of solid, 3 mm long and 2 cm thick, is held vertically inside a liquid. Calculate the rate of dissolution in g/min under natural convection conditions assuming that mass transfer is rate controlling.

Given	$D = 2 \times 10^{-6}$ cm²/s
	$\mu = 1$ poise
	Saturation conc., $C_s = 0.043$ mol/cm³
	ρ (liquid) = 2.09 g/cc
	ρ (saturated solution) = 2.20 g/cc
	Molecular wt. of liquid = 90
	$N_{Nu} = 0.545(N_{Gr} \cdot N_{Sc})^{0.25}$

Solution: We should first understand the mechanism of the dissolution process. There is a saturated solution near the solid surface. However since this layer is heavier, it tends to sink thus causing natural convection. Surface equilibrium prevails because the dissolution reaction at the interface is fast.

Using the correlation among dimensionless groups, we obtain

$$N_{\mathrm{Nu}} = 0.545(N_{\mathrm{Gr}} \cdot N_{Sc})^{0.25}$$

$$\text{i.e.} \quad \frac{h_m d}{D} = 0.545\left[\frac{g d^3 \rho^2}{\mu}\left|\frac{\Delta\rho}{\rho}\right|\frac{\mu}{\rho D}\right]^{0.25} \tag{5.61}$$

$$\text{or} \quad h_m = 0.545\left[\frac{g|\Delta\rho|}{\mu}\right]^{0.25} \cdot D^{3/4} \cdot d^{-1/4} \tag{5.62}$$

where the characteristic dimension d should equal the length axis. The total width is 2×2 cm.

The saturation concentration $C_s = 0.043$ mol/cm^3
Bulk concentration of solute $C_o = \frac{2.09}{90} = 0.023$ mol/cm^3
Hence, total dissolution rate in mol/min is given by

$$\text{Rate} = 4\left[\int_0^3 h_m (C_s - C_o)\,\mathrm{d}x\right] \times 60 \tag{5.63}$$

The integration takes into account the fact that the mass transfer coefficient changes along the length of the solid. The rate then is

$$\text{Rate} = 240 \times 0.545 \times \left[\frac{g|\Delta\rho|}{\mu}\right]^{0.25} D^{3/4}(C_s - C_o)\int_0^3 x^{-\frac{1}{4}}\mathrm{d}x \tag{5.64}$$

Taking $g = 980.6$ cm/s^2 and $|\Delta\rho| = 2.20 - 2.09 = 0.11$, the rate is calculated as 1.375×10^{-3} mol/min. In this calculation, we have ignored the edges.

5.5 Kinetic Equation in Special Cases

5.5.1 Reaction of a Solid with a Constant Overall Dimension

Consider the basic liquid phase mass transfer equation (Eq. 5.5)

$$\frac{\mathrm{d}n}{\mathrm{d}t} = \frac{V}{A} \cdot \frac{\mathrm{d}C}{\mathrm{d}t} = \frac{D}{\delta}(C_s - C_o)$$

If during the reaction, solid forms a solid reaction product or residue of the same density and the overall dimension remains unchanged, then one obtains a special situation. We have

$$\frac{dC}{dt} = k \cdot (C_s - C_o) \quad (5.65)$$

where k is a constant $\left(\frac{DA}{V\delta}\right)$.

If the fluid has a constant fresh supply, then $(C_s - C_o)$ is constant. That is

$$\frac{dC}{dt} = \text{constant} \quad (5.66)$$

The reaction then acquires the feature of a zero-order reaction.

5.5.2 Reaction of Solid in Fluid with Unchanging Composition

If $(C_s - C_o)$ only is constant, then $\frac{dc}{d_t}$ becomes proportional to surface area A dissolving. The kinetics then resembles that for phase boundary controlled reaction. This will be explained with reference to some work done on dissolution of silica in glass melts (Ray et al. 1982).

5.5.3 Dissolution of Silica During Glass Melting

Dissolution of silica is of much importance in glass melting. Accordingly, numerous laboratory studies have been reported on the mechanism of dissolution of pure silica rods and spheres and also of ordinary sand grains.

It is generally agreed that the dissolution reaction is diffusion controlled. Yet, the data often fit the phase boundary control model. This has led to controversy and ambiguity. This apparent anomaly is examined here. It is shown that under certain conditions both models may lead, approximately, to the same rate equation.

During melting of silicate glasses, a good amount of silica is consumed directly in the initial reactions which lead to meltdown. The residual silica dissolves slowly in the primary glass thus formed presumably by diffusion. The dissolution reaction, because of its prolonged nature, is of considerable practical importance in glass melting. Accordingly, much work has been reported on the subject.

Riebling (1969) studied dissolution of silica rods above 1300 °C and concluded that the reaction was diffusion controlled, the diffusion coefficient being independent of rod size below 3 mm. The diffusion mechanism has been supported by several other studies (Roborts 1959; Krieder and Cooper 1967; Cable and Martlaw 1971). Attempts have been made to use diffusion theory in predicting time for complete dissolution and estimation of melting time in the industry.

Use of diffusion theory in practical glass melting, however, is beset by problems. According to Swartz (1972), the rate of disappearance of sand grains is limited, in principle, by the diffusion kinetics of silica in the melt but in practice becomes extremely sensitive to those factors which upset silicate concentration gradients.

The diffusion theory is directly contradicted by the data reported by Preston and Turner (1940). They found that, at any time, the rate of dissolution was proportional to the surface area of silica grains exposed to attack and that the average grain diameter decreased almost linearly with time after the initial chemical reactions had been completed. Published data, Cable and Martlaw (1971), also indicate this although these were interpretated on a diffusion control model.

5.5.4 Rate Equation

The equation for dissolution of silica into surrounding melt through diffusion across a boundary layer is (Eq. 5.5)

$$\frac{\mathrm{d}C}{\mathrm{d}t} = \frac{D}{\delta}(C_s - C_o)\frac{A}{V}$$

The solution is complicated by the fact that both A and $(C_s - C_o)$ change continuously. An approximate solution is easily found if we assume that in the silica-rich primary melt $(C_s - C_o)$ essentially remains constant. Then, we have

$$\frac{\mathrm{d}C_o}{\mathrm{d}t} = \frac{\mathrm{d}W}{\mathrm{d}t} = \mathrm{k}_1 A \tag{5.67}$$

where W is the weight of silica particle.

Equation 5.67 is the basic postulate of phase boundary control mechanism. Also, if r is radius of each particle in an ensemble of n uniform particles then

$$-\frac{\mathrm{d}}{\mathrm{d}t}\left(m\frac{4}{3}\pi r^3 \rho\right) = \mathrm{k}_1 n 4\pi r^2 \tag{5.68}$$

which reduces to

$$\frac{\mathrm{d}r}{\mathrm{d}t} = \frac{\mathrm{k}_1}{\rho} \tag{5.69}$$

where ρ is the density. Equation 5.69 shows that size should decrease linearly with time as found by Preston and Turner (1940). They have also shown that the melting time, i.e. the time of complete dissolution, is proportional to the initial diameter. This can be proved as follows. Suppose that the kinetic equation is expressed as

$$g(\alpha) = k_2 t \tag{5.70}$$

where α is the degree of reaction, $g(\alpha)$ is an appropriate function, and k_2 is the rate constant. For any fixed value of α, $g(\alpha)$ is fixed. Time t^* for this is given by

$$t^* = \text{a constant}/\, k_2 \tag{5.71}$$

When reaction rate is proportional to surface area, A, then k_2 is inversely proportional to r because A is inversely proportional to the initial radius of particles if reaction rate is proportional to the surface area.

5.5.5 *Cementation Reaction*

Cementation reaction or contact reaction means precipitation of one metal by another from solution, e.g. of copper by iron power added to $CuSO_4$ solution. The reaction may be written as

$$M_1^{Z_1} + \frac{Z_1}{Z_2} \cdot M_2 = M_1 + \frac{Z_1}{Z_2} \cdot M_2^{Z_2} \tag{5.72}$$

In most cases, the kinetics of such a reaction is controlled by solution diffusion of ions across a boundary layer at the metal solution interface (Wadsworth 1969). In some cases, however, both solution diffusion and surface reaction have been found to be important. A simple rate equation applicable in many cases is

$$\frac{dC}{dt} = -\frac{CA_c \cdot D}{V\delta} \tag{5.73}$$

where A_c is the cathodic surface area and other terms have their usual meanings.

Ingraham and coworkers (1969) have shown that kinetics of cadmium cementation by zinc cylinders and plates and that of copper on aluminium discs follow a similar first-order kinetics except, sometimes, for extended periods when rate increases presumably due to increases in cathodic surface area.

5.6 Smelting Reduction Processes

In the early 1980s, worldwide over ninety-five per cent of liquid iron used to be produced through coke–sinter–blast furnace route. However, scarcity of good quality coking coal, high capital costs and environmental constraint of this route made it imperative to look at alternate ironmaking processes. The direct reduced iron (DRI) route, which emerged in the late 1960s and early 1970s, is now well established. Currently, about fifty-five per cent of iron is produced through the conventional blast furnace, whereas the DRI process accounts for another forty per cent. Smelting reduction (SR) is the other promising route. Already a few smelting reduction processes like COREX and FINEX have been commercialized, and a few plants are in operation, including in India; a few other processes have proved successful in the pilot plan scale.

In smelting reduction, a substantial part of iron is produced through high-temperature reduction of FeO in molten slag with the help of coal and oxygen injection. Iron ore is partially prereduced and the obtained prereduced material dissolved in slag for final reduction of foaming FeO slag by solid or carbon dissolved in iron.

5.6.1 *Reduction of* FeO *in Slag (Paramguru et al. 1996a, b; Sato et al. 1987)*

Conceived in the late 1930s, fundamental studies in smelting reduction started in the 1950s. Studies were carried out on reduction of pure molten iron oxide and/or slags containing different levels of wustite by solid carbon, carbon-saturated iron melt and CO gas. Systematic research over the years has now established the following:

a. Reduction by carbon-saturated liquid iron is much faster as compared to that of solid carbon (often by one order of magnitude).
b. The reduction rate is influenced by FeO content, temperature, slag height, crucible/slag interfacial area when the crucible material is graphite, slag composition—particularly basicity, presence of additive such as CaF_2.
c. Gases evolving during reduction cause foaming, and behaviour is determined by reduction rate and slag properties.
d. Reduction by solid carbon is often characterized by an induction delay, particularly when FeO level is low.
e. When reduction is by solid carbon, then the metallic particles produced initially by reduction are often surrounded by a gas halo, and there are significant amounts of such particles in the slag phase.
f. The iron produced, solid or liquid depending on the reaction temperature, eventually make a carbon-saturated liquid iron. Therefore, there can be no pure solid carbon reaction at later stages of reaction.
g. There may be gas layers enveloping metallic particles or droplets produced and also layers separating carbon surface from liquid slag thus decelerating reduction rate.

Details are available in the reviews available in the literature. As mentioned earlier, FeO reduction can be by solid carbon, solute carbon (in molten iron), CO gas, or a combination of these. Studies have been extended also to slags containing MnO, Cr_2O_3.

5.6.2 *Solid Carbon Reduction*

Different forms of carbon can be used as reductant, viz., solid graphite coke/ graphite crucible, or carbon/coke/graphite pieces or rods in crucibles of graphite or other refractory materials, e.g. alumina magnesia, non-reacting vitreous graphite, steel. Reduction rates can be calculated from the amount of CO gas evolved or analysis of slag. Values for rates of reduction by solid carbon are usually $2.1–8.2 \times 10^{-5}$ mol. FeO cm^2 s at 1400–1600 °C.

A simple scheme of reduction as follows:

$$FeO + CO(g) = [Fe] + CO_2\,(g) \quad (5.74)$$

$$CO_2(g) + C = 2CO\ (g) \quad (5.75)$$

$$\text{Overall}\ (FeO) + C\,(s) = 2CO\,(g) \quad (5.76)$$

where ( ) and [ ] denote, respectively, liquid and solid phases.

Some workers prefer to write the final product as a carbide MC_y where M is metal. If the di- and trivalent iron (or chromium) oxide are expressed as MO, then one can write the following equation.

$$(MO_x) + (y + x)\,C\,(\text{solid, solute}) = [MC_y] + x\,CO\,(g) \quad (5.77)$$

$$(MO_x) + (y + x)\,CO(g) = [MC_y] + x\,CO_2\ (g) \quad (5.78)$$

It is assumed that the carbide product is in the solid state. The $CO_2(g)$ procured reacts with carbon to produce CO (g) according to Bouduard reaction (Eq. 5.75).

Possible rate-controlling steps are the following:

a. Mass transfer of FeO (i.e. Fe^{2+} and O^{2-} ions) in slag,
b. CO_2–C reaction,
c. Gas diffusion in the gas layer separating solid carbon and liquid slag,
d. Gas–slag reaction,
e. Nucleation of iron,
f. A mixture of two or more of the above.

Owing to the complexity of the system, the rate-limiting step differs as the conditions for the reaction are changed. In high FeO systems, diffusion, chemical reaction CO_2–C each individually or a mixed control have been mentioned as rate controlling. Some have mentioned nucleation also as a possibility so as to explain the induction delays are particularly prominent at lower FeO levels.

However, generally, mass transport of FeO in slag is considered to be the rate-controlling mechanism at lower values of FeO (<10%).

5.6.3 Solute Carbon Reduction

Carbon–oxygen reaction takes place, in this case, at the slag/metal interface with diffusion steps in either phase (i.e. oxygen in slag and carbon in iron) as well as reactions occurring at the interface. Even though there may be some gas layers occasionally separating regions of metal and gas phase, the reduction by solute carbon is much faster than that by solid carbon. In the latter case, this is always a gas halo around carbon and the metal particles, and a gas layer substantially separates slag from graphite/carbon surface. There is better availability of carbon reduction when carbon is present as a solute in iron.

5.6.4 Foaming and Foaming Index

A foam is agglomeration of bubbles in a liquid matrix. Bubbles are separated from each other by thin liquid films, thus generating a structure in which relative movements is quite restricted. A foam has an extremely low liquid content and a large amount of entrapped gas as individual unconnected bubbles. The fraction of volume occupied by liquid is very small (<0.05), which means that a 1000 mm of foam may come from far less than 50 mm height of liquid. If one passes air through a column of soap solution, generally 2–3 mm of liquid creates more than 100 mm height of foam. A foam, obviously, would have a large surface area, the development of which requires surface tension to be small.

A foam is different from an emulsion. If the thickness of the interstitial liquid layers is at least commensurate with the bubble diameter, then the dispersion can be called a gas emulsion or gas dispersion. There is, however, no exact boundary separating a foam and an emulsion. If 1 cc of disperse phase contains ϕ cc of liquid and, therefore $(1-\varphi)$ cc of gas, then, customarily, the system has foam when $\varphi > 0.9$ and it has "emulsion" when $\varphi < 0.1$. The region in between is a grey area. $(1-\varphi)$ is frequently referred to as "hold-up" of gas or wetness. The ratio of foam volume to liquid volume, i.e. $1/\varphi$, called expansion factor or expansion ration. The void fraction $(1-\varphi)$ may change with foam height.

For a given cross-sectional area of the smelting reduction reactor, A, the superficial gas velocity V_g^s increases with total gas flow rate, Q_g

$$V_g^s = Q_g/A \tag{5.79}$$

Foaming index $\sum$ is defined as the ratio of the foam height $\Delta h/V_g^s$ (i.e. change in slag height due to foaming to the superficial gas velocity).

$$\sum = \Delta h/V_g^s \tag{5.80}$$

$\sum$ is essentially the time of travel of gas in the foam phase.

Some workers have done extensive studies by creating foams in $CaO–SiO_2–FeO$ system by external inert gas purging. At 1400 °C, $\sum$ decreased with increasing basicity to a value of about 1.22 beyond which it increased. The presence of P_2O_5 marginally increased foaming index, whereas CaO decreased it. Increase in CaF_2 and temperature also decreased $\sum$ because of lowering of viscosity. The relationship of $\sum$ with slag properties was expressed as follows:

$$\sum = A\mu(\gamma \cdot \rho)^{-0.5} \tag{5.81}$$

where μ, γ and ρ are, respectively, viscosity (Pa s), surface, surface tension (N m^{-1}) and density (Kg m^{-3}). A is a constant. Addition of coal or coke suppresses slag foaming as mentioned earlier. Thus, production rates limited by foaming will be more in actual bath smelting operations compared with calculated values. Pressurized operation will decrease foam height due to decrease in gas flow velocity. Both reaction kinetic and foaming behaviour are related to temperature and slag properties and, therefore, foaming index and reaction rate constant are interrelated.

The foaming index and the foam height depend on several factors, and Figs. 5.4 and 5.5 indicate some data for a particular investigation using a plasma reactor (Jouhari et al. 2000, 2001). Maximum foam height increases with initial slag height, FeO concentration in slag, and volumetric gas flow rate. These are indicated, respectively, by a, b and c in Fig. 5.4. If one considers the foaming index, then variations are as shown in Fig. 5.5. $\sum$ does not have a simple relationship with basicity, and it does not vary with crucible diameter. It increases with slag height and decreases with initial FeO constant beyond some value.

The foaming phenomenon, as expected, is related to reaction kinetics which again depends on the reductant. As has been mentioned earlier, carbon in solution (pig iron) is far more effective reducing agent than solid forms of carbon. In the industry, the product liquid iron containing solute carbon itself reduces the FeO containing slag continuously producing fresh liquid iron which dissolves carbon made available in the system. There can be considerable foaming because of high reaction rates. Therefore, some processes, e.g. Ausmelting operates is very low levels of FeO(2–4%) where foaming is less.

5.6.5 Reaction Kinetics

Kinetics of FeO reduction in slags have been studied over wide ranges of FeO concentration, basically and temperature using several laboratory techniques. In these techniques, the sample size can vary from fraction of a gram to one kilogram. The progress of reaction can be studied by determining the total amount of CO and CO_2 generated and the gas analysis, or loss in weight of slag assuming that all oxygen is lost as CO or actual analysis of FeO in slag.

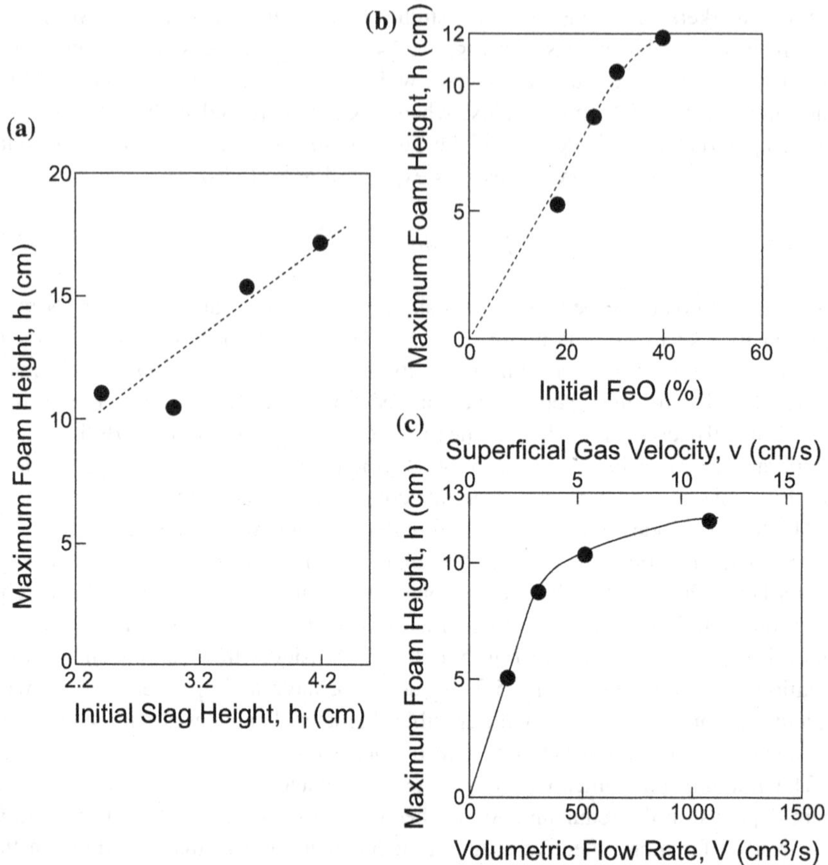

Fig. 5.4 Effects of different parameters on foaming: plot of maximum foam height against. **a** Initial slag height. **b** Initial FeO content in slag. **c** Volumetric flow rate/superficial gas velocity due to variation of % FeO in slag. (Crucible dia = 11 cm, $CaO:SiO_2 = 1.29$, temperature: 1550 ± 100 °C) (Jouhari et al. 2000, 2001)

It is now well accepted that reduction by solute carbon follows a simple first-order kinetics through the activation energy value determined may vary from one set to another. Figure 5.6 shows some results from a study using a plasma reactor set-up using nearly 0.5 kg of slag (Jouhari et al. 2000). Figure 5.6a shows that the E value is approximately 153 kJ/mol. Figure 5.6b, c indicates the importance of the ratio of slag volume and metal–slag interfacial area which controls the rate constant. Figure 5.6d shows that the rate constant increases with increase in FeO concentration which implies that the reaction is controlled by FeO diffusion in slag. This result has been substantiated by thermogravimetric analysis using 2 g samples (Paramguru et al. 1996a). However, one obtains smaller value of E ($\sim$90 kJ/mol).

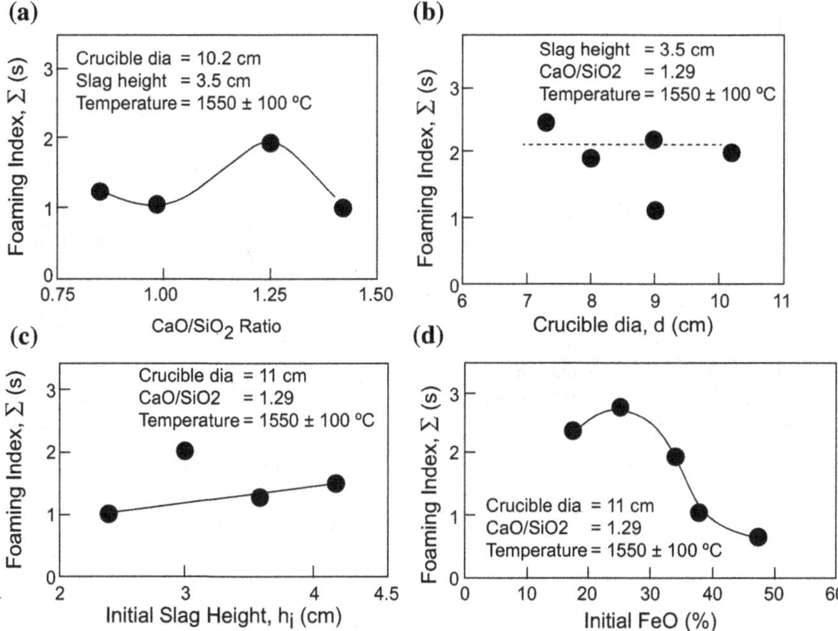

Fig. 5.5 Variation of foaming index with different parameters: **a** $CaO:SiO_2$ ratio (crucible dia = 10.2 cm, slag height = 0.05 cm, temperature: 1550 ± 100 °C); **b** crucible dia (slag height = 3.5 cm, $CaO:SiO_2 = 1.29$, temperature: 1550 ± 100 °C); **c** initial slag height (crucible dia = 11 cm, $CaO:SiO_2 = 1.29$, temperature: 1550 ± 100 °C); **d** initial %FeO in slag (crucible dia = 11 cm, $CaO:SiO_2 = 1.29$, temperature: 1550 ± 100 °C) (Jouhari et al. 2000, 2001)

When FeO in slag is reduced by solid carbon, then the reaction is characterized by an incubation period, particularly at low FeO levels in slag. The plot of ln $g(\alpha)$ vs ln t generally has a slope of about 0.9 as against 1 for first-order kinetics. The rate constant, however, remains proportional to FeO per cent in slag and the E value also remains about 90 kJ/mol.

The kinetic equation assumes the nucleation—grain growth (Johnson–Mehl) typewritten as follows (Paramguru et al. 1996b)

$$g(\alpha) = -\ln(1-\alpha)^{1/n} = k't$$

The plot of ln $g(\alpha)$ vs $\ln t$ generally has a slope of about 0.9 as against 1 for a first-order reaction. The rate constant, however, remains proportional to FeO per cent in slag, and the E value also remains around 90 kJ/mol.

Studies have shown that foaming and kinetics can be correlated using some dimensionless numbers. The Reynolds number defined as follows: (see Eq. 5.53)

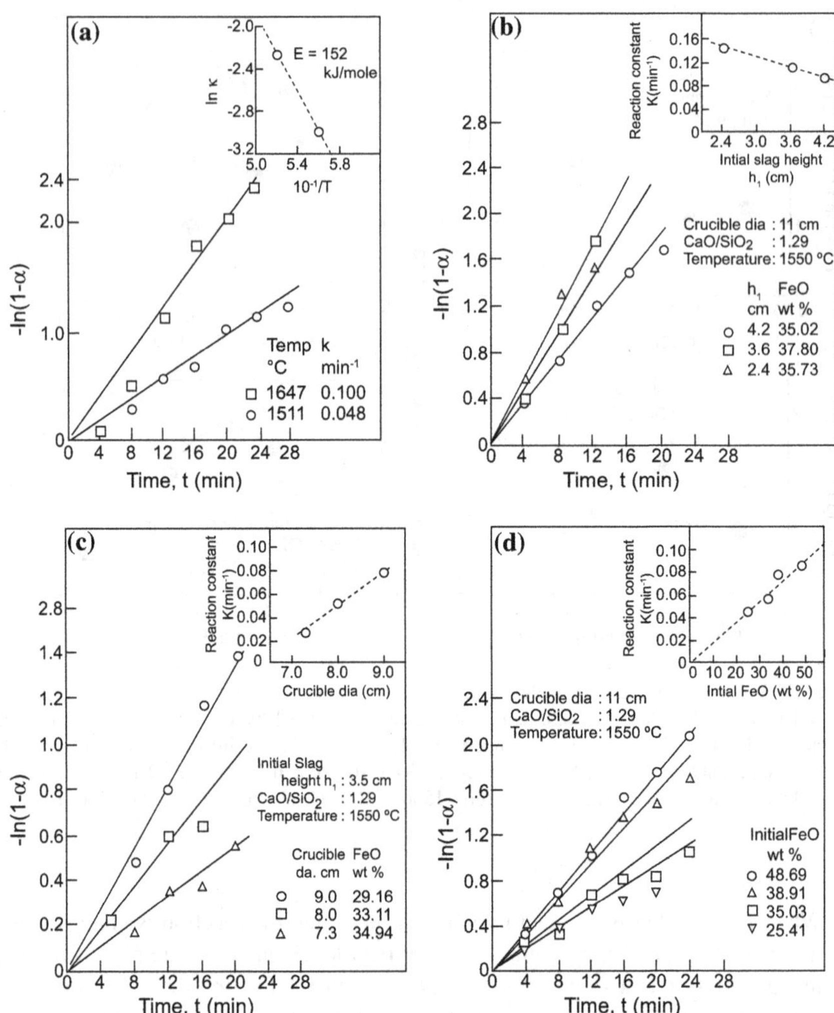

Fig. 5.6 First-order kinetic equation for solute reduction of FeO containing slag and effects of various parameters. (Results from the plasma reaction). **a** $\ln(1-\alpha)$ versus t and Arrhenius plot for estimation of apparent activation energy; **b** $\ln(1-\alpha)$ versus t plot for variation of initial slag height; **c** $\ln(1-\alpha)$ versus t plot for variation of crucible diameter; **d** $\ln(1-\alpha)$ versus t plot corresponding to data shown (Jouhari et al. 2000)

$$N_{Re} = \frac{4\rho v d}{\pi d^2 \mu} \quad \text{or} \quad \frac{\rho v d}{\mu}$$

where ρ = density, v = velocity, d = characteristic dimension.

The Morton number M is defined as follows:

$$M = \frac{g\mu^4}{\rho \upsilon^3}$$

The other dimensionless number often used in the literature is the following:

$$\pi_1 = \sum g\mu/\upsilon$$

$$\pi_2 = \frac{\rho\upsilon^3}{g\mu^4} = \frac{1}{M}$$

Except π_1, all the dimensionless numbers are defined in terms of liquid properties, whereas π_1 uses the parameter $\sum$ which must be determined experimentally.

The following correlation has been established in the literature:

$$\log \mu = A - B \log N_{Re}$$

where A and B generally have values of −3 to −5 and 2, respectively.

The v term in the Reynolds number relates to superficial velocity of CO bubbles that are produced by reduction and this relates to reaction rate. One can derive relationship such as follows (Paramguru et al. 1996a, b)

$$\log k = A + 0.5 \log \frac{(\upsilon^{3/2})}{M}$$

where A is a constant with a value around −9. This equation shows that one can estimate a rate constant value on the basis of viscosity and surface tension of slag. The latter affect foaming also and, thus, one can establish more correlations. An approximate equation is the following.

$$\log \pi = -2 \log \pi_1 + 6.34$$

From this the following, correlation is derived (Jouhari et al. 2000, 2001)

$$\sum = A'\mu(\rho\upsilon)^{0.5}$$

where A' is a constant.

Different workers have reported for A' values in range 100–600. Thus, one can estimate the foaming index also long knowing the slag properties. It should be obvious thus one can combine appropriate equations to relate foaming index with reaction rate.

5.7 Review Questions

1. Discuss whether the following statements are correct or false.
   a. Boundary layer thickness at a solid/fluid interface is independent of the fluid flow conditions.
   b. Enhanced stirring of bulk liquid accelerates solid/liquid reaction by increasing the diffusion coefficient.
   c. The limiting current during electrolysis is a diffusion current.
   d. The limiting current density is independent of the cathode surface area.
   e. Periodic current reversal increases the output of an electrolytic cell for a given input current.
   f. During acid leaching of a mineral, the leaching rate is always enhanced by intensifying stirring.
   g. Rate of leaching of gold by cyanide solution may or may not depend on the cyanide concentration depending on other factors.
   h. A fluidized bed electrode increases rate of metal deposition during electrolysis.
   i. Periodic current reversal during electrolysis increases current efficiency.
   j. Diffusion coefficient has the same dimensions as velocity.
   k. For liquid phase mass control, the time of complete dissolution of a solid plate in a liquid is infinity.
   l. Natural connection arises out of intense stirring of the liquid.
2. Draw concentration profiles for a slag–metal reaction

$$[\mathrm{A}] + (\mathrm{B}^{++}) = [\mathrm{B}] + (\mathrm{A}^{++})$$

   where [ ] and ( ) denote metal and slag phases, respectively. Assume any two rate-controlling steps.
3. What steps can be taken to increase the production capacity of an electrolytic cell? What are the principles underlying these measures?
4. The dissolution of a solid sphere in a liquid is liquid phase mass transfer controlled. It, however, follows the same kinetic equation as that for phase boundary controlled reaction. What does this imply?
5. Derive an equation to show the relationship between the limiting current density and process variables during electrolysis.
6. When a solid reacts with a gas, the reaction rate increases with gas velocity. What does this imply?
7. A spherical particle of quartz is dissolving in a large quantity of molten glass. The reaction is controlled by liquid phase mass transfer. Show that the reaction would have characteristics of interfacial reaction control.
8. The first-order kinetics for dissolution of a solid in liquid is expressed in terms of the concentration of solute. Rewrite the law in terms of fraction of solid consumed when the solid is (a) a sphere and (b) a plate.

9. Manganese is going from liquid iron to the slag phase. Draw concentration profiles at the interface assuming that Mn transfer in the metal phase is rate controlling. Calculate % Mn in the metal after 40 min if the initial content is 0.6%.

Given: Initial Mn in slag = 0%; D_{Mn}(liquid iron) = 5×10^{-5} cm^2/s; δ_{Mn}(liquid iron) = 2×10^{-3} cm; ρ (liquid iron) = 7.6 g/cc; Concentration of Mn at interface = 0.0015%; Volume liquid metal/Area of interface = 2.5 cm; Total volume of metal = 5 l.

References

Cable, M., Martlaw, D.: Glass Tech. **12**(6), 142 (1971)

Ghosh, A.: Principles of Extractive Metallurgy (Unpublished lecture notes), IIT/Kanpur (1972), Ch. 5

Ingraham, M.R., Korby, R.: ibid, **245**, 7 (1969)

Jouhari, A.K., Galgali, R.G., Chattopadhyay, P., Gupta, R.C., Ray, H.S.: Scand J. Met. **30**, 14–20 (2001)

Jouhari, A.K., Galgali, R.K., Datta, B., Bhattacherjee, S., Gupta, R.C., Ray, H.S.: Ironmaking Steelmaking **27**(1), 27–31 (2000)

Krieder, K.C., Cooper, A.C.: Glass Tech. **8**(3), 71 (1967)

Mckenzi, D.J., Ingraham, T.R.: Can. Met. Q. **9**, 443 (1970)

Paramguru, R. K., Ray, H. S., Basu, P.: Ironmaking and Steelmaking, **23**(4), pp. 328/354 (1996a)

Paramguru, R.K., Ray, H.S., Basu, P., Jouhari, A.K.: ibid **23**(5), pp. 411–415 (1996b)

Preston, E., Turner, W.C.: J. Soc. Glass Tech. **24**, 24 (1940)

Ray, H.S.: N.KH, Tumanova and S.N. Flengas. Can. J. Chem. **55**(4), 656 (1977)

Ray, H.S., Bandyopadhyay, G., Gupta, R.K.: Trans IIM **41** (1972)

Ray, H.S.: Proceeding Symposium on High Temperature Chemistry B.A.R.C., p. 213. Bombay. Jan (1982)

Riebling, E.F.: Bull Amer. Ceram Soc. **48**(8) (1969)

Roborts, A.L.: In: Kingery, W.D. (ed.) Kinetics of High Temperature Processes, p. 222. John Wiley and Sons Inc, N.Y. (1959)

Sato, A., Avagaro, G., Kanihira, K., Yashimatsu, S.: Trans ISIJ **27**, 789–796 (1987)

Swartz, E.L.: In: Pye, L.D., Stevans, H.J., La Course, W.C. (eds.) Introduction to Glass Science, p. 273. Plenum Press, N.Y. (1972)

Von Bogdandy, L., Engell, H.J.: The Reduction of Iron Ores, Springer Verlag, New York (1971), Ch. 2

Wadsworth, M.E.: Trans TMS. AIME **245**, 1381 (1969)

Wadsworth, M.E., Wadia, D.R.: Trans TMS. AIME **203**, 755 (1955)

Chapter 6
Reaction Between Two Fluids

6.1 Introduction

6.1.1 Mass Transfer Across Two-Fluid Interface

When two fluids are in contact with a well-defined interface, then there would be a boundary layer on either side in an ideal case. Consider, for example, desulphurization of sulphur-rich iron by a basic slag. The basic reaction may be written as

$$[S] + (O)^{2-} = (S)^{2-} + [O] \tag{6.1}$$

where [] and () denote, respectively, the metal and slag phases. One has to consider the diffusion flux of four species if mass transfer in liquid is rate controlling. However, the concentration profile set-up will depend on the rate-controlling step. There would be concentration profile for the species whose diffusion is rate controlling. Figure 6.1 schematically explains this. This, however, is a simplified representation.

In reality, the fluid–fluid interface may not remain static and well defined specially when the flow is turbulent at or near the interface. If a laminar flow prevails at the interface, then boundary layers are more likely to be established. If the viscosities of the two liquids are appreciably different, then the less viscous liquid may behave like a rigid solid in relation to the light liquid.

One can think of a similar situation for interaction with bubbles of gas with a liquid. Under ideal conditions, there will be a boundary layer for the gas and one for the liquid on either side of the interface. We shall first analyse some reactions based on this simplifying assumption.

H. S. Ray and S. Ray, *Kinetics of Metallurgical Processes*,
Indian Institute of Metals Series, https://doi.org/10.1007/978-981-13-0686-0_6

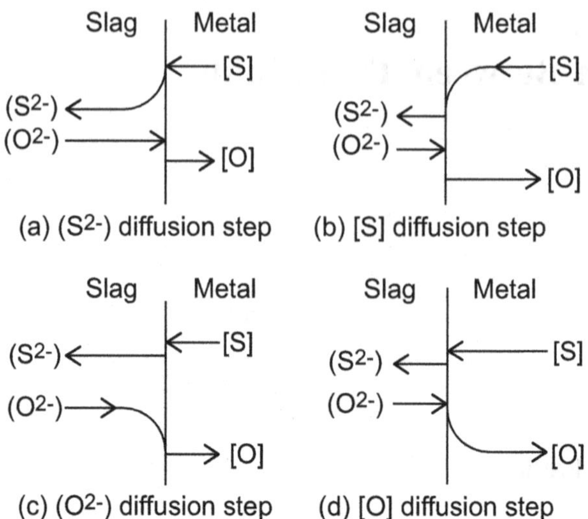

Fig. 6.1 Concentration profiles for different rate-controlling steps (a–d)

6.1.2 Slag–Metal Reaction

Consider another problem of a slag–metal interaction. A MnO-rich slag removes Si from liquid iron by the reaction

$$[\mathrm{Si}] + 2(\mathrm{MnO}) = 2[\mathrm{Mn}] + (\mathrm{SiO_2}) \tag{6.2}$$

There are boundary layers on either side of the interface, and there would be a concentration profile for a diffusion step if it is singly rate controlling. The various kinetic steps are as follows:

(a) Transport of Si in bulk metal to the interface,
(b) Transport of MnO in slag to the interface,
(c) Chemical reaction at the interface,
(d) Transport of Mn from the interface to bulk in the metal phase,
(e) Transport of SiO_2 away from the interface and into the bulk slag.

Actually, the reaction mechanism is ionic. However, this molecular representation explains the phenomenon just as well.

Suppose that the chemical reaction step is fast. The diffusion flux for any species i is given by the equation

$$J_\mathrm{i} = h_\mathrm{m} \left| \left[C_\mathrm{i}^\mathrm{o} - C_\mathrm{i}^\mathrm{s} \right] \right| \tag{6.3}$$

The rate-controlling diffusion step can be identified if values of $(C_s - C_o)$ and h_m can be estimated using the following correlations

$$N_{Nu} = \frac{h_m d}{D}$$

$N_{Nu} = B(N_{Re})^{0.5}(N_{Sc})^{0.5}$ where B is a constant.
This gives,

$$h_m = \frac{N_{Nu} \cdot D}{d} = B(N_{Re})^{0.5}(N_{Sc})^{0.5} \cdot \frac{D}{d} \tag{6.4}$$

Substituting this value in Eq. 6.3

$$J_i = h_m \Delta C = B(N_{Re})^{0.5} \cdot \left(\frac{\mu}{\rho D_i}\right)^{0.5} \cdot \frac{D_i}{d} \tag{6.5}$$

B, N_{Re} and d are common for all species. Hence, if we compare the fluxes for two solutes x and y we get

$$\frac{J_x}{J_y} = \frac{\left[\frac{\mu_x D_x}{\rho_x}\right]^{0.5} |\Delta C_x|}{\left[\frac{\mu_y D_y}{\rho_y}\right]^{0.5} \Delta C_y} \tag{6.6}$$

By substituting appropriate values, one can compare J_x and J_y and identify the slower step. If the diffusion fluxes of solutes are comparable, then we have mixed control.

6.2 Static Gas–Metal Interaction

Consider a gas atmosphere over a molten layer. In many metallurgical gas metal systems, there is diffusion of a species from the gas to the liquid or vice versa. Consider, for example, diffusion of oxygen into metal. The rate-controlling process would be diffusion of oxygen into the metal.

One can calculate oxygen dissolution rate assuming a concentration profile in the metal boundary layer in the same way as was done in Chap. 5. On the other hand, if the liquid is considered a rigid plane and the radial diffusion is neglected, then the diffusion equation is given as

$$\frac{C - C_i}{C_s - C_i} = 1 - \frac{4}{\pi} \exp\left(-\frac{D\pi^2 t}{4L^2}\right) \tag{6.7}$$

where

C concentration of oxygen in bulk metal
C_i initial bulk concentration of oxygen
C_s surface concentration of oxygen which is saturation concentration
D diffusion coefficient of oxygen
t time
L height of melt

The left-hand side equals the degree of reaction, α. Hence,

$$\ln(1-\alpha) = \ln\left(\frac{4}{\pi}\right) - kt, \tag{6.8}$$

where $k = \left(\frac{D\pi^2}{4L^2}\right)$.

If the term $\ln(4/\pi)$ can be ignored, then Eq. 6.8 is a first-order equation.

According to Eq. 6.7, because $(C_s - C_i)$ is a constant, a plot of $\ln(C - C_i)$ versus t should be a straight line, with the slope giving $D\pi^2/4L^2$. Again, from known values of C_s and C_o one can plot Dt/L^2 against t to obtain straight line passing through the origin. The slope makes possible calculation of the diffusion coefficient D. Bandyopadhyay and Ray (1972) have used these equations and approaches to analyse kinetic data on dissolution of oxygen from an oxygen–argon atmosphere into molten lead at 750 °C. Figure 6.2 shows some data reported by Bandyopadhyay and Ray (1972).

Rigorous mathematics requires that for radial diffusion to be neglected L should be zero. To approach this ideal condition, the diffusion coefficient was calculated using two different melt heights. For each set of data (corresponding to a given height of melt or a given L/d ratio, where d is a diameter), the D value was calculated. To minimize error, the correct value was obtained by extrapolation to zero value of L/d. The value so obtained matched well with value reported in the literature (Fitterer and Arckle 1969), Fig. 6.3.

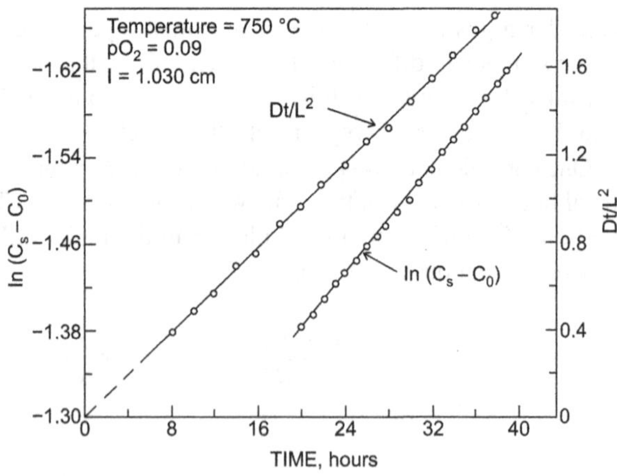

Fig. 6.2 Plots of Dt/L^2 and $\ln(C - C_i)$ versus time (Bandyopadhyay and Ray 1972)

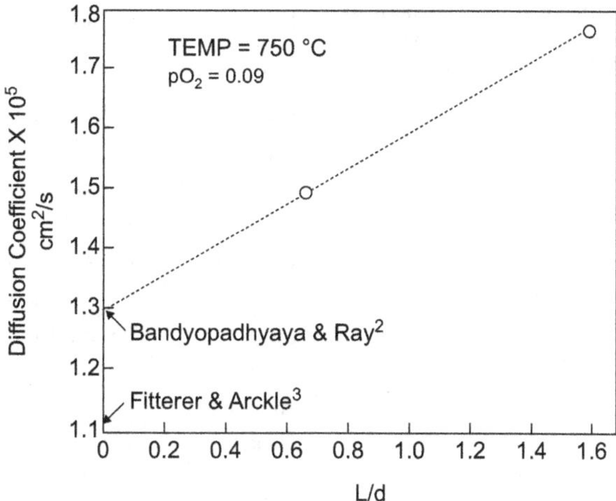

Fig. 6.3 Calculated diffusion coefficient against L/d ratio (Bandyopadhyay and Ray 1972)

Diffusion of gas from a static atmosphere, however, is very slow. In the case of dissolution of oxygen in a deoxidised metal, the oxidation rate can be enhanced manyfold by bubbling oxygen through the melt. The rate should increase with temperature, bubbling rate and height of the melt. The oxidation rate, however, appears to be independent of p_{O_2} which implies that gas phase diffusion is not rate controlling. However, during bubbling, the melt is stirred and while there is liquid phase transfer at bubble–metal interface as in the previous case, one can no longer treat the liquid as a rigid plane. One has to now invoke the concept of a liquid boundary layer around every gas bubble.

This subject is now discussed in some detail.

6.3 Bubble–Metal Interaction

6.3.1 Introduction

Bubbles play an important role in many metallurgical operations. In steelmaking, oxygen bubbles and jets are used for stirring as well as rapid oxidation of impurities by increasing the gas–metal reaction surface area. Carbon boil and lime boil involve evolution of bubbles and consequent stirring and gas–metal interaction. Many metals are purified by bubbling oxygen or chlorine through the liquid. The impurities react preferentially and are removed either as volatile products or as separate slag phase. Table 6.1 lists some metals which can be purified by bubbling oxygen or chlorine. Metals may also be degassed by bubbling a gas which itself has low solubility in the metal. Thus, oxygen and nitrogen in liquid steel are eliminated by bubbling, argon, and hydrogen in liquid copper is eliminated by bubbling carbon monoxide.

Table 6.1 Systems purified by oxygen or chlorine bubbling

Liquid metal	Impurities removed by	
	Oxygen bubbling	Chlorine bubbling
Iron	C, Si, P, Mn	Zn
Copper	Pb, As, Sb, S, Mg, Al, Fe	Fe, Pb, Zn
Lead	As, Sb, Sn, Zn	–
Zinc	Na	Pb
Bismuth	Pb, As, Sn, Zn	–
Tin	Fe, Al, Zn, Ni	–
Silver	Pb, Zn, Cu	–
Gold	Zn	Ag, Pb, Cu, Sb
Aluminium	–	Mg
Nickel	–	Cu

In metallurgical processes, bubbles bring about mass transfer by either giving some element to the melt or by taking away an element or by an exchange reaction. These reactions may take place in a homogeneous medium (metallic phase) or at a slag–metal interface.

There is considerable literature on the behaviour and reaction of bubbles in liquids at room temperature. Literature on high-temperature metallic systems, however, is very limited. This may be attributed to lack of transparency of metallic melts which make visual examination of bubbles impossible and to difficulties generally associated with high-temperature experimentation.

Studies involving interactions of bubbles with liquids are generally aimed at one or more of the following:

(a) Determination of size and shape of bubbles for various orifice diameters, flow conditions, system geometry, etc.
(b) Theoretical estimation of mass transfer coefficients.
(c) Experimental determination of mass transfer coefficients.
(d) Determination of the rate-controlling step during the interaction of bubbles with liquid systems, etc.

(b), (c) and (d) would all depend on fundamental informations concerning (a). Again, due to scarcity of high-temperature data, most of high-temperature studies depend heavily on information on low-temperature systems concerning bubble formation (a). It is generally believed that bubble formation depends on certain general principles valid universally. Some selected literature on bubble formation in low-temperature liquids is listed at the end of the chapter (Bownder and Kumar 1970; Ramakrishnan et al. 1969; Khurana 1969; Padmavathy et al. 1969; Calderbank 1969; Davenport et al. 1967).

Table 6.2 lists some of the important equations that have been proposed in the literature for analysing the nature of bubble–liquid interactions at room temperatures. The equations are for the following:

Table 6.2 Some equations proposed for analysing the nature of bubble–liquid interaction

Equation and reference		Remark
Bubble size		
$d_b = 2d_o$	Pehlke and Bement (1962)	Bubble size is obtained from orifice diameter
$d_b^3 = \frac{6d_o}{g(\rho - \rho')}$	Nanda and Geiger (1971)	Bubble size is expressed in terms of orifice diameter, surface tension, and densities
$d_b = 0.18(d_o)^{0.5}(N_{Re})^{0.33}$	Leibsen et al. (1956)	Gives bubble size in jets
Bubble surface		
Area of curved surface = 11.675 r_b^2 Area of flat surface of a cap = 9.599r_b^2	Davies and Taylor (1950) Baird and Davidson (1962)	Gives surface area for spherical cap bubbles
Bubble velocity		
$U_\infty = \frac{1}{3}\frac{g r_b^2 \rho}{\mu}$ or $\frac{2}{9}\frac{g r_b^2 \rho}{\mu}$	Strokes	Terminal velocity for spherical bubbles in terms of bubble size and the nature of the liquid
$U_\infty = 0.72\sqrt{g d_e}$	Davies and Taylor (1950)	Gives terminal velocity for spherical cap bubbles
$U_\infty = \frac{1}{2}\sqrt{g r_e}$	Collins (1967)	Gives terminal velocity for spherical cap bubbles
$U = U_\infty\left\{\frac{1}{b}\left(\frac{1-d_e}{d}\right)\right\}^{0.765}$	Uno and Kinter (1956)	Accounts for wall effects on the terminal velocity of bubble
$U_\infty = \left\{\frac{4-(\rho - \rho')g - d_b}{3\rho f}\right\}^{0.5}$	Pehlke and Bement (1962) Nanda and Geiger (1971)	Gives wall effect of terminal velocity of bubble
Equation and reference		Remark
Mass transfer coefficient		
$K_l = \frac{D}{d_e}(2 + 0.55N_{Re}^{0.5}N_{Sc}^{0.5})$	Hammerton and Garner (1954), Froessling (1938)	Liquid-phase mass transfer coefficient is expressed using dimensionless numbers
$K_l = 2(DU_\infty/2\pi r_b)^{0.5}$	Hughmark (1967), Froessling (1938), Baird and Davidson (1962)	Mass transfer coefficient is related to bubble size and velocity
$K_l = \text{const}.\, d_e^{-0.25}D^{0.25}g^{0.25}$	Baird and Davidson (1962)	Mass transfer coefficient is related to bubble size and velocity
$\frac{2K_l r_b}{D} = f\left\{\left[\frac{\mu}{\rho D}\right]^{0.5}\cdot\left[\frac{2r_b\cdot U_\infty\cdot\rho}{\mu}\right]^{0.766}\right\}$ $K_l \propto (2r_b)^{0.86}$ for large bubbles $K_l \propto (2r_b)^{0.67}$ for small bubbles	Calderbank (1969)	Relation for estimation of mass transfer coefficient

Symbols
K_l = mass transfer coefficient; U = terminal velocity of bubble; U_∞ = terminal velocity in liquid of infinite width; d = dia. of container vessel; d_b = bubble diameter; b = constant; d_e = dia. of equivalent sphere, d_o = orifice dia.; r_b = bubble radius; r_e = radius of equivalent sphere; f = friction factor

(a) Estimation of bubble size from the orifice diameter,
(b) Estimation of the velocity of the bubble,
(c) Estimation of the liquid phase mass transfer coefficient.

Table 6.3 lists some literature pertinent to the subject of interaction of bubbles with liquid metals.

Some workers (Davenport et al. 1967; Panenis and Davenport 1969) have shown that the behaviour of gas bubbles in liquid metals is similar to that of the bubbles in low-temperature liquids under certain conditions.

6.3.2 Low-Temperature Liquids

Previous studies on low-temperature liquids related to the determination of the following factors in various regimes of bubble motion: (i) bubble size and shape, (ii) bubble velocity and motion, (iii) mass transfer rates and coefficients across the gas–liquid interface and (iv) interfacial area for different configurations.

The experimental data and theoretical investigations were concerned mainly with a single bubble rising at low velocities in the laminar region. Mass transfer

Table 6.3 Literature pertinent to bubble–liquid metal interaction

Literature	Remark on topic
Richardson (1973)	Review on drops and bubbles in steelmaking
Ali et al. (1973)	Review on interaction of gas bubbles with molten metals
Bradshaw and Richardson (1968)	Experiments with mercury and aqueous solutions at room temperature and silver at 1000 °C. Experiments related to steelmaking conditions, Interaction in a homogeneous phase and at an interface
Devenport et al. (1967)	Spherical cap bubbles in mercury and silver
Panenis and Davenport (1969)	"Two-dimensional" experiments, nitrogen bubbles rising in a sheet of mercury
Guthrie and Bradshaw (1969)	Large nitrogen and oxygen bubbles in liquid silver
Lange and Coworkers (1960)	Globular and calotte shaped bubbles in liquid iron baths
Nanda and Geiger (1971)	Carbon monoxide bubbles in copper (a solid electrolyte cell method)
Bandyopadhyay and Ray (1972)	Oxygen–argon bubbles and hydrogen bubbles in molten lead oxygensystem (use of a solid electrolyte cell technique)
Ali and Ray (1974)	Oxygen–argon bubbles and hydrogen bubbles in molten lead oxygensystem (use of a solid electrolyte cell technique)
Ali et al. (1973)	Rate-controlling step during the above-mentioned interactions
Papamantellos et al. (1970)	Mathematical model for mass transfer between ascending bubbles and liquid steel

between swarms of bubbles in agitated turbulent liquids has not received much attention. A comprehensive review of the bubble–liquid interactions at room temperatures is available elsewhere (Levich 1962; Danckwerts 1970) and, therefore, this aspect has not been covered in this review. However, some basic information which is useful for the study of bubble interactions in liquid metals is provided (Ali et al. 1973; Ali and Ray 1974).

Size, shape and motion of bubbles

Several workers have studied experimentally the shape and motion of air bubbles in water and other liquids. It has been established that the shape and nature of bubble motion are determined by the nature of the liquid flow around it. It has been observed that the bubble shape is spherical when the Reynolds number (N_{Re}, defined in terms of bubble dimensions) is less than 400, oblate spheroid of varying geometric proportions when N_{Re} is between 400 and 1100, 5000, and is spherical cap when N_{Re} is above 5000. Also, when the value of the Reynolds number is several hundred, the character of the bubble motion changes, depending on whether the bubble motion radius is below or above about 0.7 mm. Smaller bubbles rise steadily in a straight line, but larger bubbles either follow zigzag motion from side to side or rise in the form of a spiral. Small bubbles appear to be spherical; larger bubbles are gradually distorted into spherical cap shape, depending on the value of N_{Re}.

Velocity of bubbles

For very small spherical bubbles, a creeping flow around the bubbles can be assumed as they rise in the liquids. Since there is no slip at the bubble, liquid interface, the normal and tangential stresses and the tangential velocities on both sides of the interface are equal and the normal components of the velocity at the interface are zero. For these conditions, there are some correlations as summarized in Table 6.2.

When a spherical bubble is moving at its terminal velocity, the drag force on the bubble must equal the force on it due to buoyancy. This equality has been used to derive the equations shown.

In the case of very small bubbles, it has been shown that because of the inherent presence of surface active agents in the liquid, internal circulation in the bubble ceases, as the surface active agents form an immobile film on the bubble surface. Hence, the bubbles behave as rigid spheres and their rising velocity is given by Stoke's law

$$U_\infty = \frac{2}{9}\frac{g r_b^2 \rho}{\mu} \tag{6.9}$$

With increase in bubble diameter, the surface active agents can no longer cover the surface completely and internal circulation occurs. Surface tension forces lose their influence further as the bubble gets distorted at larger sizes. The transition between spherical and ellipsoidal forms depends on a balance between inertial, viscous and surface tension forces for which no general relationship is yet available.

Ali et al. (1973) have shown that for a spherical cap bubble, potential flow can be assumed around the bubble. Accordingly, they derived the following expression for terminal velocity:

$$U_\infty = 0.72\sqrt{g d_e} \tag{6.10}$$

where d_e is the diameter of an "equivalent" sphere (i.e. a sphere with volume equal to that of the spherical cap).

In the transition region between various bubble shapes and in the region where the bubble is an oblate spheroid, no theoretical solutions are available for the prediction of bubble velocity.

Mass transfer between the bubble and the surrounding liquid

Several physical absorption models for mass transfer have been proposed, viz. the film model, the still surface model, the surface renewal model, etc.

In the film model (Whitman 1923), it is assumed that there is a stagnant film of thickness δ at the surface of the liquid next to the gas. The concentration of the diffusing species changes only in this stagnant film, the rate of change being given by Fick's first law. The liquid mass transfer coefficient K_1 is defined as D/δ, where D is the diffusion coefficient. The hydrodynamic properties of the system are accounted for by the parameter δ, which depends on the bubble geometry, liquid agitation and physical properties of the liquid.

The basic concepts of the still surface model are as follows. When mass transfer occurs from a solid surface to a moving turbulent liquid, only molecular diffusion occurs in a region close to the surface. Further away from the surface, transport by the turbulent eddies becomes more important. The situation may be similar in the neighbourhood of a free liquid surface. Instead of a discontinuity, as assumed in the film model, there may be a progressive transition from purely molecular transport to predominantly convective transport, as the distance from the surface increases. The proponents (Danckwerts 1970) of these concepts have shown that K_1 may be proportional to any power of D between 0 and 1.

The surface renewal models assume the replacement at intervals of elements of liquid at the surface by liquid from the interior, which has the local mean bulk composition. While an element of liquid is at the surface and is exposed to the gas, it absorbs as though it were quiescent and infinitely deep and the rate of absorption is a function of the time of exposure of the element. The replacement of liquid at the surface might be brought about by turbulent motion of the body of the liquid. This model suggests that the surface of an agitated liquid consists of elements which have been exposed to the gas for different lengths of time (or have different "ages") and which will, therefore, in general, be absorbing at different specific rates. Different versions (Guthrie and Bradshaw 1969; Large et al. 1960) of the model lead to different distributions of the surface ages about the mean value, and the liquid mass transfer coefficient is given by the following expression:

$$K_1 = (DS)^{1/2} \tag{6.11}$$

where S is a constant, which has the dimensions of reciprocal time. Here the hydrodynamic properties of the system, so far as they affect K_1, are accounted for by a single parameter S.

When mass transfer is accompanied by a chemical reaction, the situation is rather complicated.

Mass transfer studies

It has been shown that the values of mass transfer coefficients for small spherical bubbles would be those predicted by the equation (Hammerton and Garner 1954),

$$K_1 = \frac{D}{d_e}(2 + 0.55 N_{Re}^{0.5} \cdot N_{Sc}^{0.5}) \tag{6.12}$$

where N_{Sc} is Schmidt number.

Higher values are expected when internal circulation begins inside the bubbles, and velocity gradients within the liquid at the gas–liquid interface are reduced or eliminated. Hughmark (1967) has also postulated a semi-theoretical relation for small bubbles. For a sphere rising through a liquid, this theory predicts

$$K_1 = 2(DU_{\infty}/2\pi r_b)^{0.5} \tag{6.13}$$

For bigger bubbles, the dependence of K_1 on the size is not completely understood due to lack of data on distortion, flow pattern, etc., and there is no general correlation available for calculating the values of the mass transfer coefficients.

For spherical cap bubbles, Baird and Davidson (1962) have shown that K_1 for the leading surface is given by,

$$K_1 = \text{const.} d_e^{-0.25} D^{0.25} g^{0.25} \tag{6.14}$$

Calderbank found that for large bubbles, K_1 was proportional to $(2r_b)^{0.86}$ and for small bubbles ($r_b < 1$ mm), to $(2r_b)^{0.67}$.

It was shown subsequently that experimental data on liquid metals often conform to the above-mentioned theoretical expressions, which were developed primarily for low-temperature liquids.

Swarm of bubbles in a liquid

The system of a swarm of bubbles moving in a liquid is a complicated one. The bubbles may be of various sizes and may be dispersed unevenly in the liquid. Moreover, bubble coalescence and breakup may be occurring simultaneously with the rise of bubbles. Thus, any attempt at a theoretical correlation for the velocity distribution of the bubbles in a swarm and mass transfer between the liquid and the bubbles must employ many simplifying assumptions.

Measurement of mass transfer coefficient

It has been discussed earlier that at high temperatures, mass transfer of solute in the liquid metal is likely to be the rate-controlling step. Then,

$$\frac{dn_g}{dt} = K_1 a_b (C_s - C) \tag{6.15}$$

where n_g signifies the number of moles of the diffusing gas, a_b, the area of the bubble–metal interface, and C_s and C, the concentration of the solute in the metal at the interface and in the bulk, respectively. For mass transfer in most systems, the liquid metal is exposed to a gaseous atmosphere over the free surface and, therefore, the mass transfer across the free surface of the melt must also be considered. This is given by

$$\frac{dn_g}{dt} = DA_s \frac{dC_g}{dy} \tag{6.16}$$

where A_s is the area of the free surface of the melt, and dC_g/dy, the concentration gradient in the gas over the surface (y being the vertical direction). If the solute content of the gas above the free surface is the same as that in the bubble, then

$$\frac{dn_g}{dt} = \frac{D}{\delta} A_s (C_s - C) \tag{6.17}$$

For such a case, the total rate of transfer is given by

$$\left(\frac{dn_g}{dt}\right)_{\text{total}} = nK_1 a_b \Delta C + A_s (D/\delta) \Delta C \tag{6.18}$$

where ΔC denotes $(C_s - C)$ and n, the number of bubbles in the melt at any time. The value of n is given by

$$n = \frac{F\tau_r}{V} \tag{6.19}$$

where F is the flow rate of gas, τ_r, the time of residence of the bubble inside the melt, and V the volume of the bubble. The total instantaneous area na_b is given by

$$na_b = A_b = 3F\tau_r / r_b \tag{6.20}$$

The term n_g may be converted to concentration units by dividing it by the volume of the melt V_m. Thus,

$$n_g = V_m (C - C_o) \tag{6.21}$$

where C_o is the initial bulk concentration. Differentiation of Eq. 6.21 with respect to temperature gives

$$\frac{dn_g}{dt} = V_m \frac{dC}{dt} = -V_m \frac{d\Delta C}{dt} \tag{6.22}$$

Substituting for na_b from Eq. 6.20 and for dn_g/dt from Eqs. 6.22 and 6.18 becomes

$$-V_m \frac{d\Delta C}{dt} = \frac{K_1 3F\tau_r}{r_b} + \frac{D}{\delta} A_s \Delta C$$

which on rearrangement reduces to

$$\frac{d(\Delta C)}{\Delta C} = -\frac{1}{V_m}\left[\frac{D}{\delta} A_s + \frac{3K_1 F\tau_r}{r_b}\right] dt \tag{6.23}$$

This on integration yields

$$\ln\left[\frac{C_s - C}{C_s - C_o}\right] = -\frac{1}{V_m}\left[\frac{D}{\delta} A_s + \frac{3K_1 \cdot F \cdot \tau_r}{r_b}\right] t \tag{6.24}$$

This may be simplified further if the concentration of the solute in the liquid metal is small, in which case C_s becomes negligible, so that Eq. 6.24 becomes

$$\ln(C/C_o) = -\frac{1}{V_m}\left[\frac{D}{\delta} A_s + \frac{3K_1 \cdot F \cdot \tau_r}{r_b}\right] t \tag{6.25}$$

A plot of $\ln(C/C_o)$ versus t should thus result in a straight line, and K_1 can then be calculated from the slope. Since C_o is a constant, it is also possible to plot in $(C_s - C)$ against t.

The residence time τ_r can be calculated by dividing melt height h by the terminal velocity U_∞.

Bandyopadhyay and Ray (1972) have used this equation successfully in their work on oxidation of liquid lead by Ar–O_2 bubbles. Some of their data are shown in Figs. 6.4 and 6.5.

6.4 Mass Transfer as a Rate-Controlling Step

Role of Activation Energy

Chapter 1 has discussed the effect of activation energy (E) on reaction rate. It was shown that at lower temperatures reaction steps with high E values become rate controlling because the rate constants (k) are smaller. However, a higher E value

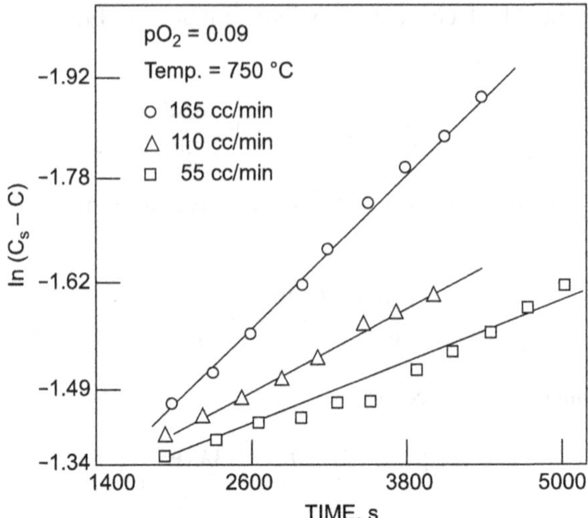

Fig. 6.4 $\ln(C_s - C)$ against t (Bandyopadhyay and Ray 1972)

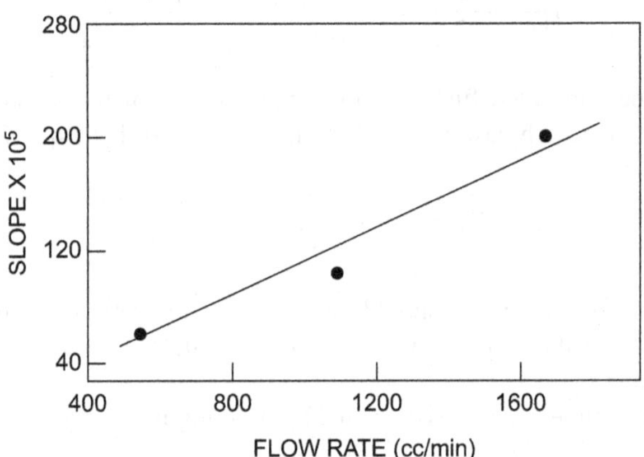

Fig. 6.5 Slope against flow rate (Bandyopadhyay and Ray 1972)

also implies a more rapid increase in k with temperature. Thus, at higher temperatures, reaction steps associated with lower activation energies are likely to become rate controlling. Generally, E values are high for interfacial reactions (10–100 kcal/mole i.e. 40–400 kJ/mole). For gas phase on diffusion, E values are usually very low (20 kJ/mole or less). Diffusion through porous solids is characterized by intermediate values. Thus, with increase of temperature different rate-controlling steps emerged. For oxidation of carbon three zones, all often encountered accordingly (Fig. 6.6).

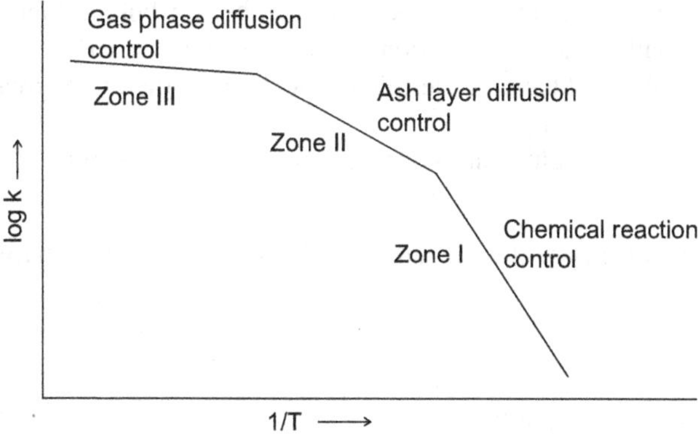

Fig. 6.6 Different zones for oxidation of carbon

Liquid phase mass transfer is also not appreciably accelerated by temperature increase. This is explained as follows. The diffusion coefficients in high-temperature liquids are comparable with those in aqueous solutions. Some values are as follows.

D in aqueous solutions $\approx 10^{-5}$ cm^2/s
D in liquid metal solution $\approx 10^{-4} - 10^{-5}$ cm^2/s
D in liquid slags $\approx 10^{-6}- 10^{-7}$ cm^2/s

The boundary layer thickness (δ) values are generally as follows.

δ in aqueous solutions $\approx 10^{-3} - 10^{-4}$ cm
δ in liquid metals and slags $\approx 10^{-2} - 10^{-4}$ cm

We, therefore, conclude that higher temperatures do not necessary imply considerable enhancement of mass transfer coefficient D/δ. Thus, mass transfer can become rate controlling at elevated temperatures and stirring can have dramatic effect in enhancing rate (up to a limit). It has, however, been pointed out in the literature that under some special conditions, the chemical reaction step can still be slow at high temperatures. This happens when some surface active impurities render considerable portions of the interface inactive, and available area is reduced. For example, presence of oxygen or sulphur in steel, both surface active, reduces the rate of absorption of nitrogen.

Example 6.1 Oxygen is bubbled through a liquid, metal so as to oxidise the same. A given extent of reaction is achieved in 1 h with bubble radius of 2 mm. How will the reaction time change if the bubble radius is reduced to 1 mm? Make suitable assumptions and approximations.

Solution
Several parameters are affected when the size is reduced. Assuming that the flow rate is unchanged, some of these are as follows:

a. For a given flow rate, there will be more number for bubbles and, though each bubble is smaller, the total reaction surface area will be larger,
b. Smaller bubbles will have less terminal velocity and, therefore, greater residence time (Eqs. 6.9 and 6.10)
c. The mass transfer coefficient may change with the bubble radius (Eqs. 6.13 and 6.14)

Assuming that (c) is negligible and that the bubbles are spherical we have from Eq. 6.9.

$$U_{\infty} = K \cdot r_{\mathrm{b}}^{2} \text{ i.e. } r_{\mathrm{b}}^{2} = K'/\tau_{\mathrm{r}}$$

where K and K' are constants.

From Eq. 6.25, ignoring the free surface

$$\frac{\tau_{\mathrm{r}}}{r_{\mathrm{b}}} \cdot t = \text{constant}$$

$$\frac{t}{r_{\mathrm{b}}^{3}} = \text{constant}; \frac{t}{1^{3}} = \frac{1}{2^{3}}; \quad t = \frac{60}{8} = 7.5\,\text{min Ans.}$$

This value should be considered as approximate because of the assumptions involved.

6.5 Bubble–Aqueous Solution Interaction

Reduction of metallic ions in solution (Meddings and Mackiw 1964; Wadsworth (1969)

Gaseous reduction of metallic ions is exemplified by the equation

$$\mathrm{M}^{n+} + \frac{n}{2}\mathrm{H}_2 = \mathrm{M} + n\mathrm{H}^{+} \tag{6.26}$$

Precipitation of a metal like Cu is more complicated in the presence of Cu^{2+} and Cu^{+} ions simultaneously, the overall reaction involving several steps. For precipitation of Ni or CO, the situation is simpler. Thus, metals readily reduce only at a solid surface, the reaction thus being strictly heterogeneous. The reduction of the amine complex is written as,

$$\mathrm{M(NH_3)}_x + \mathrm{H}_2 \xrightarrow[\text{surface}]{\text{Solid}} \mathrm{M} + 2\mathrm{NH}_4^{+} + (x-2)\mathrm{NH}_3 \tag{6.27}$$

where x is maintained between 1.9 and 2.2. At a constant hydrogen pressure, the rate of reduction remains independent of the metal concentration up to about 85% reduction. The rate equation may be written as

$$-\frac{d[Ni^{2+}]}{dt} = k \cdot A \cdot p_{H_2} \tag{6.28}$$

where A is surface area.

It is assumed that the surface of the metal powder is covered by a strongly adsorbed layer of nickel ions (or nickel amine ions), and that these are so strongly adsorbed that the number of adsorbed ions begins to decrease appreciably only where more than 85% of the ions originally present are reduced.

6.6 Review Questions

1. Discuss whether the following statements are correct or false.

   a. A given volume of reactant gas reacts better with a liquid when it is dispersed as smaller bubbles.
   b. When a liquid is stirred by gas bubbles, then mixing efficiency is governed more by total volume of gas than by size of bubbles.
   c. During slag–metal reactions, the reaction at the interface is slowed down if the interface is agitated.
   d. An interface between two liquids is a three-dimensional region.
   e. A gas bubble emerging from a nozzle immersed in a liquid usually has a diameter smaller than the diameter of the nozzle.
   f. In liquids, gas bubbles are more likely to ascend in a straight line if the liquid viscosity is lower.
   g. The extent of reaction between gas bubbles and a liquid metal depends on the total volume of gas irrespective of the number of bubbles into which the gas volume is split.

2. The dissolution of a diatomic gas in liquid iron at 1600 °C is controlled by the diffusion of the dissolved gas in the liquid metal. Calculate the rate of dissolution in g/cm^2 s from an atmosphere containing 70% Ar and 30% of the gas into an initially pure iron at 1600 °C. The pressure of gas is 1 atm. The solubility of the gas in liquid iron at 1600 °C and 1 atm partial pressure is 0.045%.

   Data: Diffusion coefficient, D = 0.0001 cm^2/s.

   Eff. boundary layer thickness δ= 0.001 cm
   At. wt. of gas = 14
   Mol wt. of Fe = 55.85
   Density of Fe = 7 g/cm^3.

3. Oxygen bubbles introduce oxygen into a molten metal during their upward passage. It may be possible for a small bubble to introduce more oxygen as compared to a large bubble. Explain how.

4. Gaseous oxygen is injected into the top of an Fe–C melt to decarburize the melt. List the reaction steps and, assuming carbon diffusion to be rate controlling, draw the carbon concentration profile. Show that the decarburization rate is approximately constant in the initial stages and calculate the rate.

   Given:

   Temperature of melt 1550 °C at which for saturation by carbon

$$[\%C] = 1.34 + 2.54 \times 10^{-3} T\,(^\circ C),$$

$$\frac{D}{\delta} = 0.5 \times 10^{-2}\ \text{cm/s}$$

   Melt is saturated by O_2 on injection and FeO precipitates as a separate phase.

$$\text{FeO (slag)} = \text{Fe} + \underline{O} \quad \log K_1 = -\frac{6320}{T} + 2.734,$$

$$\underline{C} + \underline{O} = \text{CO(g)} \quad \frac{A}{V} = 1\ \text{cm}$$

$$\log K_2 = \frac{1168}{T} + 2.07$$

   Henrian activity coefficient of $C = 1.05$
   [Hint: $K_1 = \frac{a_{Fe} \cdot a_{\underline{O}}}{a_{FeO}}\ a_o = 0.1854$

$$K_2 = \frac{p_{CO}}{a_o \cdot a_{\underline{C}}} p_{CO} 1 \therefore a_{\underline{C}}(eq) = .0154 = 1.05 \times C_s$$

$$C_s(\text{surface conc} - \text{eq})\ .01$$

5. How do size and shape of bubbles influence interfacial area and reaction time during interaction of bubbles with a liquid?
6. Liquid steel is being poured into an ingot from a ladle as a smooth stream. The liquid stream sucks in some oxygen from the atmosphere. Derive an expression to relate oxygen intake with

   (a) height and diameter of stream (h and d, respectively),
   (b) pouring rate (W) and
   (c) oxygen content of steel in ladle.

   Assume the following rate-controlling steps

(a) oxygen diffusion inside the liquid stream
(b) oxygen transfer across the gas film surrounding the stream.

   Find the oxygen content in the ingot for the following data.

$W = 1.0$ kg/s.
$d = 0.02$ m
$h = 1.0$ m
D (in Fe liquid) = $10^{-8} m^2/s$.
Oxygen saturation conc. = 0.28wt%
(O in equilibrium Fe—FeO)
Oxygen concentration in ladle = 0.01 wt%

Reference Szekely (1969).

References

Ali, A., Ray, H.S.: J. Appl. Chem. Biotechnol. **24**, 539 (1974)
Ali, A., Ray, H.S., Rao, V.Y.: J. Sci. Ind. Res. **32**(7), 336 (1973, July)
Baird, M.H., Davidson, I.F.: Chem. Eng. Sci. **17**, 87 (1962)
Bandyopadhyay, G.K., Ray, H.S.: Trans. IIM **64** (l) (1972)
Bandyopadhyay, G.K., Ray, H.S.: Met. Trans. **2**, 3055 (1971)
Bownder, B., Kumar, R.: Chem. Engg. Sci. **25**, 25 (1970)
Bradshaw, A.V., Richardson, F.D.: Kinetic aspects of bubble agitated systems. In: Proceedings of the Symposium on Chemical Engineering in Iron and Steel Industry March (1968), Institution of Chemical Engineers (1968)
Calderbank, P.H.: Chem. Eng. (CE 209) (1969)
Collins R.: Chem. Engg. Sci., **22**, 89 (1967)
Crank, J.: Mathematics of Diffusion. Oxford Clarendon Press (Chapter 6) (1956)
Danckwerts, P.V.: Gas-Liquid Reaction. McGraw-Hill, New York (Chapter 5) (1970)
Davenport, W.G., Bradshaw, A.V., Richardson, F.D.: JISI **205**, 1034 (1967)
Davenport, W.G., Bradshaw, A.V., Richardson, R.: J.I.S.I. **205**, 1034 (1967)
Davies, R.M., Tayler, G.: Proc. R. Soc. **200A**, 375 (1950)
Evans, D.J.I., et al.: Can. Min. Met. Bull. 530 (1961, July)
Fitterer, G.R., Arckle, G.F.: Paper presented in Met Soc. AIME annual meeting 1969 (Abstr. J. Met.) 20 (1968)
Froessling, N.: Beitr. Geophys. **52**, 170 (1938)
Hughmark, G.A.: Ind. Eng. Chem. Process Des. Dev. **6**, 218 (1967)
Guthrie, R.I.L., Bradshaw, A.V.: Trans. TMS-AIME **24**, 2285 (1969)
Hammerton, D., Garner, F.H.: Trans. Inst. Chem. Engrs. **32**, 18S (1954)
Khurana, A.K., Kumar, R.: Chem. Engg. Sci. **24**, 1711 (1969)
Large, W.K., Ohji, M., Papamantellos, D., Schenk, H.: Arch. Eisenhuttwes **40**, 99 (1960)
Leibsen, I., et al.: AI.Ch.E. J1. **2**, 296 (1956)
Levich, V.G.: Physicochemical Hydrodynamics. Prentice Hall, Englewood Cliffs (Chapter 8) (1962)
Meddings, B., Mackiw, V.N.: In: Wadsworth, M.E., Davis, F.T. (eds.) Unit Processes in Hydrometallurgy, p. 345. Gordon and Breech, New York, (1964)
Nanda, C.R., Geiger, G.H.: Met. Trans. **2**, 1101 (1971)
Ohji, M., Papamantellos, D., Lange, V.K., Schent, H.: Arch. Eisenhuttwes **41**, 321 (1970)
Padmavathy, P., Kumar, R., Kuloor, N.R.: Ind. J. Tech. **3**, 133 (1969)
Panenis, M., Davenport, N.G.: Trans. TMS-IME **245**, 735 (1969)
Pehlke, R.D., Bement, A.L.: Trans. TMS, AIME **224**, 1237 (1962)

Ramakrishnan, S., Kumar, R., Kuloor, N.R.: Chem. Engg. Sci. **24**, 731 (1969)
Ray, H.S., Verma, R.K.: Trans. IIM **29**(1), 59 (1976)
Richardson, P.D.: Trans. I.S.I. Jpn (First Yukawa Memorial Lecture) **13**, 369 (1973)
Szekely, J.: Trans. TMS-AIME **245**, 341 (1969)
Uno, S., Kinter, R.: A.I.Ch.E. Jl. **2**, 420 (1956)
Wadsworth, M.E.: Trans. TMS-AIME **245**, 1381 (1969)
Whitman, W.G.: Chem. Metall. Eng. **29**, 147 (1923)

Chapter 7
Nucleation and Growth

7.1 Introduction

A vast majority of metallurgical transformations take place by nucleation of stable nuclei of product phase, followed by their and growth. Roasting of sulphides, decomposition of carbonates, etc., all produce new phases. Decarburization of iron carbon melts proceeds by nucleation and growth of CO bubbles. Condensations of vapours, freezing of liquids, melting of solids, precipitation processes are all examples of nucleation and growth processes. The phase transformations that may take place when a single phase alloy α is cooled from a temperature at which it is stable to a temperature at which it becomes unstable are of considerable technological interest. Solidification and allotropic transformation are examples of polymorphous phase transformation, which involves only change of structure. In the other type of phase transformation, the compositions of parent and product phases differ. In most instances, this is associated with changes in structure as well. Only in a minority of reactions like precipitation of Guinier–Preston zones (G. P. zones), e.g. in Al–Cu alloys, the two phases have identical crystal structures and orientations.

There are two or more alternative reactions, for example, one metastable and another stable (i.e. with lower free energy) product phase, then the fastest total reaction takes place and determines the overall kinetics and the product. The initial metastable reaction considerably reduces the driving force for formation of the stable phase; as a result, the metastable product may remain for the useful lifetime.

When nucleation and growth process determine the overall kinetics, the α–t plot follows a variation of the Johnson–Mehl type. An empirical approach for analysis of such kinetic data has been discussed in Chap. 2. Growth of strain-free recrystallized grains on annealing a cold-worked material is an example of a polymorphous phase transition. In this instance, nucleation is not necessary as nuclei of low dislocation densities and correct structure pre-exist in the cold-worked material.

H. S. Ray and S. Ray, *Kinetics of Metallurgical Processes*,
Indian Institute of Metals Series, https://doi.org/10.1007/978-981-13-0686-0_7

Here, kinetics is controlled by growth, and here too, Johnson–Mehl type equations have been successful in describing the kinetics.

This chapter deals primarily with the thermally activated nucleation of product phases, which is prevalent in a majority of the technologically important metallurgical reactions. The mechanism of growth of such phases has been dealt with in other chapters. For the necessary perspective, however, various phase transformations are very briefly recounted in the rest of this section. Details may be found in textbooks and scholarly reviews in edited volumes such as Cahn and Haasen (1996) and Kostorz (2001).

7.1.1 Gibbs' Type I Transformation

Following J. Willard Gibbs, nucleation and growth transformations are classified as Gibbs' type I transformation. The nucleus has a completely different composition and/or structure from the matrix and is separated from it by an interface, which has an interfacial energy. The change in composition and/or structure is large in magnitude, but spatially localized to the region of the matrix that has transformed. The interfacial energy plays a crucial role in the nucleation and growth processes. The reaction proceeds by the movement of this interface. When the parent and product phases differ in composition, growth involves two successive steps:

(i) Long-range diffusional process, usually thermally activated, over many atomic spacings, and
(ii) Uncoordinated thermally activated jumps of atoms across the interface.

Since these are successive steps, the slowest of these two processes will be rate controlling. When the parent and product phases have identical compositions (as in solidification of a pure metal), the first step is not required, but growth is thermally activated. Such phase transformations are called "diffusive", "reconstructive" or "civilian" phase transformations.

Martensitic Phase Transformation

Originally, the name martensite referred to the hard microconstituent found in quenched steels. Many materials other than steel are known to exhibit martensitic phase transformation. In a martensitic (also called "military", "shear" or "displacive") phase transformation, the parent and product phases have identical compositions. Unlike diffusive phase transformations, growth in martensitic transformation takes place by the movement of the parent phase–martensite interface, a coordinated shear type movement of all the atoms in the interface over distances typically less than an interatomic distance. This movement of the interface produces an invariant plane strain relationship, in which one plane (the habit plane) remains undistorted and unrotated. Ideally, the interface is planar, coherent/semi-coherent, and glissile. In CO and its alloys, the martensite–parent interface for the fcc → hcp martensitic transformation is fully coherent; its movement normal to itself accomplishes the required lattice change. In contrast, in most cases including Fe–C alloys, the martensite–parent interface is semi-coherent; its movement is related to strain in a complex way, and in order that it may remain glissile during growth, auxiliary deformation processes must periodically relieve the accumulating strain. This secondary deformation leads to fine scale inhomogeneity in martensite, such as slip, twinning, or faulting observable in transmission electron microscopy, which produces the invariant plane strain condition in the macroscopic scale.

Martensitic transformation is thus equivalent to deformation of the parent crystal lattice into that of the product. In a single phase material, thermally activated grain boundary migration is a reconstructive transformation, while mechanical twinning is a displacive transformation.

On rapid quenching, a parent phase is retained in metastable equilibrium. With increasing undercooling, the driving force for phase transformation increases, and also diffusivity decreases. At sufficiently high undercooling, the driving force may become adequate to supply the distortion energy associated with the martensitic transformation. The temperature at which martensite first forms on cooling (commonly called M_s temperature) varies widely; e.g. M_s is 1200 K for ZrO_2 and <4 K for Fe–34 Ni–0.22 C (wt%) alloy. Since long-range diffusion is not involved, martensite can grow at temperatures below 100 K and at speeds higher than 1000 $m.s^{-1}$. The overall kinetics of martensitic transformation depends on both nucleation and growth, and the slower of the two dominate the overall kinetics. In most cases, the amount of martensite that is obtained depends only on the transformation temperature and not on the time at the temperature; i.e. the transformation is athermal. Slow thermal nucleation may give rise to isothermal transformation characteristics.

7.1.2 Gibbs' Type II Transformation

Gibbs' type II transformation, also called "continuous transformation", can occur only if the initial and final structures share either a common lattice, or, as in liquids and glasses, a lack of lattice. Here, a region of the matrix that is initially of uniform composition develops a composition wave that is small in magnitude, but unlike in a Gibbs' type I transformation, delocalized in space. For such composition fluctuations to be stable, uphill diffusion must be possible. In a binary solution, the thermodynamic condition for uphill diffusion is $\partial^2 G/\partial C^2 < 0$, where G is the free energy for the solution and C the solute content. The point where $\partial^2 G/\partial C^2 = 0$ is called the spinodal point. Any small perturbation in composition can be expressed as a Fourier series of sine or cosine waves with different wavelengths. In absence of any additional restrictions, of the various wavelengths initially present, the shorter wavelengths requiring diffusion over shorter distances should grow faster in amplitude. This would eventually lead to an ordered solid solution. (Note: however that many disorder $\rightarrow$ order transformations take place by the nucleation and growth route.) In spinodal decomposition, wavelengths that grow in amplitude are much larger than interatomic distances. Thus while in continuous ordering, the unlike near neighbour bonds are maximized, and in spinodal decomposition, the like near neighbour bonds are maximized.

From thermodynamics of inhomogeneous solid solutions, it has been shown that in a system that wishes to phase separate, for amplitudes to grow in time, there is a minimum critical wavelength, which represents a balance between thermodynamic driving force and diffusion distance. Although binary equilibrium diagrams indicate only a few potential examples of equilibrium spinodal decomposition of metals, there are many metastable reactions which could occur by the continuous mechanism. Spinodal decomposition is important in formation of ordered coherent precipitates.

7.2 Homogeneous Nucleation

Turnbull and Fisher (1949) first applied the homogeneous nucleation model to solidification of metals. Since then, the concept has become fundamental to the study of nucleation and growth type phase transformations and has achieved remarkable success in at least qualitatively interpreting experimental results for reconstructive phase transformations.

Fundamental to the theory of nucleation is the formation of embryos. Many embryos of the new phase form continuously by thermal fluctuations. Formation of an embryo results in local increase of free energy, associated with the interfacial energy of the product phase. In solid-state phase transformations, additional local increase of free energy may be required because of elastic strain energy:

(i) *Coherency*: For coherent precipitates, elastic strain energy arises from mismatch in lattice parameters of parent and product phases.
(ii) *Dilational (volumetric) strain energy*: When the number of atoms in the product nucleus is the same as the volume of parent matrix, it replaces (e.g. in a polymorphous transformation), dilatational strain energy results from difference in densities of parent and product phases. One example is the nucleation of less dense α–Fe from γ–Fe in pure iron. Dilatational strain energy may also result in when the number of atoms in the product phase is larger than that in the parent phase it replaces. An example is the precipitation of Zn-rich γ-brass from β-brass; since diffusivity of Zn is higher than that of Cu, the inward flux of Zn atoms exceeds the outward flux of Cu atoms, and likewise for precipitation of Fe_3C because of faster diffusion of interstitial C.

The strain energy component would be negligible for nucleations from metallic melts, at least above glass transition temperature. In this case, only interfacial energy needs to be considered.

Of the embryos formed, only those meeting a stability criterion form stable nuclei and the unstable embryos revert to the parent phase by thermal fluctuations. The stable embryos are extremely small and yet contain a few hundreds of atoms. The theory of nucleation essentially consists in deriving this stability criterion.

Types of Nucleation

Nucleation may be homogeneous, with nuclei forming at random locations in the parent phase, or heterogeneous, with nuclei forming at preferred locations like dislocations, grain boundaries, impurity atoms, crevices, container walls.

The energetic condition for heterogeneous nucleation is discussed in the next section. Heterogeneous nucleation is energetically favourable, and homogeneous nucleation is difficult to achieve. A simple example of a homogeneous nucleation could be the solidification of a pure liquid. However, even the so-called pure liquid may contain impurity, promoting heterogeneous nucleation. One way to obviate the problem is to disperse liquid droplets in another liquid medium with much lower freezing point, and then gradually cooling the whole system. Experiments have been carried out by suspending liquid metal droplets in oil, or in very low melting slag. Continuous monitoring of such systems during slow cooling has shown that the droplets solidify at different temperature in groups; a few groups solidify last after large undercooling by 200–300 K. The drops that require the largest undercooling are the ones presumably involved in homogenous nucleation. These experiments confirm that solidification by homogeneous nucleation is more difficult than by heterogeneous nucleation. In practice, solidification starts at a few degrees of undercooling because of heterogeneous nucleation.

7.2.1 Critical Size of Nucleus

Consider homogeneous nucleation of solids from melt, where the elastic strain energy can be ignored and the surface energy can be considered to be isotropic (i.e. independent of local surface orientation of the nucleus) and constant, independent of the nucleus size. The free energy change for the formation of a spherical embryo of radius r is given by

$$\Delta G = \frac{4}{3}\pi r^3 \Delta G_v + 4\pi r^2 \sigma \tag{7.1}$$

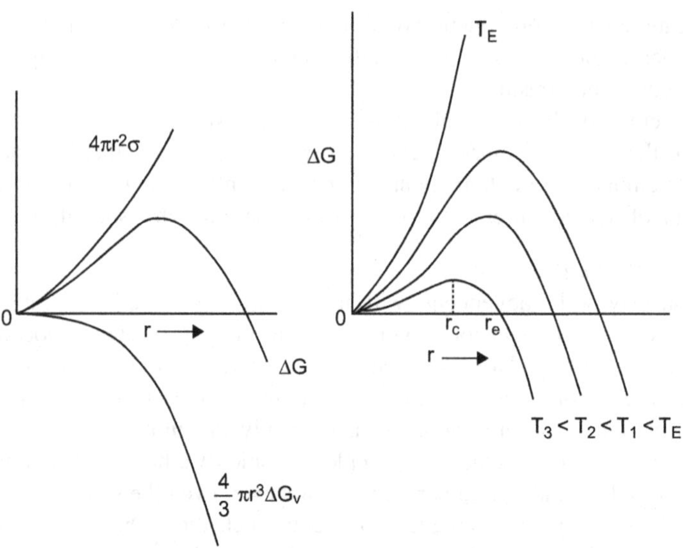

Fig. 7.1 Free energy change during the formation of a nucleus as a function of radius

ΔG_v is the "chemical" free energy change per unit volume, and σ is the surface energy. ΔG_v is zero at the equilibrium transformation temperature T_E, and it becomes increasingly negative as temperature is lowered. In Fig. 7.1, the figure on the left schematically shows the variation of ΔG and its two components with r for a temperature below T_E, and the figure on the right schematically shows the variation of ΔG for different temperatures below T_E; temperature dependence of σ has been ignored compared to that of ΔG_v.

For nucleus to be stable, $\Delta G \leq 0$, that is, $r \geq r_e$ marked in Fig. 1. Let r_c be the size of the embryo corresponding to the maximum in ΔG. Any embryo with size $r \geq r_c$ can reduce the change in free energy by growing in size, while for any embryo with size $r < r_c$, for ΔG to decrease r must also decrease; that is, these embryos are unstable and eventually disappear into the matrix.

The value of r_c is obtained as the solution of:

$$\frac{\partial \Delta G}{\partial r} = 4\pi r^2 \cdot \Delta G_v + 8\pi r\sigma = 0 \tag{7.2}$$

$$\therefore \quad r_c = -2\sigma/\Delta G_v \tag{7.3}$$

Substituting Eq. 7.3 in Eq. 7.1,

$$\Delta G^* \equiv \Delta G(r = r_c) = \frac{16\pi}{3} \cdot \frac{\sigma^3}{(\Delta G_v)^2} \tag{7.4}$$

Clearly, $r_c \rightarrow \infty$ at $T = T_E$, and as Fig. 7.1 schematically shows, r_c and ΔG^* decrease as T is progressively reduced, i.e. with increasing undercooling. If v is the volume of one atom, then the number of atoms in an embryo of radius r is obtained from $nv = 4\pi r^3$. Using Eq. 7.2, the number of atoms in an embryo of size r_c is then obtained as

$$n^* = \frac{32\pi\sigma^3}{3v(\Delta G_v)^3} \tag{7.5}$$

For solid-state transformations, dilatational strain energy can be taken into account by substituting ΔG_v in the equations with $\Delta G_v + \Delta G_v^E$, where ΔG_v^E is the increase in elastic strain energy per unit volume of the precipitate.

7.2.2 *Equation for Undercooling*

At the equilibrium temperature T_E, the free energy change must be zero, i.e.

$$\Delta G_v = \Delta H_v - T_E \cdot \Delta S_v = 0, \quad \text{i.e, } \Delta S_v = \Delta H_v / T_E \tag{7.6}$$

where ΔH_v and ΔS_v are enthalpy and entropy changes per unit volume. We assume that ΔH_v and ΔS_v are independent of temperature, and L is the latent heat. Then at any temperature T

$$\Delta G_v = \Delta H_v - \frac{T \cdot \Delta H_v}{T_E} = \Delta H_v \cdot \frac{T_E - T}{T_E} = L \cdot \frac{\Delta T}{T_E} \tag{7.7}$$

$\Delta T = T_E - T$ is the undercooling. Substituting ΔG_v from Eq. 7.7 into Eqs. 7.3 and 7.4,

$$r_c^* = -2\sigma/\Delta G_v = -2\sigma T_E / L\Delta T \tag{7.8}$$

$$\Delta G^* = \frac{16\pi \cdot \sigma^3 \cdot T_E^2}{3 \cdot L^2 \cdot \Delta T^2} = \frac{K'}{\Delta T^2} \tag{7.9}$$

Equation 7.9 clearly brings out the role of undercooling, as K' is a constant. These equations show that for $\Delta T = 0$, i.e. zero undercooling, $r_c^* = \Delta G^* = \infty$. This clearly shows that solidification is not possible without some degree of undercooling. The assumptions about ΔH_v and ΔS_v being independent of temperature should be acceptable for at least small degrees of undercooling.

Transformation hysteresis
As the derivation above shows, it is impossible for solidification process to be carried out at the equilibrium temperature T_E, and some undercooling is essential. Similar reasoning shows that some superheating must be necessary for carrying out the reverse phase transformation of melting from the solid state. Consequently, phase transformations in the two directions take place at different temperatures giving rise to transformation hysteresis, and the equilibrium transformation temperature lies somewhere in between.

7.2.3 Homogeneous Neucleation Rate

One can calculate the number of nuclei by assuming equilibrium between creation of fresh embryos and their dissolution due to instability. The stable embryos present, i.e. nuclei, are assumed not to affect the equilibrium. Let N_0 be the total number of sites for nucleation; for homogeneous nucleation, this equals the number of atoms. If N_r is the number of embryos with radius r, then in equilibrium,

$$N/N_0 = \exp(-\Delta G/K_B T) \tag{7.10}$$

and for critical size,

$$N_r^* = N_0 \cdot \exp(-\Delta G^*/K_B T) \tag{7.11}$$

A critical embryo becomes stable when it starts growing instead of dissolving. For this to happen, atoms from matrix must jump onto the surface. Let

N_s^* Number of atoms in the parent phase at the surface of embryo,
p Probability of vibration of atom in the direction of embryo,
υ Frequency of vibration of atoms,
U Activation energy for addition of the atom to embryo surface.

Then, the frequency with which critical embryos become stable is given by

$$N_s^* p\upsilon \cdot \exp\left(-U/k_B T\right)$$

Using $P = N_s^* p\upsilon$, and considering that rate of formation of all critical embryos is $\exp(-\Delta G^*/k_B T)$, the rate of formation of nuclei which continue to grow is

$$I = P \cdot \exp(-U/k_B T) \cdot \exp(-\Delta G^*/k_B T) \tag{7.12}$$

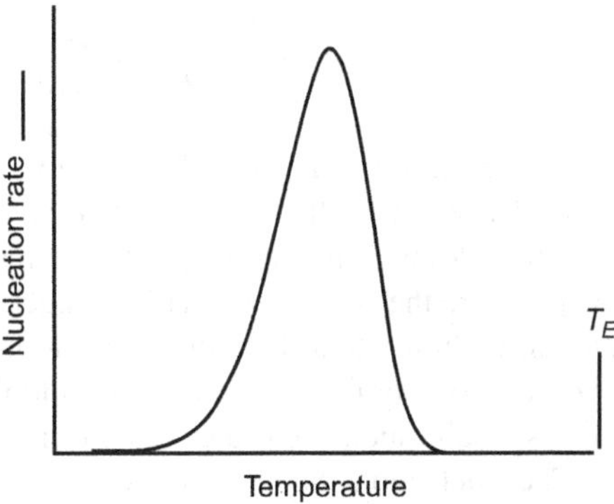

Fig. 7.2 Schematic illustration of the temperature dependence of nucleation rate, Eq. 7.13. The equilibrium transition temperature is indicated

There are two exponential terms in this equation. The first exponential term indicates the kinetic barrier, while the second one may be considered as the thermodynamic barrier, to nucleation. The role of temperature becomes explicit by substituting for ΔG^* (Eq. 7.9):

$$I = P \cdot \exp\left(-\frac{U}{K_B T}\right) \cdot \exp\left(-\frac{K'}{\Delta T^2 \cdot K_B T}\right) \tag{7.13}$$

On the right-hand side, all parameters except T and ΔT are constants. This equation shows that nucleation rate falls to zero for $\Delta T = 0$ (zero undercooling) and also at $T = 0$. It is maximum at some intermediate temperature. The temperature dependence of nucleation rate is schematically shown in Fig. 7.2.

Time for a given extent of transformation

Ignoring temperature dependence of growth of nuclei, the temperature dependence for the time t_α for a fixed extent of transformation α should be inversely proportional to the rate of nucleation. Therefore, using Eq. 7.13, t_α can be expressed as

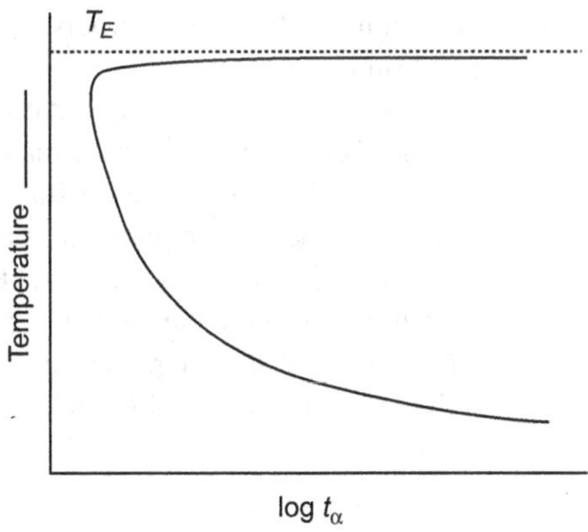

Fig. 7.3 Schematic illustration for the variation of time for a fixed amount of transformation determined from nucleation rate alone, showing the familiar shape of TTT (time–temperature–transformation) curves

$$t_\alpha \propto \frac{1}{I} = \frac{1}{P} \cdot \exp\left(\frac{U}{K_B T}\right) \cdot \exp\left(\frac{K'}{\Delta T^2 \cdot K_B T}\right) \tag{7.14}$$

Figure 7.3 schematically shows the variation. The physical interpretation of this dependence is as follows. At low ΔT values, the critical nucleus size is large, and therefore despite high atomic mobility, few nuclei form. As ΔT increases, initially decrease in the size of the critical nucleus more than compensates decrease in atomic mobility, and therefore, nucleation rate rapidly increases. Eventually however, loss of atomic mobility becomes too severe, and even though critical nucleus size is small, nucleation rate starts decreasing.

The nucleation theory discussed so far is often called the classical theory of homogeneous nucleation. Several simplifying assumptions have been made in deriving the equations:

(i) Bulk property data have been considered for ΔG_V, even though it is doubtful whether this is valid for a small cluster of atoms.
(ii) Each embryo is considered to have an atomically sharp boundary, with a well-defined, isotropic specific surface energy that is independent of curvature. For multicomponent systems, there may be a gradient in chemistry and/or order across the interface. Actually surface energy σ which plays such a critical role is itself not so well defined.
(iii) It has been assumed that atoms may join at all atomic sites on the surface, whereas in some cases, atoms may be able to add to the nucleus only at ledges.
(iv) Mutual interaction of nuclei, for example, loss of some of these by the process of Ostwald ripening, has been ignored.

In spite of these simplifications, this classical theory of homogeneous nucleation has become an important tool in understanding of the kinetics of phase transformation, because of the following reasons:

(i) It does provide qualitative, but straightforward explanation for the increase of nucleation rate by several orders for a small extent of undercooling from the equilibrium temperature T_E.
(ii) It does provide the key to qualitative understanding of the familiar C-shape of Time–Temperature–Transformation (TTT) curves (see Fig. 7.3).
(iii) It also explains why in solid-state precipitation processes it is frequently found that a metastable phase is characterized by good atomic fit with matrix and therefore low σ is the one that forms. The strong dependence of nucleation rate on σ means that a phase with low σ nucleates most rapidly and therefore that precipitates in preference to a phase that may be thermodynamically more stable, but has a higher σ.

7.3 Heterogeneous Nucleation

7.3.1 Energetics of Heterogeneous Nucleation

As indicated in Sect. 7.1, heterogeneous nucleation on preferred sites is energetically more favourable, and therefore by far more common, compared to homogeneous, random nucleation. For example, carbide precipitation during tempering of ferritic steels is found to be associated with dislocations, grain boundaries, etc. While quantitative derivation for each such instance is complex, and beyond the scope of the present chapter, the fundamental reason for this preference for heterogeneous nucleation can be understood by considering the nucleation process in solidification from melt. For this process, inclusion particles, container walls, or parts of equipment in contact with melt may act as preferred sites of nucleation. Considering the free energy change associated with the formation of an embryo, Eq. 7.1, it is clear that the presence of a foreign phase modulates the interface energy so as to facilitate heterogeneous nucleation.

Consider, for example, nucleation from melt in the presence of mould surface (M), schematically shown in Fig. 7.4. In addition to the energy of the solid–liquid interface σ_{SL} considered in homogeneous nucleation (depicted by the symbol σ in the relevant equations), two more surface tension (i.e. surface energy/per unit area) terms must now be considered:

(i) Between parent liquid and the mould, σ_{ML}
(ii) Between solid nucleus and the mould, σ_{MS}

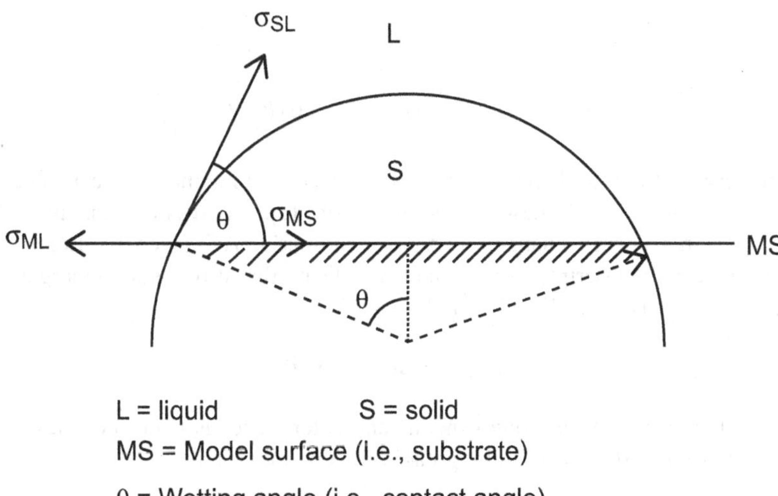

Fig. 7.4 Schematic surface tension forces on a nucleus during heterogeneous nucleation

For a nucleus that is stable, the forces acting on it must be balanced. Therefore, denoting the contact angle (also called wetting angle) as θ, from Fig. 7.4,

$$\sigma_{\mathrm{ML}} = \sigma_{\mathrm{MS}} + \sigma_{\mathrm{SL}} \cdot \cos\theta; \quad \cos\theta = (\sigma_{\mathrm{ML}} - \sigma_{\mathrm{MS}})/\sigma_{\mathrm{SL}} \tag{7.15}$$

Free energy change on nucleation

The total free energy change ΔG for the formation of a nucleus is given by the equation

$$\begin{aligned}\Delta G &= V_{\mathrm{s}} \cdot \Delta G_{\mathrm{v}} + A_{\mathrm{SL}} \cdot \sigma_{\mathrm{SL}} + A_{\mathrm{MS}} \cdot \sigma_{\mathrm{MS}} - A_{\mathrm{MS}} \cdot \sigma_{\mathrm{ML}} \\ &= V_{\mathrm{s}} \cdot \Delta G_{\mathrm{v}} + A_{\mathrm{SL}} \cdot \sigma_{\mathrm{SL}} - A_{\mathrm{MS}} \cdot (\sigma_{\mathrm{ML}} - \sigma_{\mathrm{MS}})\end{aligned} \tag{7.16}$$

Here, V_{s} = volume of the solid, ΔG_{v} = chemical free energy change per unit volume (negative), and A_{SL}, A_{MS} =, respectively, the solid/liquid and mould/solid interface areas.

In Eq. 7.16, the term involving A_{MS} reflects the fact that formation of the nucleus results in replacing A_{MS} area of mould/liquid interface with the same area of mould/solid interface. Substituting Eq. 7.15 in the above,

$$\Delta G = V_{\mathrm{s}} \cdot \Delta G_{\mathrm{v}} + \sigma_{\mathrm{SL}} \cdot (A_{\mathrm{SL}} - A_{\mathrm{MS}} \cos\theta) \tag{7.17}$$

If the nucleus has a radius r, then

$$A_{\mathrm{MS}} = \pi r^2 \sin\theta; \quad A_{\mathrm{SL}} = 2\pi r^2 \cdot (1 - \cos\theta)$$

Substituting these in Eq. 7.17,

$$\begin{aligned}\Delta G &= \left[\frac{4}{3}\pi r^3 \Delta G_{\mathrm{v}} + 4\pi r^2 \sigma_{\mathrm{SL}}\right] \cdot S(\theta) \\ S(\theta) &= (2 + \cos\theta) \cdot (1 - \cos\theta)^2 \Big/ 4\end{aligned} \tag{7.18}$$

The term $S(\theta)$ is a shape factor, and the term within the square brackets is exactly equal to that of homogeneous nucleation, Eq. 7.1. It is convenient to designate this free energy change by ΔG_{hom}. Similarly, ΔG given by Eq. 7.17 is denoted with the subscript het to explicitly indicate that it refers to heterogeneous nucleation. That is, from Eqs. 7.1 and 7.18,

$$\Delta G_{\mathrm{het}} = \Delta G_{\mathrm{hom}} \cdot S(\theta) \tag{7.19}$$

The difference between homogeneous and heterogeneous nucleation lies in the shape factor term $S(\theta)$. The limiting cases for this factor are:

(i) When $\theta = 180°$, i.e. the nucleus is detached from mould surface and nucleation is homogeneous, $S(\theta) = 1$ and $\Delta G_{\mathrm{het}} = \Delta G_{\mathrm{hom}}$ as expected.

(ii) When $\theta = 0°$, i.e. the liquid completely wets the mould surface, $S(\theta) = 0$ and $\Delta G_{het} = 0$ as expected.

For heterogeneous nucleation then,

$$180° > \theta > 0°, 1 > S(\theta) > 0$$

This shows that compared to homogeneous nucleation, free energy change is less for heterogeneous nucleation. Heterogeneous nucleation is, therefore, more prevalent.

7.3.2 Critical Nucleus Size

The variation of ΔG_{het} with r is similar to that of ΔG_{hom}, except for the constant multiplicative factor $S(\theta)$, as shown schematically in Fig. 7.5.

The critical nucleus size is obtained by setting $\partial \Delta G_{het}/\partial r = 0$ as

$$r_c = -2\sigma_{SL}/\Delta G_v \tag{7.20}$$

that is, exactly equal to that for homogeneous nucleation. The corresponding value for free energy change is obtained by substituting this value in Eq. 7.19 as

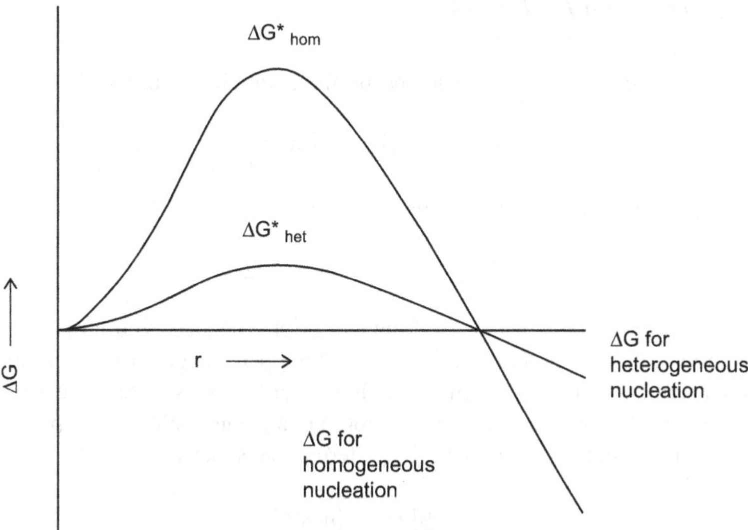

Fig. 7.5 ΔG versus r plots for homogeneous and heterogeneous nucleations

$$\Delta G_{het}^{*} = \frac{16\pi\sigma_{SL}^{3}}{3\Delta G_{v}^{2}} \cdot S(\theta) = \Delta G_{hom}^{*} \cdot S(\theta) \tag{7.21}$$

Since $\Delta G_{het}^{*} < \Delta G_{hom}^{*}$, following the derivation of Eq. 7.9, one can readily show that ΔT for heterogeneous nucleation is much lower than that for homogeneous nucleation.

7.3.3 Other Factors that Aid Nucleation

We have seen that nucleation is aided by undercooling, and by the presence of a foreign substrate that promotes heterogeneous nucleation. As Eqs. 7.20 and 7.21 show, nucleation can be aided by additional factors, which reduce ΔG_v (recall that ΔG_v is negative) and/or reduce σ_{SL}. For example, a gas or vapour bubble may be made to nucleate and grow by reduction in pressure. Nucleation and growth of CO bubbles in molten steel can be induced by dissolution of more oxygen in the bath. During deoxidation of molten steel, oxides are made to precipitate by adding deoxidizing metals such as Al, Fe–Si, V, which lower the oxygen potential.

7.4 Examples

7.4.1 Deoxidation of Steel

Consider a molten salt being deoxidized by the addition of Mn and Si.

$$\underline{\mathrm{Mn}} + \underline{\mathrm{Si}} + 3\underline{\mathrm{O}} = \mathrm{MnSiO_3} \tag{7.22}$$

Since activity of deoxidation product is unity, one can write

$$K_e = [\%\ \underline{\mathrm{Mn}} \times \%\underline{\mathrm{Si}} \times \%\underline{\mathrm{O}}^3]_e \tag{7.23}$$

K_e is an equilibrium constant, and the subscript e denotes equilibrium.

Equation 7.23 implies that if the product of the percentage terms at equilibrium exceeds a critical value K_e, then the oxide precipitates as a separate phase. In practice, however, one needs supersaturation, i.e. a greater value of the product K_s, at supersaturated state. The ratio K_s/K_e is termed as supersaturation

$$\frac{K_s}{K_e} = \frac{[\%\underline{\mathrm{Mn}} \times \%\underline{\mathrm{Si}} \times \%\underline{\mathrm{O}}^3]_s}{[\%\underline{\mathrm{Mn}} \times \%\underline{\mathrm{Si}} \times \%\underline{\mathrm{O}}^3]_e} \tag{7.24}$$

For $MnSiO_3$, the supersaturation is calculated to be between 500 and 4000 for homogeneous nucleation, whereas in practice, it is around 10–100. Thus, it has been said (Turkdogan 1973; Von Bogdandy 1965) that nucleation is heterogeneous during deoxidation by silicomanganese. Homogeneous nucleation, however, may be possible with strong deoxidizers such as Al,Zn,Ti since these are capable of imparting very high supersaturation ($K_s/K_e \approx 10^{-3}$ to 10^{-8}).

If Al, Si are used together, then perhaps Al_2O_3 precipitates first and then SiO_2 precipitates on it.

Oxide inclusions precipitate when the deoxidizer is added to the metal. In the case of most deoxidizers, the precipitation reaction is complete within a very short period and a very large number of nuclei are formed. For example, during silicon deoxidation, the number of nuclei particles reaches 3×10^7 to 10^8 per cc within a few seconds, the largest particles being of the order of 2–4 μm. Calculations have shown that after 1 s, less than 0.001% of the original oxygen remains to precipitate. Perhaps the other deoxidizers behave in the same manner. Subsequently, the oxide particles grow and the larger particles tend to float up and separate more rapidly. For each type of oxide, the growth determines the sight of the final inclusions as well as the total inclusion content of the steel. The exact mechanism is steel controversial, and it is likely that a number of mechanisms are operative at different stages of the growth. This is schematically shown in Fig. 7.6.

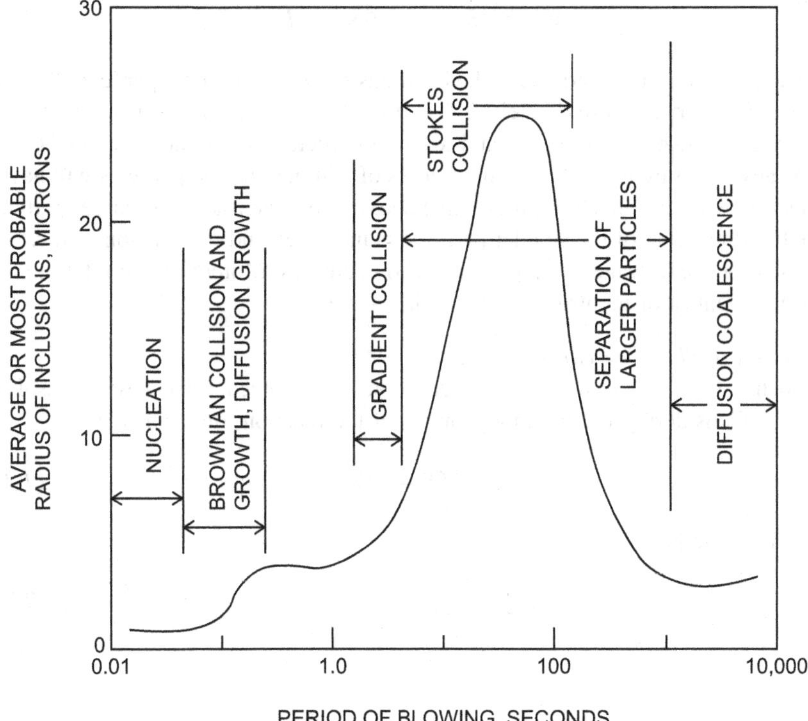

Fig. 7.6 Various mechanisms of growth of deoxidation product particles (Ray et al. 1970)

It is seen that the average inclusion size increases for about 2 min. Thereafter, a considerable proportion of the larger particles is removed, and the average size of the inclusions decreases with time. The growth phenomenon is perhaps less important here because the deoxidation products preferentially precipitate on the FeO inclusions already existing in the oxidized metal. The results of many studies mostly concern the condition on the right-hand side of the peak where escape of the larger particles is the predominant process. According to Stoke's law, the rate of rise of a suspended particle in a liquid is given by

$$V = \frac{2gr^2(d_1 - d_2)}{9\eta} \tag{7.25}$$

Here,

V The rising velocity in cm/sec,
g The acceleration due to gravity in cm/sec^2,
r The radius of the particle in cm,
d_1 The density of the liquid in g/cm^3,
d_2 The density of the suspended particle in g/cm^3 and
η The viscosity of liquid in dynes/sec sq cm or in poises.

In the case of liquid steel, this can be reduced to

$$V = 8.38 + 10^{-3}(6.94 - d)\,r^2 \tag{7.26}$$

The only important variable is thus r. It has been shown that a particle 30 μm in diameter will rise approximately 1.5 m in 30 min, while a particle of 300 μm diameter will rise 1.5 m in 21 s and a 1 mm particle will rise at the rate of 6 m/s.

At any given moment during the process of elimination, the particles will have a given size distribution. With progressive deoxidation, the fraction of larger particles is reduced and that of the smaller particles is increased and the fraction of the most probable or the average-sized particles also decreases. Moreover, the total number and the quantity of inclusion particles also decrease.

Oxidation of dissolved carbon

There has been much work on the reaction of dissolved carbon with dissolved oxygen. Considering the thermodynamics of the reaction, one can write

$$\underline{C} + \underline{O} = CO\ (g) \tag{7.27}$$

for which one has

$$\%\underline{C} + \%\underline{O} = K' p_{CO} = K \tag{7.28}$$

where constant K is a function of temperature and pressure. This equation shows that CO should generate as a separate phase if the product of concentrations of dissolved carbon and oxygen exceeds a minimum value set by thermodynamics. For actual steelmaking practice, however, there is supersaturation and the value is higher, the deviation from thermodynamic value being governed by a number of factors.

Attempts have been made to study homogeneous nucleation of CO by subjecting levitated Fe–C alloy droplets in an oxidizing gas (Distin et al. 1968; Kaplan and Philbrook 1960). At low concentration of carbon, outward diffusion of carbon is slow and carbon is mostly oxidized by inward diffusion of oxygen from the environment. CO nucleates inside the levitated droplets, and growing CO bubbles may cause the droplets to swell and burst.

A levitated droplet, however, does not necessarily imply homogeneous nucleation. Distin et al. (1968) postulated that nucleation occurs in liquid iron oxide entrapped in the melt. Kaplan and Philbrook (1960) ruled out either of the preceding postulates. According to them, cavities swept into the levitated droplet from the surface serve as nuclei for CO bubble formation.

For ordinary melts, the container walls allow heterogeneous nucleation. The slag–metal interface also serves the same purpose. Others[1] showed that when iron was melted in glazed silica crucibles, at ordinary pressures, C–O reaction requires a supersaturation of ~ 15 before CO can nucleate and evolve. (Thermodynamically, this corresponds to CO evolution at $p_{CO} \approx 15$ atm.) However, the supersaturation required is drastically reduced if the crucible wall is scratched. Pores and cavities, containing entrapped gasses, function as permanent gas pockets, and therefore, nucleation is not required at all.

Driving force for supersaturation

It has been mentioned before that CO will separate as a separate phase only if the product of concentrations of dissolved oxygen exceeds a value set by thermodynamics of Eq. 7.27. The driving force for CO nucleation is then, obviously, provided by the supersaturation, which can be expressed in terms of excess p_{CO}, ΔP.

A CO bubble inside a melt must be at an excess pressure in order to maintain mechanical equilibrium. The excess pressure is given by the equation

$$\Delta P = \frac{2\sigma}{r} \tag{7.29}$$

where r is the radius, and σ the gas/liquid surface tension. For nucleation at constant temperature

$$dG = -p dv \tag{7.30}$$

[1]Loc. Cit, Ref 6.

For critical nucleus size in homogeneous nucleation, from Eq. 7.3,

$$r_C = -2\sigma/\Delta G_V = 2\sigma/\Delta P \tag{7.31}$$

where ΔP is the equivalent excess pressure for supersaturation required for nucleation. From Eq. 7.4

$$\Delta G^*_{hom} = \frac{16}{3}\pi\frac{\sigma^3}{(\Delta P)^2} \tag{7.32}$$

Assuming r_C for CO nucleus to be 0.1 μm and σ about 1500 ergs/cm^2, ΔP has been calculated to be about 3×10^3 atm. It is impossible to have such a high degree of supersaturation. This gives an additional reason for the support of heterogeneous nucleation.

7.4.2 Segregation Roasting of Copper Concentrate (Dutta 1991)

If cupric oxide is heated to around 700 °C with small amount of NaCl and charcoal, then metallic copper deposits on charcoal particles. This is basis of TORCO (treatment of refractory copper ores) process, which has been specially designed to extract copper concentrates from oxidized copper ores, such as aluminosilicates. Normal leaching and floatation methods are unsuitable for such ores. In this process, the comminuted ore with 60% as −50 μm and 2–6% Cu is first fixed with 0.5–2% coke or coal and 0.5–1% NaCl and then heated in a segregation reactor to a temperature of 700–800 °C for 30–60 min. When the process is completed, most of the copper gets precipitated on the carbon surface and the product consists of copper-coated carbon particles. The segregates and the gangue are separated from each other by rotation. The copper is finally separated in a reverberatory smelter with 85–90% recovery.

It has been suggested that the overall reaction involves a series of intermediate reactions such as the following (Wright 1973).

$$\begin{aligned}
&2NaCl + SiO_2 + H_2O = Na_2SiO_3 + 2HCl \\
&3CuO + 3HCl + 1.5H_2 = Cu_3Cl_3(g) + 3H_2O \\
&1.5C + 3H_2O = 1.5CO_2 + 3H_2 \\
&Cu_3Cl_3(g) + 1.5H_2 = 3Cu + 3HCl \\
&2CuO + CO = Cu_2O + CO_2 \\
&Cu_2O + 2HCl = \frac{2}{3}Cu_3Cl_3\ (g) + H_2O \quad \text{etc.}
\end{aligned}$$

There are obviously several gas–solid reactions and gas phase reactions with individual steps. Sodium chloride is hydrolysed in the presence of water vapour to produce HCl gas, which reacts with cuprous and cupric oxides to provide Cu_3Cl_3 gas. Carbon–moisture reaction generates hydrogen, which reduces this chloride and regenerates HCl, the reduced copper depositing on carbon surface by nucleation and growth.

Dutta (1991) found that the system continues to produce CO_2 which can be used as an index for the progress of the reaction. The total CO_2 generated at different times at three different temperatures in a laboratory experiment is shown in Fig. 7.7. The total copper recovery was also measured after some randomly selected experiments. It was also found that recovery was almost complete at 750 °C after 75 min. Taking this point as $\alpha = 1$, the curves are replotted as α–t plots in Fig. 7.8. This figure also shows actual metal recovery. The plots are Johnson–Mehl type and can be further analysed using equation discussed in Chap. 2. In the present case, the data can be shown to follow the equation.

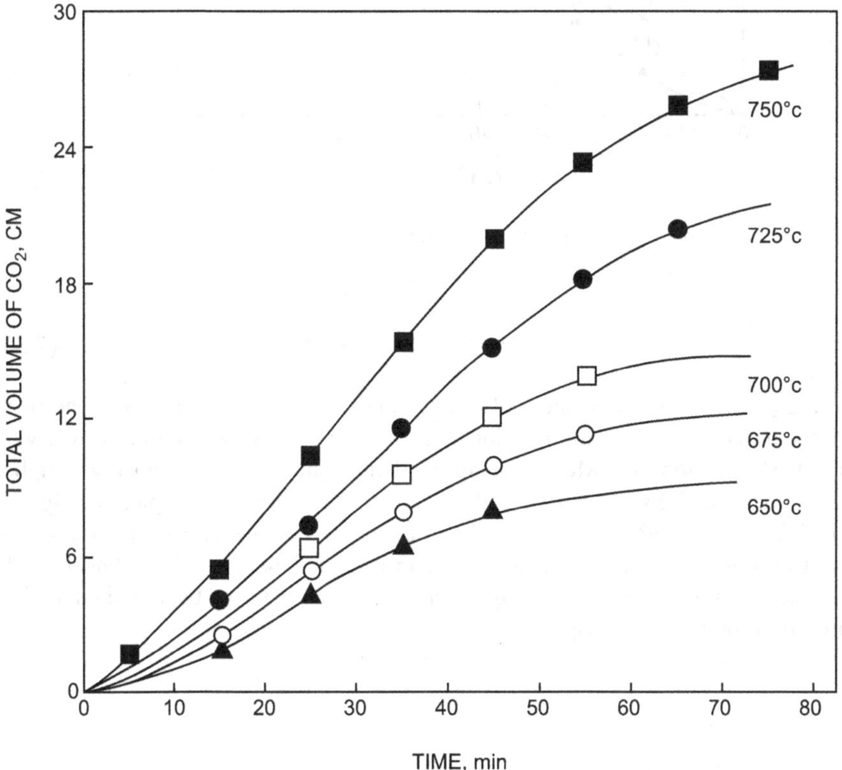

Fig. 7.7 Total volumes of CO_2 generated at different temperatures (Dutta 1991)

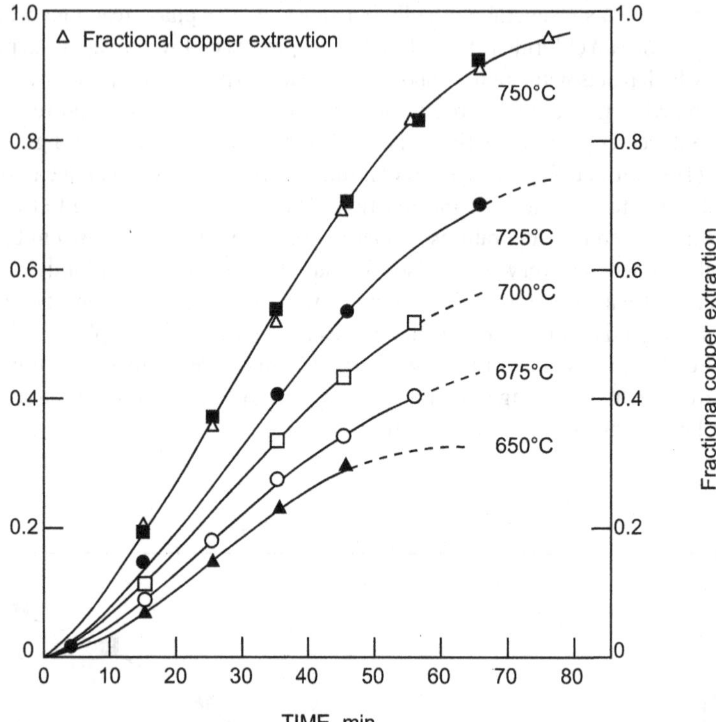

Fig. 7.8 Kinetic plot for copper extraction (Dutta 1991)

$$[-\ln(1-\alpha)]^{1/1.64} = kt \tag{7.33}$$

Figure 7.9 shows that linear plots are obtained using this equation. One may, therefore, conclude that the reaction is controlled by nucleation and growth with $n = 1.64$ in Johnson–Mehl equation. This value of n probably implies two-dimensional growth, i.e. spread of copper on a surface. The plots in Fig. 7.9 yield for the activation energy a value of about 66 kJ/mole. However, as has been discussed in Chap. 2, this is not the correct value because the Johnson–Mehl equation is not a true kinetic equation. Equation 7.33 can be modified by the procedure outlined in Chap. 2.

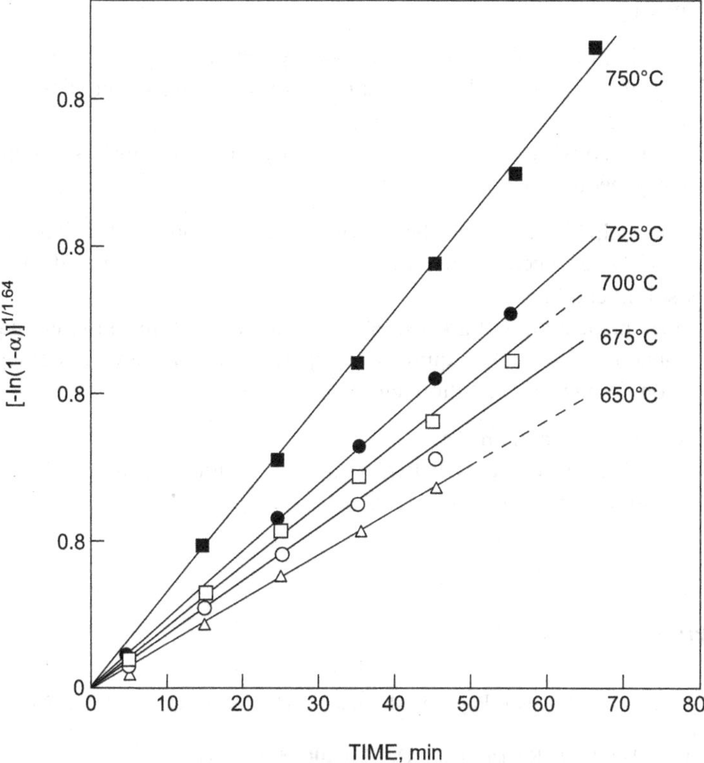

Fig. 7.9 Kinetic plots according to unmodified Johnson–Mehl equation (integrated form) (Dutta 1991)

7.5 Review Questions

1. Discuss whether the following statements are correct or false.

    a. The maximum size of solid embryos in liquid increases with the increase in undercooling.
    b. The critical size formed depends on the free energy changes involved and not the transformation temperature.
    c. At a particular temperature, particles larger than the critical nucleus are stable.
    d. There is no way one can study homogeneous nucleation during solidification of a liquid metal.
    e. For solidification, there must be an incubation period whatever be the supercooling.

2. Explain why

   a. Formation of a stable nucleus necessarily requires supercooling.
   b. The probability of dendrite formation increases with supercooling and rate of quenching.
   c. Homogeneous nucleation leads to more supercooling as compared to heterogeneous nucleation.

3. With the help of classical theory of nucleation prove that heterogeneous solidification can occur with a much less supercooling compared to homogeneous solidification.
4. Why does nucleation and growth phenomena lead to Johnson–Mehl equation?
5. Heterogeneous nucleation requires less supercooling compared to homogeneous nucleation. Which of the following explains this?

   a. The driving force is more.
   b. Specific interface energy between solid and liquid is less.
   c. Activation energy barrier is less.

References

Cahn, R.W., Haasen, P. (eds.): Physical Metallurgy, 4th edn. Elsevier Science, North Holland (1996)

Distin, P.A., Hallet, G.D., Richardson, F.D.: JISI **206**, 821 (1968)

Dutta, P.S.: Segregation roasting of copper and nickel, Ph.D. Thesis, IIT Kharagpur (1991)

Kaplan, R.S., Philbrook, W.O.: Trans. TMS AIME **245**, 2195 (1960)

Kostorz, G. (ed.): Phase Transformation in Materials. Wiley-VCH, Weinheim (2001)

Ray, H.S., Gupta, R.K., Bandyopadhyay, G.: Ind. J. Tech. **8**, 63 (1970)

Turkdogan, E.T.: Chemical Metallurgy of Iron and Steel, p. 153. Iron and Steel Institute, London (1973)

Turnbull, D., Fisher, J.C.: J. Chem. Phys. **17**, 71 (1949)

Von Bogdandy, L.: In: Elliott, J.F., Meadoweroft, T.R. (eds.) Steelmaking: The Chipman Conference, p. 156. The MIT Press, Cambridge, Massachusetts (1965)

Wright, J.K.: Miner. Sci. Eng. **5**, 119 (1973)

Chapter 8
Non-ideal Conditions and Complex Reactions

8.1 Introduction

8.1.1 Real Systems and Non-ideal Conditions

So far we have discussed various rate-controlling processes and rate equations under some ideal conditions. For example, gas–solid reactions have been discussed with reference to single spherical particles and unchanging single reaction mechanisms, whereas under real conditions, the conditions prevailing can be beset with many uncertainties and a mechanistic approach in developing kinetic models is much more difficult. Laboratory kinetic studies, carried out under controlled conditions, therefore, are often not relevant for understanding real systems.

Consider, for example, the reactions of iron oxide reduction. From a practical point of view, the large number of laboratory kinetic studies reported on iron ore reduction has had very little impact on the design and operation of actual reactors (Barner et al. 1963; Szekely and Evans 1972). This is so, because numerous factors make interpretation of laboratory data and their application rather difficult.

8.1.2 Factors Which Lead to Complications

Some of the factors which limit the relevance of the laboratory studies are as follows:

(a) *Presence of impurities*: Laboratory studies are generally carried out using pure Fe_2O_3 and carbon or single reducing gas, e.g. H_2 or CO. Real systems contain in varying quantities, considerable impurities. In fact, there can be vastly different reaction rate constants even for materials coming from adjacent ore bodies. Moreover, reduction reaction often involves a quaternary gas mixture, namely $CO–CO_2–H_2O$, in the diffusional field.

H. S. Ray and S. Ray, *Kinetics of Metallurgical Processes*,
Indian Institute of Metals Series, https://doi.org/10.1007/978-981-13-0686-0_8

(b) *Prediction of packed bed reaction rates from single sphere experiments*: Such experiments can often be in error. For example, consider decomposition of $CaCO_3$. While small single particles behave as if the reaction were controlled by a chemical step at the reaction interface, the reaction of a packed bed is controlled by heat and mass transfer (Hills 1968).

(c) *Multiparticle nature of systems*: Iron ore reduction involves multiparticle polydisperse system for which the kinetic laws may fall to hold (Herbst 1979) because of size variation among particles. The size distribution can also change during reaction due to cracking of particles (Smith and Maddocks 1969; Dollimore 1978; Hills 1978; Turkdogan 1978) or sintering of particles (Bagshaw et al. 1969). The latter may be significant enough to show a decrease in reaction rate with increase in temperature (Szekely and Evans 1970, 1971).

(d) *Non-isothermal conditions*: The vast majority of real systems involve non-isothermal conditions and, therefore, the results of the traditional isothermal kinetic experiments are unrealistic. Temperature variation during reaction may also give rise to variation in gas composition and flow rate.

(e) *Variations of pore structure*: Various workers have confirmed that gas diffusion within the porous reduction product, iron or iron oxides, plays an increasingly dominant role in controlling reaction rate (Ghosh 1979; Hills 1978). Therefore, the physical characteristics of the pores and the changes that occur in the solid during the reaction influence the course of the reaction appreciably.

During reduction of ore pellets by coal char in a rotary kiln simulator, the product was found to contain uniform distribution of metallized iron particles whose size increased with time and temperature. The effect of temperature was more drastic. Coalescence of particles let to fewer, but larger pores. As a combination of pore closure, plastic deformation of the outer iron shell and other factors the rate of reduction was found to decrease beyond 1075 °C (Morrison et al. 1978).

It should be noted that both the topochemical model and the solid diffusion model envisage decrease of the reaction rate with increase in the particle size. However, in real systems, there is an optimum value of particle size for which reaction rate is maximum. This is so because conversion is expected to increase with increase in flow rate of reducing gas, providing that the gas leaves the bed at equilibrium, and hence with increase in particle size. Conversely, for sufficiently large particles the flow rate will be high but the surface area available for reaction will limit the reduction rate. The optimum size will be somewhere in between (Barner et al. 1963; Szekely and Evans 1970, 1971). In fact, it is best to say that there exists an optimum combination of grain size, porosity and reaction temperature which would give the fastest overall reaction rate (Szekely and Evans 1970, 1971).

The reaction scheme can be complicated. For example, consider iron oxide reduction by carbon. Carbon reduction occurs through gaseous intermediates CO and CO_2. Rao (1979) has described the reaction scheme in terms of the following stages:

Initiation:

$$C(s) + \frac{1}{2}O_2(g) \rightarrow CO(g) \tag{8.1}$$

$$C(s) + Fe_xO_y(s) \rightarrow Fe_xO_{y-1} + CO(g) \tag{8.2}$$

Propagation:

$$Fe_xO_y + CO(g) \rightarrow Fe_xO_{y-1} + CO_2(g) \tag{8.3}$$

$$C(s) + CO_2(g) \rightarrow 2CO(g) \tag{8.4}$$

Termination:

$$\text{CO (g, interior)} \rightarrow \text{CO (g, inert gas phase)} \tag{8.5}$$

$$\text{CO}_2\text{(g, interior)} \rightarrow \text{CO}_2\text{(g, inert gas phase)} \tag{8.6}$$

where $x = 1$, 2 or 3 when $y = 1$, 3 or 4.

In addition, CO may also be formed by true direct reduction occurring at the points of contact between carbon and oxide particles. CO thus produced readily reacts with hematite particles.

According to Rao (1979), the rates of reduction of Fe_2O_3, Fe_3O_4 and FeO at moderately high temperatures (~1000 °C) are much greater than the rate of the Boudouard or solution loss reaction and the overall process becomes limited by the availability of CO gas according to this reaction. This, however, is not universally accepted and mass transfer of gaseous CO and CO_2 through the porous sample may also be rate controlling.

Overall extent of reaction for a porous mass

Consider gaseous reduction of a porous spherical pellet of hematite grains. As the reducing gas penetrates, its reduction potential falls and there will be no reduction beyond a certain degree of penetration at a given moment. Of course, the penetration depth will increase with time. At any given instant, the individual grains of hematite will be more reduced the nearer they are to the outer surface. In other words, the local extent or degree of reduction α' varies, from a maximum value at the outer surface to a zero value somewhere in the interior, provided the pellet diameter is sufficiently large.

Although the degree of reduction of individual grains at any given extent varies with the location, one can express an overall average value for the whole pellet. The overall extent of reaction may be expressed as a volume average as

$$\alpha = \frac{3}{R_0^3}\int_0^{R_0} R^2\left(1 - \frac{r^3}{r_s^3}\right) dR \tag{8.7}$$

where R is the distance of penetration, R_0 the radius of pellet, r is the radius of reaction front within a single grain and r_s the radius of grains, all assumed to be of uniform size.

The quantity $1 - \left(r^3/r_s^3\right)$ expresses the local extent of reaction, α'. The weighted average α is not a function of R but, of course, a function of t. The dependence can be expressed by a relationship of the type

$$g(\alpha) = kt \tag{8.8}$$

where $g(\alpha)$ would be an appropriate function of α and k, the rate constant. Equation 8.8, the kinetic law for isothermal reduction of the entire pellet as a whole, will not remain valid if pore structure and particle size vary. It is not easy to incorporate these variations into the kinetic law. However, it is possible to account for variation of temperature.

It should be remembered that there can be no theoretical estimation of rate of a reduction and actual experiments must be carried out. It may be useful to carry out tests under conditions simulated to be close to actual situations envisaged for the reactor. However, apprehensions about kinetic investigation with real systems are common. Yet kinetic laws, which are required for design of reactors, can be established only through work on real systems.

Additional discussion on some of the factors that lead to non-ideality is given later.

8.2 Different Kinetic Laws for Similar Processes

8.2.1 Introduction

Previously, it has been shown that a gas–solid reaction can be controlled by anyone of several possible rate-controlling steps. This is true even if one considers more specific examples, e.g. oxidation of metals, and leaching of sulphide minerals. Not only is the oxidation behaviour of one metal different from that of another, the reaction for the same metal may overlap and the rate equation may be combination of two or more laws. Some of these complications will now be discussed with reference to some specific examples.

8.2.2 Oxidation of Metals

Methods of investigating the growth of oxidation layers include determinations of the changes in thickness of the scale, in weight (strictly, mass) of the metal sample, or in volume of the surrounding gas. Most quantitative data have been obtained as changes in weight (Δm) per unit surface area.

Kinetic theory is concerned primarily with finding relationships between oxidation and time. A number of relationships have been discovered empirically and are to be listed here, in terms of weight increases Δm and time t.

The linear relationship,

$$\Delta m = k_l \cdot t \tag{8.9}$$

with k_l a constant, is the simplest equation that is found to express some experimental data.

The parabolic relationship

$$(\Delta m)^2 = k_p \cdot t \tag{8.10}$$

with k_p a constant represents a straight line when $(\Delta m)^2$ is plotted against t. If this line does not intersect the origin of the coordinate axes, the more general form of the parabolic equation,

$$(\Delta m)^2 = k_p \cdot t + C \tag{8.11}$$

with k_p, C constants, may be applicable.

Experimental values have occasionally been found to agree with a cubic relationship:

$$(\Delta m)^3 = k_c \cdot t. \tag{8.12}$$

And finally, there is the more general logarithmic (or exponential) relationship written as

$$\Delta m = k_e \cdot \log(a \cdot t + t_0) \tag{8.13}$$

which contains three constants: k_e, a and t_0. There is also the inverse logarithmic relationship:

$$1/\Delta m = A - k_{il} \log t \tag{8.14}$$

where A and k_{il} are constants.

Combinations of two or more of these relationships in a single oxidation–time curve are also quite common. A metal or alloy may, for instance, start to oxidize parabolically and then continue linearly. This relationship has been termed paralinear.

Figure 8.1 gives curves representing the various relationships, and these have been drawn in such a manner as to intersect at one point. It is seen that these equations cover a considerable number of possibilities which are further increased by the fact that some of the equations contain more than one constant which may vary within wide limits. If we take into account the fact that the experimental values

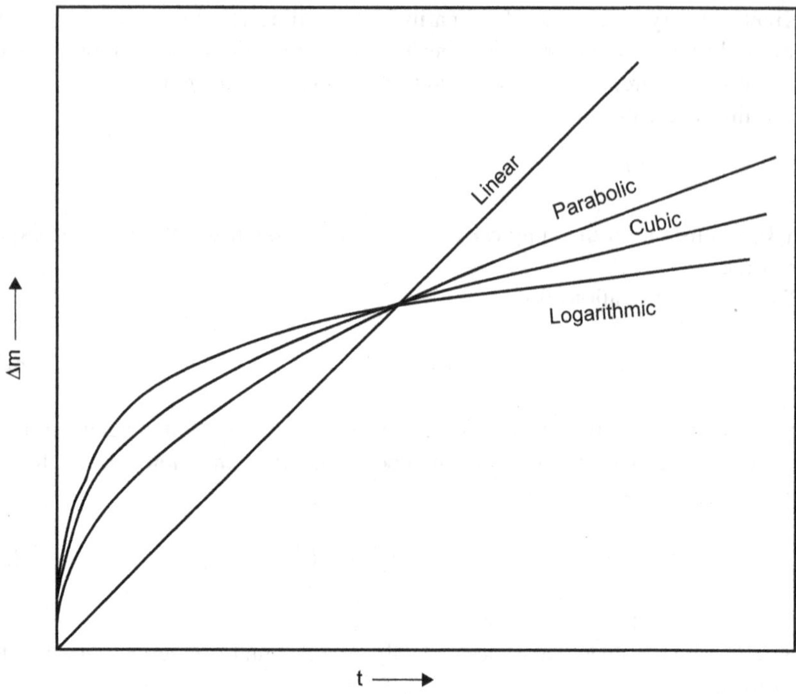

Fig. 8.1 Oxidation–time relationships

rarely agree very accurately with anyone curve, it can be concluded that one of the equations above should always be capable of expressing any regular progresses of oxidation observed experimentally. This would restrict the usefulness of the relationships considerably unless it was possible to attach a physical meaning to the respective equations.

Equations 8.9–8.14 can also be expressed in terms of thickness (y) of the surface oxide or decrease in pressure (Δp) of the gas in place of increase in weight per unit surface area (Δm). As a rule, one may convert these quantities one to another. If the gas volume (V) and the composition of the layer are accurately known, the decrease in pressure is calculated from the increase in weight by the ideal gas law.

The type of time relationship that can be applied to a given metal or alloy depends largely on the thickness of the film already formed, that is, on time and temperature, Table 8.1 shows, for a number of metals, the time relationships against temperature that have been observed experimentally.

It is seen that at low temperatures, that is, for thin oxidation films, the logarithmic and inverse logarithmic relationships prevail. The oxidation versus time curves may even appear to be asymptotic. Experimental work has shown that many metals when exposed to oxygen at room temperature oxidize rapidly at the beginning, but after a few minutes the oxidation rate drops to low or negligible values, a stable film of thickness 20–50 Å being formed.

Table 8.1 Oxidation–time relationships observed at various temperatures in air or oxygen for a number of metal (Hills 1968)

Temp °C	100	200	300	400	500	600	700	800	900	1000	1100	1200
Mg	*log*		*par*	*paralin*	*lin*							
Ca	*(log)*		*par*	*lin*	*lin*							
Ce	*log log-lin*	*lin accel*										
Th			*par*	*lin*	*lin*							
U	*par*	*paralin*	*lin-acc*									
Ti			*log*	*cubic*	*cubic*	*paralin*			*paralin*			
Zr		*log*	*cubic*		*cubic*				*cubic*	*cubic-lin*		
Nb		*par*	*par*		*paralin*		*lin*		*lin*		*accel*	*asym.*
Ta	*log inv. log*			*par*	*paralin*		*lin*		*lin*			*delayed*
Mo			*par*	*paralin*	*paralin*		*lin*		*lin*			
W				*par*	*par*		*paralin*		*paralin*		*paralin*	
Fe	*log*	*log*	*par par*		*par*		*par*		*par*		*par*	
Ni		*log*	*log*	*cubic*	*par*				*par*		*par*	
Cu	*log cubic*	*(par)*			*par*	*par*		*par*				
Zn		*log*	*log*	*par*	*par*							
Al	*log inv. log*	*(log)*	*par*		*(asym.)(lin)*							
Ge				*par*	*paralin*							

The literature on oxidation of metals is too vast to be reviewed briefly. We can only consider some specific examples. In some cases, the kinetic law is simple. In air or oxygen, oxidation of lead follows a simple parabolic law—up to the melting point (Bagshaw et al. 1969).

In oxygen, zinc oxidizes following a logarithmic law or a parabolic law (200–390 °C). Under high pressures, it may follow an equation of the form $\alpha = kt^n$ with n = 0.29–0.46 (300–400 °C, 133 mbar). Oxidation of solid nickel also follows a parabolic law at high temperatures. Oxidation of copper follows different laws at different temperatures because of change of reaction mechanism. Sponge iron samples (DRI) oxidize at low temperatures according to the logarithmic law. But some experiments by Bandyopadhyay (1988) indicate that the kinetic switches over to the first-order law, $-\ln(1-\alpha) = kt$ at elevated temperatures.

Liquid metals

Oxidation of liquid nonferrous metals in air or oxygen has been reviewed by Drouzy and Mascre (1969). They also summarized the effect of other solute metals on oxidation rate. Some of their observations are as follows.

Liquid cadmium (400–550 °C) oxidizes following the parabolic law ($E \sim 9$ kcal/mole). While 1% Pb, Bi or Cu has no effect on oxidation, only 0.05% Zn prevents oxidation completely, Zn oxidizing to ZnO according to parabolic law. Liquid zinc usually ignites at around 830 °C, but in some cases, ignition can be delayed until 900 °C. A zinc wire can be melted in its oxide sheath. Oxidation rate is practically equal in air or oxygen.

Oxidation of liquid lead in the range 500–700 °C can be described in the initial periods by a parabolic law; thereafter, a logarithmic law applies. The time for transition from the first to the second law decreases with increasing temperature. Small additions of Ag increase resistance to oxidation, whereas even small amounts ($\sim 1\%$) of Cd, Bi or Sn accelerate oxidation.

Oxidation of liquid copper is more complicated. There can be three situations, namely (a) dissolved oxygen is below the saturation limit, (b) dissolved oxygen exceeds this limit and the oxide formed is liquid and (c) the oxide formed is solid. In (a) and (b), the oxidation rate is the same and it depends on replenishment of oxygen by diffusion in air and experimental factors. In (c), the rate is lower because it is controlled by diffusion through solid oxide layer. Between 1100 and 1200 °C, the rate of weight gain per unit area is approximately constant in all cases, and the activation energy being around 20 kcal/mole.

Oxidation of zinc vapour is relevant to zinc metallurgy. It has been discussed by Clarke (1979) who highlighted its implications in the imperial smelting furnace. In the imperial smelting process, the zinc vapours are passed into a lead splash condenser, immediately after they leave the furnace. In the condenser, both the temperature of vapours and their zinc content are progressively reduced. Lowering of temperature reverses the reaction.

$$ZnO(s) + CO(g) = Zn(g) + CO_2(g) \tag{8.15}$$

Injection of preheated air at the top of the furnace results in exothermic conversion of part of CO to CO_2 and raising of gas temperature from 800 to 1050 °C. At the same time, the reoxidation temperature for the new gas mixture is raised to only 950 °C, allowing a permissible 100 °C cooling of the furnace gases before reoxidation occurs. As 95% of the zinc leaving the furnace charge is condensed, zinc adsorption into lead droplets far outweighs zinc reoxidation but still, according to Clarke, the problem is worth studying. Clarke's experimental studies indicated that, in the range 600–900 °C, oxidation of zinc in CO–CO_2–Ar mixtures was heterogeneous, requiring an existing surface on which to react. Under the reaction conditions investigated, two distinct reaction regimes were observed, the reaction rate in both cases being controlled by surface chemical reactions as opposed to gas phase diffusion. Below 800 °C, the rate was controlled by prevailing excess zinc partial pressure.

8.2.3 *Roasting of Sulphides*

Roasting of galena, represented by the equation

$$PbS + 2O_2 = PbSO_4 \tag{8.16}$$

is said (Khallafalla 1979) to follow the kinetic equation

$$1 + 2(1 - \phi\alpha) - 3(1 - \phi\alpha)^{2/3} = \frac{k}{4r_0^2}t \tag{8.17}$$

where ϕ represents the volume of $PbSO_4$ per unit volume of PbS. Gold marking experiments have shown that PbS diffuses from inside through $PbSO_4$ layer to react on the surface. If Eq. 8.17 is divided all-throughout by 3, then it assumes the form similar to the well-known Ginstling–Brounshtein product layer diffusion equation. Here, the reaction is sensitive to particle size. On the other hand, as has been said earlier, sulphation of ocean manganese nodules is independent of particle size and the rate law been given by the first-order equation. Leaching of the completely sulphated nodules in boiling water provides Ni, Cu and Co extraction above 80%. Much of the manganese but little of the iron is also extracted.

According to one source (Khallafalla 1979), oxidation of ZnS pellets (0.4–1.6 cm) in the temperature range 740–1040 °C is controlled by gaseous transport through the product layer of ZnO. For suspensions of individual ZnS particles in a fluidized bed reactor in an oxygen–nitrogen mixture, the kinetics is controlled by a surface reaction at the ZnS–ZnO interface. The reaction is

$$\mathrm{ZnS} + \frac{3}{2}\mathrm{O_2} = \mathrm{ZnO} + \mathrm{SO_2},\ \Delta H = -106\,\mathrm{kcal/mole\ ZnS}(800\text{–}1000°\mathrm{C}) \tag{8.18}$$

During reaction, the rate of diffusion across gas film enveloping the particle, that through the ZnO product layer, and the rate of reaction at the ZnS/ZnO interface must be equal. Fukunaka et al. (1976) express this as follows

$$\frac{\mathrm{d}n_{\mathrm{O_2}}}{\mathrm{d}t} = 4\pi r_0^2 k_\mathrm{g}(C - C_\mathrm{s}) = \frac{4\pi r_0 r_\mathrm{i}}{r_0 - r_\mathrm{i}} \cdot D(C_\mathrm{s} - C_\mathrm{i}) = 4\pi r_\mathrm{i}^2 k_\mathrm{r} C_\mathrm{i} \tag{8.19}$$

where

C = oxygen concentration in bulk gas
C_s = oxygen concentration at particle surface
k_g = gas film mass transfer coefficient
C_i = oxygen concentration at reaction interface
r_i = radius at reaction interface
r_0 = radius of particles
D = effective diffusivity of oxygen in ZnO shell
k_r = rate constant for interfacial reaction.

The overall rate equation is expressed as

$$\frac{\mathrm{d}n_{\mathrm{O_2}}}{\mathrm{d}t} = 4\pi r_0^2 \cdot k \cdot C \tag{8.20}$$

where k is an overall constant.

The oxidation of zinc sulphide is highly exothermic, and the temperature changes can introduce considerable uncertainty. In order to minimize the temperature rise, Fukunaka et al. (1976) in their work on batch-type fluidized bed reactor, mixed less than 1 wt% ZnS particles with fused alumina particles and the mixed particles were fluidized. A high activation energy ($\sim$75 kcal/mol) justified the postulate of chemical reaction control. Non-isothermal reaction of ZnS has also been described in the literature.

According to Rao and Abraham (1971), the oxidation of Cu_2S involves diffusion through an outer boundary layer, diffusion through the layer formed during a non-isothermal period, plus diffusion through a product layer formed during continuing reaction. The apparent activation energy after an initial period is low ($\sim$6 kcal mol) which supports diffusion mechanism. The reactions may be represented as

$$Cu_2S + 1.5O_2 = Cu_2O + SO_2 \tag{8.21}$$

$$Cu_2O + 0.5O_2 = 2CuO \tag{8.22}$$

The fact that Cu_2O layers are not built up to other than interfacial boundary layers of constant thickness is evidenced by the fact that the overall weight remains constant during the course of the reaction (Rao and Abraham 1971).

During such gas–solid reactions, the possibility of gas film diffusion being rate controlled is eliminated using high gas velocities. Thus, Ammann and Loose (1971) investigated oxidation of molybdenite (MoS_2) powders using thin layers and high gas velocities. The kinetics follows the simple chemical control equation.

Chlorination of minerals

Chlorination reactions are conveniently used in many beneficiation reactions (Ammann and Looso 1971). Chlorination of chalcocite (Cu_2S) is said to proceed sequentially as follows (Luthra et al. 1973)

$$Cu_2S \rightarrow CuS \rightarrow CuCl \rightarrow CuCl_2$$

The kinetics observed between 100 and 220 °C has been found (Bandyopadhyay 1988) to fit a cubic rate law (Ingraham and Parsons 1969)

$$\alpha^3 = k \cdot t \tag{8.23}$$

which indicates a rate lower than the parabolic law. Perhaps complication arises because of reaction of reaction of CuS with $CuCl_2$; thus,

$$CuCl_2 + CuS = 2CuCl + \frac{1}{2}S_2 \tag{8.24}$$

This reaction will lower the rate of formation of $CuCl_2$.

8.3 More Examples of Complex Processes

8.3.1 *Leaching of Sulphides*

A very large number of kinetic studies on the leaching of sulphide minerals have been reported in the literature. A few examples are considered.

Acid leaching of chalcopyrite ($CuFeS_2$) in an autoclave (160 °C) has been found to follow the chemical control equation

$$1 - (1 - \alpha)^{1/3} = kt \tag{8.25}$$

The reaction is

$$CuFeS_2 + \frac{17}{4}O_2 + \frac{1}{2}H_2SO_4 = CuSO_4 + \frac{1}{2}Fe_2(SO_4)_3 + \frac{1}{2}H_2O \tag{8.26}$$

Equation 8.25 implies, as discussed in Chap. 3, that the reaction boundary moves at a constant velocity, the rate constant increase with increasing p_{O_2} and also that the reaction rate, at any given instant, is proportional to the reaction interface area. It should be noted that similar kinetics will be observed for diffusion through a limiting boundary layer of solution of constant thickness δ at the receding mineral interface. Therefore, it is not possible from kinetics alone to determine the exact mechanism. The effect of temperature on the observed kinetics must be evaluated to distinguish between the two. The following derivation shows how a diffusion mechanism can also lead to the phase boundary control equation. We have (V_s = solution volume)

$$\frac{dn(CuFeS_2)}{dt} = \frac{4}{17}\frac{dn(O_2)}{dt} = -\frac{dn(Cu)}{dt} = -\frac{1}{V_s}\frac{d(Cu^{2+})}{dt} \tag{8.27}$$

If diffusion through a limiting boundary film, δ is rate controlling, then

$$\frac{dn}{dt} = -\frac{4\pi r^2 DC}{\sigma\delta} \tag{8.28}$$

where C is concentration of bulk solution assume to be much greater than C_s and σ is a stoichiometric factor which represents the number of moles of the diffusing species required for each mole of metal value released by the reaction.

Leaching of finely ground chalcopyrite by ferric sulphate, on the other hand, is controlled by a diffusion mechanism (Dutrizac et al. 1970). The reaction is written as

$$CuFeS_2 + 4Fe^{3+} = Cu^{2+} + 5Fe^{2+} + 2S^0$$

The kinetic equation follows the Ginstling–Brounshtein equation because the sulphur which forms is said to adhere to the mineral surface forming a tightly bonded diffusion layer explaining the exceptionally low reaction rate for the ferric sulphate leaching processes. Figure 8.2 shows some results taken from the literature (Dutrizac et al. 1970).

Pure oxide minerals which leach without forming products of reaction should follow a simple surface reaction rate control or diffusion through a limiting boundary film. Acid leaching of chrysocolla, however, follows it more complicated kinetics (Wadsworth 1979). The reaction is

$$CuO \cdot SiO_2 \cdot 2H_2O + 2H^+ = Cu^{2+} + SiO_2 \cdot nH_2O + (3-n)H_2O \tag{8.29}$$

The kinetic equation contains shrinking core model–surface reaction plus diffusion terms (β, γ are constants)

$$\left[1 - \frac{2}{3}\alpha - (1-\alpha)^{2/3}\right] + \frac{\beta}{r_0}\left[1 - (1-\alpha)^{1/3}\right] = \frac{\gamma[H^+]}{r_0^2}t \tag{8.30}$$

During leaching, the remaining silica lattice provides a diffusion barrier with a moving reaction boundary.

Similar rate equations have been found in many other leaching studies. Such mixed control kinetics is discussed in the next section.

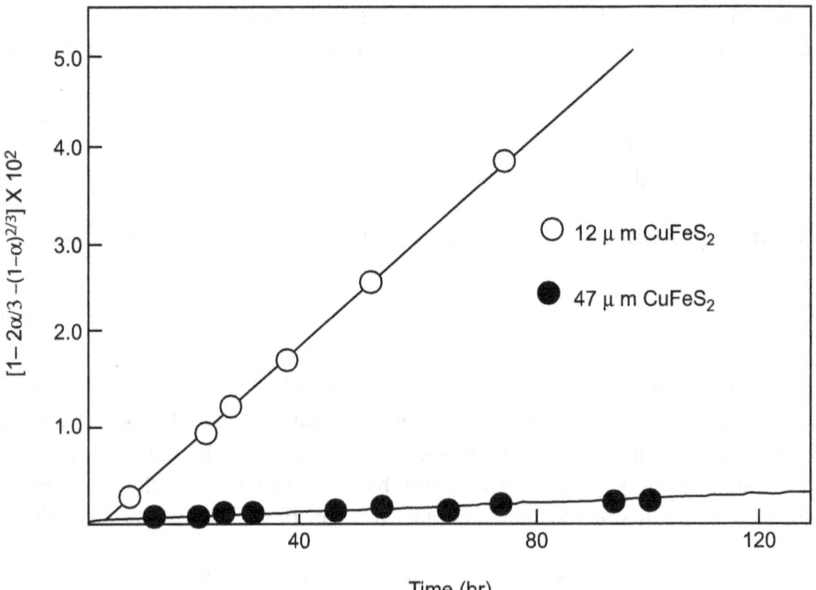

Fig. 8.2 Plot of $1 - 2\alpha/3 - (1-\alpha)^{2/3}$ versus t for monosize chalcopyrite particles (Drouzy and Mascre 1969)

Dutrizac and coworkers (Dutrizac et al. 1970) have reported kinetic studies on leaching of several synthetic minerals. They have shown that the leaching of bornite ($CuFeS_4$) below 35 °C follows a parabolic law, leaching being controlled by diffusion through a non-stoichiometric product layer ($Cu_{5-x}FeS_4$). About 40 °C, this compound reacts further forming chalcopyrite and elemental sulphur and the reaction is now controlled by aqueous diffusion through a liquid boundary layer adjacent to mineral surface. Obviously while stirring will accelerate leaching in the second case, it will not be effective in the first case. For synthetic cubanite ($CuFe_2S_3$), the kinetics is said to be linear following the overall reaction

$$CuFe_2S_3 + 3Fe_2(SO_4)_3 = CuSO_4 + 8FeSO_4 + 3S^0 \tag{8.31}$$

The activation energy in this case ($\sim$12 kcal/mol) is about twice that of the former diffusion-controlled processes.

Dissolution of metal sulphides is often governed by electrochemical mechanisms, and this has been recognized by many investigators. We can think of the overall reaction to be a summation of an anodic reaction and a cathodic reaction.

For example, anodic dissolution of ZnS may be represented as

$$ZnS = Zn^{2+} + S^0 + 2e^- \tag{8.32}$$

The cathodic reduction involves reduction of chemisorbed oxygen whose presence may vary the number of charge carriers within the zinc sulphide due to its semiconductor properties.

$$2e^- + 2H^+ + \frac{1}{2}O_2 = H_2O \tag{8.33}$$

The anodic dissolution of chalcopyrite in the presence of ferric ion is explained on the basis of diffusion of ferric iron through the deposited sulphur film resulting from the dissolution of chalcopyrite.

$$CuFeS_2 = Cu^{2+} + Fe^{2+} + 2S^0 + 4e^- \tag{8.34}$$

The depletion of ferric ion in solution is said to result in the built up of a sulphur layer of thickness which may be assumed to be directly proportional to Δn, and the amount of copper in solution. A kinetic equation based on this assumption is

$$\frac{\Delta n^2}{1 - \sigma\Delta n} - \sigma^{1/2}\left[\frac{1}{1 - \sigma\Delta n} + 4.606\,\log(1 - \sigma\Delta n) - (1 - \sigma\Delta n)\right] = kt \tag{8.35}$$

where σ is a stoichiometric factor.

Reaction of chalcocite is said to take place in two stages, and the reactions being represented as

$$\text{I.} \quad Cu_2S = CuS + Cu^{2+} + 2e^- \tag{8.36}$$

$$\text{II.} \quad CuS = Cu^{2+} + S^0 + 2e^- \tag{8.37}$$

Figure 8.3 illustrates the process.

Aqueous oxidation of PbS has been described by Eddington and Prosser (1979) to follow the electrochemical model

$$PbS = Pb^{2+} + S^0 + 2e^- \tag{8.38}$$

$$O_2 + 2H_2O + 4e^- = 4OH^- \tag{8.39}$$

It is generally believed although electrochemical reactions are important in the sequence of steps, charge transfer probably is not involved in the rate-controlling process (Wadsworth 1979; Dutrizac et al. 1970). However, preferential galvanic attack can occur when one mineral is in contact with another mineral or metal. Thus, pyrite (FeS_2) can accelerate the anodic dissolution of galena (PbS); PbS becomes the sacrificial anode. $CuFeS_2$ can react cathodically and produce Cu_2S if it comes in contact with metals like Cu, Fe, Pb, Zn having more negative potential.

$$2CuFeS_2 + 6H^+ + 2e^2 = Cu_2S + 2Fe^{2+} + 3H_2S \tag{8.40}$$

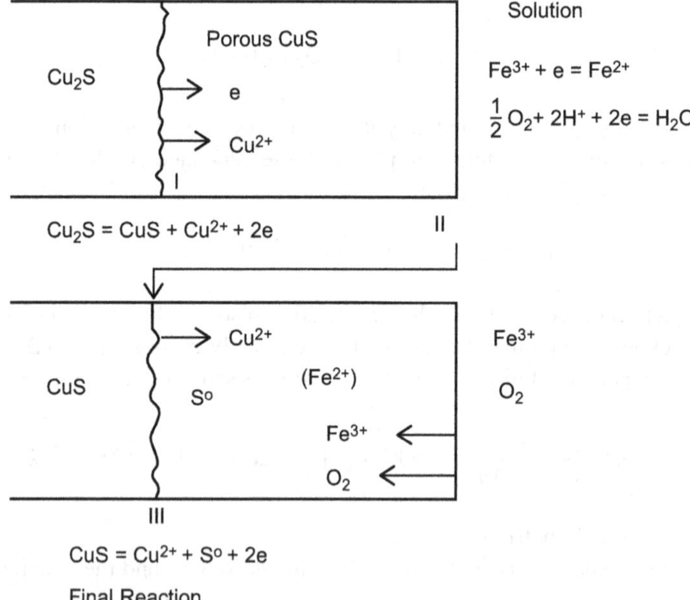

Fig. 8.3 Surface layers formed during anodic dissolution of CuS and Cu_2S

The anode in this case is induced to react anodically. For copper, the reaction is

$$2Cu + H_2S = Cu_2S + 2H^+ + 2e^- \tag{8.41}$$

8.3.2 *Ammonia Leaching of Copper Sulphides*

We now discuss in more detail one particular leaching process, viz. ammonia leaching of copper sulphides. Ammonia leaching is industrially important not only for extracting copper values from sulphide ores but also for extraction of several metals from complex sulphide (Reilly and Scotty 1977; 1978; 1984; Warren and Wadsworth 1984; Beckstead and Miller 1977a, b).

The basic electrochemical mechanism of leaching has been described by Eqs. 8.38 and 8.39. The first step in the anodic reaction is thus formation of elemental sulphur. This sulphur is subsequently oxidized to different ionic species. Thus, acid leaching of chalcopyrite dissolves copper as $CuSO_4$. During ammonia leaching, sulphate ions are produced.

However, significant quantities of elemental sulphur may form which can be recovered. Thus, during leaching of covellite (CuS) up to 60% of the sulphur can be recovered as elemental sulphur while achieving high degrees of solubilization of the covellite. However, the recovery is made possible by the presence of a non-miscible organic sulphur solvent (Reilly and Scott 1978).

Leaching of CuS (Reilly and Scott 1978)

In the presence of organic sulphur, solvent leaching of copper has been found to relate closely to sulphur recovery and the ratio of copper dissolved in aqueous phase to sulphur dissolved in organic phase remaining constant. This suggests that oxidation of sulphur is prevented. Since there is no protective coating of sulphur on the mineral surface, the interfacial chemical control equation will prevail for either species.

The reaction steps according to the electrochemical leaching model can be written as:

$$\text{a.}\quad CuS \rightarrow Cu^{2+} + S^0 + 2e^- \tag{8.42}$$

$$\text{b.}\quad Cu^{2+} + 4NH_3 \rightarrow Cu(NH_3)_4^{2+} \tag{8.43}$$

$$\text{c.}\quad \frac{1}{2}O_2 + H_2O + 2e^- \rightarrow 2OH^- \tag{8.44}$$

$$\text{d.}\quad 2NH_4^+ + 2e^- + \frac{1}{2}O_2 \rightarrow 2NH_3 + H_2O \tag{8.45}$$

$$\text{e.}\quad 2S^0 + 2OH^- + O_2 \rightarrow S_2O_3^{2-} + H_2O \tag{8.46}$$

$$\text{f.}\quad S_2O_3^{2-} + 2OH^- + 2O_2 \rightarrow 2SO_4^{2-} + H_2O \tag{8.47}$$

This mechanism allows for an effect of ammonia and oxygen on copper dissolution rates and for a dependence on the hydroxyl ion concentration of the oxidation rate of elemental sulphur, if there be no organic solvent. Apparently, the effect of oxygen partial pressure on the fractional recovery is insignificant which means that the leaching rate is not subject to rate limitations by oxygen mass transfer processes. Dissolution of Cu and precipitation of S increase with ammonia concentration.

Leaching of chalcopyrite

Although elemental sulphur has been shown to exist as a transient species, during leaching sulphur is generally oxidized to various ionic species. Even then small quantities of sulphur may remain. The leaching reaction is again said to involve an electrochemical mechanism. There is agreement that the cathodic step is the reduction of oxygen on the mineral surface.

$$O_2 + 2H_2O + 4e^- = 4OH^- \tag{8.48}$$

There is, however, disagreement about the anode reaction. According to one model (Beckstead and Miller 1977a, b), the reaction is

$$CuFeS_2 + 19OH^- = Cu^{2+} + \frac{1}{2}Fe_2O_3 + 2SO_4^{2-} + \frac{19}{2}H_2O + 17e^- \tag{8.49}$$

According to another model (Warren and Wadsworth 1984), the reaction is

$$CuFeS_2 + 4NH_3 + 9OH^- = Cu(NH_3)_4^{2+} + Fe(OH)_3 + S_2O_3^{2-} + 3H_2O + 9e^- \tag{8.50}$$

However, neither of these account for formation of elemental sulphur. Thus, a third model (Reilly and Scott 1984) has been proposed which is written as

$$CuFeS_2 + 4NH_3 + 6OH^- = Cu(NH_3)_4^{2+} + \frac{1}{2}S_2O_3^{2-} + S + Fe(OH)_3 + \frac{3}{2}H_2O + 7e^- \tag{8.51}$$

$$\frac{d\alpha}{dt} = \frac{127f}{d_0}(OH^-)^{1/2}\left(\frac{K_1 p_{O_2}}{1 + K_2 p_{O_2}}\right)^{1/2} \times \left[k_1 + k_2 + (Cu^{2+})_0 + k_2' \alpha\right]^{1/2}(1-\alpha)^{2/3} \tag{8.52}$$

where

d_0 = particle diameter
f = shape factor (actual area/area of sphere of equivalent volume)
(OH^-) = hydroxyl ion concentration
$(Cu^{2+})_0$ = initial concentration of cupric ion
p_{O_2} = partial pressure of oxygen
and, K_1, K_2, k_1, k_2 and k_2' are various constants.

For large values of $(Cu^{2+})_0$ and constant leaching conditions, Eq. 8.52 reduces to

$$\frac{d\alpha}{dt} \propto (1-\alpha)^{2/3} \tag{8.53}$$

which is identical to the equation for phase boundary control process. For small values of p_{O_2}, $K_2 p_{O_2} \ll 1$, and therefore the rate becomes proportional to $\sqrt{p_{O_2}}$, whereas for large values of p_{O_2} (>1.25 atm) where $K_2 p_{O_2} \gg 1$, $d\alpha/dt$ becomes nearly independent of p_{O_2}. The rate is inversely proportional to particle diameter and significantly sensitive to temperature (E ~10 kcal/mol) while ammonia concentration appears to have little effect on rate, the hydroxyl ion concentration is important.

In most cases, a plot of $1-(1-\alpha)^{1/3}$ versus t is linear. However, often α appears to have a positive value at $t = 0$, i.e. the line does not pass through zero. Backstead and Miller (Beckstead and Miller 1977a, b) have written the overall reaction as follows.

$$CuFeS_2 + 4NH_3 + \frac{17}{4}O_2 + 2OH^- = Cu(NH_3)_4^{2+} + \frac{1}{2}Fe_2O_3 + 2SO_4^{2-} + H_2O \tag{8.54}$$

They have shown that for dilute solids concentration, the rate is controlled by an electrochemical surface reaction. Even though stirring speed influences the rate of the electrochemical reaction, this effect is due to changes in the morphology of the hematite deposit which alters the surface reaction kinetics, rather than being indicative of mass transfer limitation. Under conditions of low stirring speeds and low oxygen pressure, the hematite reaction product often virtually stops. Thus, iron-free sulphides, e.g. chalcocite (Cu_2S), covellite (CuS), are more readily leached at lower temperatures and pressures. The same is true for bornite (Cu_5FeS_4) which presumably does not produce sufficient amounts of Fe_2O_3 because of the low iron content.

Leaching in ammonia medium has the advantage that only metals such as copper, nickel and cobalt dissolve readily and the medium has negligible solubility for iron, manganese and other elements. Ammoniacal leach liquors are also ideally suited for treatment by solvent extraction for recovery of copper values. However, large quantities of ammonium sulphate produced present a disposal problem.

Kinetics

For ammonia, leaching of chalcopyrite the liquid–solid mass transfer step is not considered rate controlling (Reilly and Scott 1977). In case of monosize particles in an intensely stirred reactor under moderate pressures and dilute solid-phase concentration, the kinetics of reaction represented by Eq. 8.54 is controlled by a catalytic electrochemical surface reaction (Beckstead and Miller 1977a, b). Often the data can be filled into the phase boundary-controlled reaction model. However, the plot of $1-(1-\alpha)^{1/3}$ versus t may not pass through the origin. This is presumably due, at least in part, to both the presence of small quantities of mineralogical forms of copper sulphides other than chalcopyrite (i.e. bornite) which dissolve more rapidly and presence of small amounts of fines not removed in the screening process (Reilly and Scott 1977).

8.3.3 Deoxidation of Steel

The presence of oxygen is known to be detrimental to the mechanical and corrosion resistant properties of steels. The oxygen content must be controlled to within reasonable limits using a suitable deoxidation process. The generally employed deoxidation processes may be classified under two heads, namely the diffusion deoxidation and the precipitation deoxidation. In the former, the oxygen in metal is diffused out, and in the latter, it is precipitated out as an insoluble oxide. Notable technological developments in diffusion deoxidation include vacuum degassing. In the precipitation deoxidation method, simple or complex deoxidizers are added to the molten metal. A portion of the oxygen in steel is combined as stable oxides, which partly float up to the surface of the bath and pass into the slag, the balance remaining in steel.

The precipitation deoxidation method is by far the most popular method of deoxidation, and a large number of investigations have been concerned with the solubility of oxygen in liquid iron in the presence of various deoxidizers. Based on these data, methods have been developed for calculating the oxygen content of commercially melted steels. In a number of cases, the kinetics of such reactions has also been studied (Szekely and Evans 1970, 1971).

It is not only important to reduce the oxygen level using a deoxidizer of a suitable strength but also to ensure that the deoxidation products, the stable oxides, are removed from the bulk of the metal to give a clean product. Thus, graphite, although it has limited affinity for oxygen, may be highly recommended in some cases. Pneumatic injection of graphite powder has been employed to produce very clean deoxidized steel. Some investigations have been directed at the kinetics of the formation and growth of the solid deoxidation products (Rao 1979; Kubaschewski and Hopkins 1953; Bagshaw et al. 1969) and the effective removal of these non-metallic inclusions from the metal bath (Ray et al. 1970, 1972; Bandyopadhyay et al. 1971).

A study of any deoxidation process necessitates the analysis of steel for oxygen. The conventional method for oxygen determination is the vacuum fusion method. The analysis, however, becomes unreliable at very low values of oxygen. Oxygen probes based on stabilized zirconia solid electrolyte have also been developed for rapid determination of oxygen (Gitterer 1967; Subbarao 1980).

It is interesting to note that, during the process of deoxidation, not only the oxygen level decreases but some other factors such as the size and density of inclusions, the carbon content, the type of inclusion, the hardness of the deoxidized solidified steel, etc., may also be affected (Sims 1954). These can be used as indirect kinetic parameters.

Ray and coworkers (1970, 1972; Bandyopadhyay et al. 1971) described a study in which they first oxidized a steel by oxygen injection and then deoxidized the steel using various deoxidizers including carbon.

Some experimentally measured (Ray et al. 1970, 1972; Bandyopadhyay et al. 1971) variations in carbon percentage, hardness and size and area density of inclusions (of cooled steel) during direct injection of oxygen are shown in Fig. 8.4. It is found that the decarburization rate is constant, whereas the indirect parameters vary in a nonlinear manner. The drop in hardness is understandable in terms of decrease of carbon content. The increase in size and density is because of ferrous oxide precipitated oxygen injection. There can be secondary precipitation of very fine FeO during cooling of oxygen saturated steel but its quantity will be much less.

During deoxidation, reactions are more complex. It is known that under thermodynamic equilibrium the product of concentrations of dissolved deoxidizer and dissolved oxygen has a fixed value.

$$\%\underline{M} \times \%\underline{O} = \text{MO} \tag{8.55}$$

Since $a_{\text{MO}} = 1$, $\%\underline{M} \times \%\underline{O}$ = constant.

For carbon deoxidation, p_{CO} may be taken as unity for 1550 °C. We have the equation (Gitterer 1967)

$$\%\underline{C} \times \%\underline{O} = 0.002 \tag{8.56}$$

Thus, as $\underline{O}$ level drops $\%\underline{C}$ can increase if free carbon is available.

Figure 8.5 shows some results of increase in carbon level during deoxidation of steel contained in a clay–graphite crucible by various deoxidizers (Ray et al. 1970, 1972; Bandyopadhyay et al. 1971). There is, obviously, forced dissolution of carbon with gradual lowering of dissolved oxygen. In the case of graphite injection, the carbon level rises drastically. In all cases, however, the rise is linear. If steel is deoxidized in a carbon-free refractory, then, of course, the carbon level cannot rise unless the deoxidizer is carbon itself. Figure 8.6 shows variations of some other parameters (Ray et al. 1970, 1972; Bandyopadhyay et al. 1971). The parameters associated with inclusions, however, vary also because of the kinetics of flotation of inclusion particles as has been discussed in Chap. 7.

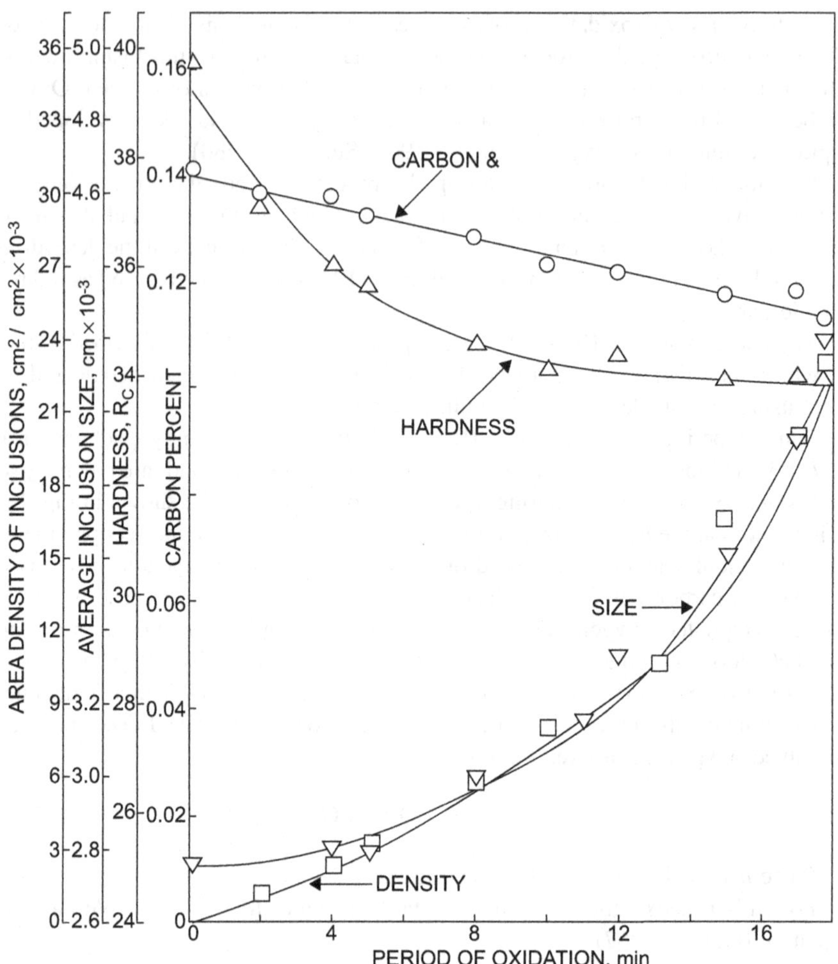

Fig. 8.4 Variations in carbon percentage, room temperature hardness, size and area density of inclusions during oxidation of steel by oxygen injection (Ray et al. 1970, 1972; Bandyopadhyay et al. 1971)

8.3.4 Slag and Glass Formation Reactions

Slag formation reactions can be very complex indeed. Consider, for example, the reaction of fluxes in a blast furnace during the descent of the charge. There can be many sequential and parallel reactions among the fluxes, the gangue and the ore. Low melting phases form first. These dissolve various compounds to form newer phases as materials are exposed to higher and higher temperatures. Thus, the slags that exist at different regions of the furnace differ remarkably in compositions. The rates of formation of phases have profound influence on the furnace operation.

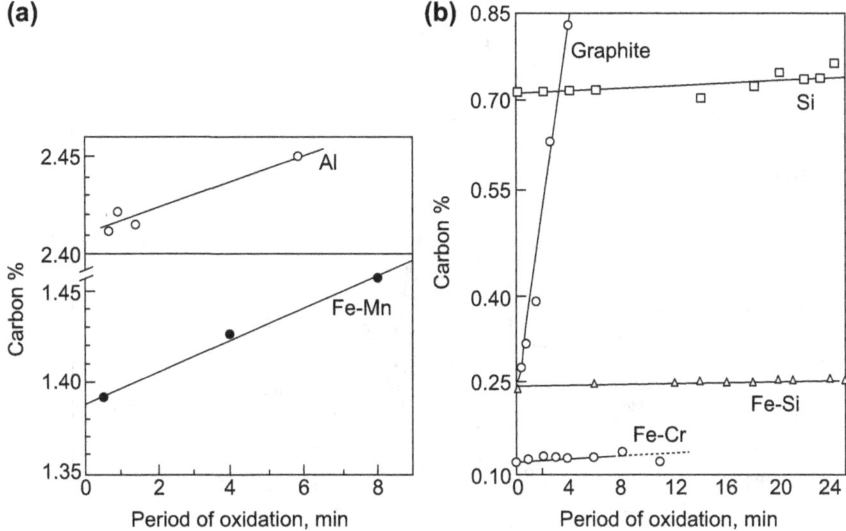

Fig. 8.5 Variations in carbon level during deoxidation of steel in clay–graphite crucible using **a** aluminium and ferro-manganeses and **b** graphite, silicon, ferro-silicon and ferro-chromium, as deoxidizers (Ray et al. 1970, 1972; Bandyopadhyay et al. 1971)

Wegman (1984) has given a general picture of slag formation along the height of blast furnace. This is shown in Fig. 8.7. The figure gives a general idea about the appearance and disappearance of phases as the charge descends in hotter zones of the furnace. Not much is known about the kinetics of these reactions. Some experimental blast furnaces have actually been frozen and samples taken from various zones analysed. Typical data for a particular furnace may be found in Wegman (1984).

Glass melting reactions

Slag formation reactions produce various glassy compositions. The reaction scheme or reaction path depends on raw materials compositions and various other factors. One can find useful information from workers in the area of glass chemistry many of whom have reported excellent studies on melting reactions which lead to commercial soda–lime–silica glasses. Some of the literature (Kautz 1969; Wilburn et al. 1965; Platt et al. 1967; Wilburn and Dawson 1972; Eitd 1976; Gibson and Ward 1940; Kroger 1948) in this field is briefly summarized.

When a glassmaking batch of raw materials is gradually heated, it first sinters and then undergoes a series of sequential and parallel reactions. At a sufficiently high temperature, some reactions seem to occur with particular rapidity resulting in the formation of primary glass and other reaction products. This causes the mass to collapse to a compact form, and the phenomenon being known as meltdown.

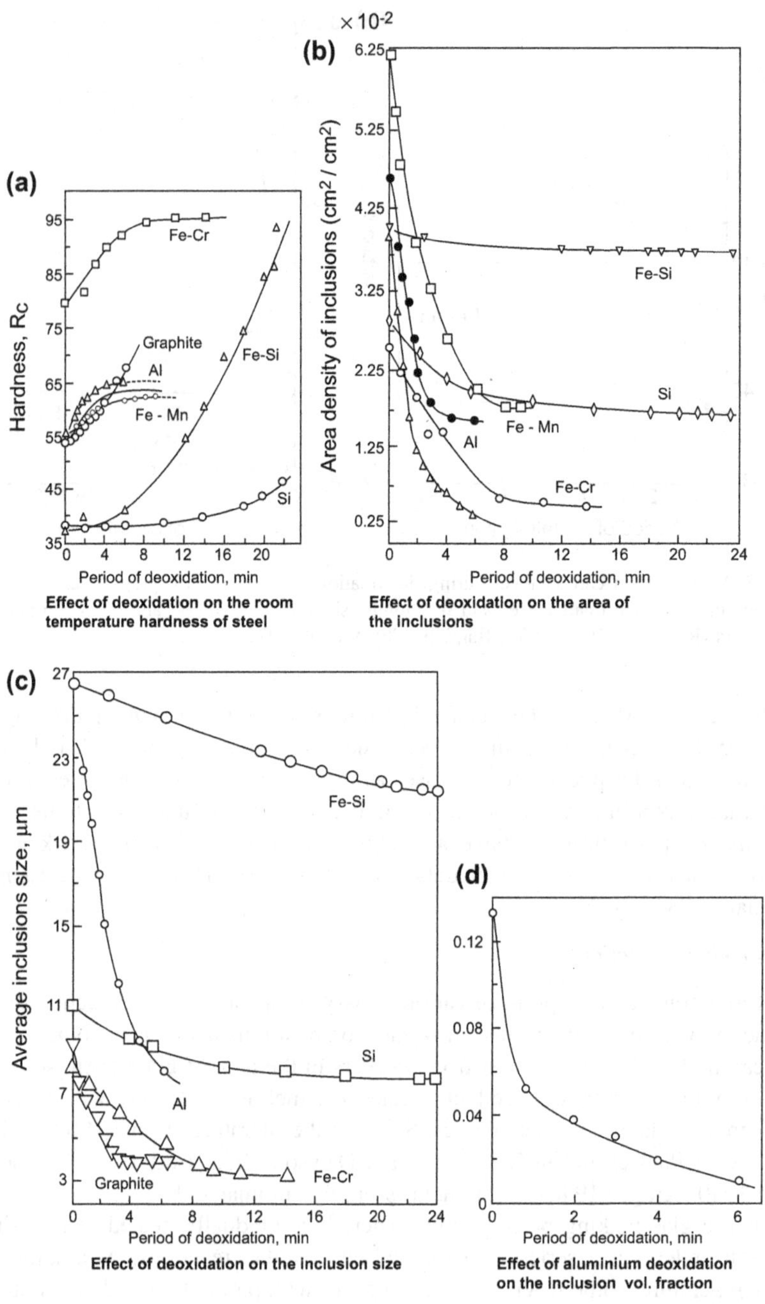

Fig. 8.6 Variations in indirect parameters during deoxidation of steel (Ray et al. 1970, 1972; Bandyopadhyay et al. 1971)

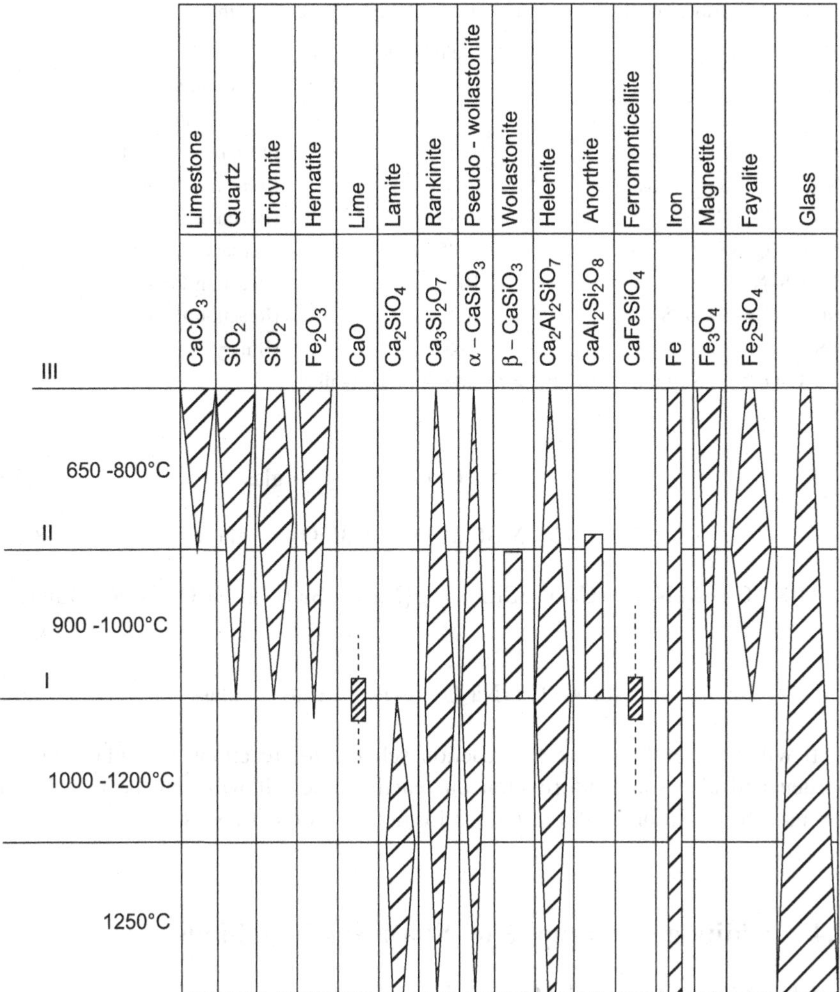

Fig. 8.7 Schematic diagram showing formation of different minerals along the height of the blast furnace (Wegmann 1984)

Table 8.2 lists the various lower melting phases that form during melting of soda–lime–silica glasses. The sequences of reactions which lead formation of the phases are as follows:

$$500\text{–}700\ °C: Na_2CO_3 + CaCO_3 = Na_2Ca(CO_3)_2 \tag{8.57}$$

$$780\text{–}870\ °C: Na_2Ca(CO_3)_2 + Na_2CO_3 + 3SiO_2 = 2Na_2O \cdot CaO \cdot 3SiO_2 \tag{8.58}$$

Table 8.2 Various phases present in Na_2O–CaO–SiO_2 phase diagram

Phases	Melting temperature	
NS–N_2CS_3	1060	(Binary eutectic)
QS–NS_2	790	(Binary eutectic)
NS_2–NS–N_2CS_3	821	(Ternary eutectic)
N_2CS_3–NC_2S_3–NC_3S_6	827	(Reaction Point)
NS_2–NC_2S_3–NC_3S_6	740	(Reaction Point)
NC_3S_6–Q–NS_2	725	(Ternary eutectic)
T–βCS–NC_3S_6	1035	(Reaction Point)
NC_3S_6–NC_2S_3–βCS	1030	(Reaction Point)
NS–NS_2	840	(Binary eutectic)

Q = CaO; N = Na_2O; S = SiO_2; Q = Quartz; T = Tridymite

$$960\,^{\circ}\text{C}: \; Na_2CO_3 + 3SiO_2 \rightarrow \text{glass} \tag{8.59}$$

$$1000\,^{\circ}\text{C}: \; Na_2Ca(CO_3)_2 + 3SiO_2 \rightarrow \text{glass} \tag{8.60}$$

$$1100\,^{\circ}\text{C}: \; 2[2Na_2O \cdot CaO \cdot 3SiO_2] = 3(Na_2O \cdot SiO_2) + Na_2O \cdot 2CaO \cdot 3SiO_2 \tag{8.61}$$

$$> 1100\,^{\circ}\text{C}: \; Na_2O \cdot SiO_2 + 2SiO_2 \rightarrow \text{glass} \tag{8.62}$$

It will be useful to have the reaction scheme for reactions in a blast furnace burden and also some information on reaction rates. It will then be possible to estimate composition of liquid phases formed at different zones.

8.4 Additional Examples of Non-ideal Conditions

8.4.1 Mixed Control Mechanism

In the previous section, we have discussed the use of a mixed control kinetic equation (Eq. 8.30) for analysis of some leaching data. A mixed control mechanism is proposed when there is no single rate-controlling step, i.e. the resistances offered by two or more steps in sequence are comparable. This situation is analogous to heat flow through multilayer insulation each layer offering some resistance and, therefore, causing a temperature drop. Figure 8.8 shows a 3-layer insulation and, the temperature drops schematically. Consider a steady-state situation. The heat flux Q is given by

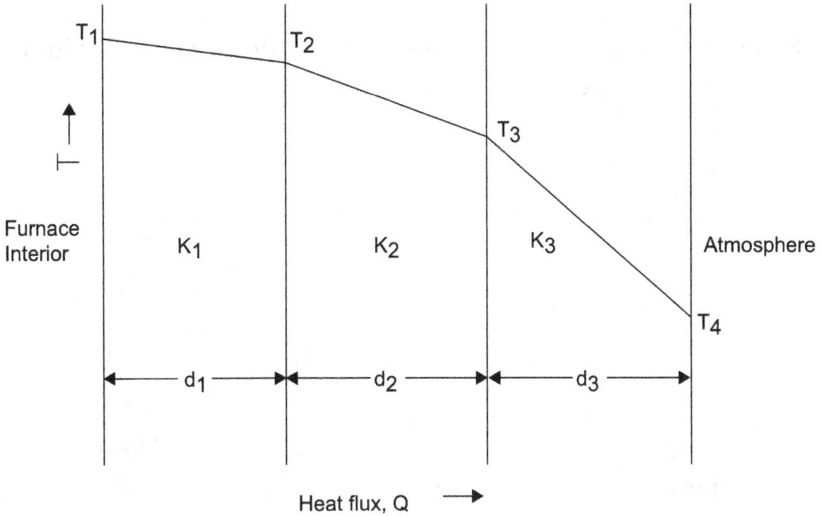

Fig. 8.8 Heat flow through a 3-layer insulation

$$Q = \frac{K_1[T_1 - T_2]}{d_1} = \frac{K_2[T_2 - T_3]}{d_2} = \frac{K_3[T_3 - T_4]}{d_3} \tag{8.63}$$

where K_1 K_2, etc., are thermal conductivities, i.e.

$$T_1 - T_2 = Q\frac{d_1}{K_1}, \quad T_2 - T_3 = Q\frac{d_2}{K_2}, \quad T_3 - T_4 = Q\frac{d_3}{K_3}$$

Adding, $T_1 - T_4 = Q\left(\frac{d_1}{K_1} + \frac{d_2}{K_2} + \frac{d_3}{K_3}\right)$,
i.e.

$$Q = \frac{T_1 - T_4}{d_1/K_1 + d_2/K_2 + d_3/K_3} \tag{8.64}$$

In Eq. 8.64, $(T_1 - T_4)$ represents the total temperature difference and the denominator in the RHS may be looked upon as the total resistance to heat flow.

In an analogous manner, the reaction rate for a fluid–solid reaction may be expressed as a ratio of total concentration difference involved to the sum of resistances offered by individual reaction steps. Consider Fig. 8.9. Bulk fluid of concentration C_1 has a concentration C_2 as it enters the product layer. At the unreacted core, the concentration is C_3 before reaction. Assume that equilibrium concentration after reaction is zero and that C_1, C_2 and C_3 remain unchanged.

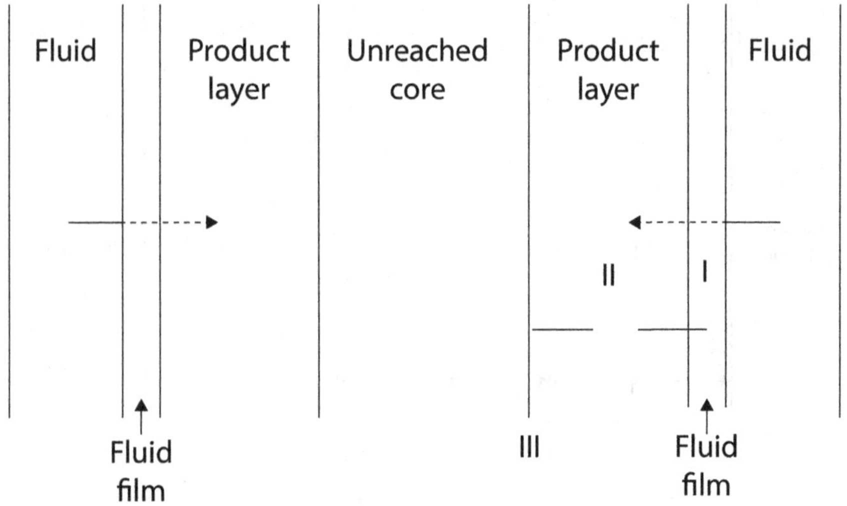

Fig. 8.9 Mixed control mechanism

In the simplest case, the solid is a flat plate with a surface area A. This area is assumed to remain unchanged as the reaction proceeds. We have the following equations

$$\begin{aligned} &\text{I Rate}: \ D_f A(C_1 - C_2)/\delta \\ &\text{II Rate}: \ D_p A(C_2 - C_3)/l \\ &\text{III Rate}: \ KAC_3 \end{aligned}$$

where D_f and D_p are, respectively, diffusion coefficients for fluid phase and product layer diffusion. δ is the boundary layer thickness in the fluid, l is the thickness of the product layer and K is a constant.

The equation analogous to Eq. 8.64 is derived as

$$\text{Rate} = \frac{C_1}{\frac{\delta}{D_f A} + \frac{l}{D_p A} + \frac{1}{KA}} \tag{8.65}$$

In the case of a spherical particle, A will not remain constant, because the area across which diffusion takes place gradually decreases. If the outer radius is r_2 and inner (core) radius is r_1, then Eq. 8.65 reduces to Eq. 8.66.

$$\text{Rate} = \frac{C_1}{\frac{\delta}{D_f 4\pi r_2^2} + \frac{l}{4\pi D_p} \cdot \frac{r_2 - r_1}{r_1 r_2} + \frac{1}{K 4\pi r_1^2}} \tag{8.66}$$

If there is no overall contraction or swelling of the particle, then r_2 remains r_0 the original particle size. We then have

$$\text{Rate} = \frac{C_1}{\frac{\delta}{D_f 4\pi r_0^2} + \frac{l}{4K D_p} \cdot \frac{r_0 - r_1}{r_0 r_1} + \frac{1}{K 4\pi r_0^2}} \tag{8.67}$$

Additivity law

There is another more novel approach for analysis of mixed control kinetics. In this approach, one first notes that resistance of every reaction step increases the total reaction time for achieving a particular degree of conversion. Accordingly, one can postulate an additivity law which essentially states that the total time is the sum of the times required by the individual steps in the absence of all other steps (Sohn 1978). Thus, for fluid–solid reactions, the time required to attain a particular value of α will be sum of the time required to attain the same α in the absence of product layer diffusion resistance plus that required under such diffusion control. Suppose we consider the reaction and diffusion steps for a gas–solid reaction, the kinetics of which are given by the following equations.

$$1 - (1-\alpha)^{1/3} = \frac{k_1}{\rho r_0} t$$

$$1 - \frac{2}{3}\alpha - (1-\alpha)^{2/3} = \frac{k_2}{r_0^2} t$$

If there is no diffusional resistance, then chemical reaction time t_c is given by

$$t_c = \frac{1 - (1-\alpha)^{1/3}}{k_1/\rho r_0} \tag{8.68}$$

If there is no resistance due to chemical reaction and the single rate-controlling step in product layer diffusion, then, for the same α, reaction time is

$$t_d = \frac{1 - \frac{2}{3}\alpha - (1-\alpha)^{2/3}}{k_2/r_0^2} \tag{8.69}$$

When both steps offer resistance together, then the actual time t is given by

$$t = t_c + t_d = \frac{1 - (1-\alpha)^{1/3}}{k_1/\rho r_0} + \frac{1 - \frac{2}{3}\alpha - (1-\alpha)^{2/3}}{k_2/r_0^2}$$

or,

$$\rho k_2[1-(1-\alpha)^{1/3}]+r_0k_1\left[1-\frac{2}{3}\alpha-(1-\alpha)\right]^{2/3}=\frac{k_1k_2t}{r_0} \tag{8.70}$$

This is the integral form of the kinetic equation.

Activation energy for a mixed control mechanism

When mixed control prevails, then obviously the apparent activation energy will be different from the activation energies associated with the individual steps. The apparent order of the reaction will also differ. This can be shown thus.

Consider two consecutive steps [l, 2] both of which are involved in rate control. During a reaction, both steps have similar rates and thus one can write

$$\text{Rate}=k_1f_1(\alpha)-k_2f_2(\alpha) \tag{8.71}$$

where k_1 and k_2 are the rate constants for the two steps and $f_1(\alpha)$ and $f_2(\alpha)$ are functions of α designating order

$$k_1=A_1\exp(-E_1/RT)$$
$$k_2=A_2\exp(-E_2/RT)$$

We have

$$\text{Rate}=A\ \exp\left(-\frac{E}{RT}\right)f(\alpha)=\sqrt{A_1A_2}\exp\left(-\frac{(E_1+E_2)}{2RT}\right)\sqrt{f_1(\alpha)\cdot f_2(\alpha)} \tag{8.72}$$

where A and E designate the apparent values of kinetic parameters under mixed control, and $f(\alpha)$ is the apparent function which designates the apparent order. It is seen that E assumes the value of the arithmetic average $(E_1+E_2)/2$. The apparent order changes depending on the form of $f_1(\alpha)$ and $f_2(\alpha)$.

Levenspiel (1962) has discussed this problem with specific reference to the effects of film diffusion on activation energy and order of reaction. He has shown that when chemical reaction at an interface is superimposed with diffusion through product layer, then the observed E becomes half of the true E value. Also, an nth order action behaves like a reaction of order $(n+1)/2$, i.e. 0 order becomes 0.5 order, 1st order remains 1st order, 2nd order becomes 1.5 order, 3rd order becomes 2nd order, and so on.

It is easy to prove these in view of the preceding analysis. One can also extend the same analysis to mixed control by a multiple of steps. In some other cases, the kinetic equation can be derived by use of proper logic. (see Example later)

8.4.2 Reactions Controlled by Reagent Supply

It is not essential for reaction rate to be always controlled by a rate-controlling reaction step. Sometimes, under a given set of circumstances, all the steps may be sufficiently rapid so that the overall reaction rate is limited only by the rate supply of reagents. Consider, for example, removal of a gas dissolved in a liquid by purging of another gas which is inert towards the liquid.

The basic mechanism of degasification is this. As the inert gas bubble enters the liquid, it begins to rise in the liquid and eventually escapes. During its passage, however, the dissolved gas diffuses into the bubble from the surrounding liquid. If this process of diffusion is very rapid, then the bubble, before its escape, takes in as much dissolved gas as allowed by thermodynamic equilibrium. In such a situation, the rate of degasification will simply depend on the rate of purging of the inert gas. The situation is analysed entirely on the basis of thermodynamics and materials' balance. Consider the following problem.[1]

Dyhydration (i.e. removal of hydrogen) of steel by purging argon gas

The process consists of bubbling argon through liquid steel containing large amounts of hydrogen. The argon bubbles pick up hydrogen from the steel, thus causing the removal. The process is not practised in large-scale operation because of the high cost of argon. However, it can be done in the laboratory. At steady state, from hydrogen balance

$$\begin{aligned}&\text{Rate of removal of hydrogen from steel } (\dot{\omega}_1,\ \text{g/s})\\&\quad = \text{Rate of transfer of hydrogen to argon } (\dot{\omega}_2,\ \text{g/s})\end{aligned} \tag{8.73}$$

$$\dot{\omega} = -\frac{\mathrm{d}(C_\mathrm{H})}{\mathrm{d}t} \times W \times 10^{-6} \times 10^{6} = -\frac{\mathrm{d}(C_\mathrm{H})}{\mathrm{d}t} \times W \tag{8.74}$$

where C_H is concentration of hydrogen at any time t from the beginning of the process in parts per million (ppm) by weight and W is the weight of steel in metric tons.

Now, we have to calculate the rate of hydrogen transfer in g/unit time assuming that the equilibrium reaction

$$2\underline{\mathrm{H}} = \mathrm{H}_2$$

always holds.

Let $\mathrm{d}v'$ be the volume (litres) of argon passed at process temperature T (Kelvin) and 1 atm pressure in an elemental time $\mathrm{d}t$. We assume that hydrogen evolution is small enough not to increase this value appreciable so that the volume of the

[1]Based on problem given in A. Ghosh, *Principles of extractive metallurgy*, Ch. 5, Lecture notes, IIT Kanpur (1969).

hydrogen–argon mixture generated is also dv'. Suppose that the pressure of hydrogen at the process temperature is p_{H_2}. Then considering hydrogen

$$\frac{p_{H_2} \cdot dv'}{T} = \frac{1 \cdot V_H}{273} \tag{8.75}$$

or,

$$V_H = \frac{dv' \cdot p_{H_2} \cdot 273}{T}$$

Here V_H denotes the volume of hydrogen at standard temperature (273 K) and pressure (1 atm). Since 22.4 l of hydrogen have mass 2 g at standard temperature and pressure (STP), the mass of hydrogen occupying volume V_H (in g) is

$$\frac{dv' \cdot p_{H_2} \cdot 273}{T} \cdot \frac{2}{22.4}$$

Now if the volume of argon is measured at STP, then dv' would have a different value, say dv, so that

$$dv' = \frac{T}{273} \cdot dv \tag{8.76}$$

Thus, the weight of hydrogen in terms of dv is $dv \cdot p_{H_2}/11.2$, and

$$W_2 = \frac{1}{11.2} \cdot \frac{dv}{dt} \cdot p_{H_2} \tag{8.77}$$

Combining Eqs. 8.73, 8.74 and 8.77

$$-\frac{d(C_H)}{dt} \times W = \frac{1}{11.2}\frac{dv}{dt} \cdot p_{H_2}, \quad \text{or} \quad -d(C_H) = \frac{p_{H_2}}{11.2W} \cdot dv \tag{8.78}$$

From Sievert's law,

$$C_H = K_H \cdot p_{H_2}^{1/2}, \quad \text{or } p_{H_2} = C_H^2/K_H^2$$

where K_H is a constant. Then

$$\frac{d(C_H)}{C_H^2} = \frac{dv}{11.2\, W K_H^2} \tag{8.79}$$

Integrating:

$$\frac{1}{C_H} = \frac{V}{11.2\, W {K_H}^2} + I \tag{8.80}$$

where I is the integration constant. Noting that at $V = 0$, $C_H = C_H^i$, we obtain $I = 1/C_H^i$. Hence,

$$\frac{1}{C_H} - \frac{1}{C_H^i} = \frac{v}{11.2WK_H{}^2} \tag{8.81}$$

Equation 8.81 will allow us to find out C_H at any value of V provided we know C_H^i, W, and K_H.

It should be noted that during the process, temperature will fall changing K_H. This effect can be taken into account if we combine heat balance.

8.4.3 Reactions Controlled by Heat Transfer

Many reactions in metallurgy are accompanied by appreciable heat effects. During the reaction, heat either flows out of the system (exothermic reaction) or into the system (endothermic reaction). Consequently, the temperature at the interior of a reaction mixture may deviate appreciably from the outer regions and the environment (Hills 1967; Rao and Abraham 1971; Themelis and Yannopoulos 1966). Thus, an external thermocouple, even if it is placed close to the sample in a furnace, reads merely the furnace temperature but not the true sample temperature. To measure the latter one must insert a thermocouple inside the sample. Figure 8.10 shows an experimental assembly used for studying sample temperature during oxidation of copper sulphide. Figure 8.11 shows the temperature data recorded during the progress of reaction for a 7 g sulphide sample pressed in a steel die about 1.22 cm dia at a pressure of 40 psi (Hills 1967). The temperature rise in the sample is obviously due to exothermicity of the reaction. For endothermic reactions, similarly, the sample temperature is likely to decrease. Figure 8.12 shows some data for sample temperature during decomposition of a spherical pellet of about 1.1 cm dia in a furnace at 863 °C (Hills 1967). The temperature probe refers to the thermocouple placed near but outside the compact.

The temperature difference between the interiors and the exterior would, of course, depend on the chemical and physical characteristics of the solid. Thus, the difference would be more pronounced if compact is larger and/or denser and if the thermal conductivity is poorer. Conversely, the difference may be small for a loose powdery sample particularly if the sample volume is also small.

The heat of reaction introduces some complications in kinetic studies.

(a) The reaction temperature becomes uncertain and although one can measure rates versus time, one may not be able to relate the measurements to a well-defined temperature.
(b) For exothermic reactions, an increase in sample temperature implies an increased rate and, therefore, a greater temperature differential. As a matter of fact, the temperature inside continues to rise until a time when increased heat

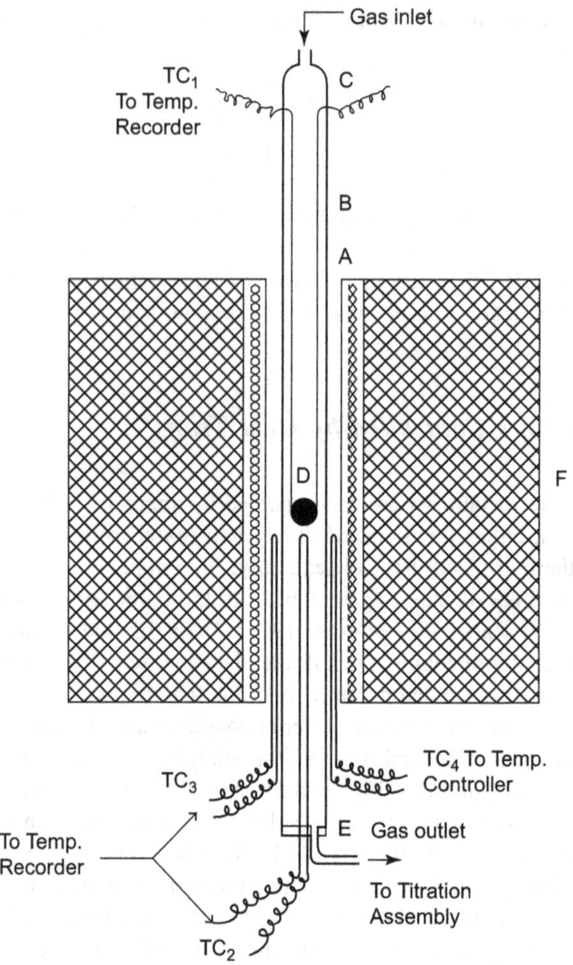

Fig. 8.10 Experimental assembly for measuring sample temperatures during oxidation of copper sulphide (Rao and Abraham 1971)

dissipation puts a stop to this rise. At the maximum difference, the rate of heat generation equals the rate of heat dissipation. Beyond this point, the inside temperature must drop again because of slowing down of reaction rate. On the whole, however, the temperature variation takes places under non-isothermal conditions although the furnace temperature may remain unchanged.

(c) Similar observations apply for endothermic reactions. In this case, however, there is a further complication. As the interior regions get cooled, the reaction rate there decreases. To keep the reaction occurring, heat must be continuously supplied from outside. Under certain conditions, this heat transfer process and the resultant reaction rate may become slower than any other reaction step. In such a situation, the reaction is said to be heat transfer controlled.

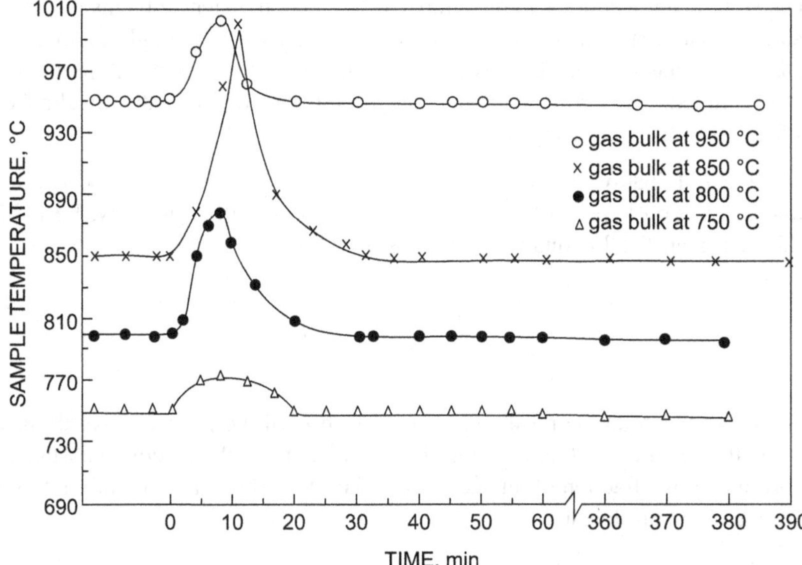

Fig. 8.11 Rise in temperature of the sample during roasting due to exothermic heat generated, at various temperatures of roasting (Rao and Abraham 1971)

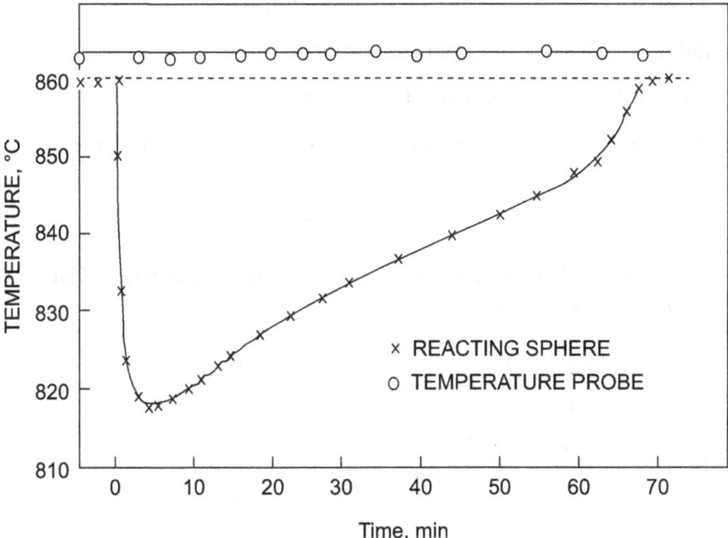

Fig. 8.12 Temperatures of calcium carbonate sample, and of an inert temperature probe, while the sample is decomposing in a furnace held at 863 °C (Rao and Abraham 1971)

There are many examples in metallurgy where heat transfer problems have to be considered. The reaction between iron oxide and carbon is strongly endothermic with the consequence that, if ore is used in a steel bath to oxidize carbon, the bath may get chilled. This will affect the rate of refining of the bath. Steelmakers, therefore, limit the use of iron ore.

Example 8.1 Tripathy, Ray and Pattanaik[2] have shown that the kinetics of solid-state reaction between Cr_2O_3 and Na_2CO_3 follows the Zhuravliv, Lesokhim and Tempel'man (ZLT) equation[3] which is as follows:

$$\left[\frac{1}{(1-\alpha)^{1/3}} - 1\right]^2 = kt$$

Show that the equation results from modification of the product layer diffusion model with the additional assumption that the activity of the reacting substance is proportional to the fraction of unreacted material. Also show that, in this case, plots of α versus $\log t$ are linear.

Solution

Consider reaction of a spherical solid particle. The reaction rate should gradually decrease because of three factors. The usual factors are:

a. A gradually decreasing reaction interface between core and product layer,
b. A gradually increasing product layer thickness,

The additional factor arising out of modification is

c. A gradually decreasing amount of unreacted material.

Therefore, the differential form of kinetic equation should be expressed as

$$\frac{\mathrm{d}y}{\mathrm{d}t} = \frac{k}{y}(1-\alpha)A$$

where y is the product layer thickness, and A is the product–core interface area. Since $y = (r_0 - r)$, we have

$$\frac{\mathrm{d}y}{\mathrm{d}t} = \frac{\mathrm{d}(r_0 - r)}{\mathrm{d}t} = \frac{k}{(r_0 - r)}(1-\alpha)\cdot 4\pi r^2$$

$$\frac{\mathrm{d}r}{\mathrm{d}t} = \frac{\mathrm{kr}^2(1-\alpha)}{r_0 - r}$$

[2] A. K. Tripathy, H. S. Ray and P. K. Pattanaik, *Trans. IIM Sec C* May–Aug (1994)C pp 49–54; *also Met. and Mat. Trans. B* Vol. **26B** (1995) pp 125–127

[3] F. V. Zhuravliv, L. G. Lesokhin and R. G. Tempel'man; *J. Appl. Chem. USSR* (English translation), Vol **21** (19)(1948), p 887; S. F. Hulbert, *J. Brit. Ceram. Soc.* Vol. **6** (1) (1969), pp 11–20.

Again, $\alpha = 1 - (r/r_0)^3, \quad r = r_0[(1-\alpha)^{1/3}]$, whence

$$-\frac{dr_0(1-\alpha)^{1/3}}{dt} = \left\{kr_0^2\left[(1-\alpha)^{1/3}\right]^2(1-\alpha)\right\}\Big/ r_0\left[1-(1-\alpha)^{1/3}\right]$$

$$\frac{d\alpha}{dt} = \frac{3k(1-\alpha)^{7/13}}{[1-(1-\alpha)^{1/3}]}$$

Differentiating ZLT equation with respect to t, one obtains

$$\frac{d\alpha}{dt} = \frac{3k}{2}\frac{(1-\alpha)^{7/3}}{\left[1-(1-\alpha)^{1/3}\right]}$$

The last two equations are identical if the constants are adjusted. Hence, the ZLT equation is consistent with the assumptions.

α *versus* $\log t$: Taking logarithm of ZLT equation

$$2\log\left[\frac{1}{(1-\alpha)^{1/3}} - 1\right] = \log k + \log t, \text{ which gives}$$

$$2\log\left[1-(1-\alpha)^{1/3}\right] - \frac{2}{3}\log(1-\alpha) = \log k + \log t$$

If we expand and neglect higher powers of α, we obtain,

$$\frac{8\alpha}{9} - \frac{7}{3} = \log k + \log t, \quad \text{i.e,} \quad \alpha = \text{const} + \frac{9}{8}\log t$$

Hence plot of α versus $\log t$ should be linear.

8.5 Kinetics of Melting of a Packed Bed

8.5.1 Development of an Elementary Model

In metallurgical, chemical and ceramic industries, there are many processes which involve melting down a bed of granular or lumpy solids. The melting of such porous beds is a slow and energy intensive step, especially when the bed has low thermal diffusivity, high latent heat of melting point. For proper control of productivity and ultimate economy, therefore, it is necessary to develop analytical and/or experimental correlations, of the melting rate with process variables. Here a simple correlation is presented from the literature (Mukherjee et al. 1983).

Some analytical methods for obtaining the temperature profile and melting rates during the melting of solid blocks have been reported in the literature (Eckert and Drake 1979). However, literature regarding melting rate in packed beds is limited (Sohn and Wadsworth 1979). The present section aims at deriving equations for melting rate on the basis of a hypothetical physical model of melting.

8.5.2 Physical Mould

It may be assumed that the surface of solids attains the furnace temperature as soon as the charge is exposed. We also assume that the furnace (and the surface) temperature remains constant. The melting front starts moving from the surface downwards. The liquid produced by the melting of the top layers will start percolating downwards through the pores and gaps. This will create a layer consisting of liquid and semi-molten solids which will grow in thickness as the melting proceeds. However, after a certain period of time further growth of this layer would be difficult because of certain factors.

(1) Melting rate would fall as the heat flux would be diminished due to the reduction in the thermal potential as the bed starts heating up.
(2) The percolating liquid would eventually freeze at cooler depths and clog the pores, thus preventing further growth of the "mushy" layer.

It is, therefore, assumed here that after a certain time the thickness of the layer reaches a limiting value. During the time required to build up this limiting thickness of the semi-molten layer, the melting rate gradually decreases reaching a terminal value after the end of the period.

After this period, with the passage of time, the top part of the semi-molten layer will become completely molten and the extra liquid produced will percolate further downwards. It is assumed now that the resultant decrease in the thickness of the semi-molten layer at the top is compensated for by the increase in thickness of this layer at the bottom, which the extra liquid produced now freshly irrigates. Thus, the thickness of this layer remains constant, and consequently, the melting rate also remains constant. The top surface of the constant thickness mushy layer remains at the furnace temperature and the bottom surface at the melting point. As the solids continue to melt and the liquid produced continue to fill up the pores and gaps, the bed height also would decrease. For a bed of finite thickness, however, some clear liquid may collect on top near the end of the melting process. It has been assumed here that the top and bottom surfaces of any clear liquid layer thus produced are at the furnace temperature. This liquid layer, thus, does not play any role in the heat transfer.

Melting rate

Let us consider the unidirectional melting (from the top to bottom) of a semi-infinite bed insulated at the sides. According to the hypothetical melting model proposed, the whole melting period may be divided into two stages: (1) period of growth of the mushy layer and (2) melting by heat transfer through a constant thickness mushy layer.

8.5.3 Period of Growth of the Mushy Layer

This is approximately determined in the following way. Assuming that the surface of the bed acquires the furnace temperature (T_f) instantaneously, an unsteady-state temperature distribution through the bed may be obtained by the finite differences method. The bed is considered to be a homogeneous body. This may be acceptable if it is assumed that the total effect of the non-uniformities in a sufficiently large granular bed becomes same in all the directions. Once the temperature profiles are thus obtained disregarding any latent heat absorption, rate of movement of the melting front may be obtained by drawing a line at the melting point parallel to the distance axis. A plot between the time steps and the corresponding bed depth which has attained the melting temperature would then give the melting rates. Such melting rates were determined under the following furnace temperature, melting temperature T_m and initial temperature T_i conditions:

$$\begin{aligned}
&(1)\quad T_f = 1800, T_i = 400,\ T_m = 1050,\ 1250,\ 1450\ ^\circ\text{C}\\
&(2)\quad T_f = 1500, T_i = 800,\ T_m = 1000,\ 1100,\ 1200\ ^\circ\text{C}\\
&(3)\quad T_f = 1600, T_i = 40,\ T_m = 1100,\ 1450,\ 1500\ ^\circ\text{C}\\
&(4)\quad T_f = 1700, T_i = 40,\ T_m = 1100,\ 1450,\ 1500\ ^\circ\text{C}\\
&(5)\quad T_f = 1600, T_i = 800,\ T_m = 100,\ 1200,\ 1300\ ^\circ\text{C}
\end{aligned}$$

Other conditions, namely the distance step ($\Delta x = 2$ cm), time step ($\Delta\tau = 60$ s) and thermal diffusivity ($\alpha = 3.33 \times 10^{-6}$ m^2/s) were unchanged in all the cases. As an example, temperature distribution and melting rates under conditions (1) are indicated in Figs. 8.13 and 8.14, respectively. It is observed from these figures that the melting rate increases initially and then decreases ultimately showing very small variation with time. Fifteen such terminal melting rates (v_u) were obtained for fifteen melting temperatures considered in the five conditions. A correlation is obtained between the melting rate v_u and the dimensionless temperature $(T_f - T_m)/(T_m - T_i)$. This dimensionless form gives the ratio of the temperature range the charge has to rise through to reach the melting point. The thermophysical properties of the bed are chosen quite with a thermal conductivity $K = 10$ w/m °C,

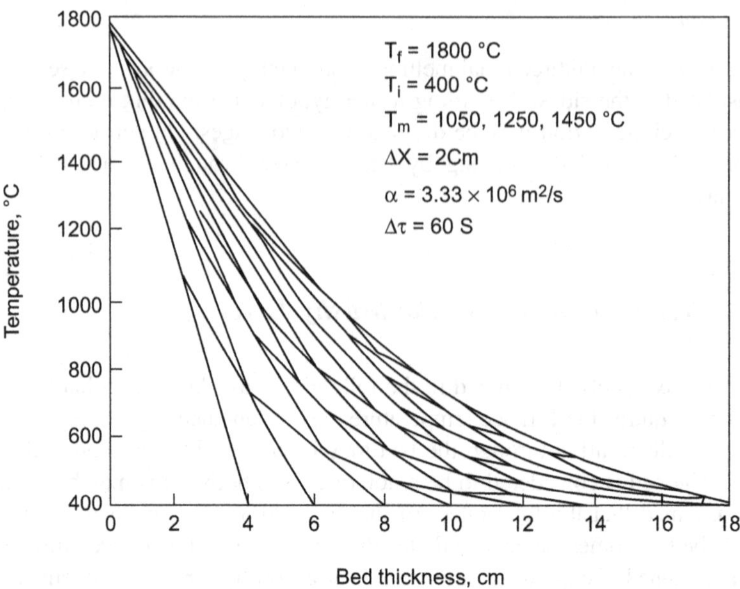

Fig. 8.13 Unsteady-state temperature distribution

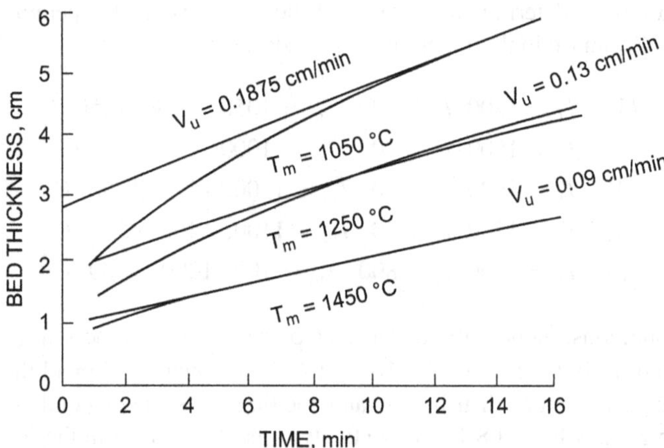

Fig. 8.14 Movement of the melting front (Mukherjee et al. 1983)

specific heat C = 1. 2 kJ/kg °C, and density $\rho = 2500$ kg/m^3, which gave a thermal diffusivity $\alpha = 3.33 \times 10^{-6}$ m^2/s. The correlation obtained by linear regression is as follows:

$$v_u = 0.166\left(\frac{T_f - T_m}{T_m - T_i}\right)^{0.649} \quad \text{cm/min} \tag{8.82}$$

Straight and log–log plots for Eq. 8.82 are shown in Figs. 8.15 and 8.16, respectively. It may be mentioned here that in the solution by the finite differences method, the distance step Δx and the time step $\Delta\tau$ are connected by the relation $\Delta\tau = \Delta x^2/2\alpha$. Analytical solution of the problem is also possible.

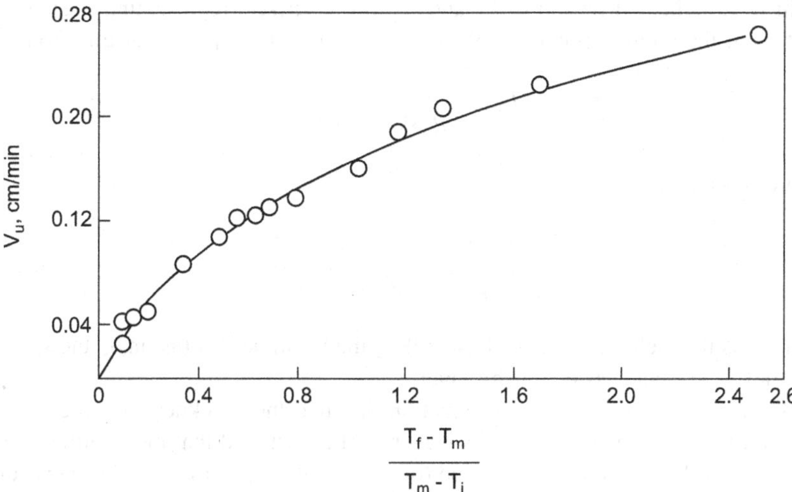

Fig. 8.15 Dependence of the melting rate in the first stage on dimensionless temperature parameter

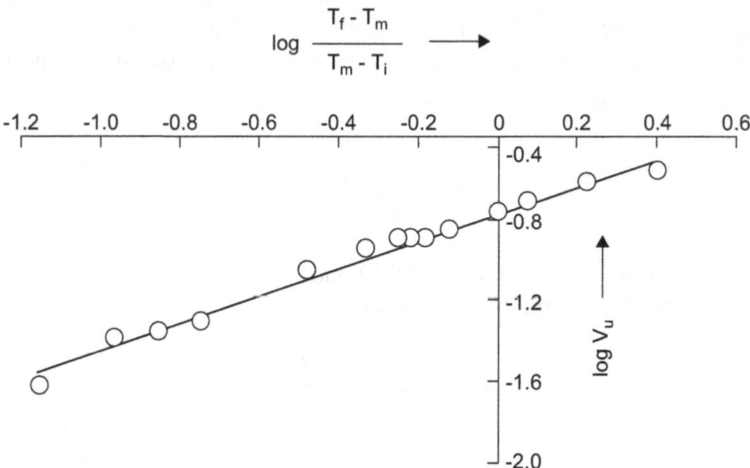

Fig. 8.16 Log–log plot of data shown in Fig. 8.15 (Mukherjee et al. 1983)

The total period of growth of the mushy layer will be determined by the thickness of this layer; this is discussed below.

8.5.4 Melting by Heat Transfer Through a Constant Thickness Mushy Layer

It is assumed here that the upper surface of the mushy layer is at the furnace temperature T_f (disregarding any temperature gradient in the liquid pool which might have collected on top of it) and its lower surface, representing the melting front, is at the melting point T_m. Writing a heat balance equation at this front,

$$\frac{K(T_f - T_m)}{l} = \rho\, v_s Q + \rho\, v_s C(T_m - T_i)$$

On rearranging,

$$v_s = \frac{K}{\rho\, l}\left[\frac{T_f - T_m}{Q + C(T_m - T_i)}\right] \tag{8.83}$$

where v_s is the melting rate (steady stage), Q the latent heat of fusion, C the specific heat and l the thickness of mushy layer.

The difficulty in using this equation is that the thickness of the mushy (semi-molten) layer is unknown. However, if it is assumed that the melting rate v_u attained near the end of the first period (of growth of the semi-molten layer) is equal to the melting rate in the second period, then the only unknown in Eq. 8.83 can be evaluated and the total duration of the period of growth of this layer can be estimated. This is done in the present case by arbitrarily choosing a value of 418.6 kJ/kg for Q. K, C and Q are kept the same as in the first period. This exercise gave a value of l ranging from 3.79 to 8.21 cm. However, a part of this layer is already built up by the time the melting rate v_u in the first period reaches a terminal value (approximately near about the 12th time step). A linear regression done between the logarithm of this part (l') and $\log\left(\frac{T_f - T_m}{T_m - T_i}\right)$ gives the relation

$$l' = 4.612\left(\frac{T_f - T_m}{T_m - T_i}\right)^{0.705} \text{cm.} \tag{8.84}$$

This dependence is shown in Figs. 8.17 and 8.18.

The total duration of the first period would therefore be

$$\tau_{tot}\, l = 12\Delta\tau + \left(\frac{l - l'}{v_u}\right) \text{min} \tag{8.85}$$

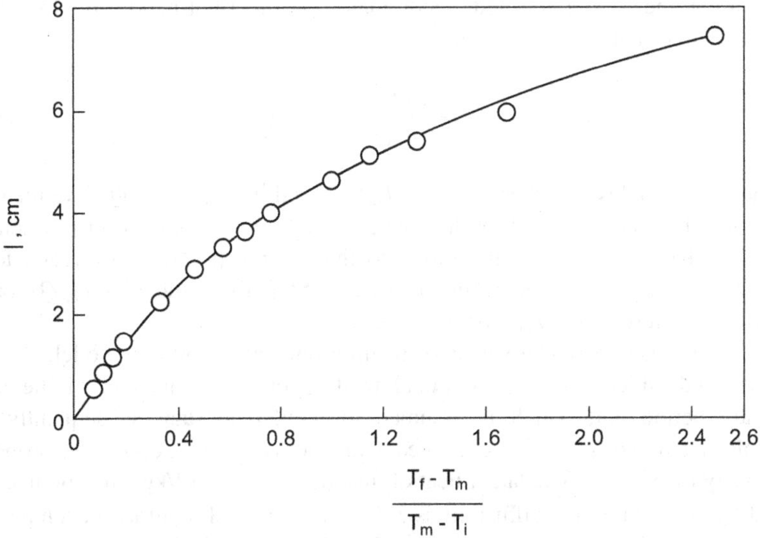

Fig. 8.17 Dependence of l' on dimensionless temperature (Mukherjee et al. 1983)

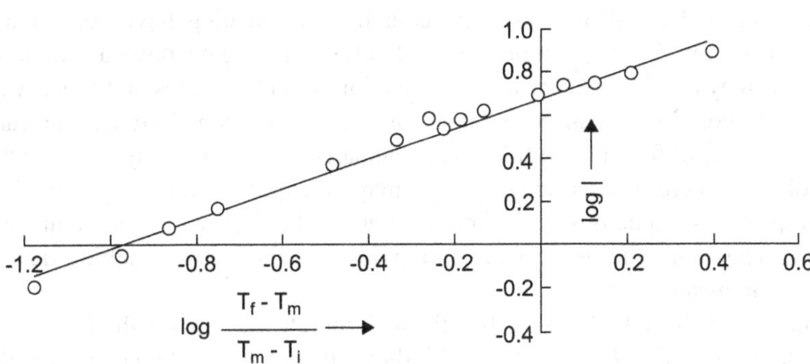

Fig. 8.18 Log–log plot for data in Fig. 8.17 (Mukherjee et al. 1983)

Since $\alpha = K/C\rho$ i.e. $K/\rho = C\alpha$, Eq. 8.83 may be written in a dimensionless form

$$v_s = \frac{C\alpha}{l}\left[\frac{T_f - T_m}{Q + C(T_m - T_i)}\right] \tag{8.86}$$

or,

$$\frac{v_s\, l}{\alpha} = \frac{C(T_f - T_m)}{Q + C(T_m - T_i)} \tag{8.87}$$

Dividing the numerator and denominator of the right-hand side by Q, the dimensionless melting rate V is given by

$$V = \frac{C(T_\mathrm{f} - T_\mathrm{m})/Q}{1 + [C(T_\mathrm{m} - T_\mathrm{i})/Q]} \tag{8.88}$$

The dimensionless numbers $[C(T_\mathrm{f} - T_\mathrm{m})/Q]$ and $[C(T_\mathrm{m} - T_\mathrm{i})/Q]$ give the ratios of the sensible heat acquired by the charge from melting point to furnace temperature and that from initial temperature to the melting point, respectively, to the latent heat. They have some similarity with Stefan's number (CT_f/Q) or the Kossovitch criteria (Q/CT_m) for melting.

To get an idea about the values of melting rates given by this model, Eq. 8.82 can be used which gives the terminal melting rate near the end of the initial unsteady period and which is assumed to remain constant subsequently. For example, let us consider a packed bed with the same thermophysical properties previously mentioned. The latent heat of fusion Q is 418.6 kJ/kg, specific heat C is 1.2 kJ/kg °C and thermal diffusivity is 3.33×10^{-6} m^2/s. For a furnace temperature of 1550 °C and initial temperature of 50 °C, a charge of melting temperature 1050 °C should melt at the rate of about 0.11 cm/min. Thus, it would take about 3 h to melt a 20-cm-thick bed.

The proposed physical model and equations for melting have been obtained under some simplifying assumptions and, thus, have some obvious limitations. However, they can be used for rough estimation of melting rates and time. A more dependable correlation can be obtained only by experimental studies on various materials under different conditions. The thickness of the mushy layer (l) which has been obtained here by assuming $v_\mathrm{u} = v_\mathrm{s}$, may be experimentally determined.

The proposed equations give some idea about the dimensionless parameters of melting. These may be used in the correlation of experimental melting data with process parameters.

Both Eqs. 8.82 and 8.83 show that the melting rate increases with $(T_\mathrm{f} - T_\mathrm{m})$ and decreases with $(T_\mathrm{m} - T_\mathrm{i})$. Equation 8.83 shows that melting rate increases with the thermal conductivity of the bed and decreases with its density, specific heat and latent heat of melting.

8.6 Review Questions

1. Discuss whether the following statements are correct or false.

    a. Specific rate constant is the rate constant in the absence of mixed control.
    b. Particle size always influences gas–solid reactions.
    c. If there is no rate-controlling step and reaction is controlled by material supply only, then the overall rate is independent of temperature.

d. When material supply controls the rate, then all individual reaction steps must be in equilibrium.
e. Exothermic reactions cannot be heat transfer controlled.
f. Sohn's additivity law finds reaction rate for mixed control mechanisms through addition of rates for individual reactions.
g. If a system is composed of a large member of particles of varying sizes, then the kinetic equation can be written by considering the arithmetic average size.
h. A series of sequential and overlapping reactions occur in a process during heating. The reaction path can be changed by changing the rate of heating.
i. In a blast furnace, the viscosity of slag decreases as it descends to regions of higher temperatures.
j. Two mechanisms, characterized by two E values, control a reaction (mixed control). The overall activation energy should correspond to the higher E value.

2. Differentiate between rate constant and specific rate constant.
3. A flat solid plate reacts with a liquid and a covering product layer is formed. The reaction is controlled by both product layer diffusion and chemical reaction at the interface. Derive the kinetic law.
4. Recommend some simple tests to identify rate control by (a) fluid phase mass transfer, (b) reaction at a phase boundary, (c) diffusion through, product layer and (d) total internal reaction.
5. Derive for a gas–solid reaction a generalized kinetic law by considering mixed control by (a) gas film diffusion, (b) product layer diffusion and (c) chemical reaction a core–product interface.
6. An inert gas is bubbled through molten steel to eliminate dissolved nitrogen. There is no rate-controlling step and every bubble of inert gas picks up sufficient nitrogen as per Sievert's law. Prove that nitrogen removal follows the equation

$$\frac{1}{C_{\mathrm{N}}} - \frac{1}{C_{\mathrm{N}}^{\mathrm{i}}} = \frac{V}{11.2W\,K_N^2}$$

where C_{N} = nitrogen concentration after passage of volume V (STP) of inert gas (ppm); $C_{\mathrm{N}}^{\mathrm{i}}$ = initial nitrogen concentration (ppm); W = weight of steel in tons.
7. Discuss the basic rationale underlying Sohn's additivity law.
8. Discuss some of the major factors because of which well-known kinetic laws for idealized systems may fail to explain the kinetic data.
9. There are four sizes of spherical solid particles with radii, 10, 15, 20 and 25 mm. 25 particles from each group make a mass which reacts with a liquid, and the reaction being chemically controlled. Write the kinetic law for the system.

10. During some reactions, e.g. oxidation of metals, the reaction mechanism depends on reaction temperature. Why?
11. During oxidation of a metallic sheet, the kinetic equation is found to confirm to the equation $\alpha^3 = kt$. Give a physical reasoning to explain the form of the equation.
12. Why are complex deoxidizers considered superior from kinetic point of view?
13. Explain how lime injection through the tuyers changes the reaction path for slag formation reactions and improves blast furnace operation.
14. In DH process of vacuum degassing, a fraction of steel is taken into a vacuum changer, degassed and then allowed to mix again with the bulk steel in a ladle. This cycle is repeated again and again until oxygen level in the bulk steel drops to a desired level.

    The process is applied to remove oxygen from a 0.3% C steel at 1600 °C using a vacuum of 10^{-3} atm. Fraction of steel drawn in each cycle is 0.05. If each cycle takes 10 min, then calculate the time required to decrease oxygen level from 0.02% to 0.001%. Given at 1600 °C and 1 atm $\%[C] \times \%[O] = 2.0 \times 10^{-3}$.

References

Ammann, P.R., Looso, T.A.: Met. Trans. **2**, 889 (1971)

Bagshaw, N.E., Feeney, J.R., Harris, M.R.: Brit. Corr. J. **4**, 301 (1969)

Bandyopadhyay, G.K., Ray, H.S., Gupta, R.K.: Met. Trans. **2**, 239 (1971)

Bandyopadhyay, A.: Investigations on kinetics and mechanism of reoxidation of direct reduced iron. Ph.D. Thesis, I.I.T. Kharagpur (1988)

Barner, H.E., Manning, F.S., Philbrook, W.O.: Trans. TMS AIME, **227**, 897 (1963)

Beckstead, L.W., Miller, J.D.: Met. Trans. B **8B**, 31 (1977)

Beckstead, L.W., Miller, J.D.: Met. Trans. B **8B**, 19 (1977)

Clarke, J.A.: Chem. Indus. **6**, 5 (1979)

Dollimore, D.: J. Thermal Anal. vol. **13**, 455 (1978)

Drouzy, M., Mascrc, C.: Met. Rev. **3**(2), 25 (1969)

Dutrizac, J.E., Macdonall, R., Ingraham, T.R.: Met. Trans. **1**, 225, 3085 (1970)

Eckert, E.R.G., Drake, R.M.: Heat and Mass Transfer. Tata McGraw-Hill Pub. Co. Ltd, New Delhi (1979)

Eddington, P., Prosser, A.P.: IMM (London) Tran. C **78**, C74 (1979)

Eitd, W.: Silicate Science, vol. 8. Academic Press (Ch. 1) (1976)

Fukunaka, Y., Monta, T., Asaki, Z., Kondo, Y.: Oxidation of zinc sulphide in a fluidizied bed, Pvt. Communication (1976)

Ghosh, A.: In: Proceedings of International Conference on Advances in Chemical Metallurgy, vol. 2, p. 21/2. ICMS 79, B.A.R.C. Bombay Jan (1979)

Gibson, G., Ward, R.: J. Amer. Ceram. **26**(7), 239 (1940)

Gitterer, G.R.: J. Metals **19**, 92 (1967)

Herbst, J.A.: In: Sohn, H.Y., Wordsworth, W. (eds.) Rate Processes in Extractive Metallurgy. Plenum, New York (1979)

Hills, A.W.D.: I. Chem. E. Symp. Ser. **27**, 27 (1968)

Hills, A.W.D. (ed.): In: Heat and Mass Transfer in Process Metallurgy, p. 39, I.M.M. (London) (1967)
Hills, A.W.D.: Met. Trans. **9B**, 21 (1978)
Ingraham, T.R., Parsons, H.W.: Can Met. Q. **B**, 29 (1969)
Kautz, K.: Glastech. Ber. **42**(6), 244 (1969)
Khallafalla, S.K.: In: Sohn, H.Y., Wadsworth, M.E. (eds.) Rate processes in extractive metallurgy (Ch. 2). Plenum Press, New York (1979)
Kroger, C.: Glasstech. Ber. **22**(5/6), 331 (1948)
Kubaschewski, O., Hopkins, B.E.: Oxidation of metals and alloys (1953)
Levenspid, O.: Chemical reaction engineering. Wiley Eastern Pvt. Ltd. Pub. Bombay (Ch. 14) (1962)
Luthra, K.L., Ray, H.S., Kumar, D.: Inst. Eng. Trans. MM **53**, 58 (1973)
Morrison, A.L., Wright, J.K., Bowling, Mc.G.: Ironmaking and steelmaking **1**, 39 (1978)
Mukherjee, A., Ray, H.S., Sen, P.K.: Trans. IIM. **36**(4 and 5), 377 (1983)
Platt, J.C., Johnes, R.H., Wilburn, F.W.: L.R.211. P.B.Ltd, Lathom, UK (1967)
Rao, V.V.V.N.S.R., Abraham, K.P.: Met. Trans. **2**, 2423 (1971)
Rao, V.V.V.N.S.R., Abraham, K.P.: Met. Trans. **2**, 2463 (1971)
Rao, Y.K.: Chem. Eng. Sci. **29**, 1435 (1979)
Ray, H.S., Bandyopadhyay, G.K., Gupta, R.K.: Trans. 11 M 41 (1972)
Ray, H.S., Gupta, R.K., Bandyopadhyay, G.: Ind. J. Tech. **8**, 363 (1970)
Ray, H.S.: Batch reactions during glassmaking, unpublished research
Reilly, I.G., Scott, D.S.: Can. J. Chem. Eng. **55**, 527 (1977)
Reilly, I.G., Scott, D.S.: Met. Trans. B **15B**, 726 (1984)
Reilly, I.G., Scott, D.S.: Met. Trans. B **9B**, 681 (1978)
Sims, C.E.: J. Metals **11**, 815 (1954)
Smith, T.J.J., Maddocks, W.R., Nixon, E.W.: Ninth Commonwealth Mining and Metallurgical Congress 1969, Mineral Processing and Extractive Metallurgy Section (1969)
Sohn, H.Y.: Met. Trans. B **9B**, 89 (1978)
Sohn, H.Y., Wadsworth, M.E. (eds.): Rate Processes of Extractive Metallurgy, p. 429. Plenum Press, New York (1979)
Subbarao, E.C. (ed.): Solid Electrolytes and Their Applications, p. 201. Plenum Press, Chicago (1980)
Szekely, J., Evans, J.W.: Chem. Eng. Sci. **26**, 1901 (1971)
Szekely, J., Evans, J.W.: Chem. Eng. Sci. **25**, 1091 (1970)
Szekely, J., Evans, J.W.: In: Soekely, J. (ed.) Blast Furnace Technology, Science and Practice. Marcal Dekker Inc. New York (1972)
Themelis, N.J., Yannopoulos, J.C.: Trans. Met. Soc. AIME 236, 414 (1966)
Turkdogan, E.T.: Met. Trans. **9B**, 1633 (1978)
Wadsworth, M.E.: In: Sohn, H.Y., Wadsworth, M.E. (eds.) Rate processes in extractive metallurgy (Ch. 2). Plenum Press, New York (1979)
Warren, G.W., Wadsworth, M.E.: Met. Trans. B **15B**, 289 (1984)
Wegmann, E.F.: A Reference Book for Blast Furnace Operators (Ch. 5). Mir Publishers (1984) (Translated from the Russian in English by V. Afanasyev)
Wilburn, F.W., Metcalfe, S.A., Warburton, R.S.: Glass Tech. **6**(4), 167 (1965)
Wilburn, F.W., Dawson, J.B.: Glass, In: R.C. Mackenzie (ed.) Thermal Analysis, vol. 2, Academic Press London (1972)

Chapter 9
Non-isothermal Kinetics

9.1 Introduction

Limitations of isothermal studies

An isothermal reactor is an abstraction. Most industrial processes involve gradual heating up of reactants with reactions proceeding under rising temperature and fluctuating temperature conditions.

Traditional kinetic studies, however, involve a series of runs carried out in isothermal environment. From the plots of fraction reacted, α, against time, t, a reaction rate constant is derived by expressing α as an appropriate function $g(\alpha)$, of t. If the experiment is then repeated at several other temperatures, one obtains the temperature dependence of the rate constant k. If the reaction mechanism is assumed to remain unchanged in the temperature range of the experiments then an Arrhenius-type plot can be obtained and the activation energy calculated from the slope of the (linear) Arrhenius plot.

The traditional approach has some limitations, which are often ignored. These may be summarized as follows.

(a) Strictly speaking, no reaction takes place isothermally because all reactions are accompanied by a heat change. For example, (exothermic) oxidation of cuprous sulphide (Ramakrishna Rao 1971) and (endothermic) decomposition of calcium carbonate (Hills 1966) cause significant changes in sample temperatures. The kinetic data, therefore, need to be analysed in terms of possible heat transfer effects due to the deviation of the temperature of the sample from that of the isothermal environment.

(b) It may not be possible to attain a thermal equilibrium with the environment without significant prereaction. In other words, there may be some reaction during the heating of the reaction system to the predetermined temperature and a part of the sample may change before the beginning of the kinetic run.

H. S. Ray and S. Ray, *Kinetics of Metallurgical Processes*,
Indian Institute of Metals Series, https://doi.org/10.1007/978-981-13-0686-0_9

(c) It may not be possible to reproduce the physical characteristics of the sample from run to run.
(d) The isothermal technique is inadequate when the product of a reaction becomes the reactant in another reaction at a higher temperature. Complex reactions involving several series, parallel, or independent overlapping reactions can hardly be studied in the traditional way.
(e) In most industrial reactors, the reacting materials are heated gradually and, therefore, isothermal conditions seldom represent a realistic situation.
(f) The kinetic parameters of the Arrhenius-type expression may not have identical values under isothermal and non-isothermal conditions. Accordingly, reaction rates determined under isothermal conditions may not be applicable for actual processes taking place non-isothermally.

Dynamic non-isothermal kinetic studies are carried out by allowing a reaction to take place at progressively higher temperatures using a well-defined temperature–time sequence, the progress of the reaction being recorded (e.g. as weight change) against temperature. This technique overcomes several limitations of the isothermal technique. It is also possible to design using the concepts of the non-isothermal technique a new class of experiments where a large mass of materials react under non-isothermal conditions in such a way that different regions react for different periods.

There is considerable interest in the development of the mathematics of non-isothermal kinetics and its application in solid-state decomposition reactions. Some go to the extent of declaring the non-isothermal determination of kinetic parameters as the only valid measurements (Draper and Mc Adie 1970).

The basic concepts of non-isothermal techniques should be applicable to all heterogeneous reactions in general, and as is shown subsequently, they allow the planning of a new class of kinetic studies, which are useful for the study of many industrial processes. This chapter gives the appropriate mathematical approach for the analysis of non-isothermal kinetic data obtained under varying conditions of temperature–time programming. It also indicates some methods of carrying out kinetic studies under non-isothermal conditions.

9.2 Procedure for Non-isothermal Studies

Thermogravimetry

For non-isothermal studies, the temperature of the reaction system should be increased in a programmed manner. In thermogravimetry (TG), the sample size is often kept very small (10–100 mg) to ensure thermal equilibrium with the furnace which is subjected to linear rise in temperature. Non-isothermal studies have been, therefore, primarily restricted to studies on pure compounds, and the method has not been extended generally to heterogeneous reactions where small sample weights are rarely representative. Moreover, the mathematical treatments have been mostly restricted to constant heating rates.

If we want to study large samples under non-isothermal conditions, then the following procedures may be adopted.

(a) The reacting system with relatively large mass may be gradually heated as in conventional thermogravimetry but using very slow heating rates to allow thermal equilibrium.
(b) The reaction mixture may be spread as a thin layer in a moving boat and the boat introduced at a steady speed into the hot zone of a furnace. The standing position of the boat rests the leading edge of the boat at the very beginning of the hot zone. As soon as the entire boat goes into the furnace, it is quickly withdrawn and the reaction mass quenched by using an appropriate technique. If no further reaction is allowed once the boat is withdrawn, then the position of a volume element along the length of the boat represents the time scale. The reaction time of the element can be calculated knowing the total length of the boat, its total travel time, and the location of the volume element. Analysis of the reaction mixture from various locations, therefore, yields kinetic data for reaction of a large number of volume elements for different periods. This technique may be called the time gradient technique or the moving boat technique. In this technique, the heating rate of a volume element will depend on its location, the actual temperature profile of the furnace, the speed of the boat, and the thermal characteristics of the reaction mixture specially if there is a thick layer of it. The heating programme can be changed widely by changing these aforementioned parameters, and the actual variation must be determined using thermocouples placed at various locations in the boat.
(c) In the third method, the reaction mixture is spread as a thin layer in a boat placed in the hot zone which has a particular built-in temperature profile. The boat is allowed to remain in the non-isothermal temperature profile for a pre-determined period and then removed and quenched to yield kinetic data for reaction of a large number of volume elements for a fixed period but at different temperatures. The run is repeated for different predetermined time values to obtain in the effect of time. This technique may be called the temperature gradient technique.

A detailed description of some of the techniques is given in the next chapter.

9.3 Kinetic Equations Under Rising Temperature Conditions

9.3.1 Reaction Rate Under Non-isothermal Condition

The theoretical treatment of kinetic data obtained under rising temperature conditions rests largely on a combination of three basic equations (Draper and Mc Adie

1970; Buker 1978; MacCullum 1127; 1971). The first is the kinetic law written in the differential form

$$\frac{\mathrm{d}\alpha}{\mathrm{d}t} = k(T) \cdot f(\alpha) \cdot \phi(\alpha \cdot T) \tag{9.1}$$

where $k(T)$ is the temperature-dependent specific rate constant, f and ϕ denote functions, t is time and T is temperature. Generally $\phi(\alpha, T)$ is assumed to be unity, and thus,

$$\frac{\mathrm{d}\alpha}{\mathrm{d}t} = k(T) \cdot f(\alpha) \tag{9.2}$$

The second equation we need is the law describing the temperature coefficient of the rate constant

$$k = A \cdot T^m \cdot \exp(-E/\mathrm{R}T) = A \cdot \exp(-E/\mathrm{R}T) \tag{9.3}$$

where E is activation energy, A the pre-exponential factor, R is the gas constant, and m is a constant generally assumed to be zero.

According to Buker (1978), the Arrhenius equation arises in gas and liquid phase reactions because of equilibrium between activated and non-activated molecules according to the Boltzmann distribution followed by formation of an activated complex from activated molecules and subsequent formation of product molecules. Solid-state decomposition reactions may not proceed by such a mechanism, and it has been claimed that the Arrhenius equation is not justified theoretically.

The third equation we need should describe the variation of the temperature T (K) with time. For a linear rate to rise

$$T = T_0 + Bt \tag{9.4}$$

where T_0 is the initial temperature (K) and B is the rate of rise of T.

Combining, Eqs. 9.2 and 9.3

$$\frac{\mathrm{d}\alpha}{\mathrm{d}t} = A \cdot \exp \cdot (-E/\mathrm{R}T) \cdot f(\alpha) \tag{9.5}$$

From Eq. (9.4), we obtain

$$\mathrm{d}T = B \cdot \mathrm{d}t \tag{9.6}$$

$$\therefore \quad \frac{\mathrm{d}\alpha}{f(\alpha)} = \frac{A}{B} \cdot \exp(-E/\mathrm{R}T) \cdot \mathrm{d}T \tag{9.7}$$

Equation 9.7 is the basic equation in non-isothermal kinetic analysis. As is obvious, several assumptions have to be made in deriving this relationship.

There has been much controversy (MacCullum and Tanner 1970; MacCullum 1971; Hill 1970; Felder and Stahel 1970; Garn 1974; Gorbachev and Logvinenko 1973) regarding the actual meaning of the term $(\mathrm{d}\alpha/\mathrm{d}t)$ since MacCullum and Tanner (1970) first claimed that this term was not identical with $(\partial\alpha/\partial t)_T$ for isothermal conditions. According to them, if

$$\alpha = f(t, T) \tag{9.8}$$

then,

$$\mathrm{d}\alpha = \left(\frac{\partial\alpha}{\partial t}\right)_T \mathrm{d}t + \left(\frac{\partial\alpha}{\partial t}\right) \mathrm{d}T \tag{9.9}$$

$$\therefore \quad \frac{\mathrm{d}\alpha}{\mathrm{d}t} = \left(\frac{\partial\alpha}{\partial t}\right)_T + \left(\frac{\partial\alpha}{\partial t}\right)_t \frac{\mathrm{d}T}{\mathrm{d}t} \tag{9.10}$$

Thus,

$$\frac{\mathrm{d}\alpha}{\mathrm{d}t} \neq \left(\frac{\partial\alpha}{\partial t}\right)_T \quad \text{unless} \quad \frac{\mathrm{d}T}{\mathrm{d}t} = 0 \tag{9.11}$$

$$\text{or,} \quad \left(\frac{\partial\alpha}{\partial T}\right)_t \cdot \frac{\mathrm{d}T}{\mathrm{d}t} = 0 \tag{9.12}$$

According to MacCullum and Tanner, therefore, Eq. 9.5 giving the temperature dependence of reaction rate cannot be written at all for non-isothermal conditions. They claimed that only isothermal methods could be used to evaluate the order of the reaction, the pre-exponential factor, and the activation energy.

Various arguments (Hill 1970; Felder and Stahel 1970; Garn 1974; Gorbachev and Logvinenko 1973) have subsequently been given to show that the above-mentioned proposition is incorrect and the chemical reaction rate is indeed given by the slope of the plot of α against t whether the temperature is varying or not. It has been shown (Felder and Stahel 1970) that the term $(\partial\alpha/\partial T)_t$ is devoid of physical meaning because temperature cannot be varied holding time constant. Moreover, even if instantaneous change were possible, $\partial\alpha$ must be zero because α cannot change instantaneously. Also, the very equation, Eq. 9.9, is incorrect only if α is a state function of the variables t and T. In reality, however, α is a path function. Gorbatchev and Logvinenko (1973) have shown that expressing α as a function of t and T leads to mathematical inconsistency.

9.3.2 Treatment of Non-isothermal Kinetic Data

Two distinct approaches may be adopted in analysing non-isothermal kinetic data.

(a) Analysis by integral methods which consider the basic differential equation such as Eq. 9.7 and obtain the kinetic constant from their integrated forms. The weight–temperature plots are sufficient, provided the temperature–time programme is known.
(b) Analysis by obtaining the data in a differential form and setting the values of $(\mathrm{d}\alpha/\mathrm{d}T)$, α and T into a suitably derived form of the expressions. This requires special derivative thermogravimetric DTG apparatus or not too accurate and lengthy procedure of graphical differentiation of TG curves.

Review of some of the methods reported in the literature has been given by several workers (Buker 1978; Sestak and Krotochvil 1973; Wendlandt 1964; Sestak et al. 1973; Satava 1973; Gyulai and Greenhow 1974; Satava and Skvara 1969). Most of these reviews give incomplete coverage of the subject and rather cumbersome mathematical approaches. Here we present some straightforward approaches based on well-defined mathematical procedures and approximations.

The analytical approach—Integration of Eq. 9.7

Suppose that the integrated form of the kinetic law as:

$$g(\alpha) = kt \tag{9.13}$$

Differentiating,

$$\frac{\mathrm{d}\alpha}{\mathrm{d}t} = \frac{k}{g'(\alpha)} = k \cdot f(\alpha) \tag{9.14}$$

Then

$$\int \frac{\mathrm{d}(\alpha)}{f(\alpha)} = \int g'(\alpha) \cdot \mathrm{d}\alpha = g(\alpha) \tag{9.15}$$

Hence integrating Eq. 9.7,

$$g(\alpha) = \frac{A}{B} \int_0^T \exp \cdot (-E/RT) \cdot \mathrm{d}T \tag{9.16}$$

It should be noted that the LHS is defined only when the integrated form of Eq. 9.13 or the analytical form of function $f(\alpha)$ is known. It should also be noted that the RHS of Eq. 9.16 cannot be integrated in finite form. Introducing a new variable

$$u = \frac{E}{RT}, \quad \text{i.e.,} \quad T = \frac{E}{Ru} \tag{9.17}$$

$$\mathrm{d}T = \frac{E}{Ru^2}\mathrm{d}u \tag{9.18}$$

$$\exp\left(-\frac{E}{RT}\right)\mathrm{d}T = \frac{E\mathrm{e}^{-u}}{Ru^2}\mathrm{d}u \tag{9.19}$$

$$g(\alpha) = \frac{AE}{BR}\int\limits_{T=0}^{T=T}\frac{\mathrm{e}^{-u}}{u^2}\mathrm{d}u \tag{9.20}$$

When $T = 0$, $u = x_0$, x_0 being infinity:

$$T = T,\ u = x, \text{ say.}$$

Doyle (1961) wrote Eq. 9.20 as

$$g(\alpha) = \frac{AE}{BR}p(x) \tag{9.21}$$

and calculated the value of the integral $p(x)$ for x values covering the range from 10 to 50. Others have also tabulated the values.

The main difficulty in applying Eq. 9.21 consists in the dependence of $p(x)$ on both temperature and activation energy. Doyle suggested a trial and error curve fitting method for the determination of activation energy. Taking logarithms, Eq. 9.21 becomes

$$B' = \log\frac{AE}{BR} = \log\ g(\alpha) - \log p(x) \tag{9.22}$$

where B' depends on the rate of heating B and the nature of the reaction (characterized by A and E) but not on the temperature. The value of $g(\alpha)$ for a given temperature can be calculated from the experimental data if the form of either $g(\alpha)$ or $f(\alpha)$ is known. Similarly, $p(x)$ for the same temperature can be found if the activation energy is known. In the trial and error method, the apparent activation energy is estimated by finding the E value for which $\log\ g(\alpha) - \log p(x)$ at different temperatures is closest to being a constant, independent of temperature. To facilitate the analysis of experimental data, a table of $\log\ g(\alpha)$ values for various mechanisms (i.e. forms of $g(\alpha)$ and a plot of $-\log p(x)$ versus T for various activation energies has been prepared by Stava and Skvara (1969) who have also discussed the use of the table and the diagram.

Approximate Solutions of the Integral

We can write, using integration by parts

$$\begin{aligned}
-\int \frac{\mathrm{e}^{-u}}{u^2}\mathrm{d}u &= \int \frac{\mathrm{e}^{-u}}{u^2}(-\,\mathrm{d}u) = \frac{\mathrm{e}^{-u}}{u^2} + \int \frac{2\mathrm{e}^{-u}}{u^3}\mathrm{d}u \\
&= \frac{\mathrm{e}^{-u}}{u^2} - \frac{2\mathrm{e}^{-u}}{u^3} + \int \frac{3!\mathrm{e}^{-u}}{u^4}\mathrm{d}u \\
&= \frac{\mathrm{e}^{-u}}{u^2} - \frac{2!\mathrm{e}^{-u}}{u^3} - \frac{3!\mathrm{e}^{-u}}{u^4} + \int \frac{4!\mathrm{e}^{-u}}{u^5}\mathrm{d}u \\
&= \mathrm{e}^{-u}\left[\frac{1}{u^2} - \frac{2!}{u^3} + \frac{3!}{u^4} + \cdots\right]
\end{aligned} \tag{9.23}$$

$$g(\alpha) = \frac{AE}{BR}\mathrm{e}^{-E/RT} \cdot \left[\frac{R^2T^2}{E^2} - \frac{2!R^3T^3}{E^3} + \frac{3!R^4T^4}{E^4} - \frac{4!R^5T^5}{E^3}\right] \tag{9.24}$$

Murray and White (1955) have examined the relative magnitudes of the terms within the brackets for thermal decomposition of kaoline using a value of 44 kcal/mole for E and 873 K for T. Taking the value of the gas constant R as approximately 2 calories, the first three terms within the brackets on the RHS of Eq. 9.24 are

$$(0.04)^2 - 2(0.04)^3 + 6(0.04)^4 = 0.0016 - 0.00013 + 0.00001$$

It may thus seem that each term is approximately one-tenth of the previous term, and the first term being dominant. Further, since the terms are alternately positive and negative, it may be justified to take the first few terms. However, one should note that after a few terms, the absolute values of the terms begin to increase again because of the factorials. This sort of function, known as asymptotic function, does not converge. However, since the positive and negative signs alternate the sum lies in between and often the first few terms can be considered for approximate solution.[1] Considering only the first term, and then only the first two terms, of Eq. 9.24 and remembering that $x = E/RT$, one obtains the following approximations, respectively:

$$p(x) = x^{-2} \cdot \mathrm{e}^{-x} \tag{9.25}$$

$$p(x) = (x-2) \cdot x^{-3} \cdot \mathrm{e}^{-x} \tag{9.26}$$

[1]Abramowtiz and Stegun (1964), Jeffreys and Swirles (1965), Chapter 17.

Doyle (1965) has listed some additional approximate relationships

$$p(x) = (x+1)^{-1} \cdot x^{-1} \cdot e^{-x} \tag{9.27}$$

$$p(x) = (x+2)^{-1} \cdot x^{-1} \cdot e^{-x} \tag{9.28}$$

These approximations reflect the best balance of accuracy and convenience in the integration of the exponential integral. As for accuracy, these approximate expressions can be compared by taking the ratio between true and approximate values of $p(x)$ for various values of x. It has been found (Doyle 1965) that Eq. 9.28 is perhaps the best form. Gorbachev (1975) has used this approximation to complete the integration of Eq. 9.20. Substitution of Eq. 9.28 and transformations yield

$$g(\alpha)\frac{AE \cdot \exp\left(\frac{E}{RT}\right)}{BR\left[\frac{E}{RT+2}\right] \cdot \left(\frac{E}{RT}\right)} = \frac{A}{B}\left[\frac{RT^2}{E+2RT}\right] \cdot \exp\left(\frac{E}{RT}\right) \tag{9.29}$$

Ozawa (1981) has suggested a method based on Eq. 9.21 for determining the activation energy. He showed that, in rising temperature thermogravimetry, the activation energy can be graphically obtained by following the course of the reaction using different heating rates. Since the LHS of Eq. 9.21 does not depend on the heating rate, it is possible to equate the RHS values for different heating rates. That is, if the weight of a reacting sample decreases to a given fraction at the temperature T_1 for the heating rate B_1, at T_2 for B_2 and so on, then

$$\frac{AE}{B_1R} \cdot p\left(\frac{E}{RT_1}\right) = \frac{AE}{B_2R} \cdot p\left(\frac{E}{RT_2}\right) = \cdots \tag{9.30}$$

He used the following approximation (Doyle 1962) for $p(x)$ reported in the literature for values of $p(x)$ 20,

$$\log p(x) = -2.315 + 0.4567x \tag{9.31}$$

Equation 9.30, therefore, can be written as

$$-\log B_1 - 0.4567\frac{E}{RT_1} = -\log B_2 - 0.4567\frac{E}{RT_2} \tag{9.32}$$

Thus the plots of log B versus $1/T$ for a given value of fraction reacted must give a straight line, the slope of which gives the activation energy. In other words, if the thermogravimetric data are plotted as fraction reacted versus $1/T$ for various heating rates, then the curves can be superimposed upon each other by shifting them along the abscissa. It is also possible to determine the form of $g(\alpha)$ or $f(\alpha)$ by comparison of experimental plots of a against $\left[\log \cdot \frac{AE}{BR}p(x)\right]$ with master plots.

Gorbachev (1976), however, considers this method mathematically unsound. Ozawa (1976) himself, in a later publication, tried to modify the method in recognition of its original limitations.

Integration Using a Graphical Approach

One of the best solutions to the problem of integration is given by Coats and Redfern (1964) who have suggested a graphical procedure based on the approximation given by Eq. 9.26.

Consider a reaction mechanism where $g(\alpha)$ has the form $(1-\alpha)^n$. (For contracting sphere model, $n = 2/3$.) We have

$$\int_0^{\alpha} \frac{\mathrm{d}\alpha}{(1-\alpha)}n = \frac{A}{B}\int_0^{T} \mathrm{e}^{-(E/RT)}\mathrm{d}T \approx \frac{ART^2}{BE}\mathrm{e}^{-(E/RT)}\left[1-\frac{2RT}{E}\right] \tag{9.33}$$

If $n = 1$, then

$$\log_{10}\left[-\ln\frac{(1-\alpha)}{T^2}\right] = \log_{10}\frac{AR}{BE}\left[1-\frac{2RT}{E}\right] - \frac{E}{2.3RT} \tag{9.34}$$

Hence, a plot of the LHS against $1/T$ gives a straight line with slope $-E/2.3R$. If $n \neq 1$, then

$$\begin{aligned} &\int (1-\alpha)^{-n}\mathrm{d}\alpha = -\frac{(1-\alpha)^{1-n}}{(1-n)} \\ \therefore\quad &\frac{1-(1-\alpha)^{1-n}}{(1-n)T^2} = \frac{AR}{BE}\left[1-\frac{2RT}{E}\right]\mathrm{e}^{-(E/RT)} \end{aligned} \tag{9.35}$$

It is easily shown that the pre-exponential term in RHS is almost constant. So a plot of $\log_{10}$(LHS) against $1/T$ gives a straight line with slope $-E/2.3R$. This approach has been used by Medek (1976), Judd and Pope (1973) and Ozawa (1973), among many others.

Analysis of non-isothermal kinetic data based on the differential values

We shall now consider some methods which aim at setting the values of $(\mathrm{d}\alpha/\mathrm{d}t)$, α, and T into some suitably derived form of the basic equation for isothermal kinetics, Eq. 9.7.

Freeman and Carroll (1958) wrote Eq. 9.5 as

$$\begin{aligned} &A\exp(-E/RT) = \frac{(\mathrm{d}\alpha/\mathrm{d}t)}{f(\alpha)}, \\ \therefore\quad &\ln A - \frac{E}{RT} = \ln\frac{\mathrm{d}\alpha}{\mathrm{d}t} - \ln f(\alpha) \end{aligned} \tag{9.36}$$

Differentiating,

$$\frac{E}{RT^2}\mathrm{d}T = d\ \ln\frac{\mathrm{d}\alpha}{\mathrm{d}t} - d\ \ln f(\alpha) \tag{9.37}$$

This equation takes a simple form in some special cases to allow a graphical approach. For example, if $f(\alpha) = (1-\alpha)^n$, then

$$\begin{aligned} \frac{E}{RT^2}\mathrm{d}T &= d\ \ln\frac{d\alpha}{\mathrm{d}t} - n\,d\ \ln(1-\alpha) \\ \therefore\quad \frac{E}{\mathrm{R}T^2}\frac{\mathrm{d}T}{d\ \ln(1-\alpha)} &= \frac{d\ \ln(\mathrm{d}\alpha/\mathrm{d}t)}{d\ \ln(1-\alpha)} - n \end{aligned} \tag{9.38}$$

Hence, a plot of $\left(1/T^2\right)\cdot(\mathrm{d}T/\mathrm{d}\ln(1-\alpha))$ against the first term in RHS should result in a straight line with slope of $-E/R$ with intercept n.

In the other approach suggested by Keatch and Dollimore (1975), Eq. 9.7 is rewritten as

$$\log\left[\frac{(\mathrm{d}\alpha/\mathrm{d}t)}{f(\alpha)}\right] = \log\left(\frac{A}{B}\right) - \left(\frac{E}{RT}\right) \tag{9.39}$$

$$\text{or,}\quad \log\left[\frac{(\mathrm{d}\alpha/\mathrm{d}t)B}{f(\alpha)}\right] = \log k = \log A - \frac{E}{RT} \tag{9.40}$$

It is thus obvious that values of A and E are obtained from a plot of $\log[(1/f(\alpha))\cdot(\mathrm{d}\alpha/\mathrm{d}t)]$ against $(1/T)$ and the value of k at a temperature T is given by Eq. 9.40 which is directly analogous to the conventional Arrhenius equation. It is thus possible to make plots of $\log k$ against $1/T$ and attempt a direct comparison with Arrhenius plot established from a series of isothermal experiments. This method of treating non-isothermal kinetic data requires knowledge of the parameters $(\mathrm{d}\alpha/\mathrm{d}T)$, $f(\alpha)$, and B, and the accuracy of determination of $(\mathrm{d}\alpha/\mathrm{d}T)$ determines the usefulness of the whole method.

9.3.3 Nonlinear Heating Programmes

So far, we have only considered a steady rate of temperature rise. It is possible to formulate the differential equation for nonlinear heating programmes.

The hyperbolic rise
A very special programme of heating which allows easy integration of the basic equation is the hyperbolic heating programme (Doyle 1962; Simon and Debreczeny 1971; Zsako 1970)

$$\frac{1}{T} = a - bt \tag{9.41}$$

where a and b are constants. At $t = 0$, $T = T_0$ and at $t = t_{max}$, $T = T_{max}$. This gives

$$a = 1/T_0, \quad b = \frac{1}{t_{max}}\left(\frac{1}{T_0} - \frac{1}{T_{max}}\right)$$

The values of T_{max} and t_{max} cannot be chosen arbitrarily as they depend on the parameters of the furnace. However, we can write

$$\begin{aligned} g(\alpha) = \int_0^\alpha \frac{d\alpha}{f(\alpha)} &= -\frac{A}{b}\int_\infty^{1/T} \exp\left(-\frac{E}{RT}\right)\cdot d\left(\frac{1}{T}\right) \\ &= \frac{AR}{bE}\cdot \exp\left(-\frac{E}{RT}\right) \end{aligned} \tag{9.42}$$

Exponential temperature rise

Suppose that a moving boat with reaction mixture is introduced into isothermal hot zone of temperature T_f. Then heating of a volume element of the reaction mixture may be expressed by

$$\frac{dT}{dt} = c\cdot(T_f - T) \tag{9.43}$$

where T is the temperature after time t in the isothermal hot zone and c is a constant. Integrating Eq. 9.43, one obtains a general case of non-programmed temperature variation which is exponential like

$$T = T_f - (T_f - T_0)\exp(-ct) \tag{9.44}$$

where T_0 is the starting temperature. It is easily seen that from Eqs. 9.43 to 9.44, $(d\ln T/dt)$ is constant (c) rather than (dT/dt) as in the linear programming. Combination of Eq. 9.43 with Eqs. 9.2 and 9.3 gives

$$\frac{d\alpha}{f(\alpha)} = -\frac{AE}{Rc}\frac{\exp(-E/RT)}{(T_f - T)}\cdot dT$$

Using the symbol $u = (E/RT)$, one obtains

$$\frac{d\alpha}{f(\alpha)} = \frac{AE}{Rc}\frac{\exp(-u)}{(T_f - T)}\cdot u^2 du \tag{9.45}$$

Equation 9.45 indicates that the mathematics of nonlinear kinetics becomes rather complicated with the exponential heating programme. It could be even more complicated in real situation where T_f is not initially constant but varies with time. If we assume that there is initially an exponential rise of T_f itself in the hot zone, the variation may be written as

$$T_f = T_{\mathrm{M}}(1 - \mathrm{e}^{-jl}) \tag{9.46}$$

where T_{M} is the ultimate maximum temperature, j a constant and l the distance from the beginning of the hot zone. If a boat containing reaction mixture enters such a zone at a constant speed v, then $l = vt$. The temperature to which a volume element is exposed at time t is then given by

$$T_f = T_{\mathrm{M}}(1 - \mathrm{e}^{-jvt}) \tag{9.47}$$

The rate of heating is given by

$$\frac{\mathrm{d}T}{\mathrm{d}t} = c(T_f - T) \tag{9.48}$$

Note that here T_f is varying, unlike for Eq. 9.43. Integration of Eq. 9.43 using the relation given by Eq. 9.47 gives the variation of T with t.

$$T = \frac{T_{\mathrm{M}}}{c - jv}\left[c\left(1 - \mathrm{e}^{jvt}\right) - jv(1 - \mathrm{e}^{-ct})\right] \tag{9.49}$$

Equation 9.49, which when plotted yields an S-shaped curve, gives a general heating programme if c is constant. For a more general case, the variation of c with temperature should be considered.

The analysis of kinetic data obtained under such heating programmes becomes complicated for the integral approach. However, analysis is made easy by the use of differential approach as discussed later.

9.3.4 *Use of a Nonlinear Heating Programme for Linear Heating*

We have seen that theoretical analysis of kinetic data obtained under increasing temperature conditions rests on the combination of three basic equations which describe the following:

(a) The kinetic law,
(b) The so-called Arrhenius-type expression,
(c) Equation describing variation of temperature with time.

Thermogravimetry (TG) is generally carried out with a constant heating rate. If the heating rate is B, as the basic equation for treatment of TG data is

$$g(\alpha) = \int_0^{\alpha} \frac{\mathrm{d}\alpha}{f(\alpha)} = \frac{A}{B}\int_0^{T} \exp(-E/RT)\mathrm{d}T$$

This exponential integral is not amenable to analytical solution. and therefore, only approximate solutions based on various approximations are possible, as has been discussed before.

A special case arises if the temperature variation is given by the equation

$$\frac{1}{T} = a - bt, \quad \mathrm{d}(1/T) = -b\,\mathrm{d}t \tag{9.41}$$

where a and b are constants. In this case, one gets the simple and exact result

$$\ln g(\alpha) = \ln(AR/bE) - E/RT \tag{9.42}$$

We will now show that under certain conditions, Eq. 9.41 can approximately describe temperature variation of the linear as well as some other forms if values of a and b are chosen appropriately. Therefore, Eq. 9.42 can be employed to analyse kinetic data (Dixit and Ray 1982).

Conversion of linear temperature rise equation to reciprocal temperature rise equation

For linear rise we have,

$$T = T_0 + Bt \tag{9.50}$$

Provided that $Bt/T_0 \ll 1$,

$$\frac{1}{T} = \frac{1}{T(1 + Bt/T_0)} = \frac{1}{T_0}\left(1 - \frac{Bt}{T_0}\right) = \frac{1}{T_0} - \frac{Bt}{T_0^2} \tag{9.51}$$

Thus, if the non-isothermal range of reaction Bt is small compared to the initial temperature, then Eq. 9.50 can be written in the form of Eq. 9.41 where $a = 1/T_0$ and $b = Bt/T_0^2$.

Matching of the actual data to the reciprocal temperature rise equation is described next, and it is shown that this is easier when the heating rate is not entirely constant but increases slightly with time as is found, in some, commercially available equipment.

Matching of data
Suppose $T_0 = 600$ K, and the heating rate is 20 K min^{-1} i.e. $B = 20$ with t in minutes. That is, the linear temperature (K) rise equation is:

$$T = 600 + 20t$$

Let us also assume that we wish to describe the temperature data in the range 900– 1100 K by a reciprocal temperature rise equation.

If we shift $t = 0$ to 900 K, then new values of T_0 and B are 900 (T_0') and 20, respectively. Thus, if we assume Eq. 9.41 to be valid then a (i.e. $1/T_0'$) equals $1/900$ and b (i.e. $B/T_0'^2$) equals $20/(900)^2$. The temperatures obtained by using these values of a and b in the equation $1/T = a - bt$ are shown in Fig. 9.1. As expected the new plot shows a gradual deviation from the linear plot at higher values of t. A better matching can be obtained by changing the value of b using a slightly lower value of B. A plot obtained by using $B = 17.5$ and, therefore, $b = 17.5/900^2$, is also shown in Fig. 9.1. Obviously, better matching can be obtained if T_0 is set at a value higher than 900 K. We may thus conclude that it is possible to approximately fit the linear data into a reciprocal temperature rise

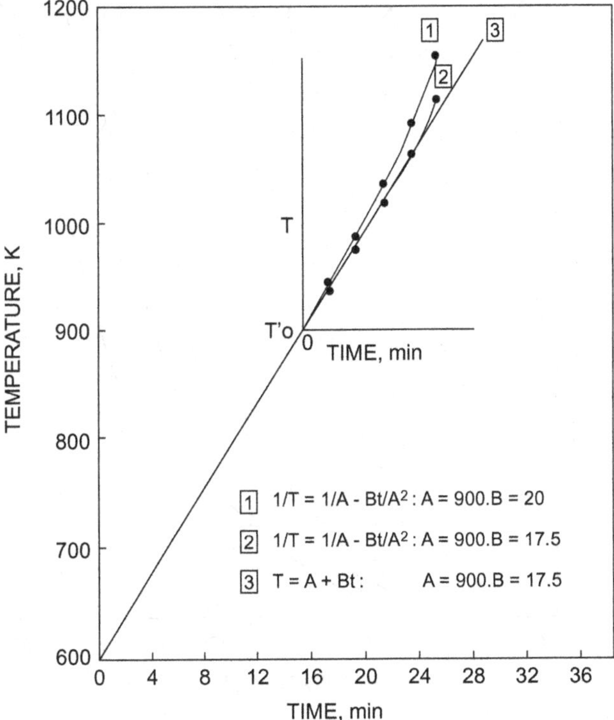

Fig. 9.1 Matching of the reciprocal temperature rise equation with constant rate of rise equation

equation, provided T_0 is high and the temperature range under consideration is small. It should be emphasized at this point that for arriving at the result (Eq. 9.42) and the evaluation of the activation energy, it is not always necessary to know the exact values of a and b. It is sufficient to establish the constraint $Bt/T_0 \ll 1$ and, therefore, ensure that there is a suitable equation. It is interesting to note that Satava (1971, 1973), using an entirely different approach, has shown that, for the correct mechanism, $\log g(\alpha)$ must be a linear function of $1/T$.

Analysis of some typical experimental data

Figure 9.2 shows some typical data for the decomposition of calcium carbonate under various conditions. These data were obtained by Dixit and Ray (1982) using

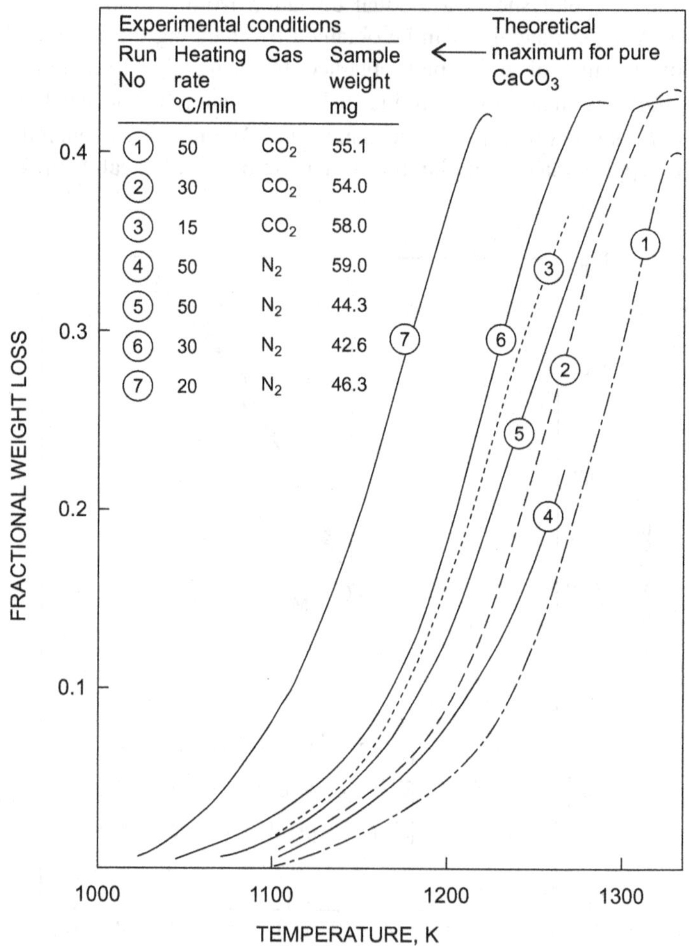

Fig. 9.2 TG data for the non-isothermal decomposition of calcium carbonate (Dixit and Ray 1982)

a Stanton Redcroft thermobalance (model TG-770), which allowed linear temperature rise at variable rates and flushing of the system with gas, if necessary. The $CaCO_3$ used was −200 mesh Ben Bennet limestone of composition CaO: 99.5–99.8 %, MgO: 0.1–0.2%, SiO_2: 0.1–0.2%, and Fe_2O_3: 0.1–0.2%. The gases used were obtained directly from the cylinder. Other experimental conditions are indicated in the figure.

The data shown in Fig. 9.2 have been analysed using the methods proposed by Coats and Redfern (1964) and the method proposed in this section. The computational results in terms of correlation coefficients for the linear plots proposed, slope and the activation energy values showed reasonably good matching. Some details of the analysis are presented in the next section.

It can be concluded that if the rate of heating is constant during increasing temperature experiments, then the temperature variation can also be described approximately by the reciprocal temperature rise equation, provided the temperature range being considered is limited and the initial temperature being considered is high. The use of the reciprocal temperature rise equation affords a simple method of analyzing non-isothermal kinetic data.

9.4 Examples of Analysis of Non-isothermal Kinetic Data

9.4.1 Decomposition of $CaCO_3$

A TG apparatus (see Chap. 10) was used for carrying out some isothermal decomposition runs using powder samples of around 50 mg. The results (MacCullum and Tanner 1970) thus obtained are shown in Figs. 9.3 and 9.4. It is seen that data for most runs conform to the phase boundary control equation. These data also yield isothermal values of E and $\log A$. The activation energy has the values of 49.7 kcal/mole in nitrogen and 57.6 kcal/mole in CO_2 (see Fig. 9.5). The value is higher for decomposition in CO_2 atmosphere.

Non-isothermal decomposition (TG experiments)

Consider the results presented in Fig. 9.2 from dynamic TG experiments for three different heating rates under gaseous environments. Runs indicating N_2 were carried out under nitrogen flowing at 6 cc/min, and runs indicating CO_2 were carried out with no flushing gas so that the samples were allowed to establish an environment of CO_2 generated by their own decomposition reactions. The results show, as expected, that decomposition is facilitated by flushing the system with an inert gas and by decreasing the heating rate. In one particular case (compare 4 and 5), there is an indication that the reaction is also aided by a decrease of the sample size.

To analyse these non-isothermal data, one can use the equation suggested by Coats and Redfern (Eq. 9.35).

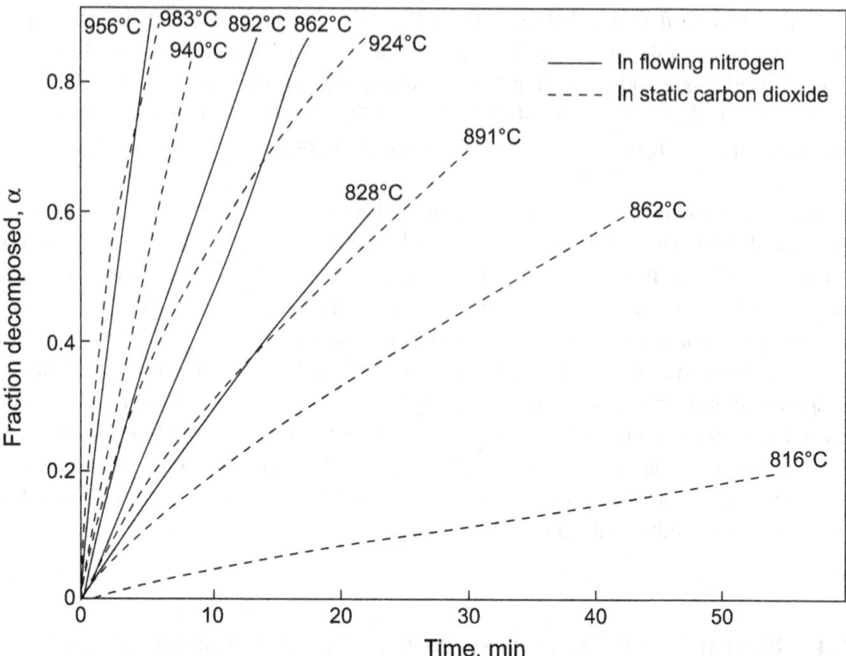

Fig. 9.3 Data for isothermal decomposition of $CaCO_3$

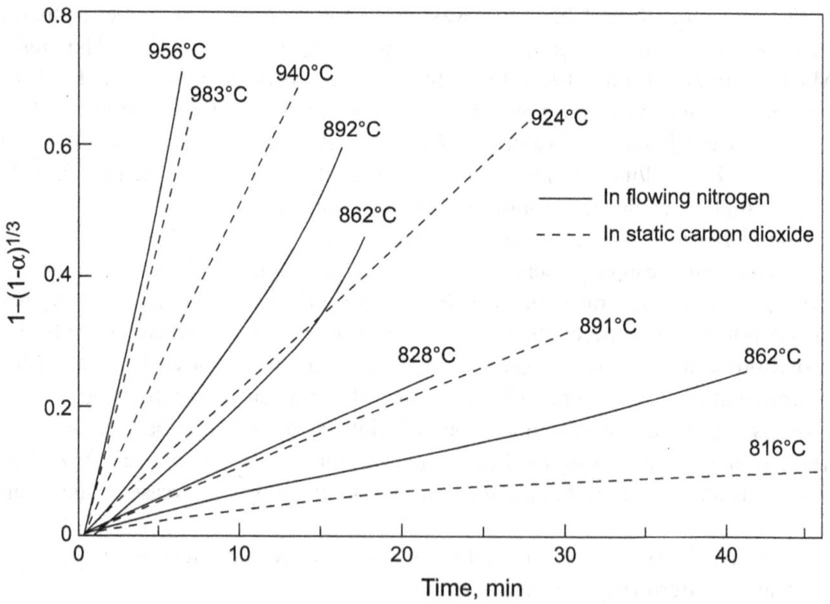

Fig. 9.4 Data of Fig. 9.3 plotted according to kinetic equation, $1-(1-\alpha)^{1/3}=kt$ (MacCullum and Tanner 1970)

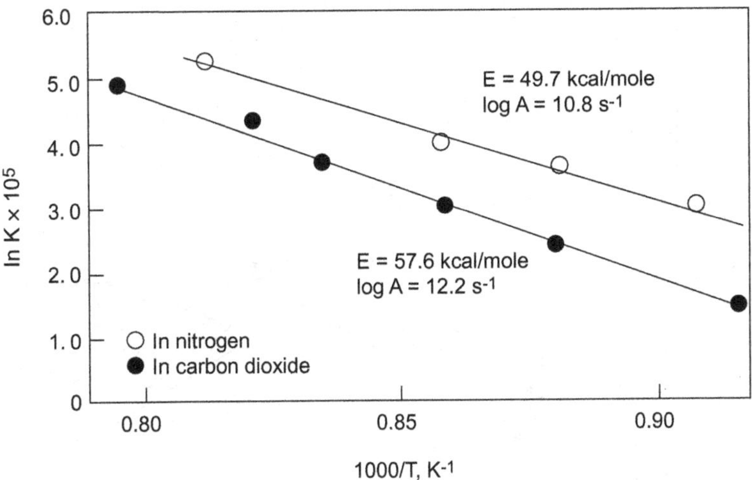

Fig. 9.5 Arrhenius plots of kinetic data for isothermal decomposition of $CaCO_3$ (MacCullum and Tanner 1970)

For the present case, we assume that the reaction is phase boundary controlled and the so-called topochemical contracting sphere model holds. According to this model,

$$g(\alpha) = 1 - (1 - \alpha)^{1/3} = k \cdot t$$

Hence, in Eq. 9.35, $n = 2/3$.

The curves shown in Fig. 9.2 have been replotted according to Eq. 9.35 to obtain the linear plots shown in Fig. 9.6. The activation energy has been calculated for two runs. Run 2 carried out under CO_2 at 30 °C/min shows a slightly higher value 58 kcal/mole of the apparent activation energy E compared to the value of 52 kcal/mole for run 6 which was carried out under nitrogen using the same heating rate. The $\log A$ values for these runs, calculated using Eq. 9.35 and appropriate values of E, R, and B, are also indicated in Fig. 9.6.

The apparent activation energy can also be obtained using a second approach described by Ozawa (Eq. 9.34). In this approach, the plots of $\log B$ versus the reciprocal absolute temperature for a given value of fraction reacted (i.e. $g(\alpha)$ constant) give a straight line, the slope of which gives the activation energy.

The $\log B$ against $1/T$ plots for three runs and for 50% reaction, i.e. $\alpha = 0.5$, are shown in the inset of Fig. 9.6. The plot is linear and the slope of the line gives for E a value of about 43 kcal/mole.

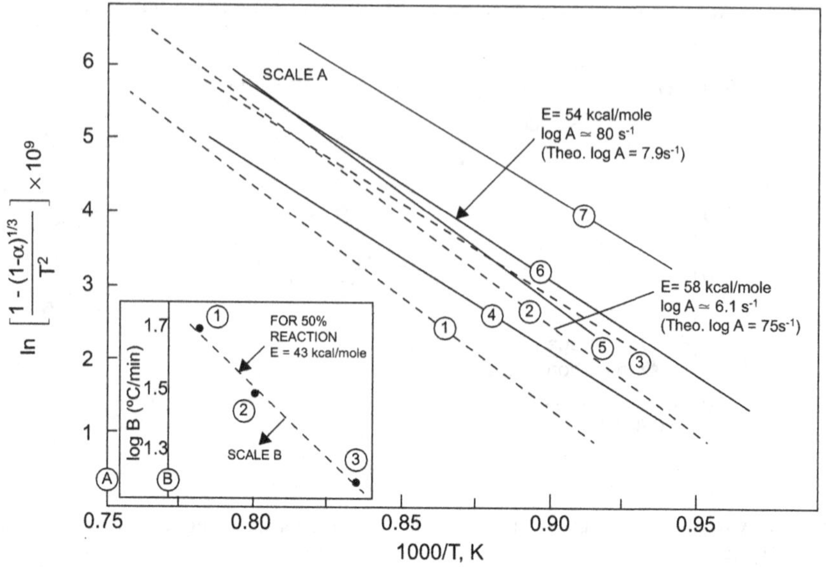

Fig. 9.6 Plots of data shown in Fig. 9.2 according to Eqs. 9.34 and 9.35 (Dixit and Ray 1982)

9.4.2 Reduction of Fe_2O_3 by CO

Figure 9.7 shows TG curves for reduction of Fe_2O_3 by carbon monoxide for three heating rates. These were obtained (Ray 1985) from the primary weight loss versus temperature plots given by a thermobalance assuming maximum possible weight loss on a stoichiometric basis (97% Fe_2O_3). The Derivative Thermogravimetry (DTG) curves for the two lower heating rates shown in Fig. 9.8 give the rates of reaction. These curves indicate the variation of the slopes of plots shown in Fig. 9.7 as a function of temperature.

The first rise in a DTG plot corresponds to the beginning of a weight loss reaction, indicating formation of Fe_3O_4 from Fe_2O_3. The second peak perhaps indicates reduction of Fe_3O_4 to iron. Beyond 560 °C, FeO becomes stable and reduction of Fe_3O_4 to iron must stop. At temperatures around 600 °C, however, this sequential reaction is characterized by many complications.

Kinetically, it is simple to consider the reduction as a simple two-phase system, oxide/metal. It has been shown (McKewan 1960; Warner 1964) that the gas–solid type of reaction takes place only at the FeO/Fe interface and is described by the topochemical reaction model. There is, however, evidence in the literature to suggest that the rate-controlling mechanism may change with temperature (Satava 1973).

Many workers have noted higher reaction rates just below 600 °C than at temperatures in the range 600–800 °C. Several reasons have been advanced for this occurrence of rate minimum (Anon 1974). This rate minimum phenomenon is clear

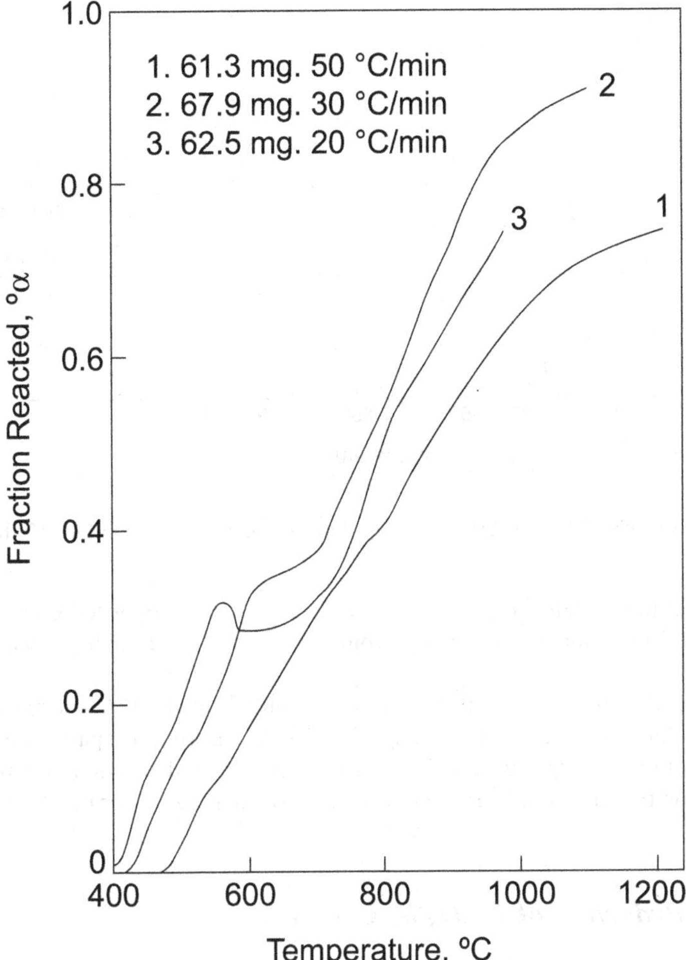

Fig. 9.7 Thermogravimetric (TG) curves for reduction of Fe_2O_3 by CO (Ray 1985)

in runs (2) and (3) shown in Figs. 9.7 and 9.8. To explain the shape of these curves more clearly, one must also consider the Budouard reaction which tends to deposit carbon.

$$2CO = CO_2 + C \tag{9.52}$$

The uncertainty of the weight loss data due to possible carbon monoxide dissociation can be examined in the light of some pertinent literature (Von Bogdandy and Engell 1971; Walker et al. 1959) on the subject. It is estimated that for a

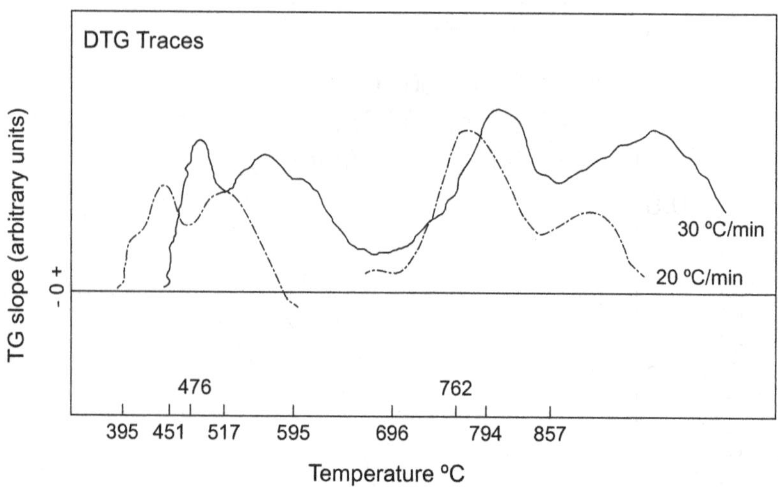

Fig. 9.8 Derivative TG plots (DTG) showing slopes of TG plots shown in Fig. 9.7 (Ray 1985)

20 °C/min rate of temperature rise, the weight of carbon deposited should be less than 5% of the sample weight. This would, of course, be lower for higher heating rate.

Figures 9.7 and 9.8 show that there is a region beyond 600 °C where the rate of reduction falls appreciably. In this range, the TG plot showed an apparent weight gain in one plot (curve 3) presumably because of some carbon deposition. It may be said that carbon introduces little error in the plot obtained at 50 °C per min heating rate.

9.4.3 Reduction of Fe_2O_3 by Carbon

The reduction kinetics in Fe_2O_3–C system depends on the way the oxide and the reductant are mixed. For example, if these are present as fine particles and coal thoroughly mixed then there is total internal reaction (Mookherjee et al. 1985). Basu et al. (Unpublished) have studied reduction of oxide–coal pellets, and the first-order equation has been found valid for isothermal reaction. This has been discussed earlier in Chap. 3. If, however, the powders are in separate layers then the reaction kinetics follow the Ginstling–Brounshtein model (Mookherjee et al. 1985). Mookherjee et al. (1985) have reported isothermal kinetic data on reduction of ore fines surrounded by char fines in small silica crucibles. Isothermal kinetic data (Fig. 9.9) follow the product layer diffusion model (Fig. 9.10a, b). In these plots, the parameter f termed degree of reduction is defined as the ratio of weight loss at given intervals to the maximum possible weight loss. The latter is obtained from the weight of oxygen in iron ore and carbon and volatiles in coal.

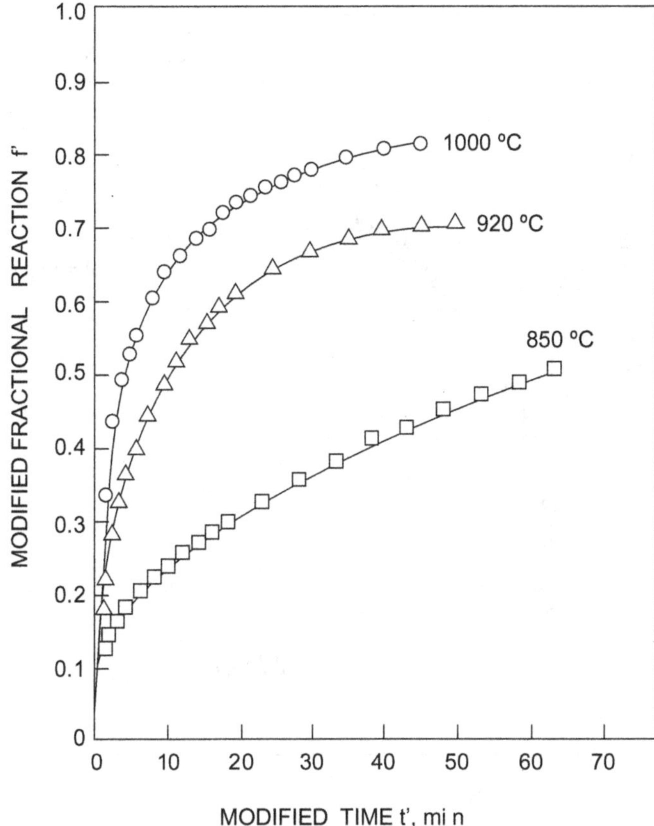

Fig. 9.9 Plot of fraction reacted against time for the ore–coal system (Mookherjee et al. 1985)

Mookherjee et al. (1985) have studied the same system of isolated layers of iron ore and char under non-isothermal conditions (Fig. 9.11). They have shown that the Ginstling–Brounshtein equation can be usefully employed in the analysis of non-isothermal kinetic data such as in the use of Coats and Redfern equation (Fig. 9.12).

9.4.4 Effect of Mass on Reaction Sequence

It is interesting to note that non-isothermal studies are not always useful in isolating sequential reactions that occur during heating. Figure 9.7 shows data on gaseous reduction of very small weights of samples. If the sample weight is sufficiently large, then reactions overlap; i.e. another sequential reaction starts before one ends. Thus, reaction steps may not be seen very distinctly. This is evidenced by the data

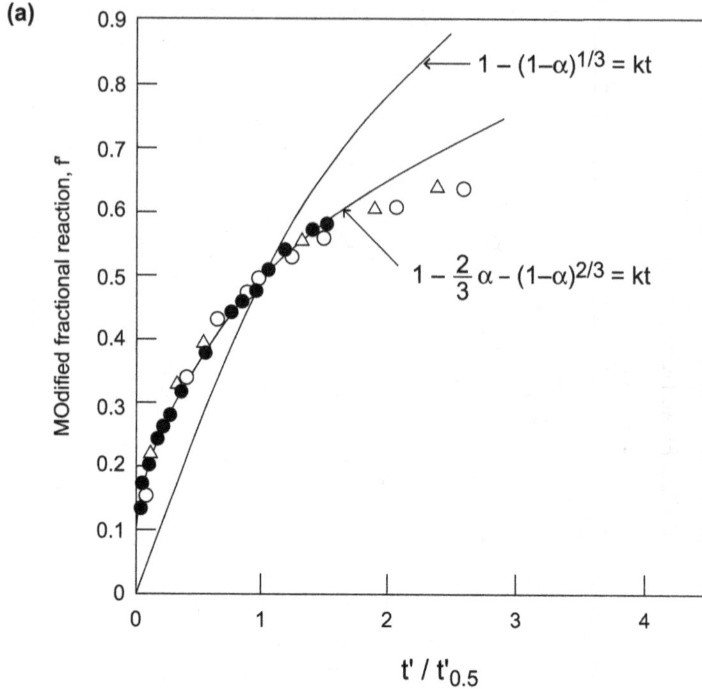

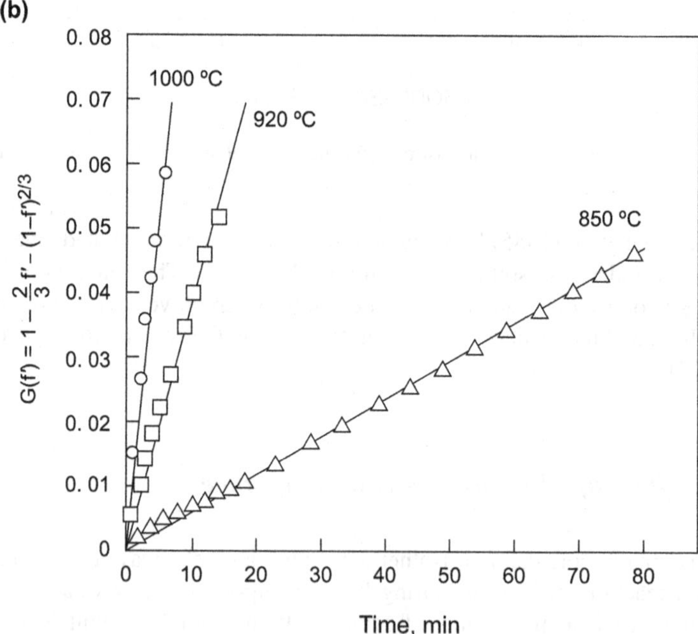

Fig. 9.10 **a** Kinetic model for reduction of ore by coal: reduced time plot for the data shown in Fig. 9.9 (Mookherjee et al. 1985). **b** Kinetic model for reduction of ore by coal: kinetic plot of the reduction of ore by coal (Mookherjee et al. 1985)

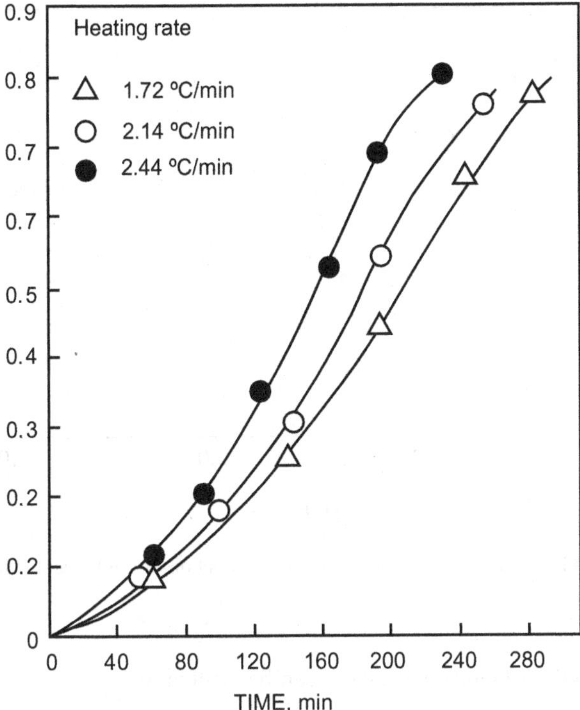

Fig. 9.11 Variation of degree of reduction under increasing temperature conditions for an ore–coal system (Mookherjee et al. 1985)

shown in Fig. 9.13, from Prakash and Ray (1987). The rate of heating here is small and yet the steps tend to overlap as in the case of high rate of heating (1) as shown in Fig. 9.7.

9.5 Kinetic Compensation Effect (KCE)

9.5.1 Variation of the So-Called Activation Energy

Inspection of data available for some solid decomposition reactions, especially carbonates, shows various values of E and A dependent upon the conditions of the material and the nature of the experiment. This has often caused arguments regarding the true or most dependable values, which should be selected from those available in the literature. The concept that a linear relationship exists between $\log A$ and E is called the kinetic compensation effect (KCE). KCE is an experimental observation and has been the subject of critical examination by several workers.

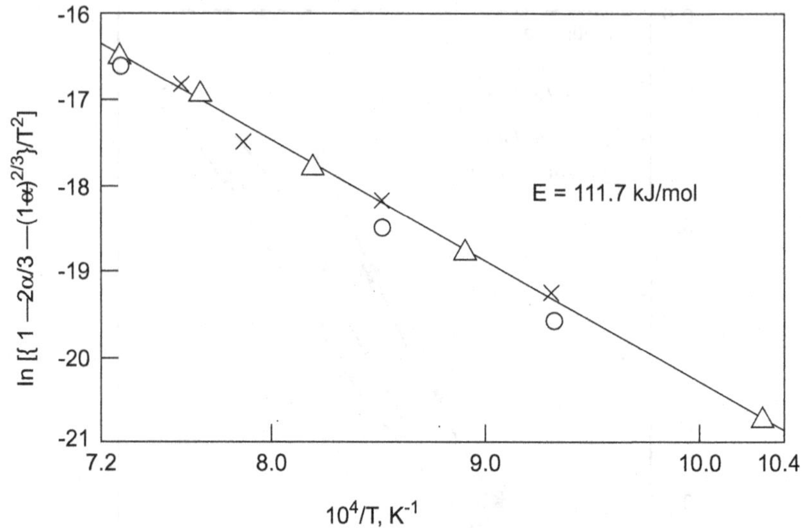

Fig. 9.12 Coats–Redfem plot of the non-isothermal reduction of ore by coal (Mookherjee et al. 1985)

The so-called Arrhenius expression can be written as

$$\log A = \frac{E}{2.3RT} + \log k$$

The equation shows that the same value of k can be satisfied by different sets of linearly related values of $\log A$ and E, provided the temperature range is limited. Thus, the kinetic parameters A and E lose clear meaning and become only formally determined values.

As an example, we can consider values reported for the decomposition of calcium carbonate. While the E values vary between 30 and 1000 kcal mole^{-1}, the corresponding pre-exponential factor varies between 10^2 and 10^{69}, but almost all values obey a linear compensation law (Gallagher et al. 1976).

It has been shown (Buker 1978) from mathematical and thermodynamic point of view that there is no fundamental error involved in extending isothermal kinetic equations to non-isothermal conditions. In some studies, where a given reaction has been examined under both isothermal and non-isothermal conditions, no significant differences were found. However, differences have been noted by some other studies.

In traditional kinetic studies, there is generally much emphasis on the determination of the activation energy because it is said to be indicative of the reaction mechanism. However, it has been pointed out (Doyle 1961; Majumdar et al. 1975) that E may not have a well-defined meaning in heterogeneous kinetics. Also, rising

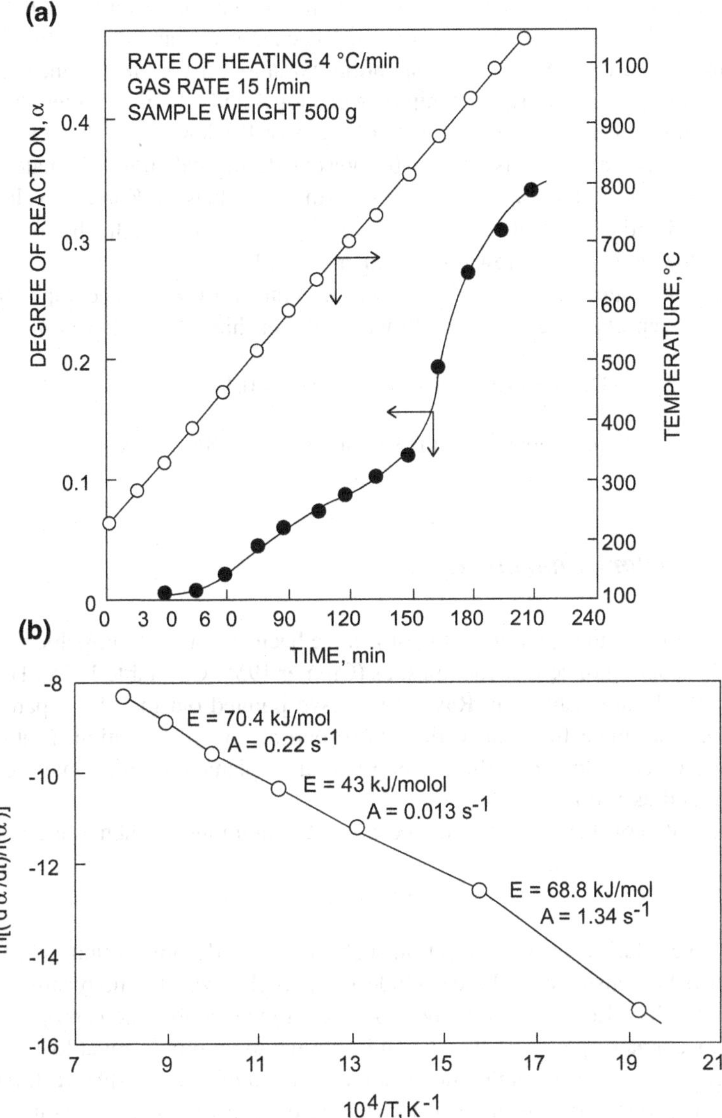

Fig. 9.13 Gaseous reduction of Kiriburu iron ore by $CO–N_2$ (30:70) under rising temperature condition (particle size −12 + 10 mm) (Prakash and Ray 1987)

temperature *E* and *A* values may not necessarily agree with the values obtained from isothermal experiments (Gorbachev 1976; Garn 1976).

In numerous heterogeneous reactions, the experimentally determined *E* has been found to vary with several factors including sample size, particle size, heating rate, the presence of impurities, even pretreatment. These factors can generally be categorized as those connected with the experimental procedure and those connected with the material. It has, however, been noted that the variation of E is sometimes accompanied by a corresponding variation of the pre-exponential factor A so that there is a linear relationship between $\log A$ and E. This is called the kinetic compensation effect (KCE) (Garn 1976; Galway Review; Gallagher et al. 1976; Zsako 1976). This effect is said to be present during calcium carbonate decomposition for which the experimentally determined values of E and A values vary between 25 and 1000 kcal/mole and 10^2–10^{69}, respectively (Gallagher et al. 1976) —the values generally following a compensation law.

For a large number of sets of $\log A$ and E values for $CaCO_3$ decomposition, the KCE has been expressed by the following relationships (Gallagher et al. 1976).

$$\text{Decomposition in } CO_2\text{:} \quad \log A = 0.175E - 2.95 \tag{9.53a}$$

$$\text{Decomposition in } O_2\text{:} \quad \log A = 0.278E - 5.98 \tag{9.53b}$$

9.5.2 *Factors Causing KCE*

The variations in the apparent value of E have been attributed to possible variations in the distribution of active reaction sites (Cremer 1955; Constable 1925). However, Lahiri (1980) and Lahiri and Ray (1982) have pointed out that the apparent variation may originate from the variations in particle size distribution, heating rate, flow rate of gas, etc. Ray (1982) has given an explanation which may be briefly summarized as follows.

If the rate constant was determined by temperature alone, then one could write

$$k = A \cdot \exp(-E/RT) \tag{9.54}$$

However, during an actual reaction such as $CaCO_3$ decomposition, other factors must also be considered. These include the particle size, thermodynamic driving force, etc. Thus, Eq. 9.54 should be modified. If the reaction rate is proportional to the surface area of particles, then k can be shown to be proportional to $1/r_0$ where r_0 is the initial radius of the particles, all assumed uniform and spherical. Again, if the rate is proportional to the thermodynamic driving force, then k would be proportional to $(C^e - C)$ where C^e is the equilibrium concentration of CO_2 at the gas/solid interface at temperature T and C is the concentration in the bulk gas at a given temperature. We can write

$$k = \frac{A}{r_0}(C^e - C) \cdot \exp(-E/RT) = \frac{A}{r_0}\left(\frac{p^e - p}{RT}\right) \cdot \exp(-E/RT) \tag{9.55}$$

where p^e is the equilibrium partial pressure at temperature T and p is the partial pressure in the bulk gas. The possible effects of several other factors can be similarly incorporated.

Consider now decomposition of $CaCO_3$ in an inert atmosphere ($p = 0$). If we assume that an equation such as Eq. 9.54 still holds and that the apparent values of the Arrhenius parameters are A' and E', then

$$k = A' \cdot \exp(-E'/\mathrm{R}T) = \frac{A}{r_0} \cdot \frac{p^e}{RT} \cdot \exp(-E/\mathrm{R}T) \tag{9.56}$$

The equilibrium CO_2 pressures over $CaCO_3$ at various temperatures are available in the literature (Hills 1967). One can thus evaluate the RHS of Eq. 9.56 for some assumed values of E and A/r_0. It has been shown (Ray 1982) that k values thus calculated follow an exponential relation, i.e. $\log k$ versus $1/T$ plot is linear. For $E = 50$ kcal/mole and $A/r_0 = 10^{14}$, E' is thus obtained as 37 kcal/mole. Equation 9.56 indicates clearly that the apparent value of the activation energy E' obtained experimentally from non-isothermal data need not be identical with the true value E.

There is often a simpler explanation for the apparent variation of the so-called activation energy. During continuous heating, there is likely to be thermal lag between the outer layers and the interior of a test sample. The sample temperature is usually read in TG by thermocouples attached to crucible exterior, and therefore, it is likely to be on the higher side. This will obviously introduce error in the reciprocal temperature axis of the Arrhenius plot and tend to reduce the calculated value of E. The thermal lag will be aggravated by increasing sample weight and particle size. The uncertainty introduced by thermal lag during heating can sometimes be removed by determining apparent E values for different heating rates and then obtaining, by extrapolation, the value for zero heating rate.

9.6 Reactions Occurring Under Fluctuating Temperature Conditions

9.6.1 Possible Aims of Studies Under Fluctuating Temperatures

Non-isothermal kinetic studies are generally restricted to studies employing thermal analysers where the sample size is small and the temperature is made to increase at constant rates. The importance of studies under increase of temperature at variable rates or under fluctuating temperatures cannot be overemphasized. There can be two main aims of such studies. Firstly, kinetic data generated under conditions of arbitrary temperature variations may be examined and analysed, for evaluating the kinetic parameters. Secondly, having evaluated these parameters through some test runs, one may try to predict the course of the reaction under different temperature–time schedules.

In real industrial situations, temperature can vary in a rather complex manner and it may not be possible to write the kinetic equation in the integrated form. In such cases, a simple differential approach should be preferable. If during the temperature variation or fluctuation the reaction is assumed to be iso-kinetic and the kinetic parameters are assumed to retain fixed values, then it is possible to analyse or, alternatively, to predict the kinetic data.

From Eqs. 9.2 to 9.3, we have

$$\mathrm{d}\alpha/\mathrm{d}t = A\,\exp(-E/RT)f(\alpha) \tag{9.57}$$

For general variation of temperature

$$\mathrm{d}T/\mathrm{d}t = f_1'(t) \tag{9.58}$$

Combining the preceding two equations

$$\left[\frac{\mathrm{d}\alpha/\mathrm{d}t}{f(\alpha)}\right] = \left[\frac{(\mathrm{d}\alpha/\mathrm{d}T)f_1'(t)}{f(\alpha)}\right] = A\,\exp\left(-\frac{E}{RT}\right) \tag{9.59}$$

Therefore, a plot of either $\ln[(\mathrm{d}\alpha/\mathrm{d}t)/f(\alpha)]$ or $\ln\left[(\mathrm{d}\alpha/\mathrm{d}\alpha)/\left(f_1'(t)/f(\alpha)\right]\right.$ against reciprocal temperature should give a straight line. The kinetic parameters E and A should be obtainable from, respectively, the slope and the intercept (Ray 1986).

Prediction of $\boldsymbol{\alpha}$ *Versus* ***t*** *plots*

To predict the course of a reaction for a known temperature–time schedule, one must first evaluate the kinetic parameters using some test runs where not only the temperature–time history should be known but α values must be determined at close time intervals experimentally. We have,

$$g(\alpha) = \int_0^{\alpha} \frac{\mathrm{d}\alpha}{f(\alpha)} = A\int_0^{t} \exp(-E/RT)\mathrm{d}t \tag{9.60}$$

The right-hand side (RHS) integral is obtained by first plotting $\exp(-E/RT)$ values at instantaneous values of time and then by graphical integration of the area covered on the t-axis using the procedure discussed. The result is multiplied by A to get the $g(\alpha)$ value from which α for the given t value is calculated. These treatments assume that the kinetic parameters remain unchanged during the entire period of reaction.

9.6.2 Graphical Integration

The graphical integration for calculation of the area covered by $\exp(-E/RT)$ values on the time axis is conveniently carried out by Simpson's one-third rule.

Consider the variation of a quantity y as a function of the independent variable x and a plot of y between the limits x_0 and x_{max}. Let us divide the total x interval into n equidistant intervals of width Δx, n being an even number. y values evaluated at x_0 and $x_0, x_1 = x_0 + \Delta x, x_2 = x_0 + 2\Delta x, \ldots, x_n = x_{max} = x_0 + n\Delta x$ are designated as $y_0, y_1, y_2, \ldots, y_n$. Then according to Simpson's one-third rule (Bajpai et al. 1978), the integral

$$I = \int_{x_o}^{x_0 + n\Delta x} y\mathrm{d}x$$

is given by

$$I = \frac{\Delta x}{3} \cdot [y_0 + 4(y_1 + y_3 + \cdots + y_{n-1}) + 2(y_2 + y_4 + \cdots + y_{n-2}) + y_n] \tag{9.61}$$

In the present case, $y = \exp(-E/RT)$ and x is time, and therefore, the desired area of the curve gives the value of $g(\alpha)/A$. The area multiplied by A gives $g(\alpha)$ value, whence α is evaluated.

9.6.3 Examples of Data Analysis

Prakash and Ray (1987) have reported some data on non-isothermal reduction of an iron ore by coal. The data are shown in Fig. 9.14a. As is seen the degrees of reduction were obtained under rising temperature conditions, but the rate of heating was not constant. Figure 9.14b shows that the data can be analysed employing the differential approach given by Eq. 9.59. Here $f(\alpha)$ has been obtained from the Ginstling–Brounshtein equation, e.g.

$$\begin{aligned} g(\alpha) &= 1 - \frac{2}{3}\alpha - (1-\alpha)^{2/3} = kt \\ f(\alpha) &= \frac{1}{g'(\alpha)} = \frac{2}{3}(1-\alpha)^{-1/3} - \frac{2}{3} \end{aligned} \tag{9.62}$$

Figure 9.14 gives for average E, a value of 75 kJ/mole and for A value of around 3 s^{-1}. Assuming that these values are operative during non-isothermal reactions in general, one can predict the course of a reaction under fluctuating temperature conditions.

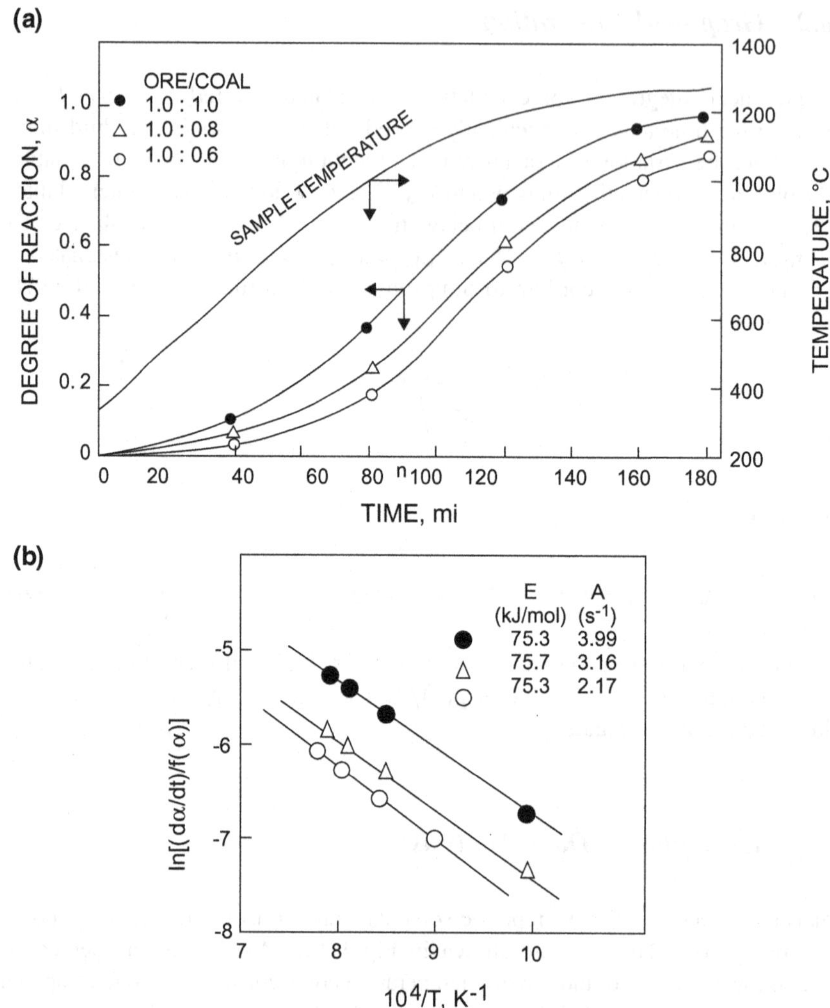

Fig. 9.14 Non-isothermal kinetic data for the reduction of Barajamda iron ore by Raniganj non-coking coal (particle size −3 + 1 mm) (Prakash and Ray 1987)

9.6.4 Prediction of $\alpha - t$ Plots

Figure 9.15 shows some data obtained in a special experiment (Prakash and Ray 1987) where the temperature of three reaction mixtures containing Barajamda ore and Raniganj non-coking coal were varied in the manner shown in the figure. The total mass of each sample was 30 g, and they were kept in a muffle furnace whose temperature was manipulated. Figure 9.15a shows plots of both T and $\exp(-E/RT)$ against t. The area covered by the $\exp(-E/RT)$ versus t plot has

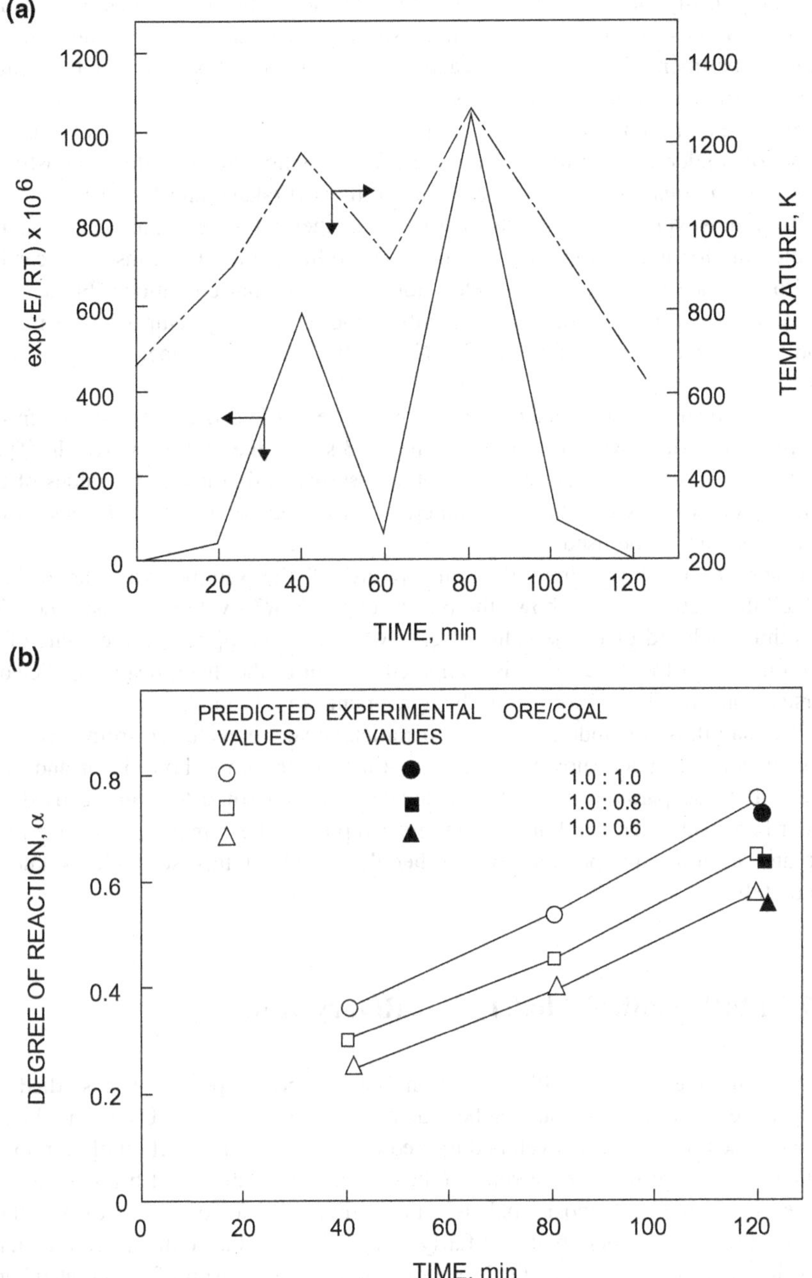

Fig. 9.15 Experimental and predicted values for the reduction of Barajamda ore by Raniganj coal under fluctuating temperatures (Prakash and Ray 1987)

been graphically integrated. Figure 9.15b shows the predicted $\alpha - t$ relation obtained by using values of kinetic parameters as shown. It also shows the α values determined experimentally at the end of each run. The agreement between the theoretical and experimental values is very close.

Some experiments were also carried out to predict the course of gaseous reduction under fluctuating temperatures. The reaction again follows Ginstling–Brounshtein equation under isothermal conditions (Prakash and Ray 1987).

It is, however, not advisable to use the isothermal kinetic parameters in the predictions for non-isothermal conditions. Therefore, some test runs were carried out to evaluate the kinetic parameters for non-isothermal conditions. Figure 9.13 presented the results of some experiments in the reducibility setup where Kiriburu iron ore was reduced by a flowing CO–N_2 mixture while the temperature was raised linearly as shown.

The α versus t plot as determined from the weight loss data (also shown in the figure) shows stage-wise reduction. Figure 9.13 shows the plot of ln $[(\mathrm{d}\alpha/\mathrm{d}t)/f(\alpha)]$ against reciprocal temperature. This plot also shows variations in the values of the kinetic parameters with change of temperature; for prediction only the values for the final stages are to be used.

Figure 9.16a shows how the temperature of the sample was varied in a reducibility setup. It also shows the plot of $\exp(-E/RT)$ values against time. The area thus enclosed gives the value of $g(\alpha)$ when it is multiplied by the value of A. The final experimental α value is compared in light of the theoretically predicted α versus t plot in Fig. 9.16b; again, the agreement is very good.

We may thus conclude that kinetic parameters can be evaluated from α versus t plots generated under known temperature–time programmes, both linear and non-linear. If these parameters are first evaluated through some test runs carried out under non-isothermal conditions, then one can predict the complete α versus t plots for any arbitrary temperature–time schedule, provided this schedule is known accurately.

9.7 Mathematical Model for a Rotary Kiln

Reduction kinetics of cold-bonded iron ore composite pellets was studied by Agarwal et al. (1996) in rotary tube reactor in laboratory scale. Using the data, a mathematical model was developed to predict the degree of reduction of composite pellets in an 8 tpt throughout rotary kiln sponge iron plant. Plant trials were conducted using these composite pellets. The predicted values of degree of reduction obtained using the model matched fairly well with the values obtained in the plant. Development of the mathematical relationship and comparison of "calculated" and "actual" values of degree of reduction are now discussed.

Isothermal reduction kinetic studies on composite pellets have been carried out under rotary kiln condition on laboratory scale to establish the kinetic law. Isothermal experiments were conducted in an electrically heated rotary tube reactor

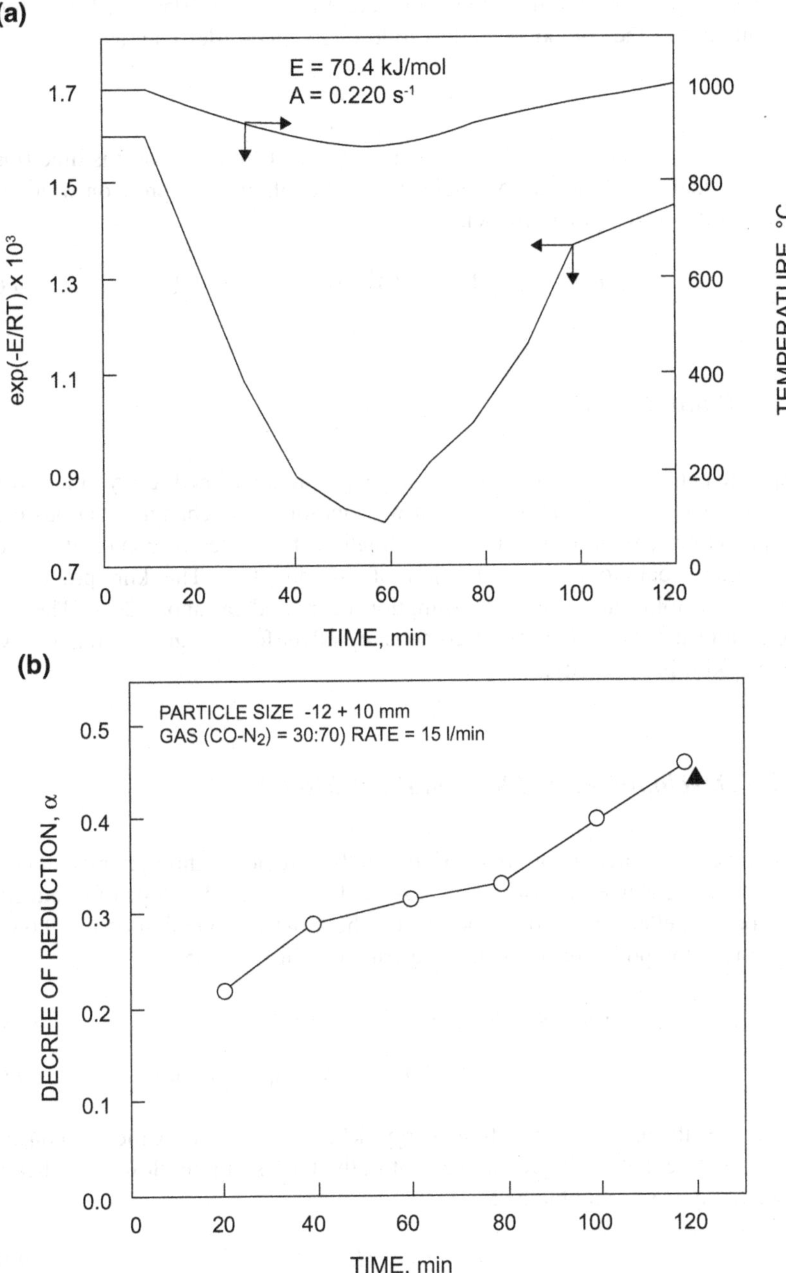

Fig. 9.16 Experimental and predicted values for gaseous reduction of Kiriburu iron ore under fluctuating temperature (Prakash and Ray 1987)

$l \approx 650$ mm, dia = 150 mm) in the temperature range 950–1050 °C. It was found that reduction of the composite pellets follows the first-order rate law

$$-\ln(1-\alpha) = kt \quad (9.63)$$

where α is the degree of reduction, k the rate constant (min^{-1}), and t is time (min).

Kinetic analysis of laboratory studies led to the following expression which will be used in the development model.

$$-\ln(1-\alpha) = 1.01 \times 10^3 t \cdot \exp(-102.5/RT) \quad (9.64)$$

9.7.1 Plant Trials

Composite pellets were reduced in a 12-m-long refractory lined rotary kiln of 8 tpd throughout capacity. The kiln operation was optimized by changing various operating parameters, such as kiln r.p.m., inclination, feed rates of composite pellets, and external coal, to get metallization of around 90%. The kiln productivity increased twofold, and energy consumption decreased by about 20%. The plant trials established the suitability of composite pellets for sponge iron making with enhanced kiln productivity.

9.7.2 Development of Mathematical Model

The degree of reduction of iron oxide can be predicted through mathematical modelling using isothermal kinetic data. A model has been developed for reduction of composite pellets and is described here. The kinetic expression of Eq. 9.63 for reduction of composite pellets can be rewritten as in Eq. 9.65

$$\text{Rate of reduction} = \mathrm{d}\alpha/\mathrm{d}t = -k(1-\alpha) \quad (9.65)$$

$$\therefore \quad \mathrm{d}\alpha/(1-\alpha) = -k\,\mathrm{d}t = -A\exp(-E/RT)\mathrm{d}t \quad (9.66)$$

Here, E is the apparent activation energy (kJ/mol) and A the Arrhenius constant.

If we assume that the degree movement in the kiln is of plug flow type, then the following expression is obtained.

$$t/\tau = x/L \quad (9.67)$$

where t (min) is the time taken by pellets to travel x m, L (m) is the length of the kiln, and τ (min) is the average residence time of charge in the kiln.

The average residence time of charge (pellets in this case) in the kiln was determined by tracer technique. Differentiating Eq. 9.67,

$$dt = (\tau/L)dx \tag{9.68}$$

Combining Eqs. 9.66 and 9.68 and integrating between the limits $x = 0$ m and $x = 12$ m, one obtains

$$\int_{\alpha=0}^{\alpha=\alpha} d\alpha/(1-\alpha) = (-A\tau/L) \int_{x=0}^{x=12} \exp(-E/RT)\, dx \tag{9.69}$$

The left-hand side of Eq. 9.69 equals $\ln(1-\alpha)$, where α is the final degree of reduction of composite pellets at the discharge end of the kiln. The right-hand side of Eq. 9.69 depends on the temperature profile of the kiln for a particular type of composite pellets.

The values of the temperature T of the charge along the length of the kiln were measured by quick response thin mineral insulated thermocouples, and these were used to calculate the term $\int \exp(-E/RT)dx$. The value of Arrhenius constant was taken as 1.01×10^3, and activation energy value was taken as 102.5 kJ/mol as found from laboratory kinetic studies. The average residence time was found to 60 min when kiln was rotating at 1.1 r.p.m. The magnetic product of the kiln was chemically analysed to obtain actual degree of reduction.

Results

Degree of reduction of composite pellets was calculated from the model as well as from the results obtained from plant trials.

The degree of reduction was calculated by solving Eq. 9.69. Two typical temperature profiles of the charge in the kiln were used as shown in Fig. 9.17. The term $\int \exp(-E/RT)$ was calculated for these temperature profiles and is plotted in Fig. 9.18. The graphical integration for the calculation of the area covered by $\{\exp(-E/RT)\}$ over the length has been calculated as per Simpson's one-third rule. The values of degree of reduction for the two temperature profiles have been calculated as follows:

$$(\alpha_A)_{\text{actual}} = 98\% \quad (\text{for temperature profile } A)$$
$$(\alpha_B)_{\text{actual}} = 95\% \quad (\text{for temperature profile } B)$$

Degree of reduction obtained from plant trials

Sponge iron produced using composite pellets was chemically analysed to find the actual value of degree of reduction. The following values were obtained:

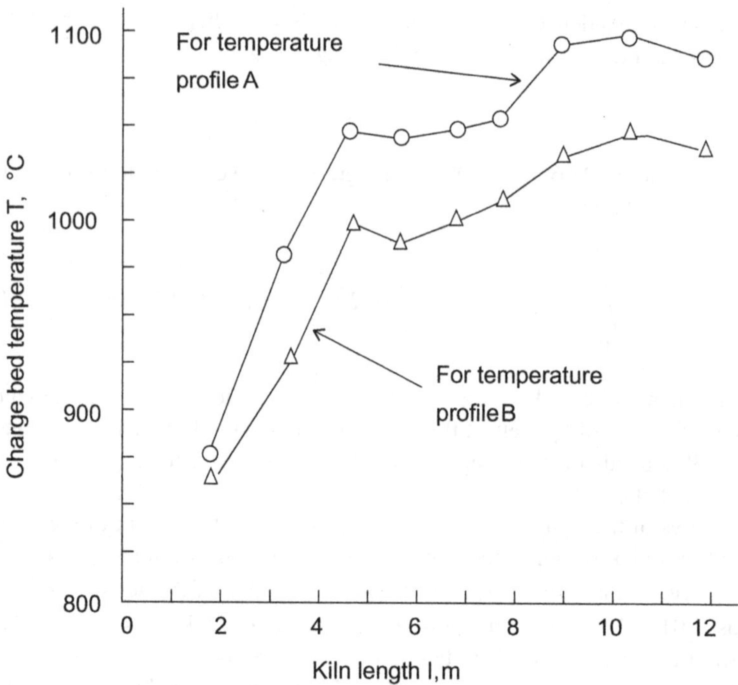

Fig. 9.17 Charge bed temperature (T) along the length of the kiln (Agrawal et al. 1996)

$$(\alpha_A)_{\text{actual}} = 93.0\% \quad \text{(for temperature profile } A\text{)}$$
$$(\alpha_B)_{\text{actual}} = 98.3\% \quad \text{(for temperature profile } B\text{)}$$

The calculated values are in good agreement with the actual values. The variation between $(\alpha)_{\text{calculated}}$ and $(\alpha)_{\text{actual}}$ is the range 5–7%. This difference may be due to the following reasons:

(i) While deriving the mathematical expressions, it was presumed that the charge movement in the kiln follows plug flow character. In actual plant situation "ideal" plug flow is not possible in any material of granular nature such as iron ore and coal. This will result in variation of residence time of charge, and hence, fluctuation in values of the degree of reduction at plant level.

(ii) In any commercial plant, as soon as sponge iron comes of rotary kiln, it encounters slightly oxidizing environment in the kiln—cooler transfer chute. This causes about 5–6% re-oxidation. Hence, the actual value is expected to be less as compared to the predicted value.

(iii) Minor variation in degree of reduction is expected due to plant factors, such as fluctuating temperature regime, even, in the most steady-state operation of the reactors.

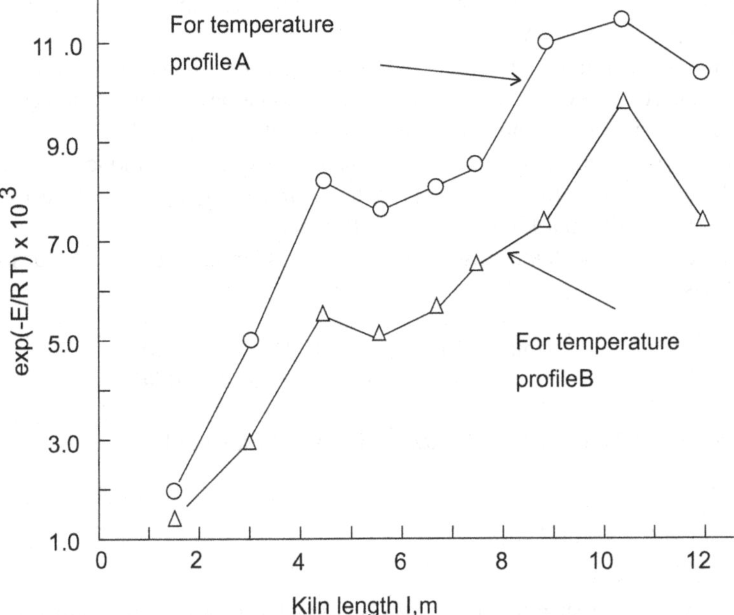

Fig. 9.18 Plot of $\exp(-E/RT)$ versus length of the kiln (Agrawal et al. 1996)

Thus, it is possible to predict the degree of reduction of composite pellets based on the laboratory kinetic data, provided that we know the temperature profile along the length of the kiln.

9.8 Review Questions

1. Discuss whether the following statements are correct or false.
   (a) During rising temperature (constant heating rate) experiments, the extent of reaction in a given time increases with heating rate.
   (b) During rising temperature processes, the rate of reaction must continuously increase.
   (c) Sintering of green compacts cannot be studied under truly isothermal conditions.
2. Discuss the conditions under which the reciprocal temperature rise equation can be used to approximately describe heating at a constant rate.
3. Of the different approaches discussed in the chapter, which is the most accurate for analysis of non-isothermal kinetic data and evaluation of activation energy?

4. The α–t plots obtained during rising temperature experiments generally are sigmoidal in nature. Why?
5. During a reaction neither temperature nor time is kept constant; i.e. both T and t vary. Discuss how one may write a kinetic equation for such a process. Make suitable assumptions and choose any mathematical approach.
6. For analysis of non-isothermal kinetic data and evaluation of kinetic parameters, both integral and differential approaches can be employed. Is it necessary to know the isothermal kinetic law in both cases?
7. In a reaction, the temperature rises at 1.72 °C/min. The $\alpha - t$ values are as follows:

t(min)	40	80	120	200	240
α	0.08	0.19	0.35	0.75	0.86

Check the data against the following isothermal kinetic laws.

(a) $\alpha^2 = kt$
(b) $1 - \frac{2}{3}\alpha - (1 - \alpha)^2/3 = kt$
(c) $-\ln(1 - \alpha) = kt$

Identify the correct mechanism by checking the best fit and evaluate the activation energy.

8. Discuss the merits and demerits of isothermal and non-isothermal experiments.

References

Abramowtiz, M., Stegun, I.A.: Handbook of Mathematical Functions. National Bureau of Standards, p. 231 (1964)
Agrawal, B.B., Prasad, K.K., Srakar, S.B., Ray, H.S.: Steel India **19**(2), 73 (1996)
Basu, P., Ray, H.S., Sarkar, S.B.: Unpublished Research
Anon.: Steelsearch 74. British Steel Corporation, p. 17 (1974)
Bajpai, A.C., Calus, J.M., Fairley, J.A.: Numerical methods for engineers and scientists, p. 258. Wiley, New York (1978)
Buker, R.R.: Thermochim. Acta **23**, 201 (1978)
Coats, A.W., Redfern, J.P.: Nature **201**, 68 (1964)
Constable, F.H.: Proc. Roy. Soc. (Lond.) **A 108**, 75 (1925)
Cremer, E.: Adv. Catal. **7,** 75 (1955)
Doyle, C.D.: J. Appl. Polym. Sci. **5**, 285 (1961); **6**, 639 (1962)
Doyle, C.D.: Nature **207**, 290 (1965)
Dixit, S.K., Ray, H.S.: Thermochim Acta. **54**, 245 (1982)
Doyle, C.D.: J. Appl. Polym. Sci. **5**, 285 (1961)
Doyle, C.D.: J. Appl. Polym. Sci. **6**, 639 (1962)
Draper, A.L., Mc Adie, H.G. (ed.): Proc. Third Toronoto Symp. Therm Anal. C.I.C. Canada, 63 (1970)
Freeman, S., Carrol, B.: J. Phy. Chem. **62**, 394 (1958)
Felder, R.M., Stahel, E.P.: Nature **228**, 1085 (1970)

Garn, P.D.: J. Thermal Anal. **6**, 237 (1974)
Gorbachev, V.M.: J. Thermal Anal. **8**, 349 (1975)
Gorbachev, V.M.: J. Thermal Anal. **10**, 191 (1976)
Gorbachev, V.M.: J. Thermal Anal. **10**, 447 (1976)
Garn, P.D.: J. Thermal Anal. **10**, 99 (1976)
Galway, A.K. (Department of Chemistry, Queen's University, Belfast BT. 95 AG, Northern Ireland (U.K.): Compensation Eff. Heterogen. Catal. (Rev.)
Gyulai, G., Greenhow, E.J.: J. Thermal Anal. **6**, 279 (1974)
Gallagher, P.K., Johnson, D.W., Jr.: Thermbchim Acta. **14**, 255 (1976)
Gorbachev, V.M., Logvinenko, V.A.: J. Thermal Anal. 193 (1973)
Hills, A.W.D.: Chem. Engg. Sci. **23**(4), 297 (1966)
Hills, A.W.D.: Trans. I.M.M. (Lond.) Sec. C. **76**, 241 (1967)
Hill, R.A.: Nature **227**, 703 (1970)
Jeffreys, H., Swirles, B.: Methods of Mathematical Physics. Cambridge, Chapter 17, Asymptotic expansion (1965)
Judd, M.D., Pope, M.I.: J. Thermal Anal. **5**, 501 (1973)
Keatch, C.J., Dollimore, D.: An Introduction to Thermogravimetry, 2nd Ed. Heyden Lond., Chapter 4 (1975)
Lahiri, A.K.: Thermochim Acta. **40**, 289 (1980)
Lahiri, A.K., Ray, H.S.: Thermochim Acta. **55**, 97 (1982)
MacCullum, J.R.: Nature **232**, 41 (1971)
MacCullum, J.R., Tanner, J.: Nature **225**, 1127 (1970)
Majumdar, S., Luthra, K.L., Ray, H.S., Kapur, P.C.: Met. Trans. **6B**, 607 (1975)
Medek, J.: J. Thermal Anal. **10**, 211 (1976)
Mookherjee, S., Ray, H.S., Mukherjee, A.: Thermochim Acta. **95**, 235, 247 (1985)
Murray, P., White, K.: Trans. Brit. Ceram. Soc. **54**, 204 (1955)
McKewan, N.M.: Trans. TMS AIME **218**, 2 (1960)
Ozawa, T.: Bull. Chem. Soc. Jap. **38**, 1981 (1965)
Ozawa, T.: J. Thermal Anal. **5**, 499 (1973)
Ozawa, T.: J. Thermal Anal. **9**, 369 (1976)
Prakash, S., Ray, H.S.: Thermochimica Acta **111**, 143 (1987)
Prakash, S., Ray, H.S., Thermochim Acta **111**, 143 (1987); ISI: J. Int. **30**(3), 183–191 (1990)
Ramakrishna Rao, V.V.V.N.S., Abraham, K.P.: Met. Trans. **2**, 2463 (1971)
Ray, H.S.: Kinetic studies under non-isothermal conditions, Part II- Methods of analysis of kinetic data. International Conference on Advances in Chemical, Metallurgy (ICMS 79) BARC, Bombay, January (1974), Proceedings, vol. II, p. 43/II/4
Ray, H.S.: J. Thermal Anal. **24**, 35 (1982)
Ray, H.S.: Thermodynamics and Kinetics of Metallurgical Processes (ICMS-81). In: Mohan Rao, M., Abraham, K.P., Iyengar, G.N.K., Mallya, R.M., IIM Calcutta (eds.), p. 185 (1985); Ray, H. S., Kundu, N.: Thermochem. Acta. **101**, 197 (1986)
Ray, H.S., Chakraborty, S.: Thermochim Acta. **101**, 131 (1986)
Satava, V., Skvara, F.: J. Amer. Ceram. Soc. **52**(11), 591 (1969)
Satava, V.: Thermochim Acta. **2**, 423 (1971)
Satava, V.: J. Thermal Anal. **5**, 217 (1973)
Sestak, J., Krotochvil, J.: J. Thermal Anal. **5**, 193 (1973)
Sestak, J., Satava, V., Wendlandt, W.W.: Thermochim. Acta **7**(5), 333 (1973)
Simon, J., Debreczeny, E.: J. Thermal Anal. **3**, 301 (1971)
Von Bogdandy, L., Engell, H.J.: The Reduction of Iron Ores. Springer, Berlin, Chapter 2 (1971)
Walker, P.L., Jr., Rakszawski, J.F., Imperial, G.R.: J. Phys. Chem. **3**, 140 (1959)
Wendlandt, W.W.: *Thermal methods of analysis*. Interscience Pub., London, Chapter 2 (1964)
Warner, N.A.: Trans. TMS AIME. **230**, 163 (1964)
Zsako, J.: J. Thermal Anal. **2**, 460 (1970)
Zsako, J.: J. Thermal Anal. **4**, 101 (1976)

Chapter 10
Thermal Analysis Techniques

10.1 Introduction

Scope for thermal analysis

Thermal analysis studies are concerned with the measurement of heat and weight changes in a system when it is heated or cooled in a predetermined manner. These studies provide insight into the physical nature of the system, mineralogical makeup and the chemical reactivity in a given environment and behaviour during heating, e.g. phase change, reactions. They also provide kinetic data for reaction rate studies. Many properties, other than weight and enthalpy, can also change during reaction and these can be employed in studying progress of reactions. However, these are generally not included within the purview of thermal analysis and are, accordingly, given special names. For example, studies on dimensional changes during reactions such as sintering are called thermomechanical analysis, studies on changes in magnetic properties are called thermomagnetometry.

Thermal analysis techniques, by consensus, now include the following methods associated with energy and weight changes.

(i) Energy change,

  (a) Heating curves and cooling curves,
  (b) Differential thermal analysis (DTA),
  (d) Differential scanning calorimetry (DSC),
  (d) Derivative DTA (DDTA),

(ii) Weight changes,

  (a) Thermogravimetry (TG),
  (b) Derivative TG (DTG).

(These can be either isothermal or non-isothermal.)

H. S. Ray and S. Ray, *Kinetics of Metallurgical Processes*,
Indian Institute of Metals Series, https://doi.org/10.1007/978-981-13-0686-0_10

In addition, thermal analysis also ordinarily includes evolved gas detection (EGD) and evolved gas analysis (EGA). Some of these are briefly discussed.

The theory of heating and cooling curves is too well known to be discussed here. Moreover, they cannot, usually, generate kinetic data being mainly restricted to measurement of temperatures for phase transformation. We will also not discuss EGD and EGA.

10.2 Differential Thermal Analysis (DTA)

10.2.1 Principles of DTA

In DTA, two small crucibles, one containing an inert reference substance like alumina and the other containing the test sample, are placed close together in a furnace and heated at a predetermined rate. There is a thermocouple in close contact with each to sense temperatures. During heating, a suitable device records the temperature of the test sample (T_S) as well as the temperature difference between the sample and the reference ($T_S - T_R$). Up to a very high temperature, no heat changes occur in alumina. Therefore, if at a temperature an exothermic reaction occurs in the test sample, the differential temperature ($T_S - T_R$) rises. Once the reaction is over, ($T_S - T_R$) must again decay to a zero value. One thus obtains an exothermic peak. For an endothermic reaction, the situation is reversed. Figure 10.1 shows schematically the DTA plots thus obtained.

The location of the peak on the temperature axis indicates the temperature at which a heat change occurs. This could be due to either a phase change or a

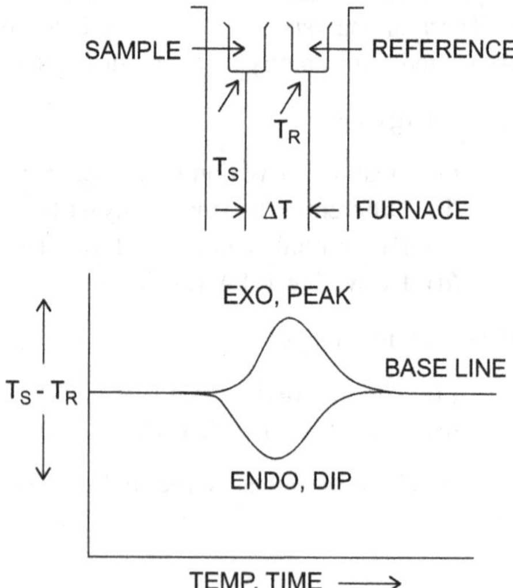

Fig. 10.1 Schematic diagram showing DTA set-up and DTA plots

reaction. If it is a phase change, the peak reappears in reverse during cooling. For reactions, this will generally not be so. Also, the area of the peak denotes the total heat effect and, therefore, is related to the total amount of reactant responsible for it. A peak, therefore, yields qualitative as well as quantitative information. Some general applications of DTA in extractive metallurgy have been reviewed by Ray and Wilburn (1982).

10.2.2 Applications of DTA

Since DTA provides a method for identification and estimation of compounds, its possibilities, when applied to the characterization of natural minerals and ores, are self-evident. For example, calcite ($CaCO_3$) shows a peak at 910 °C in air and magnesite ($MgCO_3$) shows a decomposition peak at 650 °C. These characteristic peaks can vary slightly due to the effects of other salts present. However, DTA affords a convenient method of differentiating between dolomite and mechanical mixtures of magnesite and calcite. For the latter, one would merely find the characteristic peaks of the constituents. Dolomite, however, shows the first peak at a higher temperature ($\sim$790 °C), the second peak remaining essentially unchanged. This difference is due to the fact that $MgCO_3$ is in combined form in dolomite and therefore more stable. Thus, it requires a higher temperature to decompose. However, once $MgCO_3$ decomposition is over, $CaCO_3$ is free to decompose at the normal temperature. Some forms of dolomite can show a new third peak as well, the first two being those for dolomite and the third for calcite. Figure 10.2 shows schematically the difference in DTA plots. Similarly, a mechanical mixture of Na_2CO_3 and $CaCO_3$ and the compound $Na_2CO_3 \cdot CaCO_3$ is differentiated easily. The latter shows a reversible peak doublet at around 450 °C which indicates characteristic reversible phase changes in this compound. Figure 10.3 schematically shows how different carbon forms show different peak positions and, to a lesser extent, variations of the shape of the carbon exotherm (Dixon 1976). Others (Dixon 1972) have also discussed correlation of DTA traces with grade, the approach involving pyrolysis of coal sample in its own vapour.

DTA has been used (Dixon 1972) for determination of free sulphur in the presence of sulphide and sulphate deposits. Among characteristic peaks of sulphur

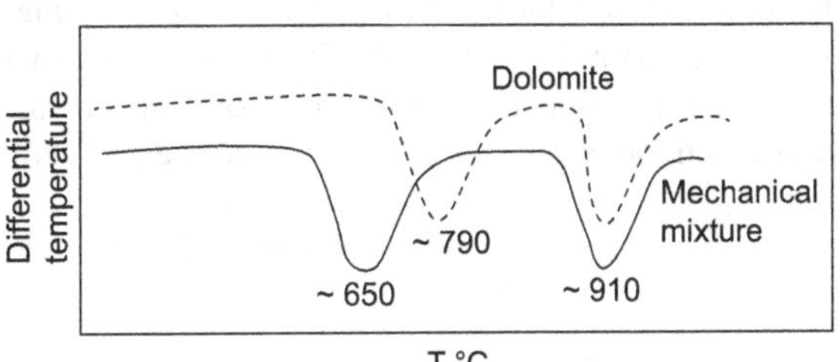

Fig. 10.2 DTA traces for dolomite ($CaCO_3 \cdot MgCO_3$) and mechanical mixture of $CaCO_3$ and $MgCO_3$ (Ray et al. 1991)

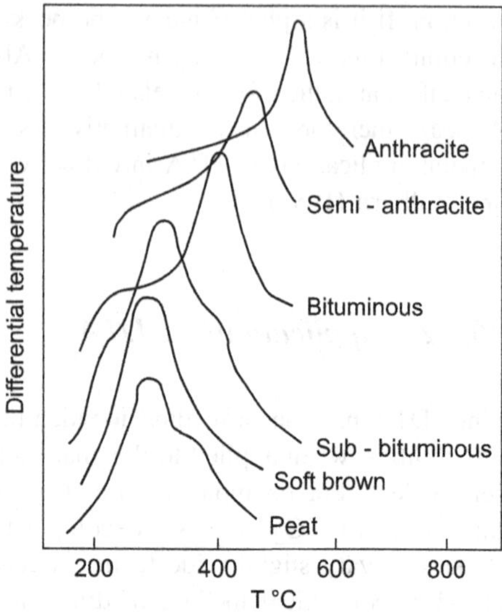

Fig. 10.3 DTA curves for various Australian coals a Welsh anthracite and an American peat (Dixon 1972)

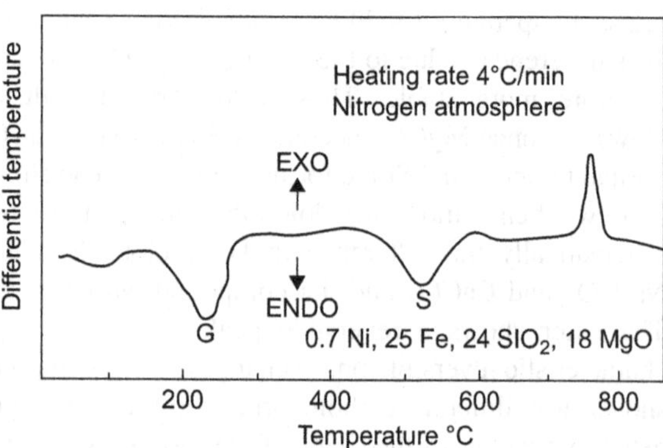

Fig. 10.4 DTA plot for a Lateritic nickel ore (Ray and Abraham 1991)

are the one at 118 °C, another at 125 °C and a third at about 190 °C. Similarly, one can distinguish between free silica (reversible peak at 573 °C) and combined silica in silicates (no such peak).

DTA carried out on lateritic nickel ores gives good indication of the expected behaviour of ore during pyrometallurgcal processing (Ray et al. 1991). A DTA plot for an ore with 1.7% Ni and 25% Fe is shown in Fig. 10.4. The first endotherm (~100 °C) is due to moisture (the ore may contain up to 40% moisture). The second endotherm G (~230 °C) is ascribed to geothite decomposition.

$$2FeOOH = Fe_2O_3 + H_2O \tag{10.1}$$

which liberates nickel oxide from the lattice. The third endotherm S (~530 °C) represents dehydration of serpentine, the NiO of which becomes active above 600 ° C. The exotherm coming next represents recrystallization of magnesium silicate from serpentine into olivine (Ni_2SO_4). At this temperature, NiO gets locked up in olivine and, therefore, it is not easy to reduce nickel above 800 °C. In some laterite ores, the magnesium silicate is in the form of talc instead of serpentine and these are easier to reduce.

A number of attempts (Kissinger 1957a, b; Read et al. 1965a, b; Ingraham and Marier 1969) have been made to evaluate activation energy of reactions and phase changes from DTA traces. This subject is discussed later. Glassy phases in minerals or reaction products can be detected as the plots show typical peak for glass annealing. Other possible applications are in determination of melting and boiling points, solidus and liquidus temperatures of slags (Verma et al. 1979). Baseline shifts can be analysed to obtain information on sintering, moisture and gas evolution, etc.

Simple and complex reactions

Of more interest in pyrometallurgy will be the use of DTA in studying reactions. The technique has been used to study the beginning and completion or reduction of a large number of oxides by hydrogen or carbon monoxide (Oates and Todd 1962; Tikkanen et al. 1963a, b). Metallothermic reductions are accompanied by large heat effects and should be specially amenable to DTA study. It should be noted that sometimes more useful information can be obtained by repeatedly heating and cooling of the test sample to determine the nature of reaction products. Thus, if one wants to study iron reduction of lead sulphide, the peak for the exothermic reaction may be so large that other heat effects are masked. However, once the reaction is over, cooling or reheating should indicate solid–liquid transformation in lead and this peak maybe used in quantitative analysis also provided suitable calibration plots are determined separately.

Sequential reactions

There is not much literature on use of DTA in the study of complex sequential reactions that are so common in pyrometallurgy. Here, we consider only two examples:

(a) *Oxidation of sulphide mixtures*:

Under controlled experimental conditions, individual sulphides, e.g. Cus, FeS, ZnS, CdS, have characteristic DTA peaks which may be attributed to sequential reactions (Sarveswara Rao and Ray 1999). For example, during oxidation of CuS the sequence of products formed is likely to be (Gray et al. 1974; Bollin 1970; Shah and Khalafalla 1971; Ramakrishna Rao and Abraham 1971) $CU_{1.8}S_1$ CU_2O a mixture of CU_2O and $CUSO_4$, CuO, $CUSO_4$ and finally CuO above about 666 °C. A DTA plot for CuS showed (Ray 1989) major exothermic peaks at 157,232 and 320 °C, and minor peaks at about 440 and 550 °C. Similarly, oxidation of FeS_2

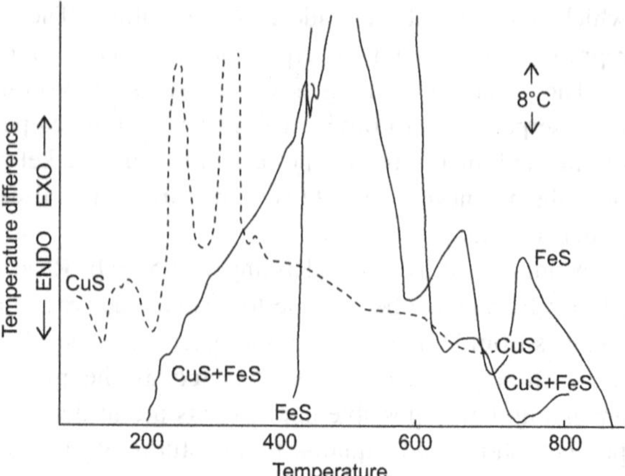

Fig. 10.5 DTA traces for CuS, FeS and CuS + FeS mixture (Sarveswara Rao and Ray 1999)

involves reactions leading to Fe_2S_3, FeS and then various oxides of iron, sulphide oxide mixture, etc. A DTA plot showed (Ray 1989) three large peaks at about 475, 651 and 736 °C. When a FeS_2 and CuS mixture was taken then, as expected, the oxidation peaks for CuS were completely suppressed. Figure 10.5 shows DTA traces for a sulphide mixture (Ray 1989) and single sulphides.

(b) *Reactions in oxide mixture*:

Glass technologists have long-studied reactions in oxide mixture to understand the sequence of products formed during glass melting. In this, DTA studies combined with TG have been very useful. To be able to interpret DTA traces for multi-component oxide mixture, one must first have DTA data on simple binaries and ternaries of the various components of the batch.

The sequence of reactions in a batch depends on conditions of test, rate of heating, sample size, particle size, gas atmosphere and flow rate, presence of impurities, etc. For example, presence of sand grains is known to accelerate decomposition of limestone. Some information concerning reactions in oxide mixtures, which may be useful in extractive metallurgy, is summarized here briefly. It should be noted, however, that all DTA data are based necessarily on fine particles.

It has been shown that in binary $CaCO_3$–SiO_2, no reaction takes place until 1000 °C above which one obtains $CaSiO_3$. If $MgCO_3$ is also present then possibly magnesium silicates also form above 1100 °C. During heating of SiO_2, Al_2O_3 and $CaCO_3$ mixtures, one may get wollastonite by about 1150 °C then anorthite, $CaO \cdot Al_2O_3 \cdot SiO_2$ which is immediately followed by eutectic melting between silica, wollastonite (Ca and/or Mg)O SiO_2) and anorthite. Silica and alumina react at temperatures above 1400 °C to form mullite $3Al_2O_3 \cdot SiO_2$ and melting takes place between mullite, anorthite and silica (tridymite). It has been observed that the temperature required to form a glassy phase may be lowered as the percentage of Al_2O_3 is increased, except in a mixture containing a low percentage of $CaCO_3$.

In glassmaking, melting is aided by the presence of large amounts of soda, Na_2CO_3, that lime reacts more easily as carbonate can be seen by the fact that melting takes place faster with $CaCO_3$ than CaO (Kroger 1948) in lime-soda-silica system. Wiburn and Dawson (1972) summarized reaction in various binaries and ternaries in lime-soda-silica-dolomite system. It was found that identification of phases was often difficult since DTA peaks were liable to shift because of a variety of factors (Lahiri and Ray 1982). However, useful conclusions could be drawn if experimental conditions were suitably standardized. It has been conclusively shown that Na_2CO_3 and $CaCO_3$ react at relatively low temperatures to form a double carbonate which in the presence of silica subsequently form a refractory phase of disodium calcium silicate. This retards glassmaking reactions. In a batch where the sodium carbonate is replaced by sodium hydroxide no such compound forms and melting proceeds uninterrupted (Kroger 1957; Kitaigo-rodskii 1958). Melting of glass can be aided (Eitel 1976; Swarts 1972; Conroy et al. 1966) by the addition of water and additives such as Na_2SO_4 which accelerate reactions through better contact among reagents.

It may be possible to make use of DTA and the DTA information on glass-making systems to understand mechanism of slag formation and factors that accelerate the reactions.

10.2.3 Quantitative DTA

Differential thermal analysis (DTA) records the difference in temperature between the test sample and a reference material as they are subjected to identical temperature regimes in an environment heated or cooled at a controlled rate. If the test substance undergoes physical or chemical changes accompanied by appreciable heat effects, then a DTA curve is obtained which shows a series of peaks in the plot between temperature difference against temperature or time. The location, size and shape of the DTA peaks are governed by a large number of factors associated with physical and chemical properties of the substance, design of the equipment, the experimental procedure adopted, etc.

Once suitable operating conditions have been established, it is seldom difficult to record the DTA curve. However, the subsequent interpretation of the peaks and baseline shifts in terms of reactions and changes in the sample can often be difficult. This is particularly true if the sample is a multicomponent system of uncertain composition, or if it contains impurities whose effect on the DTA peaks is not known. The basic theory of DTA, the factors affecting peak size, shape and location, the significance of the peak temperature, etc., have been discussed extensively in the literature (Mackenzie 1970; Garn 1965).

Although the peak height has been occasionally used in quantitative analysis of DTA curves (El Jazairi 1977), the peak area is, without doubt, the most satisfactory criterion for quantitative DTA. Total heat effect is proportional to peak area (Ray and Wilburn 1982; Smother and Yao 1958; Melling et al. 1965;

Wilburn et al. 1968; Cunningham and Wilburn 1970) and so the area can be used in quantitative analysis. The accuracy of measurement of areas of many (why many?) peaks, however, is limited by several factors. These include overlapping peaks, chemical reactions, the accuracy of area delineation and measurement, the intimacy of mixing, the extent and nature of dilution, gas evolution, the magnitude of the thermal effect concerned, the differences in thermal conductivity, particle size, bulk density, degree of crystallinity.

The area of a DTA peak, when the temperature difference ΔT is plotted against time t, is

$$\text{Area} = \int_{t_1}^{t_2} \Delta T \, \mathrm{d}t \qquad (10.2)$$

where t_1 and t_2 represent the beginning and the end of the peak along the time axis. This area depends not only on the thermal properties of the substance but also on those of the apparatus. This is so because the reaction in the sample heats or cools the sample holder as well as the reference material, in either case diminishing the magnitude of the ΔT signal. This effect is time dependent. A more rapidly heated specimen will show a higher apparent peak height because in this case the reaction is carried to completion before the heat dissipation effect becomes appreciable. The effect is minimal with isolated cups since there is no rapid heat transfer from sample to reference. When the cups are not isolated but placed in a high conductivity block, then peak height and area are reduced because of the relatively free heat exchange. However, the block permits approach of temperature equilibrium very quickly after completion of one reaction so that a subsequent reaction may be observed as separate reaction. This arrangement is ideally suited for qualitative DTA which needs sharp and isolated peaks.

Theoretical expressions of the peak area

Several workers have given the mathematical analysis of the DTA system in order to use the technique as a quantitative tool. The main aim of such analyses is to describe the shape of the DTA peak in terms of the transfer of heat from the source to the sample and of the rate of internal generation or absorption of heat by the test sample as it undergoes a physical or chemical change. A review of many mathematical models is available elsewhere (Melling et al. 1969; Wilburn et al. 1968, Cunningham and Wilburn 1970).

To eliminate the effect of the heating rate on DTA peak area, it is necessary to record the differential thermocouple signal against time (t) and not temperature. The equations that follow therefore always use "peak area" in the context of $\Delta T - t$ plot.

For a cylindrical sample placed in a holder of infinite thermal conductivity where the thermocouple is located in the centre of the sample (Fig. 10.6a), the area under a DTA curve, plotted as ΔT against time t is given by

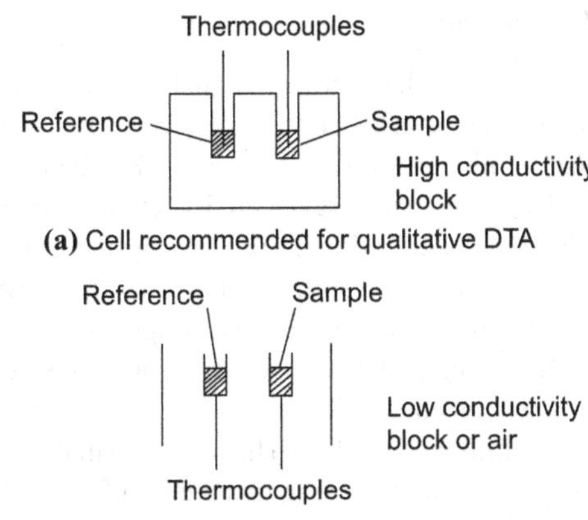

Fig. 10.6 Cell designs for qualitative and quantitative differential thermal analysis

$$\text{Area} = \frac{G \cdot m \cdot \Delta H}{k_s} \tag{10.3}$$

where m is the mass of the sample, ΔH the heat of reaction or transformation and k_s the thermal conductivity of the sample, G is a constant. This expression is, of course, independent of the heating rate. However, the parameter k_s makes the cell non-quantitative in the sense that the DTA peak area changes from one sample to another and even for the same sample, if there is variation in the thermal conductivity.

If the cups (i.e. DTA cell compartments) are isolated by a medium of low thermal conductivity, e.g. air, (Fig. 10.6b) and thermocouples are located outside, and then the expression for the area is different (Melling et al. 1969; Wilburn et al. 1968). It is written as

$$\text{Area} = \frac{m \cdot \Delta H}{2k_{\text{m}}} \tag{10.4}$$

where k_{m} is the thermal conductivity of the low conductivity separating medium. This formula is most interesting for it shows that for a sample of fixed shape and volume the area is inversely proportional to the holder conductivity. Thus, the area is directly proportional to the heat of reaction regardless of the physical properties of the sample. It is, however, important to ensure that the sample diameter is constant (i.e. the cup is unchanged) and the cup or crucible is filled to the same level every time. Because the formula relating the peak area with heat does not contain any terms associated with the physical properties of the sample being examined, the apparatus, once calibrated, may be used for quantitative analysis with widely different materials.

Determination of area

Garn (1965) has indicated methods of measuring the areas of DTA peaks. The methods of construction, indicated in Fig. 10.7, generally involve simply joining of the terminals (ends) of a peak. In such cases, the best construction method is perhaps the one suggested by Cunningham and Wilburn (1970) and shown in Fig. 10.7. The rationale underlying the construction has been discussed elsewhere (Melling et al. 1969; Wilburn et al. 1968; Cunningham and Wilburn 1970). Essentially, this method closely approximates the theoretical baseline, shown by the dotted line, and gives for area, a value very close to the area bounded by the theoretical line.

McIntosh et al. (1974) have shown that reliable linear correlation between area and weight is obtained using this method for the $\alpha - \beta$ transition in quartz (573 °C). As a second example, we can consider the transformations in the compound $Na_2CO_3 \cdot CaCO_3$. DTA plots for this compound were obtained using a DuPont calorimetric (isolated cup) cell (Mackenzie 1970).

On heating, the double carbonate Na_2CO_3 shows two distinct peaks at 407 and 450 °C. On cooling, the peaks appear at lower temperatures, namely 425 and 365 °C. The peaks are schematically shown in Fig. 10.8 which also indicates the method of construction at peak areas. It may be mentioned that the peaks are reproduced during heating and cooling cycles.

During cooling, the peaks (d) and (c) are clearly separated from one another. During heating, however, there is a slight overlap between peaks (a) and (b). Peak areas were measured using the graphical technique indicated, for a large number of samples diluted with varying amounts of Al_2O_3. Some typical calibration curves obtained are shown in Fig. 10.8. It is seen that excellent linear correlation is obtained between area and the amount of double carbonate in the test sample using both individual areas for peaks (a), (b), (c) and (d) as well as total areas (a + b) and (c + d). The regression coefficient of the linear fit is generally better than 99.5%.

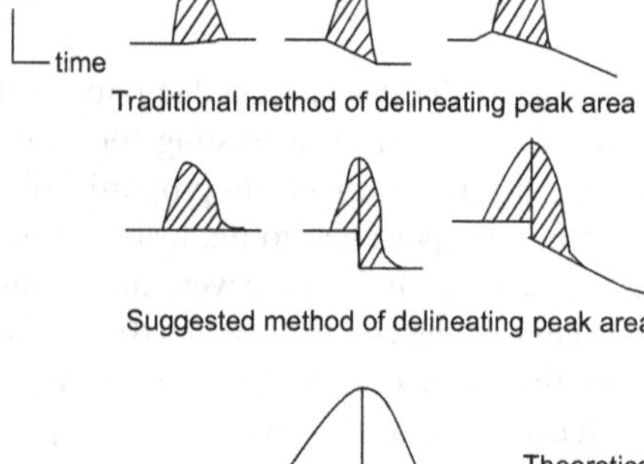

Fig. 10.7 The traditional construction and "Wilburn" construction for delineation of DTA peak area

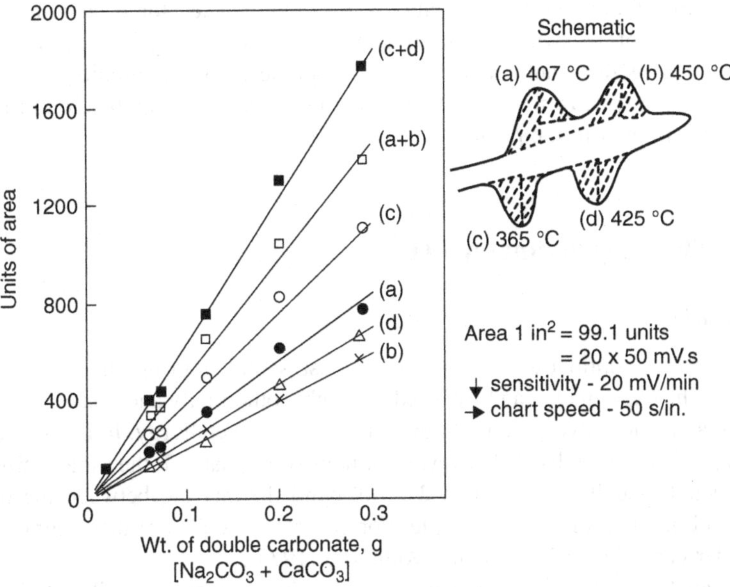

Fig. 10.8 DTA peaks for Na_2CO_3–$CaCO_3$ and calibration of peak areas against weight (Ray, unpublished research, 1978)

The graphical procedure can, therefore, be recommended for quantitative DTA work.[1]

For kinetic studies, one needs to determine the rate of generation of the peak area which can then be used as a parameter that is proportional to the rate of reaction. This can be done under both isothermal and non-isothermal conditions.

10.2.4 Differential Scanning Calorimetry (DSC)

In DTA, the peak area arises out of the combined effect of heat generation (or absorption) and heat dissipation effects. Accordingly, the area is dependent on the experimental arrangement. Therefore, for quantitative DTA, one needs a calibration curve. This is completely dispensed with in DSC. In DSC, the temperature difference ($T_S - T_R$) is all through maintained at zero by external supply of heat to the reference or test crucible to compensate for the temperature difference.

[1]For procedural details regarding the use of DTA in quantitative heat measurements see the following papers: Sarangi et al. (1991a, b, 1992)

One way of doing this is to employ one microheater for each crucible. An electrical current is switched on to generate compensating heat as and when required. This is done automatically by appropriate electrical circuits.

However, in spite of tremendous progress in this area, the application of DSC is mostly still limited to temperatures below 700 °C or so.

10.3 Thermogravimetry (TG)

TG and Kinetics

TG comprises continuous weighing of a test sample during the course of its reaction. The experiment can be carried out both isothermally and non-isothermally. Isothermal studies have been traditional in extractive metallurgy. In isothermal TG, the sample size can be large. However, for non-isothermal TG, the sample size must be restricted (usually to less than 2 g) to avoid thermal lag between the outside layers and the interiors of the sample. For general principles of thermogravimetry, the reader can refer to Keatch and Dollimore (1975).

TG can be used to study reactions or reaction sequences easily. A TG plot, however, is sensitive to experimental parameters which must be controlled. TG data, both isothermal and non-isothermal, essentially represent kinetic data. Methods of analysing these data have been discussed earlier in the book.

Derivative Thermogravimetry (DTG)

The slope of TG curves measures rate of reaction, and the continuous plot of slope with time gives DTG trace. Figure 9.9 in the previous chapter shows DTG plots for carbon monoxide reduction of a Fe_2O_3 sample. The plot shows low reaction rate increases and decreases with temperature. The peaks are associated with sequential reactions as discussed in Chap. 9.

10.4 Simultaneous DTA, TG and DTG

Principle and examples

If weight changes are also measured along with heat effects, then one knows if the latter is only due to phase transformation or a reaction involving a weight change. If there is a reaction without weight change then also DTA plot is not reversed during reheating.

Simultaneous measurements are carried out by simultaneously weighing together both crucibles of a DTA assembly, all weight changes being ascribed to weight changes in the test samples because the reference material Al_2O_3 does not lose weight on heating.

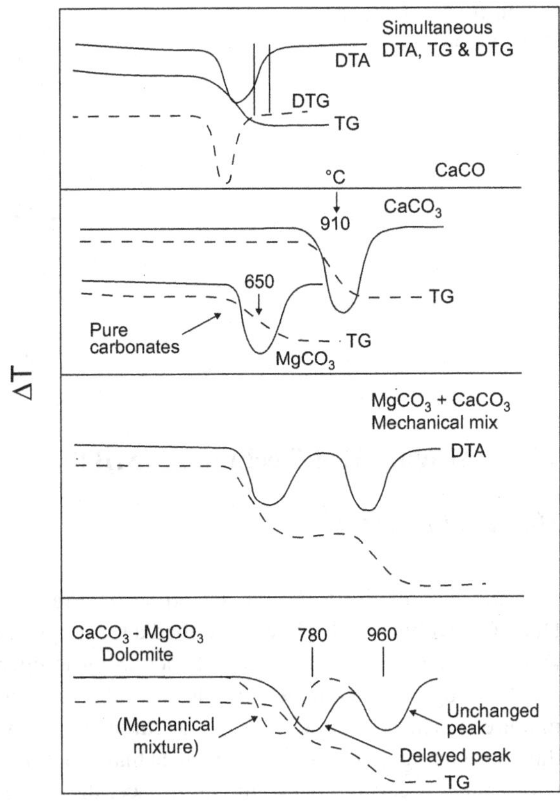

Fig. 10.9 Simultaneous TG, DTA plots of mechanical mixture of $CaCO_3$ and $MgCO_3$, and Dolomite (Schematic)

Figure 10.9 indicates, at the top, the nature of simultaneous DTA, TG, DTG plots. It also shows, schematically, the kind of plots one obtains using a mechanical mixture of $CaCO_3$ and $MgCO_3$ or these carbonates in combined form as dolomite Keatch and Dollimore (1975). Reaction can be more thoroughly studied by simultaneously collecting and analysing the gases evolved (EGA).

Figure 10.10 shows some preliminary data (Khan et al. 1984) as another example of thermal analysis using simultaneous measurements. The data are for some samples from ancient zinc retorts of Rajasthan, India. The retort contents showed hydrated sulphate $ZnSO_4 \cdot 7H_2O$. DTA endotherm and TG weight loss at around 100 °C were obviously due to dehydration. Overlapping DTA endotherms beyond 600 °C were due to melting of $ZnSO_4$ (no weight loss) and decomposition to and SO_3 (considerable weight loss). Broad depression in DTA between 340 and 360 °C could be attributed to melting of an alloy of Zn and about 5% Pb. This was substantiated by chemical analysis data. The data showed that there was some metal and $ZnSO_4$ in retort wall and slag also.

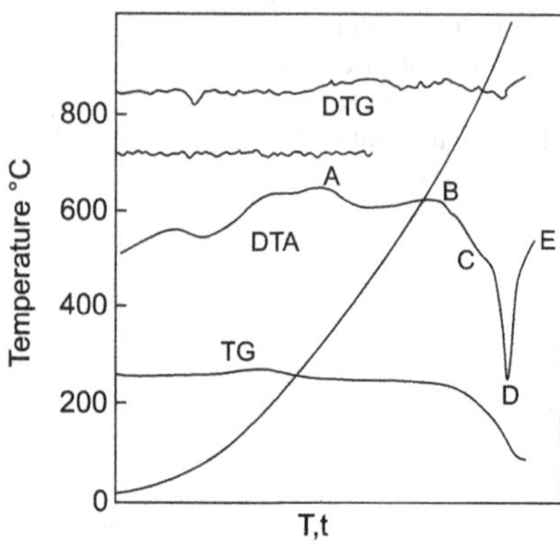

Fig. 10.10 TG, DTG and DTA of contents of an ancient zinc retort

10.5 Moving Bed Technique (MBT)

10.5.1 Principle

Ray and coworkers (Ray and Sewell 1979; Prakash et al. 1986, 1987; Ray and Dutta Chowdhury 1986) have described a technique for carrying out non-isothermal kinetic studies under conditions of increasing temperature. This technique, called the "moving bed technique", employs a tube or a capsule containing the reaction mixture which is introduced into the furnace hot zone at a predetermined speed, thus simulating material flow in an actual reaction zone (Fig. 10.11). Once the whole tube or capsule is inside the furnace, the movement is abruptly stopped and the container quickly withdrawn and cooled. Samples taken from different regions then represent the state of the system at different time–temperature conditions.

In this technique, at the starting position the leading edge of the container rests at the mouth of the furnace. As the container enters, successive volume elements are exposed to the hot zone at different intervals, heating of each element being gradual. The reaction, therefore, proceeds with both time and temperature varying. The movement of the container simulates material flow. On cooling and sampling after a run, the reaction time at any particular location can be calculated by knowing the total length of boat, its total travel time and location of the element.

The MBT recreates some features of the composite situation of a continuous (plug flow) reactor and the total sample, after removal, represents a frozen picture of the steady-state situation prevalent in the interior. A visual examination of the sample along the length can yield a great deal of useful information such as meltdown or fusion, the onset of other physical and/or chemical changes which leave visible clues, etc. The moving boat concept can, obviously, be used to model horizontal or vertical reactors (Fig. 10.12).

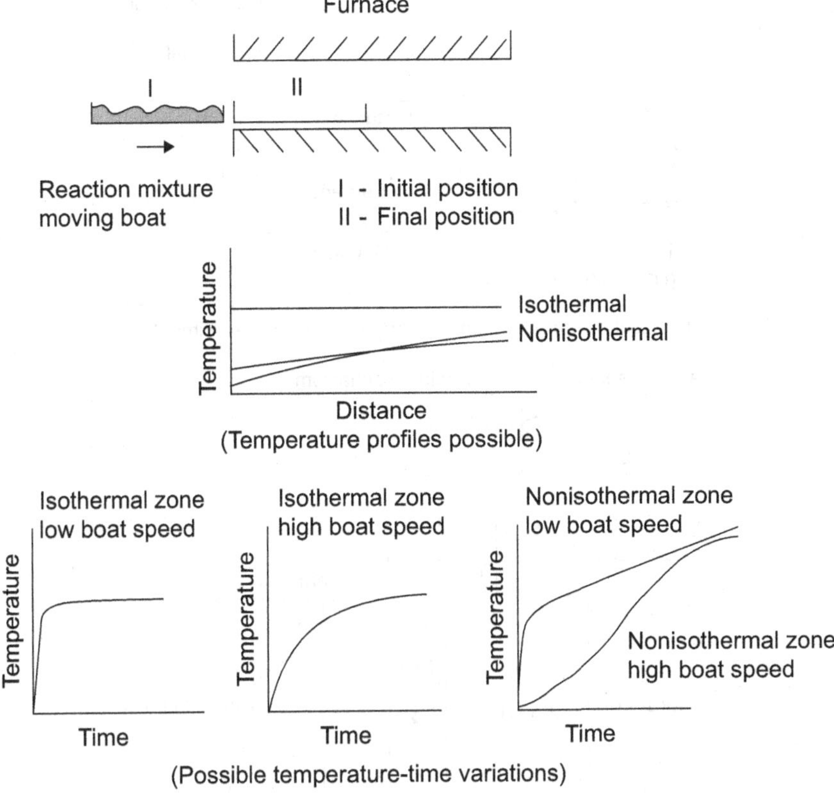

Fig. 10.11 The time gradient technique and possible temperature–time variations

10.5.2 Heating Programme

The degree of a reaction obtainable under non-isothermal conditions will depend among other factors, on the actual variation of temperature with time and, therefore, the control and measurement of this variation is of considerable importance. In the case of the moving boat, the heating of a volume element depends on the actual temperature profile of the furnace, the speed of the boat (which simulates material flow), the heat transfer coefficient and the thermophysical properties of the reaction, mixture. The heating programme can be varied by varying one or more of these parameters. The actual heating programme can be obtained by an arrangement such as the one shown in Fig. 10.13 which shows a series of thermocouples embedded in the reaction mixture. Figure 10.14 shows some typical temperature–time plots for

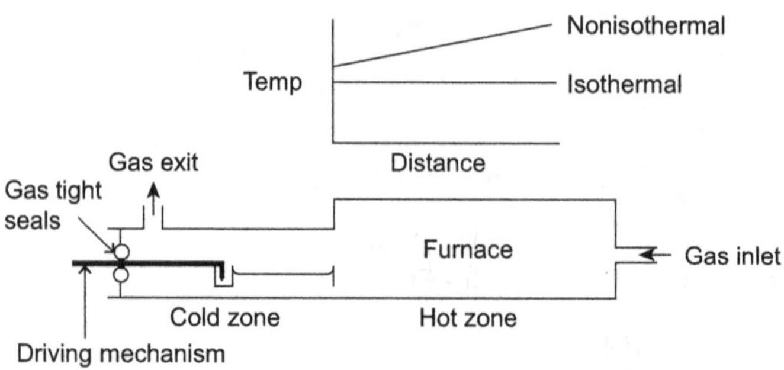

Horizontal set-up for time/temperature gradient experiments

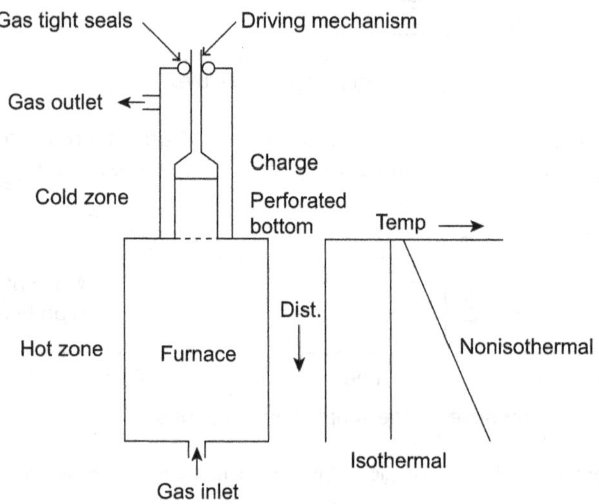

Vertical set-up for time/temperature gradient experiments

Fig. 10.12 Horizontal and vertical versions of moving bed technique

four thermocouples when a long boat was introduced into a furnace under the conditions mentioned. In this case, the heat transfer along the bed is small because successive thermocouples essentially show similar $T - t$ plots. The two cases where plots deviate substantially can be explained as due to special reasons (Ray and Sewell 1979).

A simple example

The principle of the technique would be understood more easily if we consider an example. Suppose that a glassmaking reaction mixture is laid on a horizontal platinum container and introduced into a hot zone where the mixture (typically

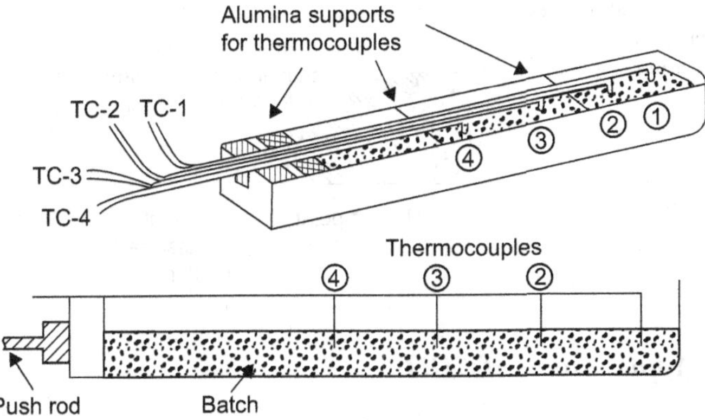

Fig. 10.13 Arrangement of thermocouples in boat

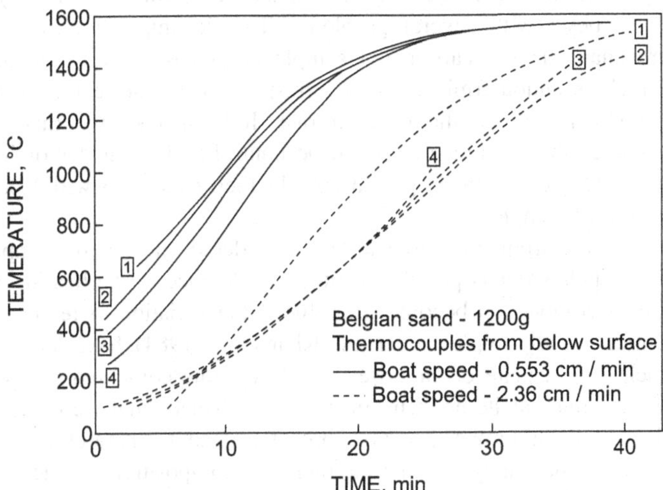

Fig. 10.14 Effect of boat speed on temperature–time plots (unpublished research by Ray 1979)

Na_2CO_3 plus $CaCO_3$ plus SiO_2 with some Na_2SO_4) melts and reacts to form glass slowly. On withdrawal and cooling, the sample would appear as shown in Fig. 10.15. The essential features visible in the sample resemble closely that which are found in actual glass melting tanks.

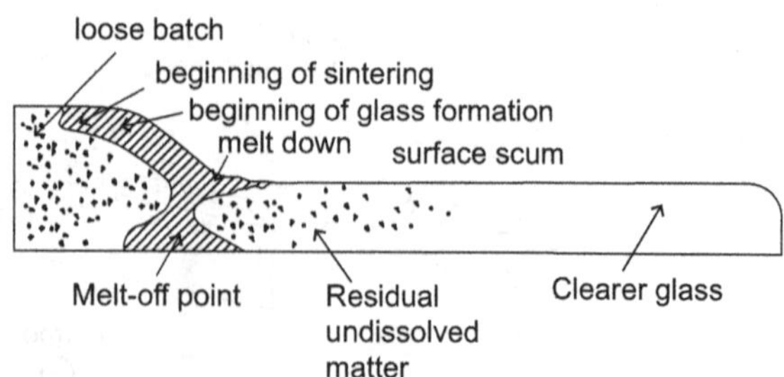

Fig. 10.15 Essential features of a glassmaking batch sample after a moving boat experiment

10.5.3 Applications of MBT

Decomposition of carbonates

The subject of proper assessment of the decomposition behaviour of limestone and dolomite samples is of considerable importance in several industries, yet proper assessment is beset with several problems. The decomposition behaviour and temperature changes in a carbonated sample are interrelated. A decomposition reaction which is endothermic tends to arrest the temperature rise, the lowered temperature slowing down the reaction rate. It is also known that for bulky, compact samples the reaction may even be controlled by transfer of heat to the cooler interior. Obviously, the importance of the thermal aspects will depend on the size and nature of sample.

When several carbonate samples are considered, no valid comparison of decomposition behaviour is possible without a reference to the heating characteristics. A direct comparison becomes possible if the variation of temperature with time is identical for all samples under test. Khan et al. (1984). have shown that for a given particle size, calcite decomposes at a lower rate compared to limestone in spite of the fact that calcite heats up more rapidly when equal weights of samples are introduced into a furnace set at a predetermined temperature. In this case, therefore, one can definitely say that the rate of decomposition of calcite would be even lower if temperatures are comparable. It is possible to ensure a uniform heating schedule in TG where the sample size is necessarily small. The restriction on the sample weight, particularly for simultaneous TG and DTA, however, involves the uncertainties of heterogeneity in test samples and sampling error. There is, therefore, the need of a method which can use relatively larger samples and, at the same time, ensure uniform heating schedules.

Presented here are decomposition data of seven limestone and dolomite samples used in a steel plant in India. The data have been obtained using the following technique on samples of fixed particle size.

(a) Measurement of rate of temperature rise when different samples of a fixed weight are suddenly introduced into a furnace maintained at a predetermined temperature.

(b) Measurement of weight loss at fixed time intervals, when samples of a fixed weight are suddenly introduced in a furnace maintained at a predetermined temperature.
(c) Simultaneous DTA, TG and DTG.
(d) The moving boat technique.

As discussed earlier, neither (a) nor (b) may allow direct comparison of dissociation behaviour. If there is no sampling error then results of (c) serve as a reliable guide. It is shown that the technique (d) perhaps yields the most reliable data. Here, we summarize a work reported by Ray and Dutta Chowdhury (1986).

Figure 10.16 outlines the MBT arrangement which was used to impose a well-defined heating programme on relatively larger samples. The set-up employs a furnace at a constant temperature and a moving boat in which is placed a series of crucibles containing the reaction mixture ($\sim$1 g). The leading edge of the boat initially rests at the furnace mouth. It is then introduced into the furnace at a constant speed using a special pushing device. As soon as the entire boat goes into the furnace, it is quickly withdrawn and the reaction mass in crucibles is allowed to cool quickly.

Under this arrangement, the crucible at the leading edge spends maximum time inside the furnace, the succeeding crucibles spending progressively lesser times, although all crucibles are individually heated in the same manner. The reaction time, of course, differs from a maximum for the crucible at the leading edge to a minimum for one at the trailing edge. The length axis of the moving boat represents time, which is determined by the total length of the boat, the total travel time and the location of the crucible. The variation of temperature with time, which is identical for all crucibles, is determined by having a thermocouple embedded in a crucible near the leading edge.

The nature of the temperature–time curve, which describes the heating schedule, depends on the steady-state temperature profile of the furnace, the speed at which

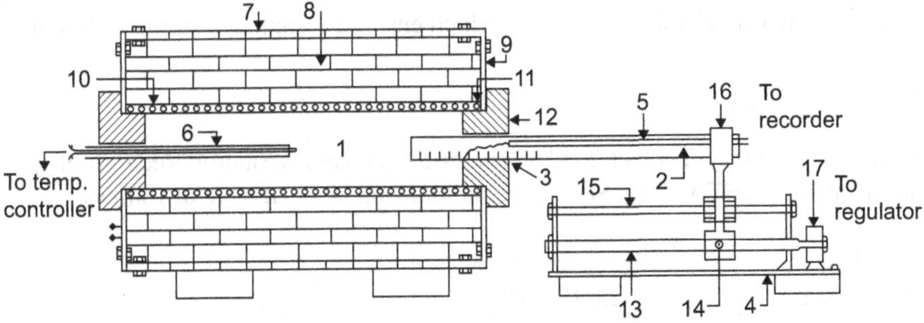

Fig. 10.16 Schematic diagram of the moving boat set-up. (1) Furnace, (2) stainless-steel channel; (3) carbonate sample, (4) moving arrangement, (5) chrome/alume thermocouplel, (6) Pt/Pt-13% Rh thermocouple, (7) furnace shell, (8) fire clay bricks, (9) sindonia board, (10) heating coil, (11) sileminite tube, (12) furnace refractory plug, (13) lead screw, (14) nut, (15) supporting shaft, (16) clamp, (17) stepping motor (Ray and Dutta Chowdhury 1986)

Table 10.1 Composition of various samples

Sample No.	Material	Source (all in India)	Composition (%)					
			CaO	SiO_2	Al_2O_3	FeO	MgO	LoI
1.	Limestone	Birmitrapur (Orissa)	45.34	8.62	2.49	1.13	4.95	38.32
2.	Limestone	Nandini (M.P.)	41.67	5.68	3.2	NA	7.14	NA
3.	Limestone	Chopan (U.P.)	42.5	11.24	2.05	1.07	4.63	NA
4.	Limestone	Satna (M.P.)	49.60	2.96	NA	NA	NA	NA
5.	Dolomite	Chopan	NA	–	7.5	–	19	–
6.	Dolomite	Birmitrapur (Orissa)	NA	3.86	1.32	NA	20.5	NA
7.	Dolomite	North Bengal	NA	2.23	NA	NA	21.0	NA

NA not available

crucibles move and the thermal properties of the sample. Accordingly, the heating programme can be changed by altering the boat speed and the temperature of the furnace hot zone.

Samples for which results are shown are described in Table 10.1.

Measurement of heating characteristics

This was carried out by placing a given weight of sample (~0.5 g) for different periods in the hot zone of a furnace maintained at a fixed temperature. The heating time required to attain the furnace temperature was measured by thermocouples embedded in the sample mass.

Isothermal decomposition experiments

These were carried out by placing a given weight of sample (~0.5 g) for different periods in the hot zone of a furnace maintained at a fixed temperature. The samples were weighed after removal from the furnace. The degree of reaction was obtained as the ratio of measured weight loss to the maximum possible weight loss observed when decomposed for 6 h at 1000 °C which ensured complete decomposition in all cases.

Thermal analysis

Simultaneous DTA, TG and DTG were carried out using a Stanton Redcroft thermal analyser (model STA-781). Standard conditions maintained were as follows:

Particle size	−90 to +45 μm
Heating rate	15 °C min^{-1}
Sample weight	~29 mg
Nature of packing	Five taps on the crucible
Gas atmosphere	As generated by the sample undergoing decomposition

Moving heat technique (MBT)

The MBT employed a horizontal furnace with a central tube of length 62 cm and internal diameter 38 mm. A narrow stainless-steel channel (70 × 2 × 1.5 cm) was used as the boat. The moving mechanism employed a lead screw principle as illustrated. A threaded shaft (39 × 1.5-cm diameter, 1.7-mm pitch) was coupled to a reversing stepping motor of rating 3-kg-cm torque, 1.25A, and 220 V AC line voltage. The transverse motion generated by a nut and plate attachment was used to push the boat into the furnace.

Figure 10.17 shows the pattern of temperature variation during heating of the samples when they were suddenly put into the hot zone. It is found that nearly 5 min elapsed before the samples came to thermal equilibrium with the furnace. For an isothermal run lasting 2 h or more, therefore, one can ignore this heating up period. Excepting in one case, the temperature variation was almost identical, presumably because the sample was rather small.

Figure 10.18 shows isothermal decomposition data for 721 °C (±4 °C). It is seen that while decomposition data for dolomite show inflections at lower temperatures, these inflections tend to disappear at higher furnace temperatures because of overlapping of the dissociation reactions for $MgCO_3$ and $CaCO_3$. The data for $CaCO_3$ approximately fit the kinetic equation for phase boundary controlled reactions, namely

$$1 - (1 - \alpha)^{1/3} = kt \tag{10.5}$$

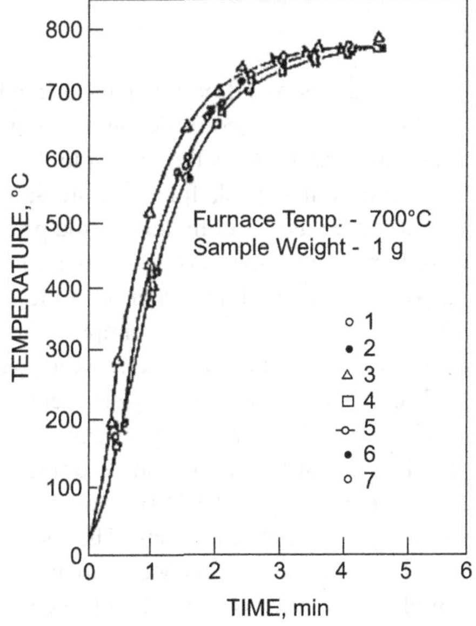

Fig. 10.17 Temperature–time plots for various samples during the heating period (Ray and Dutta Chowdhury 1986)

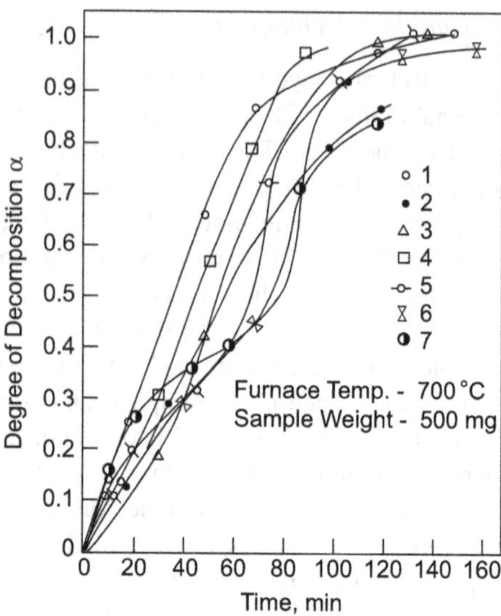

Fig. 10.18 $\alpha - t$ plots for isothermal decomposition at 721 °C (Ray and Dutta Chowdhury 1986)

where α is the degree of conversion and t is time. The slopes of the linear plots of the left-hand side of Eq. 10.5 versus t give values of the rate constant, k. These plots, which three are parallel, yield an activation energy value of around 40 kJ mol^{-1} (33 kcal mol^{-1}).

It is seen that the dissociation behaviour of different samples can be compared in terms of the $\alpha - t$ plots. We may arrange them in decreasing order of ease of decomposition as follows:

Limestone: 1,4,3,2,
Dolomite: 5,6,7.

The order is based on α values at fixed values of t. Thus, for a given t sample 1 shows a higher degree of decomposition compared to samples 4, 3 and 2.

Figure 10.19 shows the DTA plots for the various samples. These plots clearly show the double peak for dolomite samples and single peak for limestone and, therefore, are useful in indicating the presence of the mineral phases. However, the peak temperature may not serve as an index for comparison of the decomposition behaviour. This is done better with the help of the TG plots shown in Fig. 10.20. These plots identify, for example, sample 1 as the best in terms of ease of decomposition, which is in agreement with results discussed earlier.

The moving boat experiments were carried out using samples of $\sim$1 g in the crucibles lined up in the boat. For a fixed furnace temperature profile and boat speed, different limestone and dolomite samples were all found to be heated in the same manner. Figure 10.21 presents the $\alpha - t$ plots for different samples for a set of fixed experimental conditions. The measured variation of temperature with time is also shown. These plots were obtained from eight different runs, one for each sample. It is, however, possible to generate all these data from a single run if the

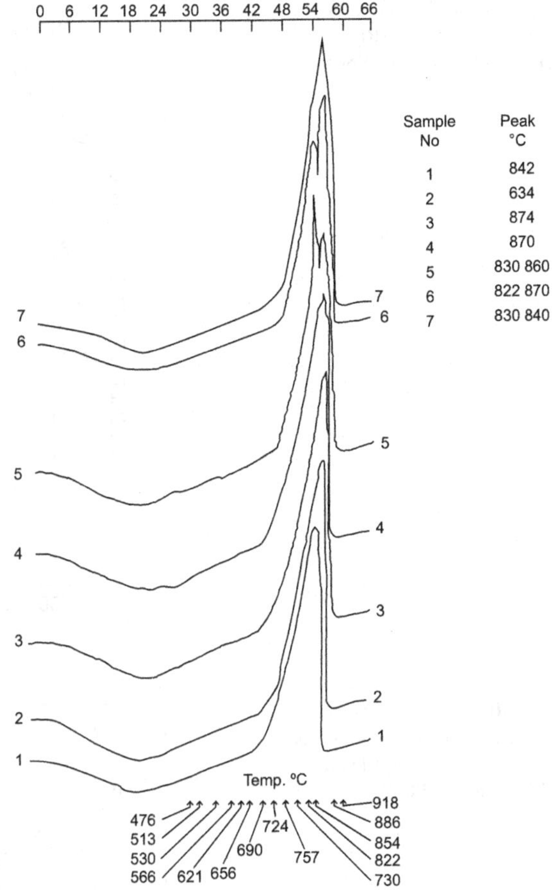

Fig. 10.19 DTA plots for different samples (Ray and Dutta Chowdhury 1986)

moving boat is wide enough to take several rows of samples. Plots given in Fig. 10.21 show that the MBT affords a convenient and effective method for establishing the difference in decomposition behaviour since the $\alpha - t$ plots are well separated.

Samples may be arranged in decreasing order of the ease of decomposition as follows:

Limestone: 1,4,3,2,
Dolomite: 5,6,7.

The orders are identical with those established through isothermal decomposition experiments discussed earlier.

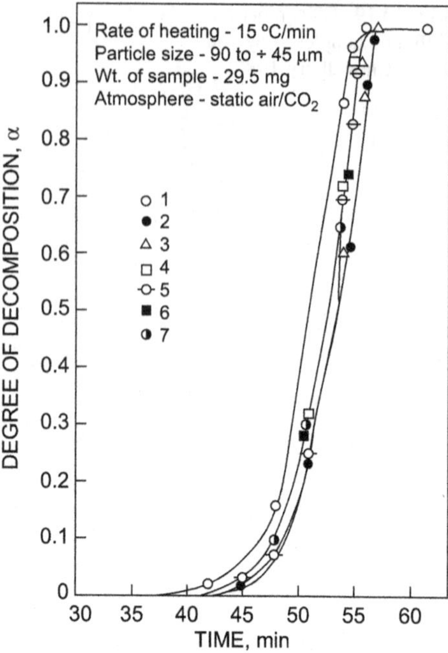

Fig. 10.20 TG plots for different samples (Ray and Dutta Chowdhury 1986)

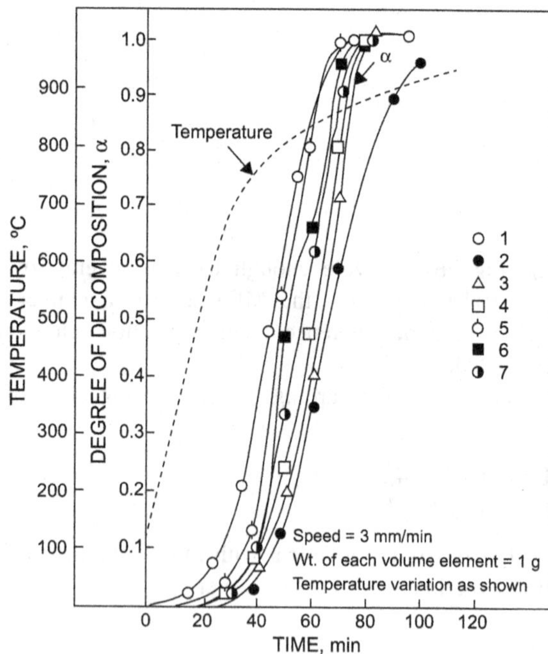

Fig. 10.21 $\alpha - t$plots generated by the moving boat technique (Ray and Dutta Chowdhury 1986)

10.5.4 Reduction of Iron Ores

Numerous standard tests have been developed to measure the reactivity of solids. All such tests have their limitations and thus there are a variety of tests for the reducibility of iron ores (Prakash et al. 1986). While some of the tests employ isothermal conditions, others employ a programmed variation of temperature. Most tests make use of static beds but a few employ moving beds.

Here, we describe the use of MBT to assess the relative reducibility of iron ores. Reduction kinetics of a number of ore–coal systems has been measured and the data compared with reducibility data obtained by other means (Prakash et al. 1986). Prakash et al. (1986) have described the reactivity of some raw materials listed in Tables 10.2 and 10.3.

Reproduced here are some typical data obtained by MBT.

The moving bed technique

The principles of the MBT have been discussed in detail earlier and only a brief outline is given here. The basic unit (Fig. 10.22) comprises a cylindrical stainless-steel capsule (0.35 m × 0.075 m) and, within this, a stainless-steel segmented cage which contains the solid reactants. The capsule is moved into the hot

Table 10.2 Analyses of iron ore samples used (wt%)

Chemical composition (wt%)	Iron ore		
	Barajamda (Bihar, India)	Kiriburu (Orissa, India)	Gadigi (Karnataka, India)
Fe (Total)	68.8	63.8	66.9
Fe_2O_3	92.0	88.1	91.7
FeO	4.3	3.6	4.0
SiO_2	0.9	1.8	1.0
Al_2O_3	1.3	2.1	1.5
CaO	0.2	0.2	0.2
MgO	0.1	0.1	0.1
Loss on ignition	1.2	4.0	1.3
S	Traces	Traces	Traces
P	0.02	0.03	0.03

Table 10.3 Proximate analysis of coal samples used (wt%)

Non-cooking coal	Chemical composition (wt%)			
	Moisture	Volatile matter	Fixed carbon	Ash
Coal A (Raniganj, W. B., India)	5.0	34.8	44.7	15.5
Coal B (Nutanganga, M. P., India)	3.0	31.8	49.6	15.6
Coal C (Raniganj, W. B., India)	7.0	31.0	44.5	17.5

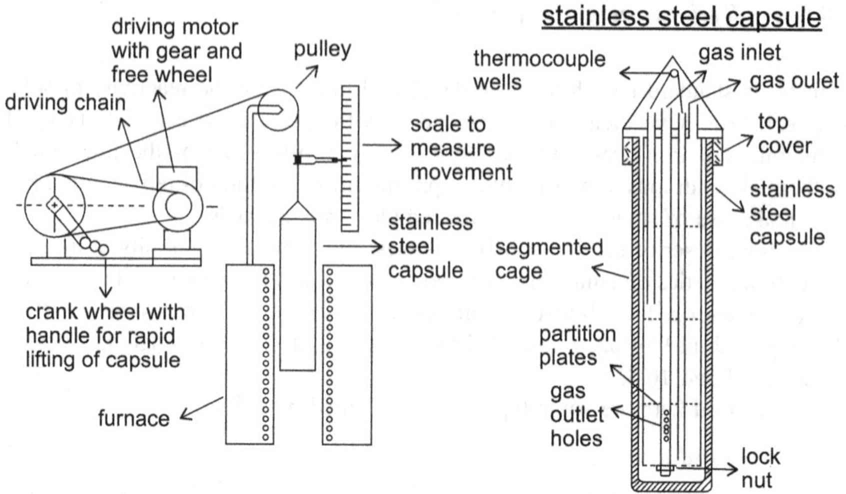

Fig. 10.22 Schematic diagram of the moving bed technique (Prakash et al. 1987)

zone of the furnace at a speed which can be varied. When the complete capsule is just inside, the movement is stopped, the capsule withdrawn and quenched with nitrogen gas. The reacted mass within the cage now represents the course of reaction in a plug flow type reactor. The volume elements near the leading edge represent the maximum time of reaction and those near the trailing edge represent almost zero reaction time. Analysis of these samples then yields kinetic data under rising temperature conditions. The heating programme that a volume element undergoes depends on the temperature profile of the furnace hot zone, the speed of the capsule, the location of the element, and the thermal characteristics of the reaction mass. The actual temperature–time $(T - t)$ plots are obtained from thermocouples embedded in the reaction mixture.

10.5.5 Standard Reducibility Test

Isothermal 1000 ± 10 °C reducibility studies were carried out by reducing large ore samples (~500 g, −12+10 mm, (i.e. the particles pass through 12 mm but are retained by a 10 mm mesh, average particle size $(12 \times 10)^{1/2}$ mm) with a carbon monoxide–nitrogen (30:70) gas mixture in a standard reducibility device (IS: 8167 1976). The ore sample was suspended from a balance and initially maintained under an inert atmosphere until it reached the temperature of the furnace. The reducing gas was then allowed to flow at a rate of 15 1/min in all cases and the loss in weight measured continuously as a function of time.

MBT data on iron ore reduction

Figure 10.23 shows results of some reducibility experiments using mixtures of three different iron ores with a fixed coal, using the moving bed set-up. The ore–coal ratio, particle size and the heating programme were unchanged in all three cases. It can be seen that, according to the MBT results, the Gadigi ore is the most easily reducible. The inset in Fig. 10.23 shows results obtained from the standard reducibility apparatus when the ores were reduced isothermally at 1000 ± 10 °C by a carbon monoxide–nitrogen mixture. According to these results, Gadigi ore is again the most reducible but the reactivity order for the other two samples is reversed. A possible explanation for the apparent discrepancy may be that the MBT results are for a system where the reduction reactions are coupled with the gasification of carbon. It is known that the latter is influenced by the nature of the ore and the metallic iron produced. This may well be different for the Kiriburu and Barajamda iron ore samples. The apparent discrepancy may also arise from the difference in the temperature conditions.

Figure 10.24 presents some MBT results on the reduction of a fixed Barajamda ore by three different coals. In all three cases the ore to coal ratio, particle size and heating rate have been kept unchanged. It is seen that the order of reactivity is A > B > C. To examine whether reduction rate is related to char reactivity, the latter was measured using a coal gasification set-up where char samples were oxidized isothermally by carbon dioxide (IS: 8167 1976). The results of measurements at 1000 °C are shown in the inset. The data show that reactivity is

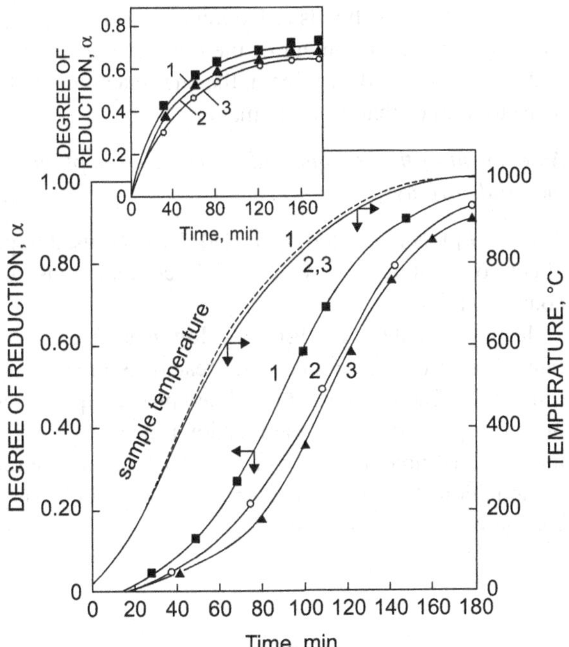

Fig. 10.23 MBT results on reduction of three iron ore samples (filled rectangle = Gadigi; filled triangle = Barajamda; filled circle = Kiriburu) by a fixed Raniganj coal. Ore to coal ratio is 1.0:0.8. Particle size is −6+3 mm. Inset: reducibility apparatus results on isothermal reduction of ore samples by a Carbon monoxide–nitrogen (30:70) mixture. Temperature, 1000 ± 10 °C; particle size, −12+10 mm; sample weight, ~500 g; gas flow rate, 151/ min. (Prakash et al. 1987)

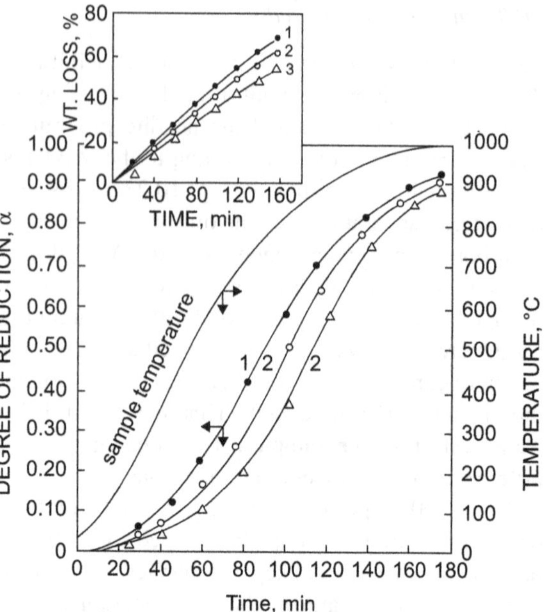

Fig. 10.24 MBT results on reduction of a fixed ore (Barajamda) by three different coals (filled circle = coal A; O = coal B; filled triangle = coal C). Ore to coal ratio is 1.0:0.8. Particle size is −6+3 mm. Inset: Data on isothermal gasification of chars made from coal samples by carbon dioxide. Temperature, 1000 °C; particle size, −10+3 mm; sample weight, ~20 g, gas flow rate, 15 ml/min (Prakash et al. 1987)

greatest for coal A and least for coal C, with coal B occupying an intermediate position. The order, which also remains unchanged at other temperatures, is the same as indicated by the MBT results. Coal A is most reactive because it has most volatile matter which aids reduction and also helps to produce a more reactive char. The volatile matter content of the other two samples, however, is nearly the same and, therefore, the difference in the char reactivity and MBT data must be attributed to factors other than volatile matter.

Assessment of throughput and energy consumption of reactors when different raw materials are used

The $\alpha - t$ plots generated by the MBT can be used to provide rough estimates of the effect of different raw material combinations on throughput and energy consumption.

It may be noted that, for plug flow reactors, the throughput, i.e. the production rate per unit cross section of the reactor is inversely proportional to the residence time of the reaction mass. For a fixed product quality, a fixed value of throughput is inversely proportional to the reaction time (t) in the $\alpha - t$ plot. Thus, if the value of the time required is t_1 for raw material combination 1 and t_2 for raw material combination 2, then the percentage change in throughput achieved because of a change of combination from 2 to 1 is:

$$100 \cdot \frac{(1/t_2) - (1/t_1)}{(1/t_1)} \tag{10.6}$$

Assessment of the changes in energy consumption because of changes in raw materials is perhaps best made with the help of isothermal $\alpha - t$ plots. It is noted that an increase in reaction temperature raises the $\alpha - t$ plots. If the data on the effect of temperature on $\alpha - t$ plots for different raw materials are available then two $\alpha - t$ plots which overlap can, possibly, be selected. The two temperatures to which these correspond indicate relative energy consumptions.

Thermal analysis studies

The MBT is essentially a non-isothermal test. To examine farther the reliability of this test three mixtures of a coal and an ore, at three different ratios, were examined by both the MBT and a Stanton Redcroft STA-780 thermal analyser. The particle and sample sizes, of course, were different. However, the heating rate employed in the thermal analyser (15 °C/min) was only marginally higher than that prevailing during the initial periods for the MBT. Figure 10.25 shows the $\alpha - t$ plots obtained by MBT and the thermal analyser.

The thermogravimetric (TG) data have been plotted in terms of fraction reacted (f) defined as the weight loss at a given time, t, divided by the maximum possible weight loss from the ore–coal sample. The denominator is obtained from the analysis of the ore and coal. It is found that although kinetic parameters used and the conditions are different, the order of reducibility remains unchanged.

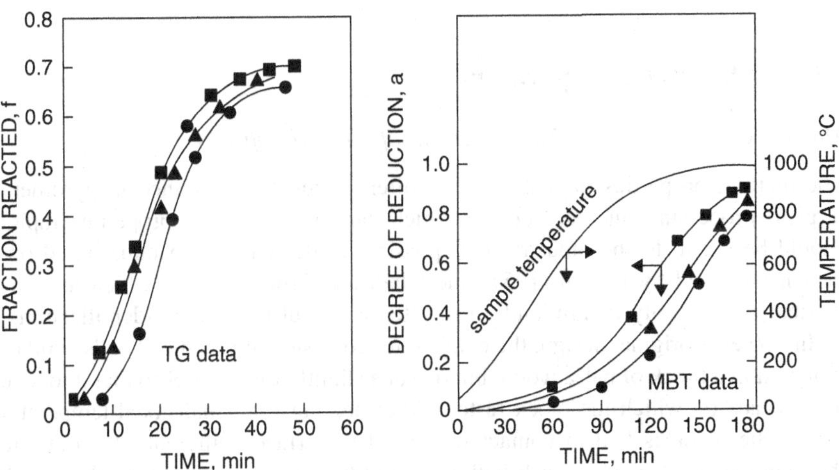

Fig. 10.25 Kinetic data for three Barajamda ore–Raniganj coal mixtures obtained using the STA-780 thermal analyser and MBT. Ore to coal ratios: filled rectangle 1.0:1.0; filled triangle 1.0:0.8; filled circle 1.0:0.6. $f - t$ plots: sample weight, ~25 mg; heating rate, 15 °C/min; atmosphere, static air; particle size −0.149+0.105 mm. $\alpha - t$ plots: sample weight, ~1400 g; capsule speed, 2.0 mm/min; particle size, −10+6 mm. (Prakash et al. 1987)

10.6 Temperature Gradient Technique (TGT)

Principles and applications

In the moving boat technique, the location of volume elements along the length indicates reaction time intervals. Thus, in a way, the technique is a time gradient technique. This is especially so if the heating up period is rapid so that volume elements essentially move in an isothermal temperature zone. In temperature gradient technique, a series of volume elements are equilibrated in a stationary boat kept in a hot zone with a built-in temperature gradient. If samples are removed after a fixed period and examined then one obtains the results of effects of different temperatures in a reaction mixture for a fixed period. One TGT experiment thus combines in the results of several isothermal experiments.

The TGT can be used in a variety of studies. Ray (1979) has used to determine the liquidus and meltdown temperatures of a number of blast furnace slag compositions. A series of small boats containing a given sample was kept in a row in a furnace with a built-in temperature profile. The temperatures at different locations were accurately measured by a series of thermocouples. After thermal equilibrium was ensured, the samples were removed and examined to identify the location where the first liquid appeared, where meltdown took place, etc. The corresponding temperatures were then obtained by noting the locations.

The same technique can be easily employed for isothermal kinetic studies for fixed reaction periods. If the experiments are carried out for 3- or 4-time intervals then one can generate a complete set of kinetic data giving effects of time and temperature, each location being used to generate isothermal kinetic data.

10.7 Thermal Comparators

Materials identification and measurement of thermal diffusivity

The thermal properties of materials are often of much importance in pyrometallurgical processing and any device which can estimate such properties rapidly should be useful to the industry. It is possible to design an apparatus based on a design proposed by Powell (1957) and Clark and Powell (1962) which measures thermal conductivity rapidly and, thus, often is useful in material identification.

In Powell's original design, the comparator consisted of two metal balls similarly mounted in a block of balsa wood, but one at a slightly lower level so that it touched any surface on which the block rested. After heating to a small fixed temperature excess the block is laid in contact with the test surface. Differentially connected thermocouples attached to each ball measured the increased rate of cooling of the ball which makes contact, and this was correlated with the thermal conductivity of the material on which the ball rested. A calibration plot made measurements more rapid and direct. In a device like this, the condition of the surface, the area of contact, etc., affect the readings and therefore these need to be standardized.

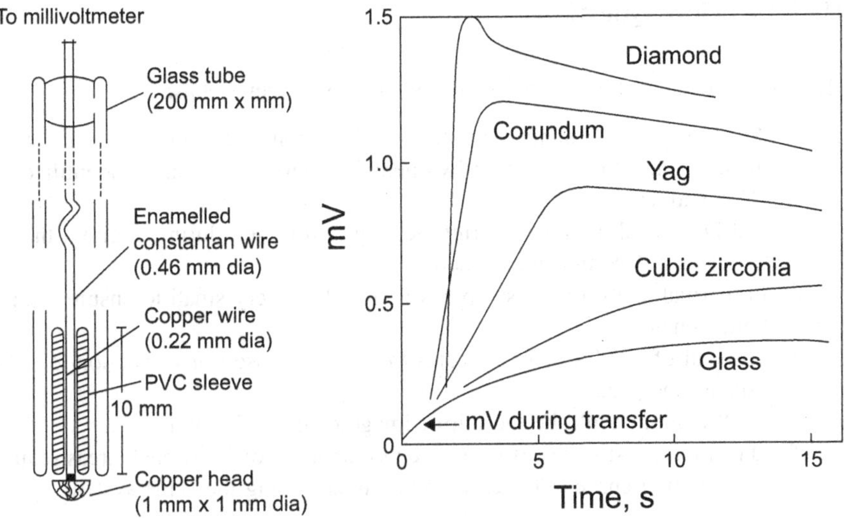

Fig. 10.26 Thermal comparator and mV-time plots (Goldsmid and Goldsmid 1989)

In addition to measuring thermal conductivity of known materials, the device can also be used in identification and sorting of metals small or large blocks and various raw materials. According to Powell, the size of the test specimen is immaterial once a certain minimum size is exceeded. Clark and Powell (1962) have given an approximate analysis of the heat flow associated with the use of the thermal comparator. They have also suggested modified forms for determining thermal conductivities of powders, for measurements made in rapid sequence or at high temperatures, etc.

Goldsmid and Goldsmid (1989) have used a similar device to differentiate between small particles of diamond and gemstones. The comparator is schematically shown in Fig. 10.26.

To use the comparator, it is first placed in a beaker of boiling water and thermal equilibrium is ensured. Subsequently, the copper head is pressed against the gemstone under test and the differential mV recorded as a function of time. Some typical plots are also shown in Fig. 10.26. Obviously, the slope of the plots in the initial periods is a direct measure of conductivity.

It is possible to devise a suitably modified form of the comparator for use on raw materials in bulk solid form or in the form of packed beds. It should be possible to use this technique also as an aid in kinetic studies in some systems where there are transformations within the bulk of a solid sample.

10.8 Review Questions

1. Discuss whether the following statements are correct or false.
   a. A equimolar mechanical mixture of limestone and dolomite cannot be distinguished from dolomite by chemical analysis but can be identified by DTA and TG.
   b. If a DTA peak obtained during heating is reversed during cooling, then it indicates a decomposition reaction.
   c. In thermal analysis, the sample weight is kept very small to ensure sample homogeneity.
   d. The end of a DTA peak indicates the end of phase transformation/reaction causing the peak.
   e. A DSC plot should be the mirror image of the DTA plot.
   f. Quantitative estimates of heat based on analysis of DTA peak areas require a calibration curve obtained by experiments using the same system.
2. A compound undergoes, on heating, an exothermic phase transformation at temperature T_1 and then an endothermic decomposition to a solid and a gas at T_2 $(T_2 > T_1)$. It also loses some moisture initially. Draw schematically *T*, DTA, TG and DTG plots for heating.
3. How can thermal analysis be useful in studying "reaction paths" when a reaction mixture undergoes complicated reaction schemes during heating?
4. What are the advantages of non-isothermal TG on isothermal TG and vice versa?
5. Draw DTA and TG plots for heating and cooling (0–900 °C) of a mixture of two carbonates. The first melts around 700 °C but does not decompose whereas the second starts solid-state decomposition at around 720 °C which is complete by about 850 °C.
6. A boat is gradually moving into a furnace hot zone so that the volume elements of a reaction mixture in the boat get heated up gradually. Show schematically plots of temperature versus distance and temperatures versus time for three different boat speeds. Explain why the order of plots in one is reverse of that in another.
7. Discuss why a TG plot is likely to shift with changes in heating rate, sample weight and particle sizes degree of packing and gas flow rate.
8. Discuss how DTA and TG plots decomposition of $CaCO_3$ will alter with increase in p_{CO_2} in the surrounding gas.
9. A coal is to be examined for proximate analysis and, thus, for determining moisture content (weight loss at 110 °C), volatile matter (weight loss up to say 700 °C after moisture removal), fixed carbon (weight loss after combustion in oxygen up to say, 1100 °C) and ash (the weight of the residue). Recommend a thermal analysis procedure for proximate analysis by a single TG experiment.
10. Why is the moving bed technique said to recreate some features of a plug flow reactor?

References

Bollin, E.M. In: Mackenzie, R.C. (ed.) Differential Thermal Analysis, vol. 1, Chapter 7. Academic Press, London (1970)
Clark, W.T., Powell, R.W.: J. Sci. Instrum. **39**, 545 (1962)
Conroy, A.R., Manring, W.H., Bauer, W.C.: Glass Ind. **84** (1966)
Cunningham, A.D., Wilburn, F.W.: Theory in Ref. I, Ch. 2 (1970)
Cunningham, A.D., Wilburn, F.W. In: Mackenzie, R.C. (ed.) Differential Thermal Analysis, vol. 1, Chapter 2. Academic Press, London (1970)
Dixon, J.P. In: Mackenzie, R.C. (ed.) Proceedings of the First European Symposium on Thermal Analysis, vol. 2, Applications, p. 390, p. 399. Academic Press, S. St. Wane (1972)
Dixon, J.P. In: Dollimore, D. (ed.) Proceedings of the European Symposium on Thermal Analysis, p. 390. Heyden (1976)
Eitel, W.: Silicate Science, vol. 8, Chapter 1. Academic Press, New York (1976)
El Jazairi, B.: Thermochimicia Acta **21**, 381 (1977)
Garn, P.D.: Thermoanalytical Methods of Investigation. Academic Press, London (1965)
Goldsmid, H.J., Goldsmid, S.E.: (School of Physics, University of New South Wales, Kensington 2033, Australia), A simple thermal comparator for testing gemstones. Private Communication (1989)
Gray, N.B., Harvey, M.R., Wills, G.M. In: Jeffs, J.N.E., Tait, R.J. (eds.) Physical Chemistry of Process Metallurgy—The Richardson Conference, p. 19. I.M.M. London (1974)
Ingraham, T.R., Marier, P. In: Schwenker Jr, R.F., Garm, P.D. (ed.) Thermal Analysis, vol. 2, p. 1003. Academic Press, New York (1969)
IS: 8167: *Methods for determination of reducibility of iron ore and sinter*, Indian Standards Institution, Manak Bhavan, New Delhi (1976)
Keatch, C.J., Dollimore, D.: An Introduction to Thermogravimetry. Heyden, New York (1975)
Kissinger, N.E.: Anal. Chem. **29**(11), 1702 (1957a)
Kissinger, H.E.: Anal. Chem. **29**, 1703 (1957b)
Kitaigo-rodskii, I.I.: Glasstech. Ber. **31**, 117 (1958)
Kroger, C.: Glasstech Ber **22**(5/6), 331 (1948)
Kroger, C.: Glasstech. Ber. **30**, 321 (1957)
Lahiri, A.K., Ray, H.S.: Thermochim. Acta **55**, 97 (1982)
Khan, D.M.A., Ray, H.S., Batra, N.K.: Trans. Ind. Inst. Met. **37**, 361 (1984)
Mackenzie, R.C. (ed.): Differential Thermal Analysis. Vol. 1: Fundamental Aspects. Academic Press, London (1970)
McIntosh, R.M., Turnock, A., Wilburn, F.W.: Trans. Brit. Ceram Soc. **73**, 117 (1974)
Melling, R., Wilburn, F.W., McIntosh, R.M.: Anal. Chem. **41**(11), 1275 (1965)
Melling, R., Wilburn, F.W., McIntosh, R.M.: Anal. Chem. **41**(10), 1275 (1969)
Oates, W.A., Todd, D.D.: J. Aus. Inst. Metals **7**, 109 (1962)
Powell, R.W.: J. Sci. Instrum. **34**, 485 (1957)
Prakash, S., Ray, H.S., Gupta, K.N.: Ironmaking Steelmaking **13**(2), 76 (1986)
Prakash, S., Ray, H.S., Gupta, K.N.: React. Solids **4**, 215 (1987)
Ramakrishna Rao, V.V.V.N.S., Abraham, K.P.: ibid. **2**, 2463 (1971)
Ray, H.S.: Met. Trans. **10B**, 677 (1979)
Ray, H.S.: Unpublished work (1989)
Ray, H.S., Dutta Chowdhury, D.: Thermochim. Acta **101**, 119 (1986)
Ray, H.S., Sewell, P.A. In: Proceedings of the International Conference on Advances in Chemical Metallurgy (ICMS-79), Bombay, Vol. **2**, 43/I/1, Jan 1979. B.A.R.C., Bombay (1979)
Ray, H.S., Wilburn, F.W.: Trans. IIM **35**, 537 (1982)
Ray, H.S., Abraham, K.P., Sridhar, R.: Nonferrous Metals Production, Ch. 7. East-West Pub., New Delhi (1991)
Read, R.L., Weber, L., Gottbried, B.S.: Ind. Eng. Chem. Fundam. **4**(1), 39 (1965a)
Read, R.L., Weber, L., Gottbried, B.S.: Ind. Eng. Chem. Fundam. **4**(1), 38 (1965b)

Sarangi, B., Ray, H.S., Tripathy, K.K., Sarangi, A.: Mater. Lett. **12**(5), 381 (1991)
Sarangi, B., Misra, S., Sarangi, A., Ray, H.S.: Trans. IIM **44**(3), 279 (1991)
Sarangi, B., Misra, S., Sarangi, A., Ray, H.S.: Thermochim. Acta **196**, 45 (1992)
Sarveswara Rao, K., Ray, H.S.: Trans. IIM **52**, 171 (1999)
Shah, D., Khalafalla, S.E.: Met. Trans. **2**, 2637 (1971)
Smother, W.J., Yao, C.: DTA-Theory and Practice. Chemical Publishing, New York (1958)
Swarts, E.L.: In: Pye, L.D., Stevens, H.J., LaCourse, W.C. (eds.) The Melting of Glass in Introduction to Glass Science. Plenum Press, New York (1972)
Tikkanen, M.H., Rosell, B.O., Wiberg, O.: Acta Chem. Scand. **17**, 513 (1963a), 521 (1963b)
Verma, R.K., Ray, H.S., Ghosh, A., Singh, R.N., Dharanipalam, S., Gupta, S.K.: Trans. IIM **32**, 232 (1979)
Wilburn, F.W., Dawson, J.B. In: Mackenzie, R.C. (ed.) Differential Thermal Analysis, vol. 2 (Chapter Glass). Academic Press, London (1972)
Wilburn, F.W., Hesford, J.R., Flower, J.R.: Anal. Chem. **40**, 777 (1968)

Chapter 11
Analysis of Kinetic Data for Practical Applications

11.1 Introduction

Since reaction rate directly relates to industrial production, kinetic data are of practical relevance. From the point of view of production, it is often important to predict:

(i) The reaction time necessary for a specific level of conversion, and
(ii) The level of conversion for a specific value for the reaction time,

with the process parameters fixed at predetermined levels. There are several quantitative approaches for obtaining the answers to these questions. These are briefly discussed in this chapter.

The preceding chapters presented systematic treatments on the various mechanisms of reactions—the "logic" of the various processes, which resulted in specific functional forms for the rate equations. The present chapter deals with the complementary question—that of systematic treatment of raw rate data generated from controlled experiments, for the purpose of answering such technological questions without specific reference to the underlying mechanisms. Systematic rate data are useful for other practical applications also. For example, with $\alpha - t$ data available, one can estimate variations in throughput and energy consumption when process parameters are changed. This is discussed towards the end of this chapter.

11.2 Traditional Kinetic Equations

Some concepts related to the traditional kinetic equations discussed so far in this book are summarized first. The idea is to outline their practical applications and also to briefly discuss some sources of discrepancies often not well understood.

H. S. Ray and S. Ray, *Kinetics of Metallurgical Processes*,
Indian Institute of Metals Series, https://doi.org/10.1007/978-981-13-0686-0_11

The functional relations that describe the rate data are known as kinetic models. About 25 kinetic models are available in the literature for routine analysis. Some common models (Mishra et al. 1995; Sharp et al. 1996; Hulbert 1969), both in integral and differential forms, are given in Table 11.1. The kinetic models in the literature are based on reaction systems with simple geometry and with additional simplifying assumptions. Yet, these have often been successfully applied for studying reaction mechanisms in many complex solid–solid, solid–gas and solid–liquid systems, which violated many of the simplifying assumptions. In most cases, the details of the nature of the systems as regards shape, size, impurity content, consumption of gas and induction period at the beginning of the reaction are not taken into account.

Often the kinetic equation to fit new data is chosen by trial and error, without any attempt to understand the relevant physico-chemical aspects of the underlying phenomena. Of the kinetic models available in the literature, the mathematically most consistent model giving the statistically best fit would generally be chosen as the applicable kinetic one. The different kinetic models are not clearly distinguished from each other even for a wide range of degree of reaction, to allow an unequivocal selection. There are examples in the literature where the kinetic model happens to fit the data very well, whereas the reaction mechanism implied by the model may be different (Sahota et al. 1987; Beretka 1984). Therefore, while fitting data to a particular kinetic model, apart from the basic assumptions inherent in the model, one must also consider possible uncertainties. Errors may be introduced by ignoring initial induction period, undesirable reactions and insufficient discrimination of models by the statistical method used.

Different forms of a particular rate equation can be derived and used where it is possible to analyse the same kinetic data in different ways and get better insight into the reaction mechanism. There is literature to show that a rate equation of one form can get transformed into another form for specific experimental conditions and assumptions.

11.2.1 Reaction Rate

Mathematically, the rate of reaction is generally expressed as a dimensionless parameter α, called the degree of reaction. It is defined as the ratio of change of a suitable parameter to the maximum change possible. In any given system, the instantaneous rate of reaction depends basically on the nature of the system, the degree of completion of reaction (α), the time of reaction t and the temperature T. For isothermal kinetics, the general expression in differential form can be expressed as

Table 11.1 Some common kinetic models in both integral and differential forms

Mechanism		
1. Deceleratory α–t curves		
1.1	*Based on geometrical models*	
	R2	Contracting area (cylindrical symmetry): $g(\alpha) = 1 - (1-\alpha)^{1/2}; f(\alpha) = (1-\alpha)^{1/2}$
	R3	Contracting volume (spherical symmetry): $g(\alpha) = 1 - (1-\alpha)^{1/3}; f(\alpha) = (1-\alpha)^{2/3}$
1.2	*Based on diffusion mechanism*	
	D1	One-dimensional diffusion: $g(\alpha) = \alpha^2; f(\alpha) = 1/2\alpha$
	D2	Two-dimensional diffusion: $g(\alpha) = (1-\alpha)\ \ln(1-\alpha) + \alpha; f(\alpha) = [-\ln(1-\alpha)]^{-1}$
	D3	Three-dimensional diffusion (cylindrical symmetry) $g(\alpha) = [1-(1-\alpha)^{1/3}]^2; f(\alpha) = 3(1-\alpha)^{2/3}\left[1-(1-\alpha)^{1/3}\right]^{-1}$
	D4	Ginstling–Brounstein (spherical symmetry): $g(\alpha) = (1-2\alpha/3) - (1-\alpha)^{2/3}; f(\alpha) = 3/2\left[(1-\alpha)^{-1/3} - 1\right]^{-1}$
		Carter–Valensi model $g(\alpha) = (1+(z-1)\alpha)^{1/3} + (z-1)(1-\alpha)^{2/3}$
1.3	*Based on order of reaction*	
	F1	First order: $g(\alpha) = -\ln(1-\alpha); f(\alpha) = (1-\alpha)$
	F2	Second-order chemical reaction: $g(\alpha) = (1-\alpha)^{-1}; f(\alpha) = (1-\alpha)^2$
	F3	Third-order chemical reaction: $g(\alpha) = (1-\alpha)^{-2}; f(\alpha) = (1-\alpha)^3$
2. Acceleratory α–t curves		
PI		Power law: $g(\alpha) = \alpha^{1/n}\ f(\alpha) = n\alpha^{(n-1)/n}$
EI		Exponential law: $g(\alpha) = \ln\alpha\ f(\alpha) = 1/\alpha$
3. Sigmoidal α–t curves		
A2		Avrami–Erofeev[a] nucleation and growth (n = 2): $g(\alpha) = [-\ln(1-\alpha)]^{1/2}; f(\alpha) = 2(1-\alpha)\ [-\ln(1-\alpha)]^{1/2}$
A3		Avrami–Erofeev[a] nucleation and growth (n = 3): $g(\alpha) = [-\ln(1-\alpha)]^{1/3}; f(\alpha) = 3(1-\alpha)\ [-\ln(1-\alpha)]^{2/3}$
A4		Avrami–Erofeev[a] nucleation and growth (n = 4): $g(\alpha) = [-\ln(1-\alpha)]^{1/4}; f(\alpha) = 4(1-\alpha)[-\ln(1-\alpha)]^{3/4}$
B1		Prout–Tompkins: $g(\alpha) = \ln[\alpha/(1-\alpha)]; f(\alpha) = \alpha(1-\alpha)$

[a]See Johnson–Mehl equations (Sect. 2.5.2) and Eq. 2.39

$$\mathrm{d}\alpha/\mathrm{d}t = f(\alpha, t, \text{ nature of the system}) = k \cdot f(\alpha) \tag{11.1}$$

The integral form for Eq. 11.1 can be expressed as follows.

$$g(\alpha) = k \cdot t \tag{11.2}$$

The rate constant k is a function of temperature and nature of system, but independent of time. The nature of system implies both the test sample and the environment and thus includes chemical composition, presence of impurity, physical factors such as shape and size of particles, composition and pressure of gas in the environment. As indicated above, $g(\alpha)$ and $f(\alpha)$ are functions of α and k.

Based on the shape of the $\alpha - t$ curves, the kinetic models can be classified as acceleratory, sigmoidal or deceleratory type. The decelerating group is further subdivided according to the controlling factor that causes deceleration. A broad classification of the kinetic models along with the shape of $\alpha - t$ has been presented in Table 11.1. Numerous examples of both decelerating and sigmoid type of standard $\alpha - t$ relations are available in the literature (Brown and Galaway 1979). The $\alpha - t$ plots can also be used for a comparative assessment of energy consumption and throughput of an industrial system; this is discussed subsequently.

For an ideal fit of any kinetic model, the $\alpha - t$ plot should pass through the origin, i.e. $\alpha = 0,\ t = 0$. However, both positive and negative deviations at origin, schematically shown in Fig. 11.1a, b respectively, have been reported in the literature. Whether the induction period is genuine or an artefact because of experimental limitations can be found out by using an appropriate function $g(\alpha)$ that yields a linear plot for $g(\alpha)$ against t. Non-zero intercepts on the abscissa in such plots, as shown schematically in Fig. 11.2a, b, indicate that the induction period is genuine. A positive deviation (Fig. 11.2a) may arise because the reaction has already progressed to some extent in the starting material. A negative deviation (Fig. 11.2b) may indicate a warm-up period. For subsequent analysis, it is

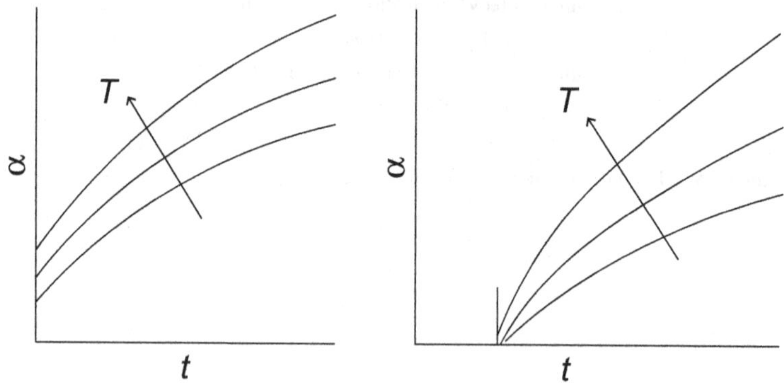

Fig. 11.1 Schematic plots of α versus time t showing positive (**a**) and negative (**b**) deviations at the origin

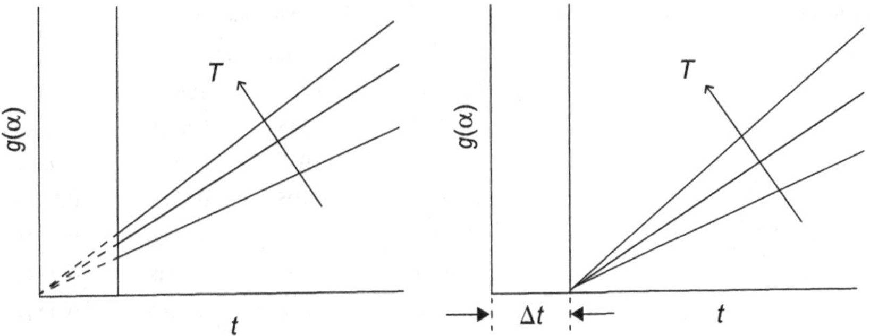

Fig. 11.2 Schematic plots of $g(\alpha)$ versus time t showing positive (**a**) and negative (**b**) deviations at the origin

necessary to correct the reaction time t, by either adding or subtracting this induction period Δt, as appropriate. These are schematically illustrated in Fig. 11.2a, b, respectively.

It is suggested that if the plot of $g(\alpha)$ versus t does not pass through origin, then the data can be verified by a log–log plot. Additional complication in establishing the zero time arises when isothermal conditions are not maintained throughout the experiment. Arbitrarily fixing the zero time in such cases is, however, not justified. Very often, computer software used for analysis of kinetic data does not take this point into account. In such cases, the reduced time method results in serious errors in the analysis of kinetic data.

11.2.2 Identification of Rate Equations

To identify the rate equation, a given set of values of fraction transformed α versus t are usually analysed using the following:

i) Test of linearity by plotting $g(\alpha)$ versus t using graphical analysis,
ii) Linear regression analysis, and
iii) Comparison of slopes of α versus reduced time $t/t_{0.5}$ with standard reduced time plots.

The details of these methods for specific mechanisms have been given in the previous chapters.

The identification of the kinetic model constitutes an important step in the analysis of high-temperature kinetic data on many metallurgical reactions. In many cases, a preliminary identification is possible using the reduced time plots. However, their applications in metallurgy have been limited.

For fixed experimental conditions, the kinetic models used can also be distinguished by comparing the correlation coefficients from the corresponding

Table 11.2 Correlation coefficient values for various α by isothermal TG

No	$f(\alpha)$	Isothermal temperature (°C)			
		185	190	194	197
1	α	0.9701	0.9545	0.9329	0.9080
2	α^2	0.9943	0.9945	0.9888	0.9857
3	$\alpha^{1/2}$	0.9414	0.9105	0.8789	0.8386
4	$\alpha^{1/3}$	0.9292	0.8919	0.8566	0.8121
5	$\alpha^{1/4}$	0.9226	08818	0.8448	0.7984
6	$1-(1-\alpha)^{1/2}$	0.9762	0.9656	0.9492	0.9291
7	$1-(1-\alpha)^{1/3}$	0.9780	0.9689	0.9542	0.9356
8	$[-\ln(1-\alpha)]$	0.9814	0.9749	0.9633	0.9477
9	$[-\ln(1-\alpha)]^{3/2}$	0.9656	0.9508	0.9326	0.9052
10	$[-\ln(1-\alpha)]^{1/2}$	0.9552	0.9348	0.9128	0.8792
11	$[-\ln(1-\alpha)]^{1/3}$	0.9430	0.9162	0.8901	0.8504
12	$[-\ln(1-\alpha)]^{1/4}$	0.9363	0.9059	0.8776	0.8351
13	$(1-\alpha)[\ln(1-\alpha)]+\alpha$	0.9946	0.9961	0.9937	0.9919
14	$[1-(2/3)\alpha]-(1-\alpha)^{2/3}$	0.9945	0.9964	0.9950	0.9936
15	$[1-(1-\alpha)^{1/3}]^2$	0.9942	0.9966	0.9971	0.9960
16	$\ln[\alpha/(1-\alpha)]$	0.9246	0.8900	0.8611	0.8145
17	$[1/(1-\alpha)]-1$	0.9891	0.9884	0.9842	0.9763
18	$\ln\alpha$	0.9009	0.8490	0.8061	0.7562

regression analyses of the kinetic data. For example, Beretka (1984) and Baretka and Lesko (1979) studied the reaction kinetics of some solid–solid reactions where the data were found to apparently fit a number of kinetic models. The correlation coefficients obtained by the regression analysis and mean residual square methods, however, distinguish the various models. For example, Pravakaran et al. (1995) studied the thermal analysis of nitrosamines at various temperatures. Table 11.2 gives an idea of using the correlation coefficients for distinguishing the reaction mechanisms. The best linearity with correlation coefficient of ~ 0.996 was obtained for the Jander equation, No 15 in this table.

11.3 Transformation of One Model to Another

Under certain assumptions, a particular kinetic equation can be transformed into another one. Note, for example, that all the diffusional models in Table 11.1 are derived from Jander equation. In such cases, the data meeting the assumptions can be analysed by different forms of the kinetic equations. Some examples are given below.

Example 11.1 The kinetic functions for Jander and Ginstling–Brounstein (GB) models are given, respectively, by

$$\text{Jander model}: \; g(\alpha) = \left\{1 - (1-\alpha)^{1/3}\right\}^2 \tag{11.3}$$

$$\text{GB model}: \; g(\alpha) = \left\{1 - (1-\alpha)^{1/3}\right\}^2 \tag{11.4}$$

For both these models, $g(\alpha) \approx \alpha^2/9$ for small values of α. So, for small values of α, the kinetic data can be analysed using the following equation:

$$\alpha^2/9 = k \cdot t \tag{11.5}$$

$$\text{i, e.,} \quad \alpha^2 = 9kt = k^+ t \tag{11.6}$$

where $k^+ = 9k$, and k is the true rate constant. Equation 11.6 is parabolic law for reactions controlled by diffusion through product layer on a flat plate (two-dimensional diffusion).

Example 11.2 Bandyopadhyaya et al. (1990) studied both low- and high-temperature reoxidation of sponge iron (direct reduced iron, DRI). At low temperature, the process obeys a direct logarithmic rate law as follows:

$$W = k^* \ln(at + 1) \tag{11.7}$$

where k^* is an empirical constant, and W is the weight gain. At higher temperatures, the reaction follows the common first-order kinetic equation:

$$\mathrm{d}\alpha/\mathrm{d}t = k(1-\alpha) \tag{11.8}$$

Both these equations, however, become identical if a simple assumption is made. The basic assumption while formulating the logarithmic law is that the rate of oxygen absorption per unit area $\mathrm{d}W/\mathrm{d}t$ is proportional to number of pores present per unit area N; i.e. N and the rate of reaction are related by an equation of the form

$$-\mathrm{d}W/\mathrm{d}t = -k''N \tag{11.9}$$

where k'' is a proportionality constant. At high temperatures, DRI being a very porous material, the number of pores available per unit area is proportional to the unreacted material W''. Therefore, Eq. 11.9 can be written as

$$-\mathrm{d}W/\mathrm{d}t = -k_1''k_1^*W'' \tag{11.10}$$

where k_1^* is a proportionality constant. The fraction conversion α is proportional to W, and W'' is proportional to $(1-\alpha)$. Thus

$$W = k_2^* \alpha \tag{11.11}$$

$$W'' = k_3^*(1 - \alpha) \tag{11.12}$$

where k_2^* and k_3^* are two other proportionality constants. Equation 11.8 is now obtained by combining Eqs. 11.10 and 11.12. This shows that both the laws for DRI oxidation can be derived from the basic equation by applying the appropriate conditions.

Example 11.3 Some solid-state decomposition reactions follow the well-known law for autocatalysis expressed by Prout–Tompkins equation

$$\mathrm{d}\alpha/\mathrm{d}t = k\alpha(1 - \alpha), \quad \text{or} \quad (\mathrm{d}\alpha/\mathrm{d}t)/[\alpha(1 - \alpha)] = k \tag{11.13}$$

In the range of α from 0.3 to 0.7, $\alpha(1 - \alpha)$ varies between 0.21 and 0.25. Assuming $\alpha(1 - \alpha)$ to be essentially constant, Eq. 11.13 reduces to Eq. 11.14:

$$\begin{aligned} \mathrm{d}\alpha/\mathrm{d}t &= \text{constant} \cdot k, \quad \text{or} \\ \ln(\mathrm{d}\alpha/\mathrm{d}t) &= E/RT + \text{constant} \end{aligned} \tag{11.14}$$

So the kinetic parameter E for autocatalytic reaction can be approximately determined by plotting $\ln(\mathrm{d}\alpha/\mathrm{d}t)$ versus $1/T$ for α within the range of 0.3–0.7. The same data can also be analysed in a different way for estimating the E value (Garbachev 1976).

Example 11.4 Consider the illustrative Example 8.1 at the end of Sect. 8.4. The reaction under consideration is the so called solid-state reaction between Cr_2O_3 and Na_2CO_3. In reality, however, the reaction involves oxygen and it should be written as follows:

$$Cr_2O_3(s) + Na_2CO_3(s) + \frac{3}{2}O_2(g) = 2Na_2Cr_2O_4(s) + 2CO_2(g) \tag{11.15}$$

Figure 11.3 indicates some data from the literature to show the influence of temperature and oxygen partial pressure on the kinetics which follow the Zhuravlev, Lesokhin and Templeman (ZLT) equation (Tripathy et al. 1994a; Tripathy et al. 1994b). Figure 11.3a, b shows the validity of the ZLT equation. The Arrhenius plot, shown in Fig. 11.3c, gives the E value of 72 kJ/mole. Figure 11.3d shows that the rate constant depends directly on the partial pressure. It is shown in Sect. 8.4 that ZLT equation can be altered to a logarithmic form. The plots are shown in Fig. 11.4.

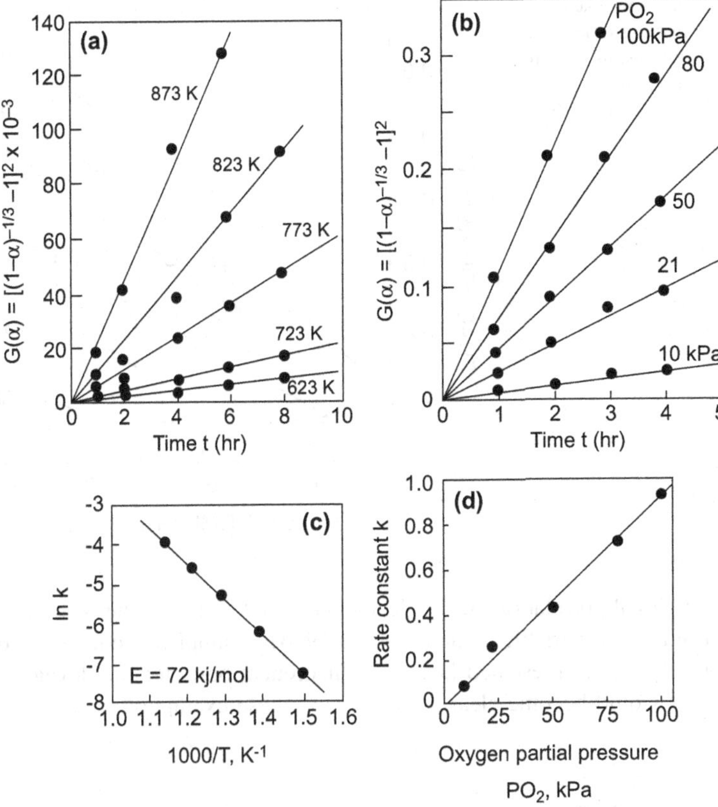

Fig. 11.3 Influence of temperature and oxygen partial pressure on Na_2CO_3–Cr_2O_3 reaction. **a** Integral function of fraction reacted $G(\alpha)$ plotted against time t. Conditions: Na_2CO_3/Cr_2O_3; ratio: 1.4; temperature T: 673–873 K; atmosphere: air. **b** Integral function of fraction reacted $G(\alpha)$ plotted against time t using ZLT model. Conditions: Na_2CO_3/Cr_2O_3; ratio: 1.4; particle size: 63–90 μm; temperature T: 873 K; p_{O_2}: 10–100 kPa. **c** Logarithm of reaction rate constant k as function of reciprocal temperature, $1/T$. **d** Variation of rate constant k with oxygen partial pressure p_{O_2}. Conditions: Na_2CO_3/Cr_2O_3; ratio: 1.4; particle size: 63–90 μm; temperature T: 873 K (Haque et al. 1992)

11.4 Empirical Correlations Based on Rate Data

A kinetic equation can be used to estimate the degree of conversion under various conditions. In the industry, an operator often needs to know the time for a given degree of conversion under a fixed set of process variables. Although this can be estimated from kinetic equations, often it is useful to have some empirical correlations for ready reference. This section discusses how experimental kinetic data can be used to establish some useful correlations (Tripathy and Ray 1995).

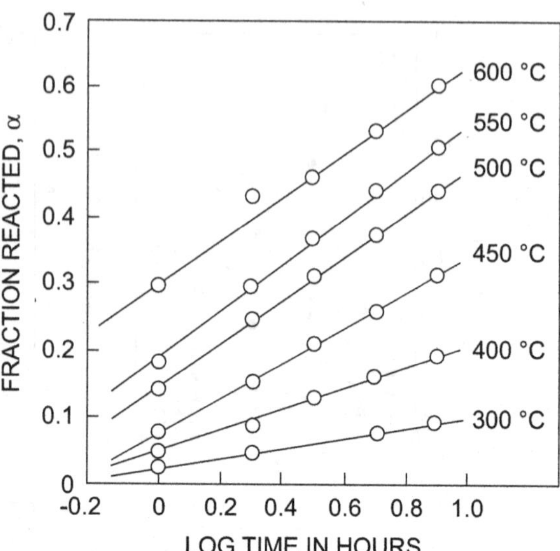

Fig. 11.4 Fraction reacted α plotted against log time (hours) for the results shown in Fig. 11.3a (Tripathy et al. 1994; Tripathy and Ray 1995)

The empirical correlations to be developed should express the reaction time as function of process variables (Tan and Ford 1984). Examples of equations for three different systems are presented here. The theoretical basis for the development of correlations would become clear as we discuss these examples.

11.4.1 Alkali Roasting of Chromium Oxide

The work reported by Tripathy, Ray and Pattanayak (1994a, 1994b) has been cited in Example 11.4 in Sect. 11.3 Equation 11.15 expresses the reaction of Cr_2O_3 with Na_2CO_3 in presence of oxygen. Table 11.3 shows the range of data employed in kinetic studies and also defines the notations for the variables. The $\alpha - t$ plots were determined under different conditions varying only one parameter at a time to obtain the experimental results. The empirical correlations were to be obtained to express time for a given α value as a function of the variables when the other parameters are fixed as listed in Table 11.4.

Table 11.3 Ranges of experimental conditions for "solid-state" reaction in the system Cr_2O_3–Na_2CO_3

Variable	Range	Normal value
Temperature (T)	673–873 K	873 K
Cr_2O_3/Na_2CO_3 ratio (m)	0.36–0.73	0.73
p_{O_2} (p)	21–100 kPa	21 kPa

Table 11.4 Range of solid-state experimental conditions for the system Cr_2O_3–Na_2CO_3

Parameters studied	Range	Values of fixed parameters
Temperature (T)	673–873 K	$m = 0.73$; $p = 21$ kPa
Cr_2O_3/Na_2CO_3 ratio (m)	0.36–0.73	$T = 873$ K; $p = 21$ kPa
p_{O_2} (p)	21–100 kPa	$T = 873$ K; $m = 0.73$

Effect of temperature (T)

The time t_α required for various degrees of conversion α in the range of 0.2–0.5 were obtained from plots as shown in Fig. 11.5. $\ln t_\alpha$ values are plotted against reciprocal temperature in Fig. 11.6. Using linear regression analysis, the activation energy for the alkali roasting of Cr_2O_3 for α in the range of 0.2–0.6 was determined to be 68 kJ/mol over the temperature range 673–873 K. The time t_α (in minutes) required for conversion at temperature T are therefore given by the following equations:

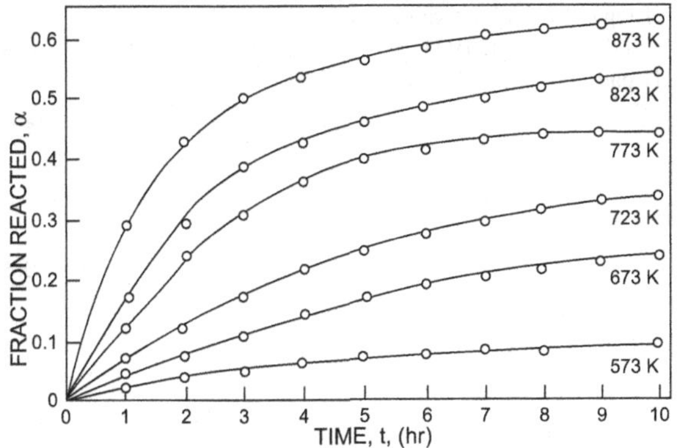

Fig. 11.5 Fraction reacted α as a function of time t. [Cr_2O_3:Na_2CO_3 = 1:1.4 (wt/wt); atmosphere: air.] (Tripathy et al. 1994; Tripathy and Ray 1995)

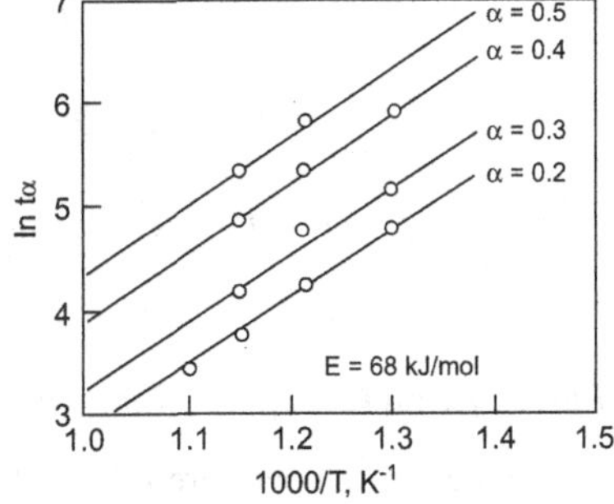

Fig. 11.6 Logarithm of time for a fixed value of α versus reciprocal temperature (Tripathy et al. 1994; Tripathy and Ray 1995)

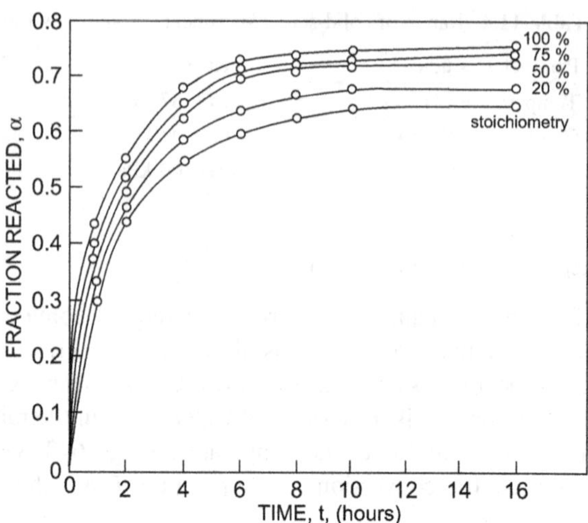

Fig. 11.7 Effect of excess flux Na_2CO_3 on fraction reacted [$T = 873$ K; $p_{O_2} = 21$ kPa.] (Tripathy et al. 1994; Tripathy and Ray 1995)

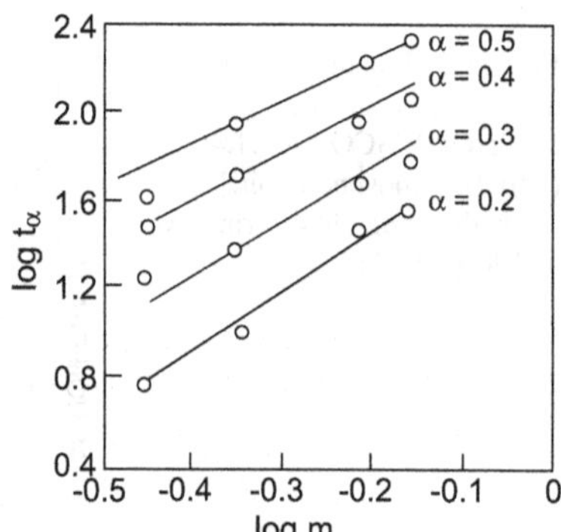

Fig. 11.8 $\log t_\alpha$ as a function of $\log m$ for various values of α (Tripathy et al. 1994; Tripathy and Ray 1995)

$$t_{\alpha=0.2} = 9.48 \times 10^{-5}\ \exp(68000/\mathrm{RT}) \tag{11.16a}$$

$$t_{\alpha=0.3} = 20.00 \times 10^{-5}\ \exp(68000/\mathrm{RT}) \tag{11.16b}$$

$$t_{\alpha=0.4} = 36.60 \times 10^{-5}\ \exp(68000/\mathrm{RT}) \tag{11.16c}$$

$$t_{\alpha=0.5} = 115.60 \times 10^{-5}\ \exp(68000/\mathrm{RT}) \tag{11.16d}$$

Effect of Cr_2O_3/Na_2CO_3 *Ratio* (m)

A reaction temperature of 873 K was selected for studying the effect of Cr_2O_3/Na_2CO_3 ratio. The gaseous atmosphere was air. The t_α values were obtained from the $\alpha - t$ plots for reactions in systems containing different Cr_2O_3/Na_2CO_3 ratio

(Fig. 11.7). Figure 11.8 shows the linear $\log t_\alpha - \log m$ plots, which are expressed by the following empirical equations.

$$t_{\alpha=0.2} = 2.75 \cdot m^{2.29} \tag{11.17a}$$

$$t_{\alpha=0.3} = 10.0 \cdot m^{2.25} \tag{11.17b}$$

$$t_{\alpha=0.4} = 22.9 \cdot m^{2.26} \tag{11.17c}$$

$$t_{\alpha=0.5} = 45.7 \cdot m^{2.25} \tag{11.17d}$$

For α in the range from 0.2 to 0.5, the value for the exponent of m in these equations varies over the small range of 2.25–2.28. For the integrated rate equation to be developed subsequently, an average value of $m = 2.26$ will be used.

Influence of oxygen partial pressure (p_{O_2})

The $\alpha - t$ plots for different oxygen partial pressures p_{O_2} are shown in Fig. 11.9. Plots of $\ln t_\alpha$ against $\ln p_{O_2}$ are shown in Fig. 11.10. Here, the reaction temperature was 873 K, and the Cr_2O_3/Na_2CO_3 ratio was 0.73.

These linear plots are described by the following relations.

$$t_{\alpha=0.2} = 9.73(p_{O_2})^{-0.97} \tag{11.18a}$$

$$t_{\alpha=0.3} = 15.80(p_{O_2})^{-0.96} \tag{11.18b}$$

$$t_{\alpha=0.4} = 25.10(p_{O_2})^{-0.97} \tag{11.18c}$$

$$t_{\alpha=0.5} = 36.30(p_{O_2})^{-0.97} \tag{11.18d}$$

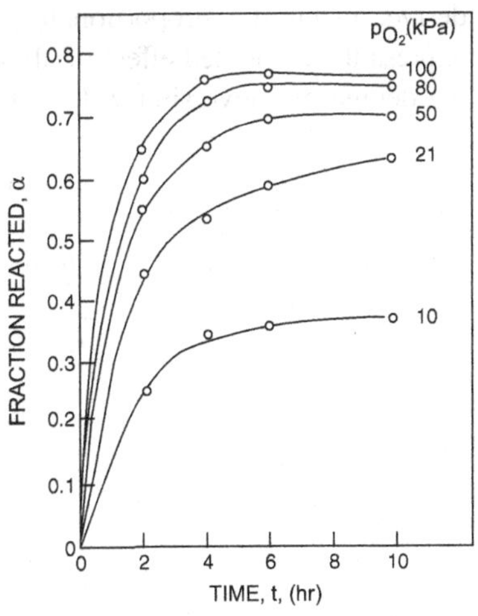

Fig. 11.9 Effect of oxygen partial pressure on fraction reacted, α, as a function of time [$T = 873$K; $Cr_2O_3/Na_2CO_3 = 1/1.4$] (Tripathy et al. 1994; Tripathy and Ray 1995)

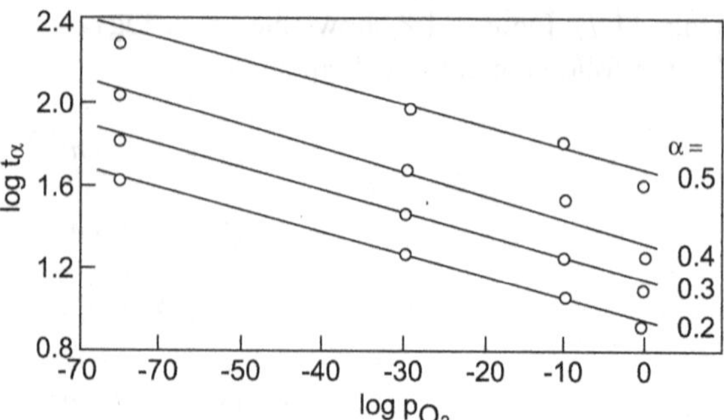

Fig. 11.10 Logarithm of time $\log t$ as a function of logarithm of oxygen partial pressure $\log p_{O2}$ (Tripathy et al. 1994; Tripathy and Ray 1995)

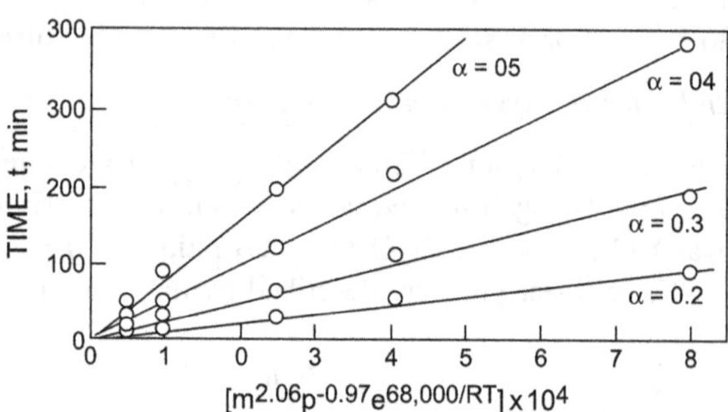

Fig. 11.11 Time required t as a function of β (Tripathy et al. 1994; Tripathy and Ray 1995)

For the integrated rate equation development, the value of the exponent is taken as −0.97.

Integrated empirical equation

From the correlations given by Eqs. 11.16a, 11.16b, 11.16c, 11.16d, 11.17a, 11.17b, 11.17c, 11.17d and 11.18a, 11.18b, 11.18c, 11.18d, it is now possible to derive, using the proportionality principle, general empirical equations which express the combined effect of all the three process variables for specific values for the degree of conversion α (0.2, 0.3, 0.4 or 0.5).

$$t_{\alpha=0.2} = 2.5 \times 10^{-3}\beta \tag{11.19a}$$

$$t_{\alpha=0.3} = 31.60 \times 10^{-3}\beta \tag{11.19b}$$

$$t_{\alpha=0.4} = 210.50 \times 10^{-3}\beta \tag{11.19c}$$

$$t_{\alpha=0.5} = 1917.69 \times 10^{-3}\beta \tag{11.19d}$$

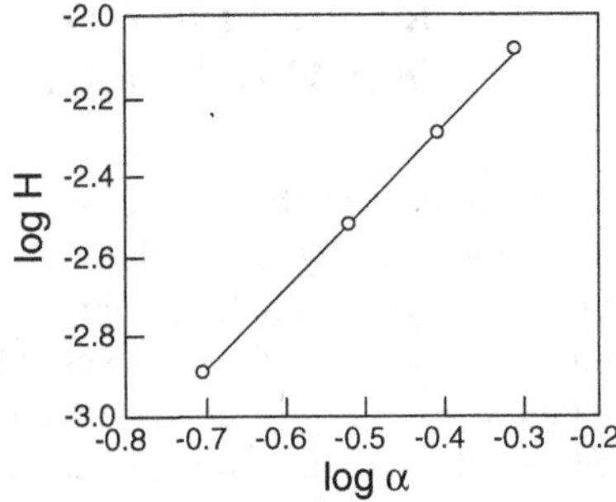

Fig. 11.12 Logarithm of slope, H, plotted against logarithm of α (Tripathy et al. 1994; Tripathy and Ray 1995)

where

$$\beta = m^{2.26}\left(p_{O_2}\right)^{-0.97}\exp\left(\frac{68000}{RT}\right) \tag{11.19e}$$

Figure 11.11 shows the plot of t_α (α from 0.2 to 0.5) against β for all the experimental results. The slopes of these lines are designated H.

In an attempt to generalize the integrated rate equations, values of $\log H$ were plotted against $\log \alpha$. These also give a good straight line fit (Fig. 11.12). The slope of this line is 3.57×10^{-2}. The correlating integrated equation can, therefore, be written as follows:

$$H_\alpha = 3.60 \times 10^{-2}\ \alpha^{2.05} \tag{11.20}$$

The most general empirical equation can, therefore, be written in the form:

$$t_{\alpha=0.2\,\text{to}\,0.5} = k\alpha^{2.05}m^{2.26}\left(p_{O_2}\right)^{-0.97}\exp\left(\frac{68000}{RT}\right) \tag{11.21}$$

Figure 11.13 shows plots of time against $\alpha^{2.05}\beta$ corresponding to all the observed values obtained in the above four series of experiments. The value of k is 3.57×10^{-2}. The overall correlation is thus expressed by the following simple and convenient empirical integrated rate equation:

$$t_{\alpha=0.2\,\text{to}\,0.5} = 3.57 \times 10^{-2}\alpha^{2.05}m^{2.26}\left(p_{O_2}\right)^{-0.97}\exp\left(\frac{68000}{RT}\right) \tag{11.22}$$

11.4.2 Reduction of Iron Ore Fines by Coal Fines in a Mixed Packed Bed System

In the context of the kinetics of direct reduction of iron ore, the term "degree of reduction" usually means the ratio of oxygen removed from the feed (iron ore) to the total oxygen initially combined with iron in the ore. Haque et al. (1992). carried out systematic experiments to establish a quantitative relationship between the time required for a given degree of reduction and six relevant process variables. An empirical correlation for a mixed system was derived which can be used conveniently for a large range of α (0.6–0.9) to determine the time required for reduction of iron ore as a function of temperature, coal/ore ratio, average particle size, bed depth, reducibility of iron ore and reactivity of the coal. For a particular combination of ore and coal, they also obtained an overall correlation, similar to that shown in Fig. 11.13. This is shown in Fig. 11.14.

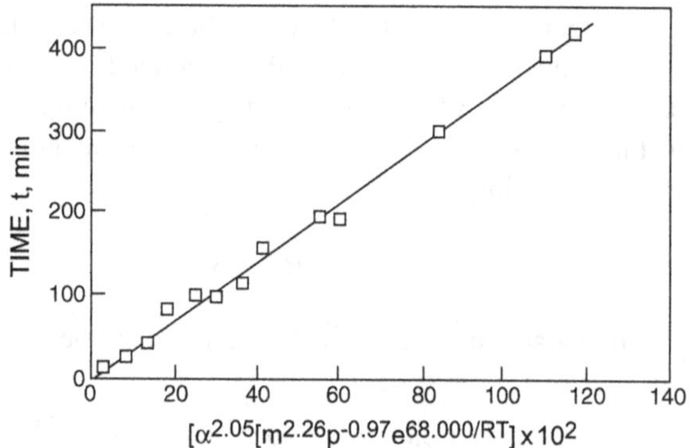

Fig. 11.13 Time required, t, as function of $\alpha^{2.05}\beta$ (Tripathy et al. 1994; Tripathy and Ray 1995)

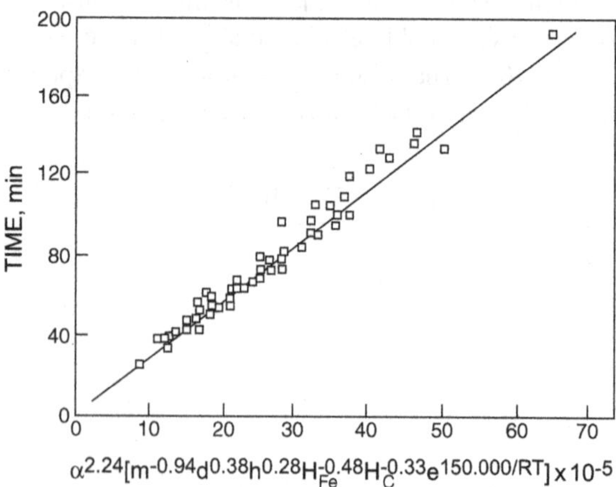

Fig. 11.14 Overall correlation for reduction of Khandband iron ore with Parascole coal (Haque et al. 1992) [m = ore/coal ratio, d = particle size (mm), h = bed depth (mm), H_{Fe} = reducibility of ore, H_c = reactivity of coal]. For compositions of ore and coal, see the original paper

11.5 Design of Experiments

In kinetics studies, parameter such as recovery is frequently desired to be expressed in terms of reaction time, temperature and other process variables. In the classical approach, one generally studies the effect of variation of one parameter on recovery, keeping the other variables constant. This approach requires a large number of experiments for the formulation of a general equation. In contrast, statistical design of experiments relies upon a limited number of experiments not only to generate a general empirical equation, but also to analyse the data statistically for a given confidence limit. Here, experiments are carried out according to a predetermined test matrix of fixed values of variables, and also a set of experiments at base level to study reproducibility. In this section, the philosophy and the basic theories are first discussed. This is followed by a discussion on the method of statistical design of experiments, with reference to actual examples.

The following sketch schematically shows how experimental design fits into a typical process improvement programme. Design of experiments is primarily based on a system of inductive reasoning. Briefly, the procedure is as follows:

a. Prepare a list of potentially important variables in the process.
b. Change the variables in a prescribed pattern.
c. Observe response on a given quantity, e.g. degree of reaction.
d. Express relationship between control and response variables.

There are several benefits from this technique. The most important one is that it can give more information per experiment than in the classical approach where one factor is varied at a time. A second benefit is an organized approach towards collection and analysis of information. The third advantage is the assessment of information reliability in the light of variation in experimental data. In addition, it is possible to get quantitative estimates for the interactions among the experimental variables.

11.5.1 Factor Variables

Factors, or the experimental variables, are controlled by the experimenter. The factor level is the value or the setting of a factor during an experimental run. Like responses, these can be classified, according to the measurement scale, as quantitative or continuous (temperature, time, etc.), or as qualitative or categorical (catalyst type, solvent, etc.). The latter are difficult to work with since the measurement scale has no natural ordering.

When different measurement scales are available, it is essential to select the one most appropriate for the problem. For example, hydrogen ion concentration can also be expressed as pH. If there are two or more factors, functional combinations of the factors may be more significant than the individual ones. For example, instead of independently manipulating the initial concentrations of reactants A and B, it might be more sensible to think in terms of reactant ratio, A/B, or total reactants, A + B.

The number of potential factors is usually large at the start of the investigation, but during the variable screening phase, only the significant or influential factors are taken into consideration for further investigation.

11.5.2 Organization of Experimental Design

Consider a system in which there are three major independent variables or factors. One desires to measure the effects of a shift from the low level to the high level of each of the three variables on the process. The possible combinations of three variables (designated here as M, T and R) at two levels consist of $2^3 = 8$ experimental runs. Fitted into a balance block, these runs assume a basic order, Fig. 11.15. The subscript 1 refers to the low level and subscript 2 the high level for each independent variable or factor within the low- and high-level grouping of R. The other variables, M and T, are balanced as to one level.

One can, therefore, estimate the effect of raising the level of R from R_1 to R_2 as follows:

$$\begin{aligned}\text{Effect of R} = &\ \text{Results from Runs}\ (3+4+7+8)/8 \\ &- \text{Results from Runs}\ (1+2+5+6)/8\end{aligned}$$

Similarly, the effects of M and T can be calculated as follows.

$$\begin{aligned}\text{Effect of M} = &\ \text{Results from Runs}\ (5+6+7+8)/8 \\ &- \text{Results from Runs}\ (1+2+3+4)/8\end{aligned}$$

$$\begin{aligned}\text{Effect of T} = &\ \text{Results from Runs}\ (2+6+4+8)/8 \\ &- \text{Results from Runs}\ (1+5+3+7)/8\end{aligned}$$

The average effect, observed with the other variables in their several combinations, is a more valid estimate of the effect of a change in operating level of a variable than the point estimate obtained by holding the other variables constant.

Fig. 11.15 Eight run block for a three-factor experiment

	R_1		R_2	
	T_1	T_2	T_1	T_2
M_1	1	2	3	4
M_2	5	6	7	8

Suppose effect of three factors is to be studied on product yield (Y, %), namely temperature (z_1) in the range of 100–200 °C, pressure (z_2) in the range of 2×10^5 Pa to 6×10^5 Pa and the residence time (z_3) in the range of 10–20 min.

For any factor z_j ($j = 1, 2, \ldots k$), we define

$$z_j^0 = \frac{z_j^{\max} + z_j^{\min}}{2}; \quad \Delta z_j = \frac{z_j^{\max} - z_j^{\min}}{2} \tag{11.23}$$

For example, for temperature, $z_1^0 = 150$ °C, $\Delta z_1 = 50$ °C for the levels chosen above. The point with coordinates (z_1^0, z_2^0, z_3^0) is called the centre point of design, or the basic level. Δz_j is the unit or interval of variation along the z_j axis. It is usual to pass from $z_1, z_2, \ldots z_k$ coordinate system to a new dimensionless system of coordinates $x_1, x_2, \ldots x_k$ defined by the following relation:

$$x_j = \frac{z_j - z_j^0}{\Delta z_j}$$

In this dimensionless coordinate system, the upper and lower levels are at +1 and −1, respectively. The centre point of design is the origin with coordinates $(0, 0, 0)$. The number of possible combinations N of the three factors ($k = 3$) is $N = 2^k = 8$. The design matrix can be reduced to coded scale. The values of the natural and coded scales, along with recovery, are shown in Table 11.5. The coded design of Table 11.5 can be portrayed geometrically as a cube (Fig. 11.16) whose eight corners represent the eight sets of experimental conditions. Table 11.6 shows the design matrix for the full 2^3 factorial design augmented with a column of dummy variables ($x_0 = 1$), and three base level experiments.

Using the design presented in Table 11.6, the regression equation including two or three factor interactions will be as follows:

Table 11.5 Experimental matrix

N	Temperature		Time		Pressure		Y
	Natural scale (°C)	Coded scale x_1	Natural scale (min)	Coded scale x_2	Natural scale (Pa)	Coded scale x_3	
1	100	−	20	−	10	−	(2) Y_1
2	200	+	20	−	10	−	(6) Y_2
3	100	−	60	+	10	−	(4) Y_3
4	200	+	60	+	10	−	(8) Y_4
5	100	−	20	−	30	+	(10) Y_5
6	200	+	20	−	30	+	(18) Y_6
7	100	−	60	+	30	+	(8) Y_7
8	200	+	60	+	30	+	(2) Y_8

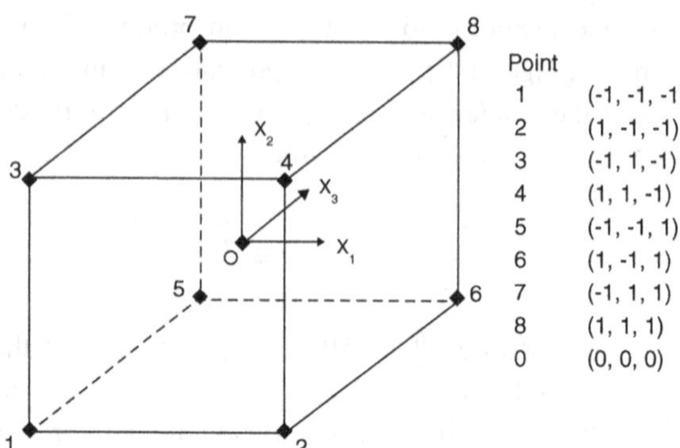

Fig. 11.16 Full factorial design at two levels for three factors

Table 11.6 Experimental matrix for 2^3 full experimental design

N	x_0	x_1	x_2	x_3	x_1x_2	x_1x_3	x_2x_3	$x_1x_2x_3$	Y
1	+	–	–	+	+	+	+	–	2
2	+	+	–	–	–	–	+	+	6
3	+	–	+	–	+	+	–	+	4
4	+	+	+	+	–	–	–	–	8
5	+	–	–	+	–	–	–	+	10
6	+	+	–	–	+	+	–	–	18
7	+	–	+	–	–	–	–	–	8
8	+	+	+	+	+	+	+	+	12
9		0	0	0					8
10		0	0	0					9
11		0	0	0					8.8

$$Y = b_0 + b_1x_1 + b_2x_2 + b_3x_3 + b_{12}x_1x_2 + b_{23}x_2x_3 + b_{31}x_3x_1 + b_{123}x_1x_2x_3 \quad (11.24)$$

where b_{12}, b_{23} and b_{31} are the second-order interaction terms, and b_{123} is the third-order interaction term. The coefficients in a regression equation like Eq. 11.24 represent joint estimate of effects and give no idea of their reliability or precision of an estimate. The precision is generally stated in the form of a confidence level (usually expressed as a percentage, say 90, 95 or 99%). The significance of the coefficients can be tested with student's t statistic.

For this purpose, it is first necessary to carry out replicate experiments at base level (the "origin", with all variables at 0 level), as indicated by test numbers 9, 10 and 11 in Table 11.6. The response standard deviation S_0 or error estimate is then calculated from the equation

$$S_0 = \left[\frac{\sum (Y_i - \bar{Y})^2}{r - 1}\right]^{1/2} \tag{11.25}$$

In this equation, r is the number of base level experiments, $\upsilon = r - 1$ is the degrees of freedom of error estimate, and $\bar{Y} = \sum Y_i/r$ is the average response. For a significance level of a (i.e. for $1 - a$ confidence interval, expressed in percentage) with degrees of freedom υ, the value of the student's t statistic $t_a(\upsilon)$ is determined from standard student's t distribution table. For a coefficient to be significant at $1 - a$ confidence level, its magnitude must be larger than $t_a(\upsilon) \cdot S_{b_j}$ where S_{b_j} is computed from the equation

$$S_{b_j} = S_0/N^{1/2} \tag{11.26}$$

N is the number of combinations used in factorial design, as defined earlier.

For the data presented in Tables 11.5 and 11.6 for example, the regression equation including two- and three-factor interaction terms is derived to be:

$$Y = 8.5 + 2.5x_1 - 0.5x_2 + 3.5x_3 - 0.5x_1x_2 - 1.5x_2x_3 + 0.5x_3x_1 + 0.11x_1x_2x_3 \tag{11.27}$$

Using the above procedure, it has been shown that the values of the coefficients b_2, b_{12}, b_{13} and b_{123} are insignificant, and the corresponding terms can be deleted from the equation. After this deletion, the regression equation becomes as follows

$$\hat{Y} = 8.5 + 2.5x_1 + 3.5x_3 - 1.5x_2x_3 \tag{11.28}$$

Here, $\hat{Y}$ is the modified response variable.

Examples The following examples will show how the factorial design of experiments is used to derive an equation for actual experiments. The experiment matrix is based on two levels and four factors, so that a total of $2^4 = 16$ experiments are necessary.

Na_2CO_3 *roasting of* Cr_2O_3

Tripathy et al. (1994), Tripathy and Ray (1995) have reported on a work on Na_2CO_3 roasting of Cr_2O_3. The effects of four process variable, namely temperature, time, amount of Na_2CO_3 addition and partial pressure of oxygen, were studied in this investigation. The range of variables selected was as follows:

Temperature (x_1): 573–873 K,
Time (x_2): 1–10 h,
Na_2CO_3 addition (x_3): 0–100% excess of stoichiometric requirement, and
Oxygen partial pressure (x_4): 10–100 kPa.

Table 11.7 Design matrix of two-level experimental design for Cr_2O_3 roasting

Expt. no.	x_1	x_2	x_3	x_4	Y
	Temp (°C)	Time (hour)	Na_2CO_3 addition (%)	Oxygen partial pressure (kPa)	Recovery %
1	+	+	+	+	84.32
2	–	+	+	+	31.10
3	+	–	+	+	40.09
4	–	–	+	+	10.23
5	+	+	–	+	74.08
6	–	+	–	+	21.00
7	+	–	–	+	33.45
8	–	–	–	+	08.50
9	+	+	+	–	53.89
10	–	+	+	–	10.83
11	+	–	+	–	20.01
12	–	–	+	–	04.50
13	+	+	–	–	49.20
14	–	+	–	–	08.33
15	+	–	–	–	27.00
16	–	–	–	–	03.25
17	0	0	0		28.85
18	0	0	0		29.13
19	0	0	0		28.73
20	0	0	0		28.80
21	0	0	0		29.30
22	0	0	0		29.03

With these ranges, the base levels are as follows: temperature—723 K, time—5.5 h, Na_2CO_3 addition—50% excess of stoichiometric requirement, oxygen partial pressure—55 kPa. Table 11.7 shows the design matrix in coded scale along with recoveries. Based on this design matrix, a regression equation for Y, chromite conversion (recovery) as a function of the process parameters (in their coded scale) was developed. The regression equation was subjected to various statistical tests, such as student's t test at 95% confidence level, to delete the insignificant coefficients and Fischer's test at 95% confidence level to test its adequacy.

The regression equation, after these statistical tests, takes the following form.

$$\hat{Y} = 29.98 + 17.76x_1 + 11.60x_2 + 1.98x_3 - 7.86x_4 + 6.01x_1x_2 + 3.17x_2x_4 + 4.73x_1x_2x_3 + 0.69x_1x_3x_4 + 0.43x_1x_2x_4 - 0.95x_1x_2x_3x_4 \quad (11.29)$$

where $\hat{Y}$ is the modified response variable.

In Eq. 11.29, the magnitudes of the various coefficients reflect the relative importance of the corresponding parameters. Thus, individually, temperature is the most important parameter, followed by duration, oxygen partial pressure and excess Na_2CO_3 addition. In addition, there are strong interactive influences, for example, of temperature with both time and oxygen partial pressure, and of time–temperature–excess Na_2CO_3 addition. Changing one variable with respect to another will have a considerable effect on the conversion.

Segregation roasting of copper concentrate

The principle of design of experiments has also been used for studying segregation roasting of copper concentrate (Datta et al. 1992). The four main parameters (factors), their base levels and ranges were as follows (Table 11.8):

The temperature, time, charcoal and sodium chloride additions were varied in steps of 50 °C, 20 min, 2 and 1%, respectively, to arrive at upper (+) and lower (−) levels of experimental conditions.

Sixteen experiments were conducted based on experimentation at two levels for four parameters. Six base level experiments were also conducted to estimate the error of variance. Table 11.9 lists the actual and coded values of the variables. Table 11.10 shows the complete design matrix based on the coded scale. It also shows the actual percentage yield of copper, *Y*, which is the response variable.

Table 11.8 Parameters, their base levels and ranges for the experimental design study on segregation roasting of copper concentrate (Tripathy 1994b)

Parameter			Base level	Range
Temperature	(x_1)	:	750 °C	700–800 °C
Time	(x_2)	:	40 min	20–60 min
Charcoal addition	(x_3)	:	6%	4–8%
Sodium chloride addition	(x_4)	:	2%	1–3%

Table 11.9 Actual and coded values of the variables for roasted copper concentrate

	x_1 temp (°C)	x_1 coded value	x_2 time (min)	x_2 coded value	x_3 charcoal (pct)	x_3 coded value	x_4 sodium chloride (%)	x_4 coded value
Upper level	800	+	60	+	8	+	3	+
Base level	750	0	40	0	6	0	2	0
Lower level	700	–	20	–	4	–	1	–

Table 11.10 Design matrix based on the coded scale for roasted copper concentrate

Expt. no.	x_1	x_2	x_3	x_4	Y (% recovery of metal)
1	+	+	+	+	52.0
2	−	+	+	+	74.8
3	+	−	+	+	46.6
4	−	−	+	+	58.6
5	+	+	−	+	40.6
6	−	+	−	+	54.0
7	+	−	−	+	31.8
8	−	−	−	+	35.4
9	+	+	+	−	60.0
10	−	+	+	−	65.6
11	+	−	+	−	45.5
12	−	−	+	−	47.7
13	+	+	−	−	24.0
14	−	+	−	−	21.7
15	+	−	−	−	26.4
16	−	−	−	−	19.1
17	0	0	0	0	55.4
18					53.9
19					55.0
20					56.0
21					56.5
22					54.8

Based on the results shown in Table 11.10, the following regression equation was developed for the pct. of metal recovery *Y*.

$$\begin{aligned} Y = 44 &- 3.11x_1 + 5.10x_2 + 12.37x_3 + 5.24x_4 \\ &- 1.81x_1x_2 - 2.2x_1x_3 + 1.65x_2x_3 + 1.03x_2x_4 - 3.58x_3x_4 - 3.34x_1x_4 \\ &+ 0.0218x_1x_2x_3 - 2.37x_2x_3x_4 - 0.009x_1x_2x_3 - 0.75x_1x_2x_4 \\ &- 0.16x_1x_2x_3x_4 \end{aligned} \tag{11.30}$$

From the results from the six base level experiments shown in Table 11.10, with degrees of freedom $\upsilon = r - 1 = 6 - 1 = 5$, using Eq. 11.25, S_0 is calculated as 0.92. Now, with $N = 16$ for the trials, from Eq. 11.26, S_{b_j} is obtained as 0.23. For a significance level of $a = 0.05$ (that is, for $1 - a = 0.95$, i.e. 95% confidence limit) with $\upsilon = 5$, student's *t* distribution table gives $t_a(\upsilon) = 2.57$. Therefore, for a coefficient to be significant at 95% confidence level, its magnitude must be larger than 2.57 $S_{b_j} = 0.59$. Using this criterion to delete the coefficients which are not significant, the regression equation becomes as follows.

$$\hat{Y} = 44 - 3.11x_1 + 5.10x_2 + 12.37x_3 + 5.24x_4 - 1.81x_1x_2 - 2.2x_1x_3 + 1.65x_2x_3 + 1.03x_2x_4 - 3.58x_3x_4 - 3.34x_1x_4 - 2.37x_2x_3x_4 - 0.75x_1x_2x_4 \tag{11.31}$$

11.5.3 Planning Factorial Experiments

The eventual success or failure of the experimental design hinges on correctly identifying all the important factors. One must be guided by experience, tempered possibly by budgetary limitations, in making this selection. If it is discovered after starting the experimental campaign that an important factor has not been but taken into consideration, the entire planning may be completely disrupted, and findings so far may be completely negated.

For meaningful results from the factorial design experiments, the variables must be accurately measured and also controlled precisely at the desired levels of +1 or −1, and 0 for the replicate base level experiments. If the variables can be accurately measured, but cannot be precisely controlled, then a much more difficult method of analysis, namely multiple regression, must be used.

The selection of the levels for the variables also requires careful consideration. After the first set of experiments, the coefficients of certain variables may be found to be small. This may indicate that these variables are unimportant. This may also indicate that the level chosen for such a variable is already optimal and therefore without effect, but only in a small neighbourhood, or the level may be a provisional optimum corresponding to the specific selections for the levels of the other variables. In such a situation, for these variables, the levels should be shifted and the scales increased in the subsequent experiments. If the variables are, in fact, without effect, then the coefficients will continue to be small. If the scales originally chosen were too small, the new coefficient will be larger. If the levels were at provisional optima, the coefficients will appear significant in the next experiments.

References

Brown, M.E. Galaway, A.K.: Thermochim. Acta **29**, 129 (1979)

Bandyopadhyay, A., Ganguly, A., Prasad, K.K., Sarkar, S.B., Ray, H.S.: React Solid. **8**, 77 (1990)

Beretka, J.: J. Amer. Ceram. Soc. **67**(9), 615 (1984)

Beretka, J., Lesko, J.: Thermochim. Acta **81**, 21 (1979)

Datta, P., Ray, H.S., Jena, A.K.: Trans. Inst. Min. Met. Soc. C **101**), C171 (1992, Sept–Dec)

Datta, P., Ray, H.S., Tripathy, A.K.: Application of statistical design of experiments in process investigations. In: Tripathy, A.K., Datta, P., Ray, H.S. (eds.) Kinetic Modeling in Quantitative Approaches in Process Metallurgy, pp 297–325. Allied Pub. Ltd., New Delhi (1995)

Garbachev, V.M.: J. Thermal Anal. **9**, 451 (1976)

Haque, R., Ray, H.S., Mukherjee, A.: Ironmaking Steelmaking **19**(1), 31–36 (1992)
Hulbert, S.F.: J. British Ceram. Soc. **6**(1) (1969)
Mishra, P., Paramguru, R.K., Ray, H.S.: In: Tripathy, A.K., Datta, P., Ray, H.S. (eds.) Kinetic Modeling in Quantitative Approaches in Process Metallurgy, pp. 297–325. Allied Pub. Ltd., New Delhi (1995)
Prabhakaran, K.V., Bhide, M.M., Kusian, E.M.: Thermochim. Acta **249**, 249 (1995)
Sahota, N.S., Gupta, K.N., Ram, M.: Trans. Ind. Inst. Metals **40**, 415 (1987, Oct)
Sharp, J.S., Brindley, G.W., Achar, B.N.N.: J. Amer. Ceram. Soc. **49**(7), 379 (1996)
Tan, T.C., Ford, J.D.: Metall. Trans. B **15B**, 919 (1984)
Tripathy, A.K., Ray, H.S.: Quantitative approach by empirical correlation in process metallurgy. In: Tripathy, A.K. Datta, P., Ray, H.S. (eds.) Kinetic Modeling in Quantitative Approaches in Process Metallurgy, pp 326–347. Allied Pub. Ltd., New Delhi (1995)
Tripathy, A.K., Ray, H.S., Pattnayak, P.K.: Trans. Inst. Min. Met. Sec. C **123**, C149 (1994a)
Tripathy, A.K., Ray, H.S., Pattnayak, P.K.: Trans. Ind. Inst. Met. **47**(6), 365 (1994b)

General reading

Khanazarova, S., Kafarov, V.: Experimental Optimization in Chemistry and Chemical Engineering. MIR Publishers, Moscow (1982)
Jakkiwar, M.S., Tupkary, H.R., Dokras, V.M.: Trans. Ind. Inst. Met. **30**(3), 348 (1980)

Chapter 12
Kinetics of Plastic Deformation

12.1 Introduction

Study of the kinetics of plastic deformation is of considerable technological significance: useful shapes are imparted to metallic materials by plastic deformation, and the useful life of a high-temperature component may be limited by plastic deformation to unacceptable levels, or even fracture. Becker (1925) and Orowan (1934) first suggested that the nonlinear resistance to plastic deformation of solids derives from thermally activated processes at local stress concentrations. Extensive experimental and theoretical researches carried out since then have elucidated various facets of the kinetics of plastic deformation. The relevant literature is vast, and continually expanding. It is now well established that for technologically important regimes, dislocation glide over slip planes causes plastic deformation in metallic materials. Depending upon stress and temperature, grain boundary sliding and/or stress-directed vacancy flow may also contribute to permanent deformation. The basic processes of interactions of dislocations and other crystal defects like vacancies, solutes, grain and twin boundaries, precipitates are now well understood.

Plastic deformation kinetics can be studied from different perspectives. The perspective for this chapter is phenomenological modelling for the kinetics of plastic deformation in polycrystalline structural materials for technological applications. Accordingly, the discussions on the mechanistic aspects are kept to a minimum. Finite element (FE) formulations incorporating such phenomenological models are being extensively used for simulating metal-forming operations and predicting deformation of structural components in service. Such modelling is considerably complicated because of the following factors:

- Distribution, interaction and migration of crystal defects are locally inhomogeneous. When deformation is treated as homogeneous, the "representative volume element" specified (e.g. volume of an element in FE formulation) or implied in discussions must be sufficiently large for statistical averaging to be meaningful.

H. S. Ray and S. Ray, *Kinetics of Metallurgical Processes*,
Indian Institute of Metals Series, https://doi.org/10.1007/978-981-13-0686-0_12

- With strain increasing beyond some level, increasing inhomogeneity of deformation manifests as formation of deformation bands, evolution of deformation texture, dynamic recrystallization, etc. Quantitative modelling for these phenomena is considerably complex.
- Depending upon the temperature and deformation rate, different rate-controlling mechanisms may become dominant.
- Wave propagation (dynamic) effects must be considered for deformation at high strain rates, say $\sim 10^3\ s^{-1}$ and higher.
- Since plastic deformation and the damage processes leading to fracture act in a synergistic manner, ideally a kinetic model should simultaneously consider deformation and fracture.

Because of these complexities, the practical approach is to seek a phenomenological model that yields technologically acceptable results for the specific material and deformation regime under consideration. The present chapter aims at introducing some of the basic concepts involved in developing such models. In common with the major body of literature, deformation and fracture are treated separately. The present chapter focuses on deformation. The next chapter deals with high-temperature (creep) fracture where thermal activation plays a key role. For simplicity, discussion in both these chapters is restricted to homogeneous, uniaxial, quasi-static deformation where dynamic effects may be ignored. Thus, the scope of the present chapter is the regime prior to onset of instability in a tension or compression test, and the primary and secondary regimes in a conventional creep test. The scope of the next chapter is essentially the tertiary creep regime.

The treatment in these two chapters is generally prescriptive—many important results are stated without citing the original literature. These may be found in the references listed at the end of this chapter. An introductory level of acquaintance with crystal defects and basic mechanical tests is assumed.

The reader may note that although there may not be many specific references to the details discussed in the previous chapters, the concepts developed here, based on the foundation of rate equations for thermally activated processes, often flow directly from those developed earlier.

For uniaxial deformation, the total longitudinal strain (ε_t) equals the sum of the its elastic (ε_e), anelastic (ε_a), and plastic (ε) strain components:

$$\varepsilon_t = \varepsilon_e + \varepsilon_a + \varepsilon$$

Anelastic strain plus plastic strain constitute inelastic strain. Elastic strains are typically up to $\sim 10^{-3}$ for deformation at ambient temperatures, and lower at elevated temperatures. Anelastic strains are similar or somewhat higher in magnitude. In this and the next chapter, generally consistent with most of the literature on phenomenological modelling for plastic deformation and fracture at elevated temperatures, anelastic strain is ignored (except in Hart's model discussed in

Sect. 12.5.3), and so is elastic strain in many situations involving large strains. This is quite acceptable for the strain ranges of interest in this chapter, except for relatively fast transients.

A note on the notations used

Notations in this and the next chapter are generally consistent with the physical and mechanical metallurgy literature, which occasionally differ from those in the process metallurgy literature. For example, Q (with appropriate subscripts, where required) is used for activation energy, while E represents Young's modulus. The reader may please take note of this. Also, in this and the next chapter, dot on top of a variable indicates its absolute differential with time t; e.g. $\dot{\varepsilon}$ is used to represent plastic strain rate. As elsewhere in this book, Δ before a variable indicates change in its magnitude, and δ its (infinitesimally) small increment.

12.2 Basic Approaches in Deformation Modelling

Consider an interrupted tensile test at ambient temperature, where a specimen of say well-annealed polycrystalline Al is loaded in tension in a testing machine, deformed well into the plastic deformation regime, fully unloaded and then again reloaded into the plastic regime, all at the same cross-head speed. Figure 12.1 schematically shows the applied stress—total strain plot that would be obtained in such a test employing the usual load and strain resolutions. If the variation of the gauge length of the specimen can be ignored (i.e. for small deformation levels), then constant cross-head speed corresponds to constant total strain rate, and ignoring anelastic strain, $\dot{\varepsilon}_t = \dot{\varepsilon}_e + \dot{\varepsilon}$. This simple test actually involves several "instantaneous" jumps in $\dot{\varepsilon}_t$. The start of the test involves a jump in $\dot{\varepsilon}_t$ from zero to the value corresponding to the imposed cross-head speed, and the intermediate unloading corresponds to jump in $\dot{\varepsilon}_t$ from a constant positive to constant negative value. In reality, an $\dot{\varepsilon}_t$ jump can never be instantaneous because of the finite mechanical inertia of the testing system. Careful measurements with high data resolutions have shown (Ray and Rodriguez 1992) that for a so-called instantaneous jump in $\dot{\varepsilon}_t$, $\dot{\varepsilon}_t$ actually varies continuously from one constant level to the other. Depending upon the stiffness of the specimen + testing machine assembly, typically $\sim$25 to 100 ms or more may be required for an $\dot{\varepsilon}_t$ transient in a stiff system. In a plot like Fig. 12.1, these continuous transients are not discernible because of inadequate data resolutions. In literature, often these transients are ignored and the $\dot{\varepsilon}_t$ jumps considered instantaneous. This is acceptable in most instances, except when the processes during this small transient period become important; Sect. 12.5.4 discusses one such example

Consider the two continuous loading segments in Fig. 12.1, both for the same $\dot{\varepsilon}_t$ level. Each loading segment has an initial linear elastic loading regime, $\dot{\varepsilon}_t = \dot{\varepsilon}_e$. With increasing stress, $\dot{\sigma}$ starts decreasing, signifying decreasing $\dot{\varepsilon}_e$ because of plastic flow at increasing $\dot{\varepsilon}$. Beyond small strain levels, $\dot{\varepsilon}_e$ and its variation with deformation becomes sufficiently small, and $\dot{\varepsilon}$ may be considered to be more or less constant.

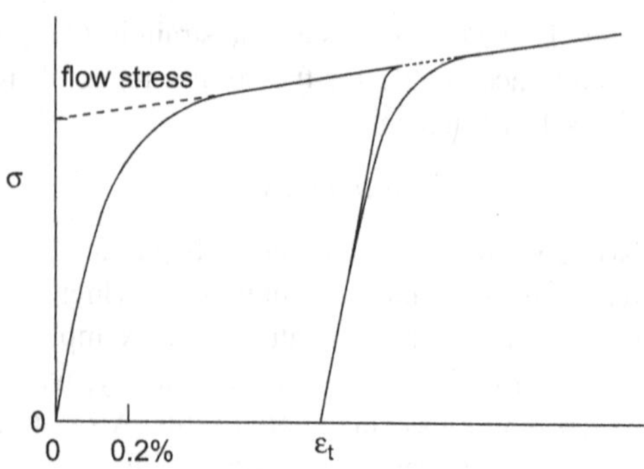

Fig. 12.1 Schematic variation of applied stress with strain in an interrupted tension test

In Fig. 12.1, the dashed line marked "Flow Stress" joins the corresponding regimes. Clearly, there are many transients in the evolution of defect substructure with plastic deformation. The dashed line in Fig. 12.1 represents the slowest of these transients. It is necessary to resort to back-extrapolation to establish the variation of this flow stress with strain for the regimes corresponding to the faster transients. Frequently, as in this chapter, it is this slowest transient that is of interest in deformation modelling. Basic approaches for such modelling are discussed in this section.

12.2.1 Micromechanical Modelling

Consider a distribution of dislocations, which continues to be spatially homogeneous (in a statistical sense) during deformation. Dislocations are propelled on the glide planes by the resolved shear stress τ, and their motion contributes to plastic shear strain, γ. For dislocations with density ρ and magnitude of Burgers vector b moving an average distance x with average speed $\bar{v}$, Orowan's equation gives the shear strain as $\gamma = \rho bx$ and shear strain rate $\dot{\gamma} = \rho b\bar{v}$. For cases where the flow stress in shear τ is solely controlled by dislocation–dislocation interactions, the following relation is found to hold over large ranges of ρ, with quite generality irrespective of the details of dislocation arrangements (Kocks and Mecking 2003):

$$\tau = \alpha\mu b\sqrt{\rho} \tag{12.1a}$$

(μ = shear modulus, α a factor.) This is valid for pure FCC and HCP metals and BCC metals at high enough temperatures where Peierls–Nabarro stress (the lattice friction stress opposing dislocation glide) becomes negligibly small. For example, for Cu at ambient temperatures $\alpha \approx 0.5 - 1$. The value of α depends on the definition of ρ adopted (dislocation intersection density, dislocation line length per unit volume, forest dislocation density, etc.), and value of μ used (as appropriate for a single crystal, or a suitably averaged value for polycrystals). Compared to the level of accuracy in α measurement, its variation with change in dislocation distribution (from uniform to cellular; see Sect. 12.3.1), which is within a range of $\sim 10\%$, may be ignored. α decreases exponentially with increasing T (and decreasing $\dot{\gamma}$), a characteristic feature of a thermally activated process. Kocks and Mecking (2003) compiled the data for variation of α with homologous temperature T/T_M for a number of a number of FCC metals (T_M is the melting point in absolute scale). Al shows the strongest temperature dependence: at $T/T_M \sim 0.5$, the value of α is ~ 0.7 times its value at absolute zero. For Ag at $T/T_M \sim 0.5$, α is ~ 0.9 times its value at absolute zero. The data for the other FCC metals investigated are between these two limits.

Other non-dislocation contributions to flow stress, which do not influence the thermal activation process, may be lumped into a friction or back stress τ_0. Thus, a more generalized formulation is:

$$\tau = \tau_0 + \alpha(\dot{\gamma}, T) \cdot \mu b\sqrt{\rho} \tag{12.1b}$$

Kinetic study with Eq. 12.1a or Eq. 12.1b as the starting point will involve formulating the function $\bar{v} = \bar{v}(\tau, \rho, T)$, and the evolution of ρ with deformation, i.e.

$$\dot{\gamma} = \dot{\gamma}(\tau, \rho, T) \tag{12.2a}$$

$$\dot{\rho} = \dot{\rho}(\tau, \rho, T) \tag{12.2b}$$

In writing Eqs. 12.2a–12.2b, geometrical relations are necessary for deriving γ and $\dot{\gamma}$ from ρ and $\bar{v}$ (see below). It is also be necessary to recognize that only a fraction of the dislocations present may actually be gliding. For the resultant formulation,

- The single structural variable ρ is used—its current value contains the entire history of deformation so far.
- The mechanical variables are identified τ, $\dot{\gamma}$ and T.
- Equations 12.2a–12.2b are the rate equations that describe the current deformation response and structure evolution.

This simple formulation is an example of a micromechanical model using a single substructural parameter ρ, the dislocation density.

Taylor orientation factor

Equations 12.1a–12.1b or Eqs. 12.2a–12.2b are basically derived for single crystals deforming in a single slip system, and appropriately, the relevant parameters are the resolved shear stress τ and the shear strain γ. For a single crystal deforming in multiple slip systems, or deformation of polycrystalline aggregates, it is natural to express the relevant equations in terms of normal stress σ and longitudinal plastic strain ε. The conversions require an average orientation factor (the Taylor orientation factor, or simply the Taylor factor) $\bar{M}$: $\sigma = \bar{M}\tau$, $\varepsilon = \gamma/\bar{M}$. Rigorous analysis to determine $\bar{M}$ for a polycrystalline aggregate is far from trivial, because it must take into account the slip systems, the statistics of the initial distribution of sizes and orientations of the grains, and gradual evolution of deformation texture with plastic deformation. For textureless FCC materials with a large number of randomly oriented grains, $\bar{M}$ can be taken as 3 for tensile deformation and 1.8 for shear deformation.

12.2.2 Yield-Based Modelling

As mentioned earlier, the two loading segments in Fig. 12.1 involve transition from purely elastic to elastic–plastic deformation. Yet, no obvious threshold stress for onset of plastic deformation (yield stress) is discernible for these segments. In fact, unique yield stress can be identified only in materials showing yield point phenomenon. In other materials, the engineering yield stress (also called 0.2% proof stress) is defined by convention as the stress for a small plastic strain of 0.002, i.e. the intersection of the plot with the computed elastic loading line passing through the point marked as 0.2% in the abscissa in this figure. Yield stress thus defined may not match with the flow stress for this strain level for the slowest transient, obtained by back-extrapolation of the dashed line in Fig. 12.1.

Consider the unloading schematically shown in Fig. 12.1. Apparently instantaneously upon the reversal of the direction of cross-head motion, the slope of the stress–strain curve changes to an essentially constant value that is consistent with purely elastic deformation of the specimen. With a micromechanical model (e.g. the one described in Sect. 12.2.1), the interpretation for such an apparently elastic unloading is as follows. On affecting the $\dot{\varepsilon}_t$-jump, plastic deformation continues, albeit with sharply decreasing rate as stress decreases. This is because plastic strain rate is a strong function of stress. Therefore, the deviation of the actual unloading line from that for purely elastic unloading also sharply decreases as unloading progresses and in effect may not be discernible without high resolution for data measurements.

This (near-) elastic unloading has, however, led to the alternative yield-based approach. In this approach, for uniaxial deformation, the flow stress for a given strain defines the threshold (i.e. the "yield") stress for plastic deformation at this strain (and strain rate). For continued plastic deformation, applied stress equals the current flow stress; further deformation is completely elastic if stress falls below this level, e.g. on affecting the unloadings in Fig. 12.1. For multiaxial deformation, the stress state must lie on the current yield surface for plastic deformation, and for stress states within the yield surface, deformation is purely elastic.

Critical experiments contradict the yield-based approach and support the micromechanical interpretation (see Sect. 12.5.4). Nevertheless, yield stress (or for multiaxial loading, yield surface) is a very useful concept, and yield-based approach is acceptable for many engineering applications.

12.3 Hardening and Recovery

Continued plastic deformation may lead to direction-independent isotropic hardening, and also kinematic hardening which reflects directionality of plastic flow properties. In cyclic deformation, kinematic hardening manifests as the well-known Bauschinger effect. In the yield surface-based approach, isotropic hardening corresponds to expansion of the yield surface with its origin unchanged, while kinematic hardening results in shift of the origin of the yield surface without change in its size. General deformation, comprising both isotropic and kinematic hardening, results in both expansion of the yield surface and shift of its origin. The sources of isotropic and kinematic hardening are summarized in the following for the necessary perspective (Kocks and Mecking 2003; Miller 1987).

12.3.1 Isotropic Hardening

FCC materials are widely used as high-temperature structural materials. Kocks and Mecking (2003) summarized the development of dislocation substructure in FCC materials with increasing strain. The sequence is dictated by stacking fault energy (SFE) Γ, or probably more accurately, by the normalized SFE $\Gamma/\mu b$, which is inversely proportional to stacking fault width. At low deformation levels, dislocation density is low, and extensive cross slipping of screw dislocations promotes a more or less random arrangement of dislocations. At higher deformation levels, the dislocation distribution becomes non-uniform. Regions of tangled dislocations with low densities ("cells") are enclosed by high densities of tangled dislocations ("cell walls"). For Cu polycrystals isothermally deformed in tension at temperatures covering a wide range, the flow stresses at identical strain hardening rate[1] were different and so were the details of cell structures. Yet, both cell diameter and average dislocation spacing scaled with flow stress. For Cu polycrystals deformed to 10% strain at different temperatures, mean dislocation spacing in the slip plane ($=\sqrt{2/\rho}$) was proportional to the average cell size. These observations combined with Eq. 12.1a indicate that in pure FCC materials, where dislocation interactions are the only strengthening mechanism, flow stress should be inversely proportional to cell diameter. With continued deformation, cell walls steadily sharpen and cell interiors are continuously cleaned up. Eventually, subgrains form with sharp

[1]The terms strain hardening and work hardening are used to refer to deformation-dependent strengthening. Often the two terms are used synonymously, and this practice has been adopted in this chapter also. In some contexts, however, it is necessary to distinguish between the two terms, depending upon whether the increase in yield stress (for uniaxial deformation), or in the radius of yield surface (for multiaxial deformation, e.g. of a von Mises material), is considered to be a function of incremental plastic strain (strain hardening), or incremental plastic work per unit volume accumulated over the deformation path (work hardening).

boundaries and essentially dislocation free interiors. The subgrains remain roughly equi-axed and maintain a misorientation angle of $1^{\circ}-2^{\circ}$. From an extensive compilation of data for subgrain sizes mostly at high temperatures, it has been concluded that normalized flow stress (i.e. flow stress/shear modulus) is inversely proportional to average cell wall spacing.

The evolution of dislocation substructure in FCC metals and alloys with other crystal structure follows similar pattern, as modulated by the number of slip systems available, SFE, solute content, precipitates, etc. At high enough temperatures, BCC materials behave like FCC metals and the strength is governed by dislocation–dislocation interaction. Pure BCC metals generally possess rather high SFE. Therefore, the work hardening is in many cases comparable to that of Al for similar values for normalized flow stress and T/T_{M}. The situation becomes complex at lower temperatures because lattice resistance (Peierls–Nabarro stress) is generally high in a BCC material. The number of easy slips systems is limited in HCP lattice. In polycrystalline HCP materials, where several different crystallographic slip modes (which often have quite different strengths) must be activated to maintain continuity of deformation, the strength is generally controlled by the hard modes of deformation. Furthermore, because of the high internal stress levels set up in HCP materials, twinning as an additional deformation mechanism comes into play in these metals much more frequently than in cubic metals.

Thus, a single phase material may show a combination of forest dislocation, cell and subgrain strengthening. The predominant source of strengthening depends on the temperature and strain rate; also, the contributions appear to evolve at different rates. While at a given temperature pure metals may show notable subgrain strengthening, some solid solution alloys are dominated by forest dislocation strengthening.

Some isotropic strengthening mechanisms are essentially independent of deformation level. Two such mechanisms are the lattice friction (Peierls–Nabarro stress) and strengthening from solutes. Solutes enhance strength in essentially two different ways: general lattice strengthening and solute drag on dislocations due to formation of solute atmosphere (Cottrell atmosphere) around the core of mobile dislocations, which is responsible for dynamic strain ageing (DSA) discussed in Sect. 12.8. Solute addition also significantly modulates SFE; decrease of SFE increases the work hardening capacity and strengthening behaviour. Thus, solutes may contribute also to deformation-dependent strengthening. Grain boundaries, precipitates and dispersoids also contribute to isotropic hardening.

12.3.2 Kinematic Hardening

There are a number of mechanisms of kinematic hardening in metals and alloys. These include bowing of dislocation between obstacles, dislocation pileups at obstacles, curvature of subgrain walls and elastic mismatch between different grains of a polycrystal. Upon strong change of direction of straining, cellular dislocation structure may tend to dissolve, while the overall dislocation density tends to remain constant—dislocations unpile and also start untangling at the cell boundaries. In the presence of a non-deformable phase, substantial back stresses are generated in the matrix, all of one sign, and they cause an effect that looks like a "permanent" lowering of flow stress in the reverse direction; only near steady state does the flow stress catch up. Each of these mechanisms creates an internal stress. This internal stress averages to zero over the volume of interest, but slows down dislocations that are moving in same direction as the recent previous straining directions (strain hardening), and accelerates dislocations moving in the opposite directions (strain softening). Some prior history of straining is required for these mechanisms. In contrast, a non-random crystallographic texture contributes to intrinsic anisotropy.

12.3.3 Strain Hardening and Recovery Processes

Strain hardening in absence of recovery processes

The deformation dependence of crystallographic strain hardening $\theta_\tau \equiv (\mathrm{d}\tau/\mathrm{d}\gamma)_{\dot{\gamma},T}$ can be derived by considering dislocation storage with deformation and Eq. 12.1a (Kocks and Mecking 2003). The rate of accumulation of dislocations with shear strain can be expressed as

$$\mathrm{d}\rho/\mathrm{d}\gamma = 1/(b\Lambda)$$

where Λ is a mean free path of dislocations, the (average) area of glide plane swept per unit increase of dislocation length. Differentiating Eq. 12.1a and substituting the above expression:

$$\frac{\tau}{\mu}\cdot\frac{\mathrm{d}\tau/\mathrm{d}\gamma}{\mu} = \frac{\tau}{\mu}\cdot\frac{\theta_\tau}{\mu} = \frac{(\alpha b)^2}{2}\cdot\frac{\mathrm{d}\rho}{\mathrm{d}\gamma} = \frac{\alpha^2}{2}\cdot\frac{b}{\Lambda} \tag{12.3a}$$

If Λ is constant, then from Eq. 12.3a, $\tau \cdot \theta_\tau$ is constant for a given temperature. For example, suppose dislocations go through the entire grain and get stored at the boundary without any storage in grain interior. Then, Λ equals grain size, and $\tau \cdot \theta_\tau$ would be independent of deformation level, but would increase with decreasing grain size. If $\tau \cdot \theta_\tau$ is constant, for deformation of polycrystals at a given temperature:

$$\sigma/\mu \propto \sqrt{\varepsilon} \tag{12.3b}$$

This is the Taylor work hardening relation. With solution and/or precipitation strengthening contributing a deformation independent "friction" or "back" stress σ_0, Eq. 12.3b is modified to:

$$(\sigma/\mu - \sigma_0/\mu) \propto \sqrt{\varepsilon} \tag{12.3c}$$

If on the other hand, gliding dislocations are arrested at the dislocation network,Λ is proportional to dislocation spacing (or, dislocation network size), $\Lambda \propto \rho^{-1/2}$. Substituting τ from Eq. 12.1a in Eq. 12.3a:

$$\frac{\theta_\tau}{\mu} = \frac{(\alpha\mu)^2}{2} \cdot \frac{b}{\Lambda} \cdot \frac{1}{\tau\mu} = \frac{\alpha}{2} \cdot \frac{1/\sqrt{\rho}}{\Lambda} \tag{12.3d}$$

Therefore, if $1/\sqrt{\rho} \propto \Lambda$, for a given temperature θ_τ is constant and $\tau \cdot \theta_\tau \propto \tau$.

For polycrystal deformation, it is convenient to define (uniaxial) strain hardening as $\theta \equiv (\mathrm{d}\sigma/\mathrm{d}\varepsilon)_{\dot{\varepsilon},T} = \bar{M}^2 \cdot \theta_\tau$. Figure 12.2 schematically shows a plot of $\sigma \cdot \theta$ against σ for a polycrystalline FCC metal (see Narutami and Takamura 1981). The positive offset on the ordinate for the extrapolation indicates contribution from Taylor work hardening (Λ is constant; Eq. 12.3a). The initial linear increase corresponds to work hardening contribution when principle of similitude applies ($\Lambda \propto \rho^{-1/2}$; Eq. 12.3d). At higher stress levels, $\theta \cdot \sigma$ decreases (i.e. Λ increases) with increasing deformation; this indicates increasing importance of recovery mechanisms.

Static and dynamic recovery

Static annealing of a heavily cold-worked material leads to considerable reduction in dislocation density, and therefore loss of strength. Such a recovery process is called static recovery. Static recovery is a thermally activated process—its driving force is the stored energy of dislocations. The process involves unzipping of attractive dislocation junctions releasing dislocations, followed by the climb of dislocations to favourable glide planes, dislocation glide and reduction in dislocation density by mutual annihilation. During deformation, in addition to static recovery, dynamic recovery may come into play. Kocks and Mecking (2003) cite an interesting example of dynamic recovery. A texture-free powder metallurgical Cu polycrystal sample was plastically deformed at 77 K. It was then strained at room temperature. This resulted in considerable reduction in flow stress subsequently measured at 77 K. Merely keeping the sample at room temperature for the same period did not produce this softening. Dynamic recovery processes, such as

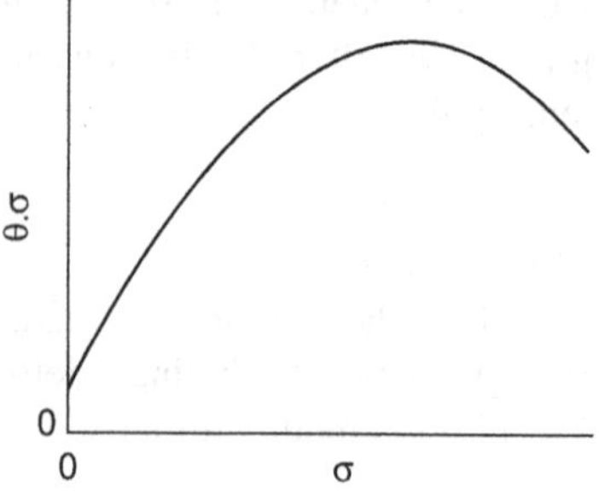

Fig. 12.2 Schematic variation of $\theta \cdot \sigma$ with σ for single phase polycrystalline FCC material at relatively low temperatures, including extrapolation to $\sigma = 0$

cross-slip leading to annihilation of opposite-signed dislocations passing sufficiently close to each other, result in softening of structure because of loss and rearrangement of dislocations. Dynamic recovery is also thermally activated and, as the name indicates, requires ongoing plastic deformation.

Evolution of deformation resistance has thus three components: dislocation storage, dynamic recovery and static recovery. Activation energies for dynamic recovery processes are lower than that for static recovery. Therefore, these processes predominate at (not too) low and intermediate T/T_M levels, while static recovery becomes increasingly important as T/T_M increases. The temperature dependence of static and dynamic recovery processes are much stronger compared to that of α (Sect. 12.2.1). Therefore, in the context of deformation-dependent strengthening, geometrical-statistical dislocation storage can be considered to be essentially athermal. This is the basis of the athermal work (or strain) hardening—thermally activated recovery approach frequently met in the literature. In this context, the term work (or strain) hardening refers actually to the (effectively) athermal geometrical-statistical storage of dislocations and not the parameters θ (or θ_τ) defined above which reflect the contributions from both dislocation storage and recovery processes.

A matter of terminology

In part of the technology literature, (i) plastic deformation is considered to be rate-insensitive deformation; (ii) the term viscoplasticity refers to rate-sensitive plastic deformation; (iii) the term creep refers to deformation at high homologous temperatures, with creep rupture lives of say $\sim 10^3$ h or more; and (iv) it is sometimes implied that creep is time-dependent permanent deformation. These descriptions must be considered as operational. Plasticity in this terminology corresponds to deformation at low T/T_M levels where geometrical-statistical storage of dislocations dominates, and recovery effects are weak enough to be ignored. Similarly, viscoplasticity refers to deformation at higher T/T_M levels where dynamic recovery effects are stronger, and creep refers to deformation at still higher T/T_M levels where static recovery dominates. From a mechanistic point, any implication that time is the control variable in driving creep deformation is not tenable. Also, from the perspective of state variable approach discussed in Sect. 12.4, the term creep can be considered to refer to deformation at constant stress (which may be extended to include the "non-ideal" condition of constant load) irrespective of the temperature or time to specimen rupture.

12.4 The State Variable Approach

State variable approach has been described as an "Article of Faith" in the literature on deformation kinetic modelling. In this approach, it is postulated that independent of the deformation path, a set of material parameters Ψ that quantifies the current microstructure and deformation substructure, along with the appropriate thermomechanical variables, provides a complete description of the current deformation response. Consider monotonic uniaxial deformation. From the micromechanical formulation in Sect. 12.2.1, $\dot{\varepsilon}$, σ, T are the appropriate thermomechanical parameters of deformation. Then, the constitutive equation for inelastic flow is a function of the form

$$\dot{\varepsilon} = f(\sigma, T; \Psi) \tag{12.4a}$$

The following functional form gives the evolution of structure parameters:

$$\mathrm{d}\Psi/\mathrm{d}t = \mathbf{G}(\sigma, \dot{\varepsilon}, T; \Psi) \tag{12.4b}$$

Note that Eq. 12.4b actually represents a set of (coupled) first-order differential equations, one for each of the elements in the set Ψ. Solution of Eq. 12.4b would describe evolution of Ψ with deformation. By definition, path dependence is excluded in Eqs. 12.4a–12.4b. Such path-independent formulations are called Mechanical Equations of State (MEOS), in the sense that the current "state" (the values for the mechanical and structure parameters) completely defines the current mechanical response, and it is not necessary to invoke any history dependence.

It is to be emphasized that the state variable approach itself does not prescribe the elements of Ψ or the thermomechanical parameters of deformation ($\dot{\varepsilon}$, σ, T in Eqs. 12.4a–12.4b). For a micromechanical (or physical) model, these are defined suitably, in terms of specific measurable components of the deformation substructure and microstructure, e.g. dislocation density, or inter-particle spacing in a precipitation-strengthened alloy. In contrast, in a phenomenological MEOS, the components of Ψ are not be quantified in terms of specific measurable features of the microstructure and deformation substructure. Nevertheless, mechanistic insight plays a significant role in identifying the relevant thermomechanical parameters and defining Ψ and its time dependence in any efficient phenomenological MEOS (see Sect. 12.9).

Example 12.1 Suppose the constitutive equation for plastic flow is given as follows:

$$\dot{\varepsilon} = \dot{\varepsilon}_0 \cdot (\sigma/\psi)^m \tag{12.5}$$

In this equation, (i) $\dot{\varepsilon}_0$ (with dimensions of strain rate) is a constant; (ii) the scalar variable ψ (with dimensions of stress) represents the deformation substructure; and (iii) $m = (\partial \ln \dot{\varepsilon}/\partial \ln \sigma)_{\psi,T}$ depends only upon T, but not on deformation level.

Some authors use the term "similitude" to indicate applicability of single parameter descriptions of the substructure. This must be distinguished from the more restrictive "principle of similitude" introduced by D. Kuhlmann-Wilsdorf, which requires proportionality of mean free path of dislocations and their spacing, and thus strictly applies for geometrical-statistical storage of dislocations (see Eq. 12.3d) (Kocks and Mecking 2003).

Obviously, a single parameter description of the current structure, as in Eq. 12.5, would prove inadequate if kinematic hardening need be considered, or even for monotonic loading, when the faster transients are also of interest. For the slowest transient in monotonic loading (recall the discussion on Fig. 12.1), a single substructural parameter might be adequate for correlating the plastic flow properties if these are determined by, for example, forest dislocation density or subgrain/cell size (see Sect. 12.3.1). In such a case, ψ becomes identified with a specific independently measurable feature of the deformation substructure.

Phenomenologically, a single parameter description of structure is viable when apparently significant changes in the substructural features only marginally modulate the flow stress (recall the insensitivity of α to the dislocation distribution, uniform to cellular, Eq. 12.1a), or when more than one significant features of the substructure change during deformation, but in more or less a linked manner over the regime of deformation of interest. In fact, there is a substantial contraction of information in shifting from metallographic description of the substructure to its characterization by mechanical properties. It transpires that a single parameter description is adequate over some technologically important regimes of deformation.

m in Eq. 12.5 can be generalized as

$$m \equiv (\partial \ln \dot{\varepsilon} / \partial \ln \sigma)_{\Psi,T}$$

Its reciprocal is the strain rate sensitivity of flow stress $(\partial \ln \sigma / \partial \ln \dot{\varepsilon})_{\Psi,T}$. This is an "engineering" definition. In Thermally Activated Strain Rate Analysis (see Sect. 12.5, Eq. 12.14b), the relevant parameter is

$$S \equiv (\partial \tau / \partial \ln \dot{\gamma})_{\Psi,T}$$

S may be called the (thermodynamic) strain rate sensitivity of flow stress. Note that in analysing test data, the notation $S \equiv (\partial \sigma / \partial \ln \dot{\varepsilon})_{\Psi,T}$, which is $\bar{M}$ times the above definition is also used. The operative definition should be clear from the context. The postulate that m is independent of the deformation level is discussed in Example 12.3 (Sect. 12.5.1).

Consider isothermal deformation, and let the current value of the single structure parameter be ψ. If Eq. 12.5 is the applicable constitutive equation, then a straight line AA with slope m in Fig. 12.3 schematically shows the corresponding iso-structural $\ln \sigma$–$\ln \dot{\varepsilon}$ relation. The current deformation state is depicted by the point O, which must lie on the line AA. The structure evolution rate $\dot{\psi}$ for this state has a unique value (cf. Eq. 12.4b), irrespective of the testing mode. Therefore, after a small time interval δt, the structure parameter has the unique value $\psi + \dot{\psi} \cdot \delta t$, the line BB. For Eq. 12.5, the line BB is parallel to the line AA and has slope m. It schematically shows the corresponding iso-structural $\ln \sigma - \ln \dot{\varepsilon}$ relation. The new deformation state must lie somewhere on the line BB. Its exact location depends on the testing constraint: it will be given by the point O_T for a tension test (constant $\dot{\varepsilon}$), the point O_C for a creep test (constant σ), and the point O_R for a stress relaxation

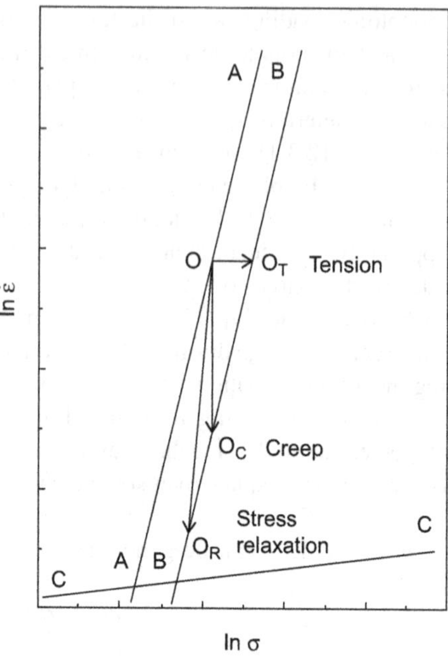

Fig. 12.3 Schematic illustration of the principle of the state variable approach using a single structural parameter for the constitutive equation given by Eq. 12.5

tests, in which cross-head motion of the testing machine is arrested so that $\dot{\varepsilon}+\dot{\varepsilon}_e=0$, and both $\dot{\varepsilon}$ and σ change, but in a coupled manner (see Sect. 12.5.3). Other states on this line BB could be reached using other deformation constraints. From Eq. 12.5, at constant temperature,

$$(1/m)\cdot(-\mathrm{d}\dot{\varepsilon}/\dot{\varepsilon})_{\sigma}=(\mathrm{d}\sigma/\sigma)_{\dot{\varepsilon}}=\ \mathrm{d}\psi/\psi \tag{12.6}$$

This equation provides the link between work hardening in tension test ($\dot{\varepsilon}$ constant) and deformation in creep test (σ constant).

The points on the schematic line CC in Fig. 12.3 correspond to the solutions of Eq. 12.4b for the specific value $\dot{\psi}=0$. (No significance need be attached to CC being drawn as a straight line in this figure.) Each point on this line corresponds to a different value of ψ. The line CC divides entire diagram into two regimes. For deformation states above this line, $\dot{\psi}\geq 0$; the structure gradually hardens during deformation, and the "normal" transients would be expected in creep tests (normal primary creep with $\dot{\varepsilon}>0$, decreasing with increasing ε or t) or in tension tests (positive work hardening rate θ, which decreases with increasing strain). For deformation states below this line, $\dot{\psi}\leq 0$, the structure gradually softens, and "inverted" transients would be expected. The line CC represents steady state of deformation, also called saturation state. The steady state determined for a specific model is actually an (extrapolated) mathematical steady state, specific to the model. Whether these states will actually be reached in the course of deformation will depend upon how faithfully the model mimics the reality.

ε and t as state variables of deformation

Consider plastic strain ε as the single (i.e. scalar) variable representing substructure, i.e. $\sigma = \sigma(\varepsilon, \dot{\varepsilon}, T)$. Nominally, this is an MEOS. It leads to a formulation:

$$\mathrm{d}\sigma = (\mathrm{d}\sigma/\mathrm{d}\varepsilon)_{\dot{\varepsilon},T} \cdot \mathrm{d}\varepsilon + (d\sigma/\mathrm{d}\dot{\varepsilon})_{\varepsilon,T} \cdot \mathrm{d}\dot{\varepsilon} + (\mathrm{d}\sigma/\mathrm{d}T)_{\varepsilon,\dot{\varepsilon}} \cdot \mathrm{d}T \tag{12.7}$$

The first term on the right-hand side of Eq. 12.7 can be determined from tension tests, the second term from strain rate jump or stress relaxation tests, and the third term by abruptly changing the test temperature during a tension test.

Now, Eq. 12.4a allows a quantitative "mechanical" definition for the structure when it is a scalar: it is the flow stress at some predetermined reference $\dot{\varepsilon}$ and T. Suppose a specimen is tensile tested under reference conditions well into the plastic deformation regime to allow sufficient work hardening, completely unloaded, fully annealed at an elevated temperature, and then again tensile tested at the reference conditions. It is well established that the intermediate thermal annealing leaves ε (virtually) unchanged and yet leads to reduction in flow stress at the reference conditions. Again, if well-annealed specimens are deformed at different strain rates and temperatures to a prefixed level of ε, unloaded, and then the flow stresses determined at reference conditions, it is found that these flow stresses are different. In these two examples, the flow stresses at the reference condition are different for identical ε levels. Clearly then, ε is path-dependent and does not qualify as a state variable of deformation. Therefore, Eq. 12.7 does not represent a proper MEOS.

The examples above actually reinforce the point that history variables ε and t can enter MEOS only in differential form. Thus, in the context of MEOS, ε and accumulated plastic work density are to be interpreted, respectively, as $\int \dot{\varepsilon} \cdot \mathrm{d}t$ and $\int \sigma\dot{\varepsilon} \cdot \mathrm{d}t$, the integrations being carried out over the deformation path, uniaxial work hardening rate θ is to be interpreted as $\theta = (\dot{\sigma}/\dot{\varepsilon})_{\dot{\varepsilon},T}$, etc. The terms strain hardening and work hardening are also to be interpreted accordingly.

Example 12.2 Equation 12.5 envisages that the entire applied stress is available for propelling the dislocations forward. In complex engineering alloys, the effective stress propelling the dislocation can be taken to be $\sigma - \sigma_b$, where σ_b is an athermal "back stress", with contributions from precipitation strengthening, solution strengthening beyond a few atom per cent, etc. (This linear superposition of stresses is discussed in more detail in Sect. 12.5.2.) Then, Eq. 12.5 is modified to

$$\dot{\varepsilon} = \dot{\varepsilon}_0 \cdot [(\sigma - \sigma_b)/\psi]^m \tag{12.8}$$

As σ_b is athermal, its temperature dependence is given by that of μ. Also, suppose σ_b is constant (e.g. the precipitate size remains constant) during the relevant period of deformation. Suppose strain rate jump tests are performed in the course of a tension test, and deformation at the new strain rate continued to an extent sufficient for determining the jump in stress level $\Delta\sigma$ corresponding to jump in strain rate by $\Delta\dot{\varepsilon}$ at constant ψ by back-extrapolation corresponding to the slowest

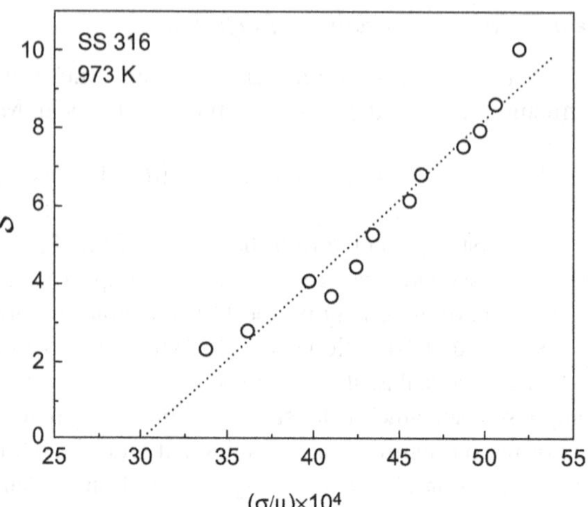

Fig. 12.4 Example of a Haasen plot. Constant nominal strain rate values employed were 3.17×10^{-4} and $3.17 \times 10^{-5}\ s^{-1}$. The dashed line shows the linear least square fit of the data, corresponding to $\sigma_b/\mu \approx 3.10^{-3}$ and $m = 24.07$ (Ray 1984)

transient (cf. Fig. 12.1; back-extrapolation for $\dot{\varepsilon}$-jump tests (Ray et al. 1993) in the presence of dynamic strain ageing is schematically shown in Fig. 12.14). Then

$$\mathcal{S} = \Delta\sigma/\Delta \ln \dot{\varepsilon} = (1/m) \cdot (\sigma - \sigma_b) \tag{12.9}$$

A plot of $\mathcal{S}$ against σ is called the Haasen plot; an example is shown in Fig. 12.4. $\mathcal{S}$ is the experimentally determined value for the (thermodynamic) strain rate sensitivity of flow stress S defined in Example 12.1. From a plot like Fig. 12.4, both m and σ_b can be determined. σ_b thus determined can also be compared with theoretical estimates, if available.

12.5 Thermally Activated Strain Rate Analysis (TASRA)

Jerky glide versus viscous glide of dislocations

When a dislocation moves in a crystal, it meets obstacles on the glide plane. If the resolved shear stress applied is less than the glide resistance offered by the obstacle, then the dislocation is arrested at the obstacle and must await thermal activation to overcome the obstacle. Thermally activated release of dislocations from such obstacles results in jerky glide. For applied stresses above the maximum glide resistance in the glide plane, the dislocation experiences linear viscous drag because of interaction with lattice vibrations (phonons), and for lower temperatures, electrons. In this viscous continuous glide regime, dislocation speed is proportional to stress. As stress further increases and dislocation speed approaches elastic shear wave speed, radiative processes become increasingly important, and the rate of increase of speed with stress gradually decreases. It is the jerky glide regime that is of interest in plastic deformation kinetics of structural components and is studied in

Thermally Activated Strain Rate Analysis (TASRA). The literature on the theory and application of TASRA is quite extensive. Kocks et al. (1975), Taylor (1992), Diak et al. (1998) and the literature cited therein may be studied for detailed expositions on these and related topics.

Barriers to dislocation motion

The nature of the obstacles in jerky glide depends on the type of glide resistance variation. Obstacles with high interaction energy with a dislocation (say ~5 eV) cannot be overcome by thermal activation. These are termed athermal obstacles and are characterized by a stress field extending over large distances. For example, strengthening by relatively large precipitates is athermal (see Example 12.2). A thermal obstacle on the other hand has a short-range character, with interaction energies of about $0.01-0.1\ \mu b^3$ (0.04–0.40 eV). These obstacles can be overcome by thermal activation by lattice vibrations (recall that $k_B T - 1/40$ eV at 300 K; k_B is the Boltzmann constant).

The Peierls-Nabarro barrier and cross-slipping of screw dislocations are examples of linear barriers. For these obstacles, the energy required to simultaneously overcome the barrier along the entire length of the dislocation is very high. Therefore, a bulge is first nucleated and subsequently expanded along the barrier. For a discrete obstacle on the other hand, the extent of its stress field in the slip plane of the dislocation is much smaller than the length of the bowing out segment of dislocation between two such obstacles. Particles, voids of sufficiently small size, precipitates, solute atoms (in dilute solutions) are examples of discrete obstacles. If their stress fields do not overlap, they can be treated as individual barriers. For a rigid barrier, for example, a precipitate particle embedded in the matrix, the line glide resistance versus distance profile of the obstacle is independent of the local stress. On the other hand, for a deformable barrier (a gliding dislocation cutting through and deforming a forest dislocation), the glide resistance–distance profile depends upon stress. The obstacle could be repulsive or attractive in nature and would affect the equilibrium position from which thermal activation takes place.

12.5.1 TASRA Formulation

Thermally activated glide of dislocations

The derivations below follow the literature already cited, notably Kocks et al. (1975). Consider a dislocation moving under a resolved shear stress τ, i.e. force τb per unit length, on a slip plane. Figure 12.5a schematically shows the variation of glide resistance force $\tilde{\tau} b$ per unit length along the reaction path (See Chaps. 1 and 2) as a dislocation moves from left to right in this diagram. The appropriate coordinate for the reaction path is the area swept out by dislocations a as it moves on the glide plane along the reaction path. The maximum glide resistance is $\hat{\tau} b$, with $\hat{\tau} > \tau$ for jerky glide. The dislocation is under a positive driving force $(\tau - \tilde{\tau} > 0)$ in

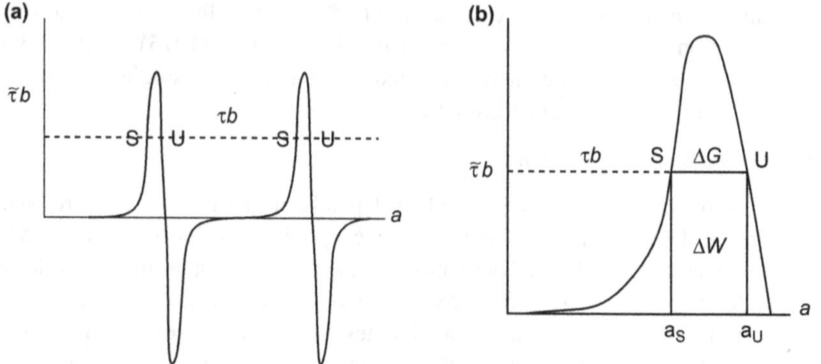

Fig. 12.5 **a** Glide resistance for a dislocation line on a slip plane. **b** Energetics of thermal activation of a dislocation (after Kocks et al. 1975)

some region of the glide plane, but would be in a state of stable equilibrium at the point marked by S in Fig. 12.5a, where $\tau = \tilde{\tau}$. The next location along the reaction path where $\tau = \tilde{\tau}$ is marked U; here the dislocation would be in unstable equilibrium. If thermal fluctuations transferred the dislocation from S to U, it would then again be under positive driving force and glide to the next stable equilibrium position with speed determined by viscous drag. The energetics of the activation process in equilibrium is shown in Fig. 12.5b. The area between S and U under the glide resistance ($\tilde{\tau}b - a$) diagram is the Helmholtz free energy change ΔF. ΔF is the sum of the work done by the applied constant τ during activation, called activation work ΔW, and the activation free enthalpy, or the Gibbs free energy of activation ΔG corresponding to $\tilde{\tau} > \tau$ which must be supplied by thermal activation.

Let a_S and a_U be, respectively, values for a at S in U in Fig. 12.5a, and $\Delta a = a_U - a_S$. a_S, a_U and Δa would in general depend upon τ (for deformable obstacles). Then

$$\Delta F = \int\limits_{a_S(\tau)}^{a_U(\tau)} \tilde{\tau} b \cdot \mathrm{d}a = \Delta W + \Delta G \tag{12.10a}$$

where

$$\Delta W = b \cdot \tau \cdot (a_U(\tau) - a_S(\tau)) = b \cdot \tau \cdot \Delta a(\tau) \tag{12.10b}$$

$$\Delta G = \int\limits_{a_S(\tau)}^{a_U(\tau)} (\tilde{\tau} - \tau) \cdot b \cdot \mathrm{d}a = \int\limits_{\tau}^{\tilde{\tau}_{max}} b \Delta a(\tau) \cdot \mathrm{d}\tilde{\tau} \tag{12.10c}$$

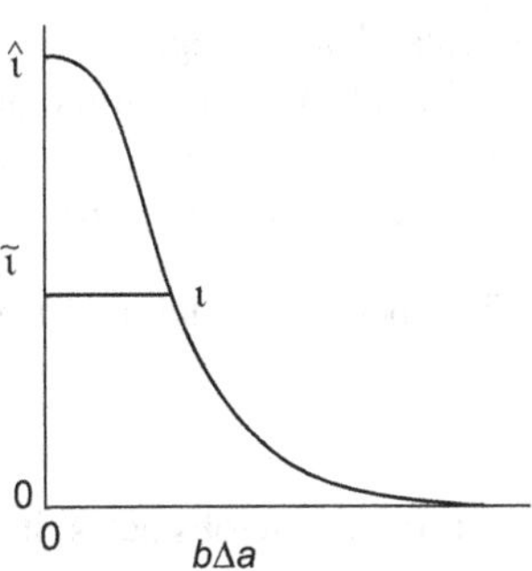

Fig. 12.6 Schematic plot of glide resistance stress $\tilde{\tau}$ against activation area multiplied by Burger's vector, $b\Delta a$ (after Kocks et al. 1975)

Figure 12.6 schematically shows a profile of stress-activation area plot with axes $b\Delta a$ and $\tilde{\tau}$, obtained from the glide resistance plot. In this plot, the horizontal line shows the applied shear stress τ. The total area under the plot equals ΔG^0 while the area under the curve above the horizontal line is ΔG (see Eq. 12.10c). It is the stress-activation area plot, and not the glide resistance profile, that can be determined experimentally.

The rate of an equilibrium fluctuation with energy greater than a given value ΔG at a given temperature is given by the product of a Boltzmann term involving ΔG, and a frequency factor υ_G:

$$P_t = \upsilon_G \cdot \exp(-\Delta G/k_B T)$$

Considering dislocation as taut vibrating elastic strings and with different additional assumptions, several theoretical estimates of υ_G have been derived in the literature. A simple, reasonable estimate is $\upsilon_G \leq 10^{11}\ s^{-1}$. Returning to Fig. 12.5a, if the characteristic time the dislocation waits at S is considerably larger than the time of flight for the dislocation from U to the next stable equilibrium S, then

$$\dot{\gamma} = \gamma_0 P_t = \gamma_0 \nu_G \cdot \exp(-\Delta G/k_B T) = \dot{\gamma}_0 \cdot \exp(-\Delta G/k_B T) \tag{12.11}$$

γ_0 is the crystallographic shear strain resulting when all the mobile dislocations accomplish one thermal activation. γ_0 may be viewed as the product of the number of activable sites and the (average) area swept by a dislocation after a successful activation. Equation 12.11 is applicable for $\Delta G > > k_B T$. When $\Delta G \sim k_B T$, the velocity of a dislocation is determined by dynamic effects and not by rates of thermal release from obstacles. On the other hand, if $\Delta G/k_B T$ is very large, it may be necessary to consider the backward jumps as well. Also, the total number of dislocation segments in a position to be activated, or the total number of activated events that a single segment undergoes, must be sufficiently large so that the statistical averaging implicit in the Arrhenius equation becomes meaningful.

Thermal activation parameters

The operational definition of activation area is

$$A \equiv -(1/b) \cdot (\partial \Delta G/\partial \tau)_T \tag{12.12}$$

For discrete obstacles, its relation with the true activation area Δa is given by the equation

$$A = \Delta a - \int_0^{\hat{\tau}} (\partial \Delta a/\partial \tau)_T \cdot \mathrm{d}\tilde{\tau}$$

The second term on the right arises because of the dependence of the glide resistance on the stress in case of deformable obstacles. Only A can be determined from experiments, and a model is needed to determine Δa from this. It has been shown that for randomly dispersed discrete deformable obstacles,

$$\Delta a = \frac{3}{2} A$$

The other parameter of interest for an obstacle is the value for activation energy at zero stress,

$$\Delta G^0 = \int_0^{\hat{\tau}} b \cdot A(\tau) \cdot \mathrm{d}\tau \tag{12.13}$$

For thermal activation at constant stress, the activation entropy ΔS and activation enthalpy ΔH are defined as

$$\Delta S \equiv -\left.\frac{\partial \Delta G}{\partial T}\right|_\tau, \quad \Delta H \equiv \Delta G + T\Delta S$$

It can be shown that

$$\Delta H = -T^2 \left.\frac{\partial(\Delta G/T)}{\partial T}\right|_\tau = k_\mathrm{B} T^2 \left.\frac{\partial(\ln \dot{\gamma})}{\partial T}\right|_\tau$$

For linear elastic obstacles, it is commonly assumed that the temperature dependence of glide resistance is given by that of μ. That is, the scaled activation energy $g \equiv \Delta G/(\mu b^3)$ is independent of temperature. Physically, this assumption implies that only the long-wavelength limit of the phonon spectrum is important, and local changes of vibrational frequency at dislocation cores, jogs, or kinks, or at surface steps left by cutting are ignored.

The effective stress formalism

In general, the obstacle profile on a glide plane may be expected to include both athermal obstacles with large range but relatively low peak (shear) stress τ_μ, superimposed on short-range barriers which can be thermally activated. Therefore in effect, for the dislocations to be mobile at all, $\tau \geq \tau_\mu$; τ_μ is called the athermal component of flow stress. When dislocation interactions are the only source of strengthening, τ_μ can arise from long parallel dislocations (accumulated during deformation) opposing dislocation glide. For elastic obstacles, τ_μ/μ is independent of temperature. The stress $\tau^* = \tau - \tau_\mu$ aids the thermal activation and is called the effective stress or thermal component of flow stress. For the effective stress formalism, the thermal activation parameters discussed above are to be written in terms of τ^*. The total stress and effective stress formalisms are equivalent in linear elasticity, and the choice is a matter of convenience.

Determining thermal activation parameters

For further discussion, it is convenient to use uniaxial normal stress, and longitudinal strain and strain rates. Taylor factor $\bar{M}$ is used for the necessary conversions. With these variables, the strain rate Eq. (12.11) becomes:

$$\dot{\varepsilon} = \varepsilon_0 \nu_G \cdot \exp(-\Delta G/k_B T) = \dot{\varepsilon}_0 \cdot \exp(-\Delta G/k_B T) \tag{12.11a}$$

ε_0 is the (longitudinal) strain resulting when all the mobile dislocations accomplish one thermal activation. In a material with a statistical distribution of obstacles, the microstructure is locally inhomogeneous, comprising of thermally hard and soft spots. Plastic flow is activated in a soft spot at a lower stress rather than in a hard spot. The statistical distribution of the stresses required to activate these spots is reflected in deformation dependence (usually, stress dependence) of $\dot{\varepsilon}_0$. For deformation at low homologous temperatures (T/T_M up to ~0.4 or more for FCC metals; see Example 12.3), variation in $\dot{\varepsilon}_0$ may generally be ignored compared to that of the exponential term. Often a power law dependence of $\dot{\varepsilon}_0$ on stress proves adequate.

The experimentally determined parameters are the apparent activation enthalpy Q, and the apparent activation area $\mathcal{A}$. In applied stress formalism, these are defined by the following equations:

$$\begin{aligned} Q &= k_B T^2 \cdot (\partial \ln \dot{\varepsilon}/\partial T) \\ &= -k_B T^2 (\partial \ln \dot{\varepsilon}/\partial(\sigma/\bar{M}))(\partial(\sigma/\bar{M})/\partial T) \end{aligned} \tag{12.14a}$$

$$\mathcal{A} = (k_B T/b)(\partial \ln \dot{\varepsilon}/\partial(\sigma/\bar{M})) = (k_B T)/(bS) \tag{12.14b}$$

This brings out the significance of the (thermodynamic) strain rate sensitivity of flow stress S (Example 12.1). $\mathcal{A}$ equals the activation area A defined in Eq. 12.12 only if $\dot{\varepsilon}_0$ is independent of stress. Writing $\Delta G = \Delta H - T\Delta S$,

$$\Delta H = Q - k_B T^2(\partial \ln \dot{\varepsilon}_0/\partial T) \tag{12.14c}$$

For inelastic obstacles, $\Delta S = 0$; when obstacles interact elastically with dislocations, $\Delta S \neq 0$. For inelastic obstacles obeying Friedel statistics, it has been shown that

$$\Delta G = \Delta H = Q - (T/\mu)\cdot(\mathrm{d}\mu/\mathrm{d}T)\cdot\mathcal{A}\cdot(\sigma/\bar{M}) \tag{12.15}$$

If $\dot{\varepsilon}_0$ is constant or at most a function of (σ/μ) only, then for linear elastic obstacles

$$\Delta G = \frac{Q + (T/\mu)\cdot(\mathrm{d}\mu/\mathrm{d}T)\cdot b\cdot\mathcal{A}\cdot(\sigma/\bar{M})}{1-(T/\mu)\cdot(\mathrm{d}\mu/\mathrm{d}T)} \tag{12.16}$$

Once ΔG is determined, $\dot{\varepsilon}_0$ can be determined from Eq. 12.11a, whence its postulated σ-dependence can be verified.

It will be noted that differential tests are required for determining the activation parameters $\mathcal{A}$ and Q (or the corresponding parameters $\mathcal{A}^*$ and Q^* defined in an analogous manner in the effective stress formalism). Analysis of data from strain rate jump tests has been introduced in Example 12.2; A can be determined from these tests reasonably accurately, provided the stress transient on affecting $\dot{\varepsilon}$-jump is not too extensive for meaningful back-extrapolation. Extrapolation may prove problematic for tests at relatively high temperatures or low strain rates because of extensive transients. Stress relaxation tests have also been widely used to determine $\mathcal{A}-\sigma$ profile; this is discussed subsequently in Sect. 12.5.3. In principle, it should be possible to determine A by noting the jump in $\dot{\varepsilon}$ on affecting a σ-jump in the course of a creep test. In practice, computing $\dot{\varepsilon}$ for new stress for the same substructure may be difficult because of the extensive ε-transients following the jump in σ in a typically high-temperature low-stress creep test.

Experimental determination of Q requires temperature jump tests. Because of large thermal inertia in typical testing systems, the transition period between two constant T-levels is invariably much larger than, e.g. that in a σ-jump or $\dot{\varepsilon}$-jump test. For intermediate and elevated temperatures, this period may be sufficiently large for significant changes in deformation substructure by the process of thermal recovery, which precludes meaningful analysis of data. One way to reduce such changes is to perform temperature jumps to progressively lower temperature levels. Once $\mathcal{A}-\sigma$ profile and Q are experimentally determined, these can be used to determine A and ΔG, verifying the postulated σ-dependence of $\dot{\varepsilon}_0$ in the process. These can be compared with theoretical models to identify the rate-controlling mechanism.

Table 12.1 Characteristics of some thermally activated deformation mechanisms (Rodriguez and Ray 1988)

Mechanism	A/b^2	Other characteristics
Peierls–Nabarro	1–100	A is independent of strain
Point defect interaction	1–100	A dependent upon point defect concentration, and on strain only if the point defect density is altered by strain
Dislocation interaction	100–10,000	A decreases and σ increases with increasing strain
Climb of edge dislocations	1	ΔH^0 equals the activation enthalpy for self diffusion (in alloys, mean activation enthalpy for alloy diffusion)
Cross-slip of screw dislocations	10–100	A decreases with increasing σ; many theoretical models available
Movement of jogged screw dislocations	10–1000	Climb of edge jogs on screw dislocations rate controlling. ΔH^0 same as that for climb
Dislocation dipoles	>1000	A decreases as the dipole density decreases with strain

Table 12.1 summarizes the characteristics for some of the thermally activable processes (Rodriguez and Ray 1988). When alternative deformation mechanisms are available, the step with the lowest activation energy is rate controlling. The case for multiobstacle deformation is discussed in Sect. 12.5.2. Transition in rate-controlling deformation mechanism is dictated by the activation energy values. The activation energy is high for climb-controlled deformation. Therefore at lower temperatures, an alternative mechanism (with stress-dependent activation energy) becomes rate controlling. For such a mechanism, for a given $\dot{\varepsilon}$, as test temperatures are increased, stress decreases and activation energy increases. When activation energy matches that for climb, climb-controlled dislocation mechanism becomes rate controlling.

For elastic obstacle, it is convenient to use dimensionless variables $g \equiv \Delta G/(\mu b^3)$ and σ/μ, because $g - \sigma/\mu$ profile is independent of temperature:

$$g/g^0 = \left(1 - \left(\tau/\hat{\tau}\right)^p\right)^q = (1 - (\sigma/\hat{\sigma})^p)^q \tag{12.17}$$

where $g^0 \equiv \Delta G^0/(\mu b^3)$, $\hat{\sigma}$ is uniaxial mechanical strength of the obstacle in polyslip (for elastic obstacles, $\hat{\sigma}/\mu$ is independent of temperature), and the parameters p and q determine the shape of the stress-activation energy profile. The constants $p = 1/2$, $q = 3/2$, or $p = 3/4$, $q = 4/3$, have been used to correlate deformation data with stress-dependent activation energy.

Example 12.3 Suppose activation area is inversely proportional to stress, $A = c'/\tau$ where c' is the proportionality constant. Then from Eq. 12.13,

$$\Delta G = \int_{\tau}^{\widehat{\tau}} b \cdot A(\tau) \cdot \mathrm{d}\tau = \int_{\sigma/M}^{\widehat{\sigma}/M} b \cdot \frac{c'}{(\sigma/\bar{M})} \cdot \mathrm{d}\left(\frac{\sigma}{\bar{M}}\right) = b \cdot c' \cdot \ln\left(\frac{\widehat{\sigma}}{\sigma}\right) \qquad (12.18a)$$

$\widehat{\sigma}$ is the mechanical strength of the obstacle (see Eq. 12.17). Substituting ΔG from Eq. 12.18a in Eq. 12.11a,

$$\dot{\varepsilon} = \dot{\varepsilon}_0 \cdot \left(\sigma/\widehat{\sigma}\right)^p \quad \text{with} \quad p = bc'/k_B T \qquad (12.18b)$$

If the variation of $\dot{\varepsilon}_0$ with temperature and stress is ignored, then the formulation reduces to Eq. 12.5 (Example 12.1). m in Eq. 12.5 is seen to be given by $m = bc'/k_B T$, the scalar structure parameter ψ is identified with $\widehat{\sigma}$, and $\dot{\varepsilon}_0$ is constant as postulated for Eq. 12.5. Extension of this analysis to the flow kinetics given by Eq. 12.8 (Example 12.2) is straightforward. If in Eq. 12.18b, $\dot{\varepsilon}_0 \propto \sigma^n$, n a constant, then

$$(\partial \ln \dot{\varepsilon}/\partial \ln \sigma)_T = n + m = n + (bc')/(k_B T)$$

Therefore, for variation in $\dot{\varepsilon}_0$ to be negligible compared to the thermal activation term,

$$m = (bc')/(k_B T) \gg n \qquad (12.18c)$$

This condition is met at sufficiently low temperatures, but will be increasingly violated when temperature increases beyond certain level. This explains why in general for deformation at lower temperatures, $\dot{\varepsilon}_0$ in Eq. 12.11a may be treated as essentially independent of deformation. Also, the strong stress dependence of ΔG in Eq. 12.18a indicates that this mechanism is unlikely to be rate controlling at higher T/T_M levels, and an alternative mechanism is likely to take over as rate controlling.

Cottrell–Stokes law

Equation 12.18a is a statement of "Cottrell–Stokes law" of deformation (see Nabarro 1991). Cottrell–Stokes law has been reported (Kocks 1976) to apply for Al, Cu and AISI type 304 austenitic stainless steel for T/T_M levels up to, respectively,0.64, 0.35 and 0.51. This law has been experimentally derived and has been defined in terms of the parameters m (as already defined in Example 12.1) and m^*:

$$\left.\frac{\partial \ln \sigma}{\partial \ln \dot{\varepsilon}}\right|_{\Psi,T} \equiv \frac{1}{m}; -\left.\frac{\partial \ln \sigma}{\partial \ln T}\right|_{\Psi,\dot{\varepsilon}} \equiv \frac{1}{m^*} \qquad (12.18d)$$

When Cottrell–Stokes law applies, m and m^* are independent of the level of deformation, i.e. the set of structure parameters Ψ. Experimentally, m is found to vary roughly linearly with $1/T$ in agreement with Eq. 12.18b, and also weakly with $\dot{\varepsilon}$; on the contrary, m^* depends only weakly on T. Note that from Eq. 12.18d, $m/m^* = Q/k_\mathrm{B}T$.

Cottrell–Stokes law has been stated also in other forms, which are mutually consistent within experimental errors. For example, in the effective stress formalism, it has been stated that if Cottrell–Stokes law applies, then $A^* = c'/\tau^*$ with c' a proportionality constant. Going through the steps used for the applied stress formalism, the plastic flow equation is now derived as:

$$\dot{\varepsilon} = \dot{\varepsilon}_0 \cdot \left(\frac{\sigma^*}{\widehat{\sigma}^*}\right)^m = \dot{\varepsilon}_0 \cdot \left(\frac{\sigma - \sigma_\mu}{\widehat{\sigma} - \sigma_\mu}\right)^m \quad \text{with} \quad m = bc'/k_\mathrm{B}T \tag{12.18e}$$

In Eq. 12.18e $\widehat{\sigma}^*$ is the effective mechanical strength of the obstacle.

12.5.2 Superposition of Different Types of Obstacles

For multiobstacle systems, analytical forms for superposition are available for some limiting cases; for other cases, in general, a numerical model is required. Consider two sets of obstacles, indicated by subscripts 1 and 2 in the following. Linear superposition applies to a system that contains both strong and weak obstacles, with density of weak obstacles considerably larger than that of strong obstacles. If a dislocation waiting between two strong obstacles bows under an applied stress, it interacts with the larger density of weaker obstacles. The effective glide resistance is the linear sum of the glide resistance of the two kinds of obstacles (the linear superposition rule):

$$\tau = \tau_1 + \tau_2 \tag{12.19a}$$

Invoking Eq. 12.14b and ignoring variation in $\dot{\varepsilon}_0$

$$1/A = 1/A_1 + 1/A_2 \tag{12.19b}$$

Superposition of solute hardening on forest hardening is expected to follow the linear superposition law, because dislocation–dislocation interaction is stronger than dislocation–solute interaction, and dislocation density, even after work hardening is usually smaller than the solute concentration. (Note, however, that beyond a few per cent of concentration, solute strengthening becomes athermal.) Square superposition applies to a system consisting of two sets of obstacles with similar strengths τ_1 and τ_2 but different planar densities. Consequently, the obstacle densities add linearly. The superposition rules for glide resistance and activation area are now:

$$\tau^2 = \tau_1^2 + \tau_2^2 \tag{12.20a}$$

$$(1/A)^2 = (1/A_1)^2 + (1/A_2)^2 \tag{12.20b}$$

The case for linear superposition of stresses is considered further in the following (Kocks et al. 1975; Diak et al. 1998). Consider a general case of linear superposition of glide resistances of solutes (σ_{sol}), precipitates (σ_p), grain boundaries (σ_g), and other sources of strengthening which can be thermally activated (σ_t), and forest dislocation interactions (σ_d). Then

$$\sigma = \sigma_d + \sigma_0 = \sigma_d + \left(\sigma_p + \sigma_g + \sigma_{sol} + \sigma_t\right) \tag{12.21a}$$

σ_d evolves during deformation; σ_0 is the sum of all the other contributions which do not vary with deformation level. For a well-annealed material with low initial forest density, σ_0 can be taken to be the back-extrapolated flow stress for $\varepsilon = 0$. Again, grain boundaries and precipitates are athermal obstacles, while the other obstacles can be thermally activated. Then, the overall (thermodynamic) strain rate sensitivity is obtained as follows:

$$\begin{aligned} S = \left(\frac{\partial \sigma}{\partial \ln \dot{\varepsilon}}\right)_{\Psi,T} &= \left(\frac{\partial \sigma_d}{\partial \ln \dot{\varepsilon}}\right)_{\Psi,T} + \left(\frac{\partial \sigma_0}{\partial \ln \dot{\varepsilon}}\right)_{\Psi,T} \\ &= \frac{(\sigma - \sigma_0)}{m_d} + \left(\frac{\partial \sigma_0}{\partial \ln \dot{\varepsilon}}\right)_{\Psi,T} \end{aligned} \tag{12.21b}$$

In this equation, m_d is the reciprocal of the engineering strain rate sensitivity for the forest dislocations strengthening component alone. If Cottrell–Stokes law is assumed to apply for forest hardening, then m_d is independent of deformation level. The corresponding relation is schematically shown in Fig. 12.7. The plot of S against $(\sigma - \sigma_0)$ is a straight line with slope $1/m_d$, and intercept of this line

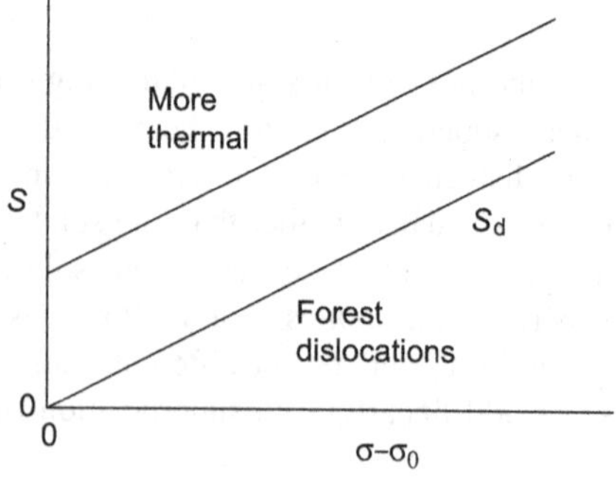

Fig. 12.7 Schematic Haasen plot for linear superposition of stresses

(from back-extrapolation) on the ordinate for $\sigma - \sigma_0 = 0$ is a measure of the total thermally activable component of the flow stress arising from sources other than forest dislocations. If $\sigma_0 = 0$ (cf. Example 12.1) or $\sigma_0 > 0$ is fully athermal (arising out of grain boundary and/or precipitation strengthening alone; cf. Example 12.2), then the line passes through origin of Fig. 12.7 (see also Fig. 12.4). Let m_{sol} and m_{t} be the reciprocals of (engineering) strain rate sensitivities for the σ_{sol} and σ_{t} components, respectively. Then, the expression for the (thermodynamic) strain rate sensitivity for σ_0 in Eq. 12.21b is obtained as:

$$\left(\frac{\partial \sigma_0}{\partial \ln \dot{\varepsilon}}\right)_{\Psi,T} = \frac{\sigma_{\text{sol}}}{m_{\text{sol}}} + \frac{\sigma_{\text{t}}}{m_{\text{t}}} \qquad (12.21\text{c})$$

Similar analyses may be carried out for the case of superposition of squares of stresses (Kocks et al. 1975; Diak et al. 1998). Such analyses show that for explicitly characterized microstructure, the parameter S together with the σ could in principle separate the relative contributions from different strengthening sources (Diak et al. 1998).

12.5.3 Stress Relaxation Testing

A stress relaxation test is actually a special type of strain rate jump test. In this test, a specimen is isothermally deformed at constant total strain rate (this can be achieved using the gauge extension for test control) to some preselected plastic strain level, and then the total strain rate is "instantaneously" changed to zero. Writing total strain as sum of elastic and plastic strains, $\varepsilon_{\text{t}} = \varepsilon + \sigma/E$ where E is the Young's modulus, during stress relaxation

$$\dot{\varepsilon}_{\text{t}} = \dot{\varepsilon} + \dot{\sigma}/E = 0, \;\; \therefore \;\; \dot{\varepsilon} = -\dot{\sigma}/E \qquad (12.22\text{a})$$

More convenient is to carry out stress relaxation tests by arresting actuator motion. In this case, Eq. 12.22a is modified to

$$\dot{\varepsilon}_t = \dot{\varepsilon} + \dot{\sigma}/\mathcal{M} = 0, \;\; \therefore \;\; \dot{\varepsilon} = -\dot{\sigma}/\mathcal{M} \qquad (12.22\text{b})$$

$\mathcal{M}$ is the equivalent elastic modulus for specimen + machine; machine here means the elements of the load train between the fixed and moving actuators, excluding the specimen. $\mathcal{M} < E$, and stiffer the testing system, closer does $\mathcal{M}/E$ approach 1. As Eqs. 12.22a–12.22b show, during stress relaxation, elastic strain is converted to plastic strain, with consequent decrease in stress level. During stress relaxation, load decreases at decelerating rate as schematically shown in Fig. 12.8. In a plot like Fig. 12.8, the sharp transient associated with $\dot{\varepsilon}_{\text{t}}$ gradually decreasing from a constant level to zero over a transient period of $\sim$50 to 150 ms is not

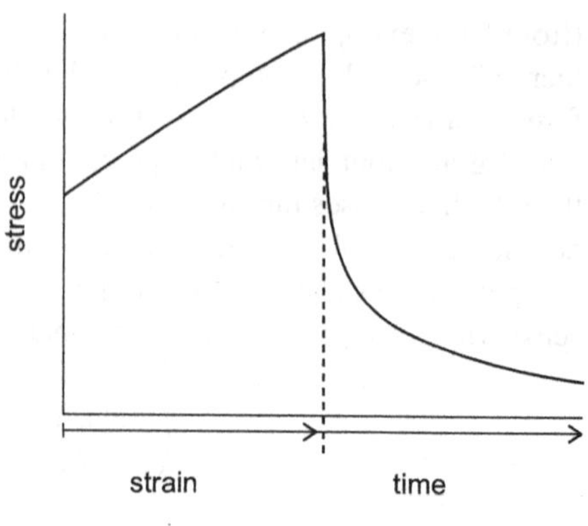

Fig. 12.8 Schematic illustration of variation of stress when a tension test is interrupted by a stress relaxation test. The stress–strain and stress–time segments are scaled differently for easy visualization

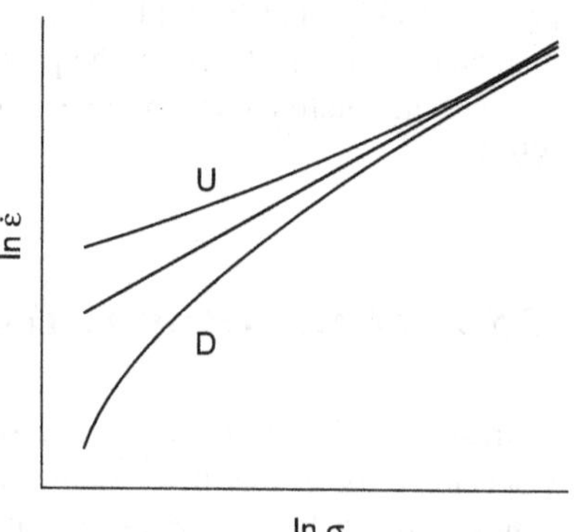

Fig. 12.9 Schematic illustration of three types of stress relaxation plots

discernible because of inadequate data resolution. Higher the values of m of the material and $\mathcal{M}$ of the testing configuration, sharper is the initial drop in load on affecting strain rate jump, and higher is the time resolution needed to determine the load transient.

Usually data from a stress relaxation test is presented in the form of a plot of $\ln \dot{\varepsilon}$ against $\ln \sigma$ (or equivalently, $\ln(-\dot{\sigma})$ against $\ln \sigma$). Figure 12.9 schematically shows the three type of stress relaxation plots observed. The D-type (curved downward) stress relaxation plot is typical of precipitation-strengthened alloys, at relatively low T/T_{M} levels. There are, however, many instances where the stress relaxation plot is U-type, i.e. curved upward. U-type plots are typically observed at intermediate and high T/T_{M} levels; one example is high purity Al at ambient temperatures.

Results from high-temperature stress relaxation testing are useful in engineering design such as high-temperature bolting–bolting stress gradually relaxes with sustained exposure to elevated temperatures. Stress relaxation tests have proved invaluable in TASRA studies, for reasons that can be anticipated from the micromechanical interpretation (Sect. 12.2.2) of tensile unloading (Fig. 12.1).

At the commencement of stress relaxation, σ and therefore $\dot{\varepsilon}$ equals that for the immediately preceding iso-$\dot{\varepsilon}_t$ deformation. Since $\dot{\varepsilon}$ strongly depends on σ, during relaxation, $\dot{\varepsilon}$ continues to decrease rapidly, though at a decelerating rate. Thus, the total strain accumulated even over extended periods of stress relaxations is very small. (Plastic strain accumulated during prolonged stress relaxation may be significant in a soft testing system, with $\mathcal{M}/E$ significantly less than the ideal value of 1). If this near constancy of ε can be taken to indicate essentially iso-structural deformation, then the stress relaxation data can be used to compute activation area, over even a few decades of $\dot{\varepsilon}$. If, however, dynamic (and static) recovery becomes significant during stress relaxation, or when dynamic strain ageing becomes important for the $\dot{\varepsilon}$ range covered during relaxation (see Sect. 12.8), the near constancy of ε during stress relaxation no longer means iso-structural deformation, and therefore, interpretation of relaxation plots requires additional considerations.

In contrast to the micromechanical interpretations, in athermal yield-based formulation (Sect. 12.2.2), plastic strain during stress relaxation must be attributed solely to recovery effects; this is dealt with in Sect. 12.5.4.

Scaling relation and model of Hart (1976)

An interesting feature of stress relaxation testing is a scaling relation—a series of $\ln \dot{\varepsilon}$ versus $\ln \sigma$ plots for stress relaxations from different (initial) ε levels can be superimposed onto a single plot by translating parallel to a line with a fixed slope (oblique translation convergence). This is schematically shown in Fig. 12.10. This scaling relation was first experimentally established by Hart and coworkers using data from high precision ambient temperature stress relaxation tests on high purity Al carried out for extended durations; the stress relaxation plots were U-type.

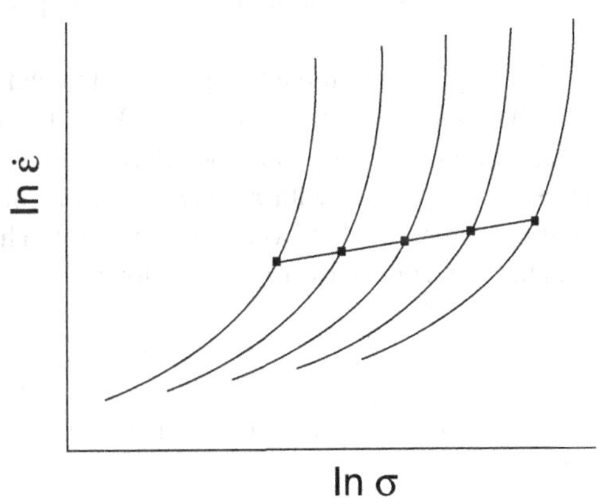

Fig. 12.10 Schematic illustration of scaling relation in stress relaxation—the stress relaxation plots superimpose when translated parallel to the axis along the straight line (oblique translation convergence)

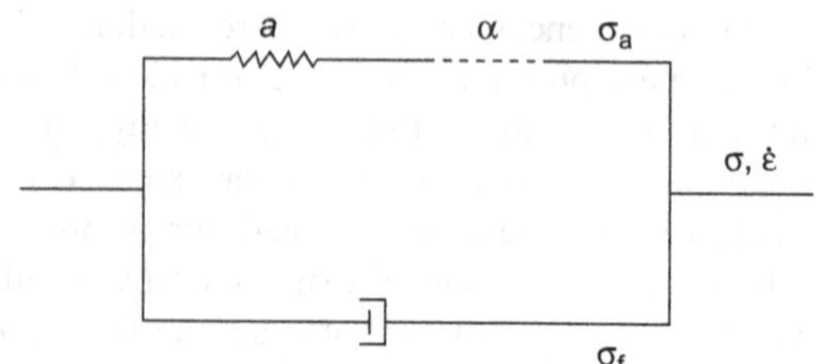

Fig. 12.11 Hart's rheological model

Subsequently, it was reported to obtain for a number of materials tested at different temperatures by many authors. Experimental results suggest that the scaling relation is probably an approximate one. Apparent exception to scaling relation is when two successive stress relaxations are carried out without allowing sufficient plastic deformation for the intermediate tensile segment for the "regular" deformation behaviour (corresponding to the slowest transient; cf. Fig. 12.1) to set in.

For isothermal deformation, increment in strain $d\varepsilon$ can be expressed by a general differential form

$$d\varepsilon = a_1 d\ln\sigma + a_2 d\ln\dot{\varepsilon}$$

where a_1 and a_2 are appropriate functions of σ and $\dot{\varepsilon}$. In general $(\partial a_1/\partial \ln\dot{\varepsilon})_{\sigma} \neq (\partial a_2/\partial \ln\sigma)_{\varepsilon}$ as ε is path-dependent. Hart argued that according to the theory of Pfaffian differential equations, the above equation can be integrated by using an appropriate integrating factor F such that $F \cdot d\varepsilon = d\sigma_H$ is a perfect differential; i.e. the parameter $\sigma_H(\ln\sigma, \ln\dot{\varepsilon})$ is path-independent (a state variable). σ_H is called the hardness parameter; of course, it has nothing to do with indentation hardness. During a stress relaxation test, the path-dependent ε remains constant, and so must the state variable σ_H.

The rheological model of grain matrix deformation proposed by Hart (1976) is schematically shown in Fig. 12.11 and the relevant equations are given in Eqs. 12.23a–12.23b.[2] This model conceives of an elastic + plastic element under stress σ_a in parallel with an anelastic element under stress σ_f and incorporates the scaling behaviour for iso-σ_H deformation. The equations relevant for plastic flow kinetics for constant σ_H are as follows:

$$\begin{aligned} &\sigma = \sigma_a + \sigma_f, && \dot{\varepsilon} = \dot{a} + \dot{\alpha} \\ &\sigma_a = \mathcal{M}a, && \dot{\varepsilon} = \dot{\alpha}^*(T) \cdot (\sigma_f/\mu)^M \\ &\ln(\sigma_H/\sigma) = (\dot{\varepsilon}_H/\dot{\alpha})^{\lambda}, && \dot{\varepsilon}_H = (\sigma_H/\mu)^{\alpha'} \cdot f \cdot \exp(-Q/RT) \end{aligned} \tag{12.23a}$$

[2]Equations 12.23a–12.23b and Fig. 12.11 generally retain the original notations of Hart, which differ from the other sections of this chapter. Interfacing, however, should not be a problem. Reader may please make a note of this.

The deformation dependence of the hardness parameter is given by the following equation:

$$\mathrm{d}\ln\sigma_{\mathrm{H}}/\mathrm{d}t = \Xi(\sigma, \sigma_{\mathrm{H}}) \cdot \dot{\alpha} - R(\sigma_{\mathrm{H}}, T) \quad (12.23\mathrm{b})$$

The first term on the right-hand side reflects work hardening and dynamic recovery, and the function $R(\sigma_{\mathrm{H}}, T)$ represents static recovery.

The values for the various material parameters are: $\mathcal{M} \approx \mu$, $M \approx 9$, $\alpha' \approx 4$ to 5, $\lambda \approx 0.15$, $f =$ a characteristic vibration frequency, and Q equals the activation energy for self (or alloy-) diffusion. λ and α' values are apparently independent of other material variables. α' is very close the typical value of the stress exponent for elevated temperature power law (PL) creep in FCC materials (to be discussed in Sect. 12.6). σ_{H} is considered to characterize strong barriers to dislocations (e.g. low angle grain boundaries or cell walls). During stress relaxation, these strong obstacles are considered to be unaffected, but climb-controlled dislocation activity takes place at the dislocation tangles between these strong obstacles, which are weaker barriers to deformation. Subsequently, many authors contributed to the application of this model, and its extension to incorporate grain boundary deformation.

During a stress relaxation test, $\mathrm{d}\varepsilon = 0$, i.e. $\mathrm{d}\sigma_{\mathrm{H}} = 0$, and therefore σ_{H} and $\dot{\varepsilon}_{\mathrm{H}}$ remain constant. From Eq. 12.23a:

- $(\mathrm{d}\ln\dot{\varepsilon}/\mathrm{d}\ln\sigma)_{t=0}$ at the start of a relaxation depends upon σ_{H} .
- For extended stress relaxation, slope of the stress relaxation plot is predicted to decrease, approaching a constant limiting value α' which is independent of test temperature, and also apparently similar for different materials.
- A series of $\ln\dot{\varepsilon}$ vs. $\ln\sigma$ plots for different constant σ_{H} values should superimpose by parallel translation on a line with slope α', i.e. in the schematic plot 12.10, the slope of the line is α'.

Consider an alternative interpretation for U-type stress relaxation, with the constitutive equation for plastic flow given by Eq. 12.18b, i.e. a single structure parameter-based model. With this equation, with $\widehat{\sigma}$ constant, the stress relaxation plot is linear with slope equal to $m = c/k_{\mathrm{B}}T$, see also Fig. 12.3. If, however, dynamic recovery comes into play during stress relaxation (because of continually decreasing $\dot{\varepsilon}$ as stress relaxation continues), $\widehat{\sigma}$ would continually decrease from its initial value. For this model:

- In a rigid testing system, $(\mathrm{d}\ln\dot{\varepsilon}/\mathrm{d}\ln\sigma)_{t=0}$ equals the value of m for the test temperature, independent of the level of plastic deformation. This is in contrast to the point A above. (In a non-rigid testing system with $\mathcal{M}/E<1$, $(\mathrm{d}\ln\dot{\varepsilon}/\mathrm{d}\ln\sigma)_{t=0}$ may show a deformation dependence which, however, should be small for typical hard testing systems.)
- From Fig. 12.3, $(\mathrm{d}\ln\dot{\varepsilon}/\mathrm{d}\ln\sigma)$ should gradually decrease yielding a U-type relaxation plot. However, with increasing time for stress relaxation, the relaxation plot will asymptotically approach the line CC in Fig. 12.3 for steady-state deformation. Thus, $(\mathrm{d}\ln\dot{\varepsilon}/\mathrm{d}\ln\sigma)_{t\to\infty}$ equals the stress exponent for steady-state

creep for the test temperature. As discussed in Sect. 12.6, this exponent depends upon temperature (and stress), and approaches a constant "universal" only in the limit of high temperature and low stresses corresponding to power law (PL) creep.

- However, with such a formulation, it is necessary to add a plausible formulation for the kinetics of dynamic recovery, which would be consistent with the scaling relation, or at least an acceptably close approximation to it.

Several authors have critically compared these aspects by careful and precise experiments, theoretical modelling for dynamic recovery with Eq. 12.18b for the iso-substructural plastic flow kinetics, and also alternative TASRA-based modelling for inhomogeneous dislocation structure using more than one structure parameter. $(\mathrm{d}\ln\dot{\varepsilon}/\mathrm{d}\ln\sigma)$ decreases sharply early during stress relaxation. Therefore, determining $(\mathrm{d}\ln\dot{\varepsilon}/\mathrm{d}\ln\sigma)_{t=0}$ accurately is difficult both because very high data resolutions are required, and the finite transition time required for $\dot{\varepsilon}_t$ to change from a constant level to zero can no longer be ignored (see Sect. 12.5.4). So the experimental studies have focused mainly on the other two points. For example, Rohde et al. (1981) carried out carefully controlled ambient temperature creep and stress relaxation tests on Al for extended periods. The results were generally consistent with predictions from Eq. 12.18b and Fig. 12.3 as listed above. Even with stress relaxations over extended periods, the limiting slope $(\mathrm{d}\ln\dot{\varepsilon}/\mathrm{d}\ln\sigma)_{t\to\infty}$ determined was 12. This value is considerably larger than the value of $\alpha' \approx 4-5$ for the Hart model, but very close to 14, the value for the stress exponent for steady-state creep of Al at ambient temperature determined from accurate creep tests. Scaling relation (or a reasonably close approximation to it) has been derived by some authors by modelling for dynamic recovery during stress relaxation; see, e.g. Rohde et al. (1981); accurate prediction for the shape of the stress relaxation plot by such modelling has proved more difficult.

12.5.4 *Time-Independent Deformation (TID) Model*

The Time-Independent Deformation (TID) model (Alden 1982) is an yield-based model (Sect. 12.2.2). It postulates that plastic deformation is athermal, and the temperature dependence of overall deformation arises solely from thermally activated recovery processes that lead to loss of dislocations and their rearrangement into softer configurations. In the absence of recovery processes, plastic deformation can be sustained only if $\dot{\sigma} > 0$. Alden's expression for strain rate is:

$$\dot{\varepsilon} = \frac{(\dot{\sigma} + r_{\mathrm{A}})}{\theta_{\mathrm{y}}} \cdot A_{\mathrm{r}} \text{ for } \dot{\sigma} > 0; \quad \dot{\varepsilon} = \frac{r_{\mathrm{A}}}{\theta_{\mathrm{y}}} \cdot A_{\mathrm{r}} \text{ for } \quad \dot{\sigma} \leq 0 \tag{12.24}$$

Here θ_y is the athermal work hardening rate in the absence of recovery (see Sect. 12.3.3), r_A a (thermally activated) recovery parameter, and A_r a free area function. In the TID model then, $\dot{\sigma}$ is a state variable of deformation, whereas in the TASRA model, it is not. The dichotomy can be resolved only by carefully designed experiments.

Consider a stress relaxation test interrupting a tension test, with its "instantaneous" change in $\dot{\sigma}$ from positive (for the tensile segment) to negative (for the stress relaxation segment). If TID model is valid, then the ratio of $\dot{\varepsilon}$ immediately after and before the transition from tension to relaxation, $\mathcal{R} = r_A/(r_A + \dot{\sigma}^-) < 1$, while for TASRA model, the ratio $\mathcal{R} = 1$; here $\dot{\sigma}^-$ is the magnitude of the stress rate just prior to the start of relaxation. Several papers reported $\mathcal{R} < 1$ for a number of materials and test conditions, which supports TID formulation, and contradicts the TASRA. Similarly, "instantaneous" stress jump in a creep test from a constant level σ_1 to a level $\sigma_2 > \sigma_1$ (this includes start of a creep test, because here $\sigma_1 = 0$) has been reported to lead to a loading strain, i.e. apparently an instantaneous increase in plastic strain. This would be consistent with TID formulation but not with TASRA.

Subsequent research, however, showed that such conclusion was based on inadequate appreciation of the role very sharp transients associated with the finite time (e.g. typically 5–100 ms for a stress relaxation test) required for the actuator speed to change from one constant level to another for transition from tension/compression to stress relaxation (see Sect. 12.2). Therefore in a stress relaxation test, it is necessary to set very high resolutions for recording both stress and time. Also, for determining the "ideal" value for $\dot{\sigma}$ corresponding to a truly "instantaneous" jump in actuator speed from a fixed level to zero, it is necessary to resort to back-extrapolation from data beyond this transient regime. Tests and analyses where these two key factors have been properly taken into account yielded $\mathcal{R} = 1$ as predicted by the TASRA formalism. Similarly, careful analysis of data from creep tests involving change in stress (this includes start of a creep test) support the TASRA, and not the TID formulation. Method of analysis for the back-extrapolation in case of a stress relaxation test, and also citations to the relevant literature for stress relaxation tests and stress jump tests in creep may be found in Ray et al. (1993).

12.6 Steady-State Creep

Elevated temperature creep has been extensively studied because of its significance for the service life of high-temperature components. There are several excellent reviews and monographs on creep deformation mechanisms (Bendarsky et al. 1985; Poirier 1985; Cadek 1988). This section deals with the steady state of isothermal creep deformation (constant σ), where $\dot{\varepsilon}$ attains a constant value (designated $\dot{\varepsilon}_s$) and continues in this state until damage processes manifest in the form of accelerating $\dot{\varepsilon}$. The σ-dependence of $\dot{\varepsilon}_s$ at a constant T is often expressed by an empirical power law relation, called the Norton equation, $\dot{\varepsilon}_s \propto \sigma_s^n$ (see Eq. 13.1); in this equation, the exponent n is called the Norton stress exponent. (For convenience, the subscript s has been used for the stress corresponding to steady state.) In typically complex structural alloys, an extended steady regime of deformation is often not found; in such situations, the minimum creep rate $\dot{\varepsilon}_m$ is used in place of $\dot{\varepsilon}_s$ in the Norton equation. Norton equation is reasonably accurate for data from relatively narrow ranges of stress, but proves inadequate when data from wider stress ranges are considered.

While most of studies on steady state have been carried out using creep testing, the existence of a steady state, or it being independent of deformation mode, is a corollary of the state variable approach, independent of any specific model or the number and nature of the components in Ψ, Eq. 12.4b (recall Example 12.1 and Fig. 12.3). In reality, a steady state may be said to have been effectively reached if the important substructural features evolve at negligible rates. There are very few direct experimental verifications in support of the experimentally determined steady state being independent of deformation mode, even though this concept occupies a key position in formulating phenomenological MEOS.

12.6.1 Power Law (PL) Creep

At sufficiently low stresses (σ_s/μ below $\sim 5 \times 10^{-6}$) $\dot{\varepsilon}_s$ is found to vary linearly with σ_s/μ (Harper–Dorn creep), and is ascribed to climb-controlled creep under conditions where dislocation density does not change with stress. Stress-directed flux of vacancies from grain boundaries perpendicular to the applied stress axis to boundaries parallel to the stress axis can also give rise to permanent deformation. This phenomenon (Nabarro–Herring creep) manifests at low stresses (σ_s/μ below $\sim 10^{-4}$) and elevated temperatures. In Nabarro–Herring creep, the diffusion path is through the lattice, and

$$\dot{\varepsilon}_s \propto \left(\sigma_s b^3 D_v\right) / \left(k_B T d^2\right)$$

$D_v = D_0 \exp(-Q_{SD}/k_B T)$ is the lattice vacancy diffusivity, and here d is the grain size. The activation energy thus equals that for self- (or in substitutional alloys, alloy-) diffusion Q_{SD}. Grain boundary diffusion predominates at lower temperature (Coble creep), resulting in

$$\dot{\varepsilon}_s \propto \left(\sigma_s b^4 D_{gb}\right) / \left(k_B T d^3\right)$$

D_{gb} is the grain boundary diffusion coefficient, with activation energy that for grain boundary diffusion. For both these mechanisms, the stress exponent equals unity.

For higher stress levels, several mechanisms, collectively called dislocation creep mechanisms, with climb as the rate-controlling step, have been proposed in the literature. These models differ in details, but lead to the steady-state creep equation with the general form:

$$\dot{\varepsilon}_{\mathrm{s}} = A_{\mathrm{C}} \cdot \left(\frac{\mu b}{RT}\right) \cdot f\left(\frac{\sigma_{\mathrm{s}}}{\mu}\right) \cdot D_0 \cdot \exp\left(-\frac{Q_{\mathrm{SD}}}{RT}\right) \tag{12.25a}$$

Here A_{C} is a constant, and $f(\cdot)$ an appropriate function of its argument. The different models differ in A_{C} and the function $f(\cdot)$. One such model is discussed in Example 12.4. The primary support for climb-controlled creep mechanism stems from the experimental observation that in a wide range of materials, the activation energy for creep $Q_c \approx Q_{\mathrm{SD}}$. For intermediate to high stresses and temperatures above $\sim 0.5T_{\mathrm{M}}$, theoretical models for climb-controlled creep predict power law dependence consistent with the equation:

$$f(\sigma_{\mathrm{s}}/\mu) = (\sigma_{\mathrm{s}}/\mu)^{n_{\mathrm{PL}}} \tag{12.25b}$$

with $n_{\mathrm{PL}} = 3$, the so-called universal exponent. It has been argued that in case of FCC materials a universal exponent of 4 is more appropriate. Experimental results with metals usually yield n_{PL} values ~ 4 to 5. This regime is aptly called Power Law (PL) creep regime.

Example 12.4 Consider a homogeneous dislocation network with average link length $\bar{l}$. Friedel (1964) showed that for climb-controlled coarsening of such a network,

$$\dot{\bar{l}} \propto D_{\mathrm{v}} \cdot \left(\frac{\mu b^3}{k_{\mathrm{B}}T}\right) \cdot \left(\frac{1}{\bar{l}}\right) \cdot c_{\mathrm{j}} \tag{12.26a}$$

the concentration of dislocation jogs $c_{\mathrm{j}} \approx 1$ for well deformed material; other terms have their usual significance. Since $\bar{l} \propto 1/\sqrt{\rho}$, using Eq. 12.1a

$$\sigma \propto \mu b/\bar{l} \tag{12.26b}$$

Differentiating Eq. 12.26b, and substituting Eq. 12.26a:

$$\left(\frac{\partial \sigma}{\partial t}\right)_{\mathrm{r}} = \left(\frac{\partial \sigma}{\partial \bar{l}}\right) \cdot \dot{\bar{l}} \propto -\frac{\mu b}{\bar{l}^2} \cdot D_{\mathrm{v}} \cdot \left(\frac{\mu b^3}{k_{\mathrm{B}}T}\right) \cdot \frac{1}{\bar{l}} \tag{12.26c}$$

Substituting Eq. 12.26b in Eq. 12.26c,

$$\left(\frac{\partial(\sigma/\mu)}{\partial t}\right)_{\mathrm{r}} = -A_{\mathrm{f}} \cdot \frac{D_{\mathrm{v}}}{b^2} \cdot \left(\frac{\mu b^3}{k_{\mathrm{B}} T}\right) \cdot \left(\frac{\sigma}{\mu}\right)^3 \tag{12.26d}$$

A_{f} is an appropriate constant. Adding a constant athermal work hardening contribution (because of geometrical-statistical storage of dislocations) when similitude applies (cf. Eq. 12.3d)

$$\frac{\mathrm{d}(\sigma/\mu)}{\mathrm{d}t} = \left(\frac{\partial(\sigma/\mu)}{\partial t}\right)_{\mathrm{wh}} + \left(\frac{\partial(\sigma/\mu)}{\partial t}\right)_{\mathrm{r}}$$

i.e.

$$\frac{d(\sigma/\mu)}{\mathrm{d}t} = \left(\frac{\theta_0}{\mu}\right) \cdot \dot{\varepsilon} - A_{\mathrm{f}} \cdot \frac{D_{\mathrm{v}}}{b^2} \cdot \left(\frac{\mu b^3}{k_{\mathrm{B}} T}\right) \cdot \left(\frac{\sigma}{\mu}\right)^3 \tag{12.26e}$$

For steady state, setting $\mathrm{d}(\sigma/\mu)/\mathrm{d}t = 0$ in Eq. 12.26e,

$$\dot{\varepsilon}_{\mathrm{s}} = A_{\mathrm{f}} \cdot \frac{D_{\mathrm{v}}}{(\theta_0/\mu) \cdot b^2} \cdot \left(\frac{\mu b^3}{k_{\mathrm{B}} T}\right) \cdot \left(\frac{\sigma_{\mathrm{s}}}{\mu}\right)^3 \tag{12.26f}$$

The result is the expression for power law creep, Eqs. 12.25a–12.25b, with $n_{\mathrm{PL}} = 3$, the universal value. This is an early model for power law creep, and is an illustration of athermal work hardening–thermal recovery approach.

12.6.2 Power Law Breakdown

At higher stresses, Power Law Breakdown (PLB) is observed: with increasing stress (and decreasing temperature) the stress exponent continually increases. Increasing contributions from dislocation pipe diffusion with decreasing temperature can account for increase in stress exponent by only 2 from the n_{PL} value. There is no unanimity about the deformation mechanism for creep in the power law breakdown (PLB) regime in single phase materials; see Ray et al. (1992a) for a summary review, and for references. The criterion for the transition from PL to PLB behaviour has also attracted attention. If the PLB is considered only as significant deviation from PL behaviour, then for a number of FCC metals, the modulus-reduced stress for the transition varies over a small range, $(\sigma_{\mathrm{s}}/\mu)_{\mathrm{PLB}} \approx 5x10^{-4}$. If instead onset of PLB is linked to a critical value of $\dot{\varepsilon}_{\mathrm{s}}/D_{\mathrm{eff}}$ (D_{eff} is the effective diffusivity for lattice + dislocation pipe diffusion), then for the same metals, the $(\sigma_s/\mu)_{\mathrm{PLB}}$ computed showed a variation of about 3 orders.

Climb-controlled PLB creep

An empirical equation, which has proved very popular for correlating steady-state data over extensive ranges covering both PL + PLB regimes, is given by the following equation:

$$\dot{\varepsilon}_s = A_g \cdot \left[\sinh\left(\frac{\alpha_g \sigma_s}{\mu}\right)\right]^n \cdot \exp\left(-\frac{Q_c}{RT}\right) \tag{12.27}$$

A_g, α_g and n are constants, and Q_c is the activation energy as before. Equation 12.27 may be compared with Eqs. 12.25a–12.25b. For Eq. 12.27, PL behaviour is reclaimed at low stresses ($\alpha_g(\sigma_s/\mu) < 1$) with $n = n_{PL}$ and $Q_c = Q_{SD}$. In effect, Eq. 12.27 with these features does not specify the mechanism, but considers that:

- Creep in PLB regime is climb controlled as in PL regime.
- The pre-exponential factor is independent of temperature, and accounts for the observed variation in the Norton stress exponent in the PLB regime.

A practical limitation of this empirical formulation is: a single parameter α_g determines both the stress for PL to PLB transition, and the curvature of the strain rate–stress plot for the PLB regime. This and similar equations have been used also for correlating PLB and also hot deformation data from metal-forming operations.

PLB creep and back stress

In complex engineering alloys strengthened by solutes (beyond a few per cents of concentration) and precipitates, PLB behaviour can be expected to result from modification of PL creep by the athermal back stress σ_b from these sources (cf. Example 12.2 and Sect. 12.5.2). Substitution of $\sigma - \sigma_b$ for σ in the expression for PL creep (Eq. 12.25b) and combining with Eq. 12.25a leads to the equation

$$Z \equiv \left[\left(\frac{\dot{\varepsilon}_s k_B T}{\mu b^3}\right) \cdot \exp\left(\frac{Q_{SD}}{k_B T}\right)\right]^{1/n_{PL}} = A_Z \cdot \left(\frac{\sigma_s}{\mu} - \frac{\sigma_b}{\mu}\right) \tag{12.28}$$

with A_Z is independent of stress. Z is the Zener–Hollomon parameter. The corresponding construction is shown schematically shown in Fig. 12.12. In this diagram, the parallel lines marked T_1, T_2, T_3 are for the corresponding temperatures, $T_1 < T_2 < T_3$; the line C, with a small positive offset on the abscissa is common to all temperatures. For creep at T_1, at relatively high stresses, the precipitates are relatively stable, and contribute to a constant back stress, say $(\sigma_b/\mu)_1$. From Eq. 12.28, the corresponding variation of Z with (σ_s/μ) is linear with an offset on the abscissa equal to $(\sigma_b/\mu)_1$. For a higher temperature T_2, the $Z - (\sigma_s/\mu)$ data for the high stress regime result in a parallel line, although with a lower offset on the abscissa $(\sigma_b/\mu)_2 < (\sigma_b/\mu)_1$, consistent with reduced resistance to deformation because of larger precipitate size at a higher test temperature. The high stress behaviour at different temperatures thus differs in the offsets on the abscissa.

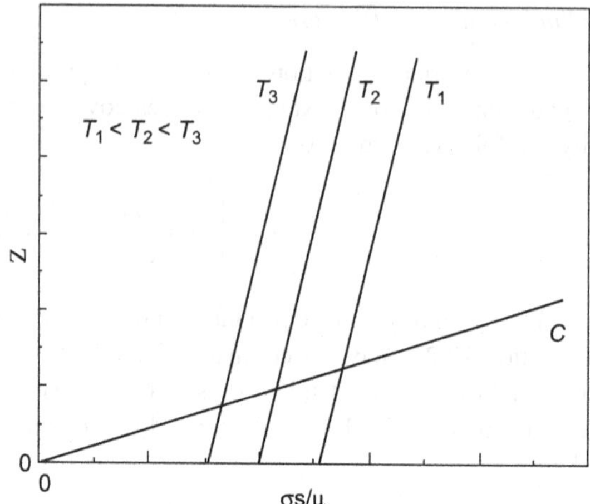

Fig. 12.12 Schematic illustration of combining power law creep with athermal back stress effects. The lines marked T_1, T_2, T_3 are for the corresponding temperatures, and the line C is common to all temperatures

However, irrespective of temperature, for sufficiently low stresses, the time to reach steady state is sufficiently high for extensive precipitate coarsening to occur. The back stress then is ~ 1 MPa (which can be ignored for technological applications), corresponding to climb of dislocations at large particles. The line C in Fig. 12.12 shows this schematically. Thus for any constant creep temperature T_1, steady-state data covering a wide range of stress would show an essentially bi-linear behaviour, falling on the lines marked C and T_1, with a smooth curvilinear transition in-between. Such bi-linear behaviour with curvilinear transition in-between has been found for many engineering alloys; for such analyses, $n_{\mathrm{PL}} = 3$ (or 4 for FCC metals) and known values of Q_{SD} are used in Eq. 12.28. More complex variations have been reported for some complex alloys.

Back stress-modulated PLB creep with activation energy equal to that for climb has been considered for pure metals also. Here, σ_{b} in Eq. 12.28 is considered to arise from inhomogeneity of dislocation configuration, and therefore σ_{b} varies with σ_{s} (see Ray et al. 1993; Bendarsky et al. 1985; Poirier 1985 for details).

Stress-dependent activation energy for the PLB regime (Mecking et al. 1986)

A possible formulation for the PLB regime with stress-dependent activation energy emerges form the analysis of tensile steady-state deformation data for Al, Ni, Cu and Ag single crystals in $\langle 111 \rangle$ orientation (Mecking et al. 1986). These materials cover a wide range of stacking fault widths: $\Gamma/\mu b$ value ranges form 0.23 for Al to 0.0018 for Ag. The steady-state deformation data were determined from extrapolation of work hardening behaviour (see Fig. 12.3, and also Sects. 12.7.2, and Appendix). By graphical analyses, these data could be correlated with the following functional relation:

$$\ln \dot{\varepsilon}_{\mathrm{s}} = A_{\mathrm{M}} + n \cdot \ln(\sigma_{\mathrm{s}}/\mu) - \frac{\mu b^3}{\mathrm{R}T} \cdot g \tag{12.29}$$

(A_{M}, n constants). Two alternative methods of analysis were considered (Mecking et al. 1986), as described in the following.

- When n was assumed to be independent of material (i.e. "universal"), then the derived g-$\ln(\sigma_{\mathrm{s}}/\mu)$ plots for these materials could be brought into coincidence to a "master plot" by scaling both g and σ_{s}/μ with material dependent parameters t^* and s^* respectively. n thus determined was 3.7, close to the universal value of 4 for PL creep for FCC materials. But the computed A_{M} varied slightly with material. Mecking et al. (Mecking et al. 1986) argued that both t^* and s^* depend only upon $\Gamma/\mu b$. In this interpretation, the results for Cu, Ni and Ag were not consistent with onset of climb-controlled creep for T/T_{M} up to 0.75, much above the value of $\sim$0.5 that is normally assumed.
- The constants in Eq. 12.29 could also be adjusted so that the results would be consistent with onset of climb-controlled creep at T/T_{M} lower than 0.75. However, in this case: (i) the superimposition of scaled stress–activation energy profile over the entire data range had to be abandoned; (ii) the computed n varied with $\Gamma/\mu b$, form 3.7 for Al to 4.6 for Ag; but (iii) yielded a constant A_{M} for all materials.

In any event, a change in mechanism to diffusion control below about $2T_{\mathrm{M}}/3$ was not compatible with the data for either method of analysis.

It has been shown (Ray et al. 1992a) that the master plot for stress-activation energy profile obtained from the first method of analysis could be described by the following empirical relation

$$g = g^0 \cdot \left[1 - \left(\frac{\tau_{\mathrm{s}}/\mu}{h}\right)^p\right], \quad p = 0.936921. \tag{12.30}$$

In Eq. 12.30, g^0 and the scaled obstacle strength h are material constants (cf. Eq. 12.17), which depend (via t^* and s^* respectively) on $\Gamma/\mu b$. For Eq. 12.30:

- $g \to 0$ for $(\tau_{\mathrm{s}}/\mu) \to \left(\hat{\tau}_{\mathrm{s}}/\mu\right)$.
- Cottrell–Stokes law obtains at the high stress limits, consistent with the experimental results for FCC metals.
- g asymptotically approaches g^0 as $\ln(\tau_{\mathrm{s}}/\mu) \to -\infty$.
- If onset of PL regime with decreasing stress is considered to correspond to g attaining a nearly constant value,[3] the stresses for $g = 0.94g^0$ (only 6% decrease

[3]Another criterion for the transition from PLB to PL regime is the condition for $(\partial \ln \dot{\gamma}_{\mathrm{s}}/\partial \ln \tau_{\mathrm{s}})_T = (\partial \ln \dot{\varepsilon}_{\mathrm{s}}/\partial \ln \sigma_{\mathrm{s}})_T \approx 4$, the "universal" value for the stress exponent for PL creep in FCC materials.

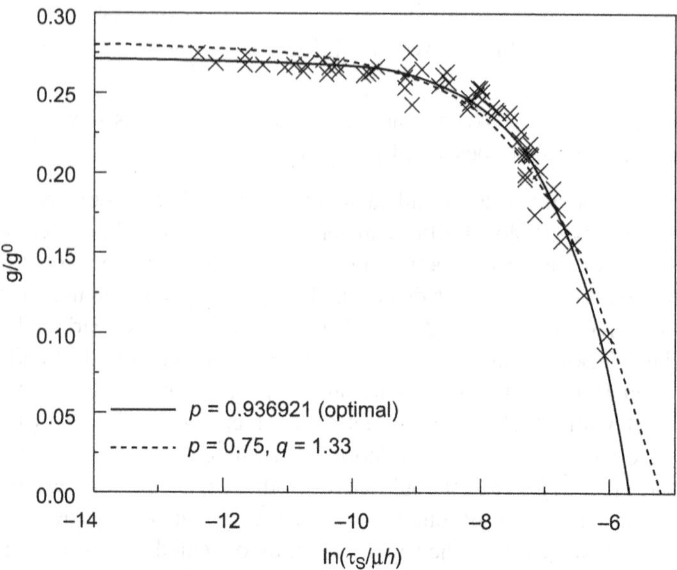

Fig. 12.13 Fit of data from "master plot" of Mecking et al. (1986) to Eq. 12.17 with the typically assumed values of $p = 0.75$ and $q = 1.33$, and Eq. 12.30 (for optimal fit) (Ray et al. 1992a)

from g^0) agree reasonably well with the $(\sigma_s/\mu)_{PLB} \approx 5 \times 10^{-4}$, indicated earlier. In this interpretation however, this stress would systematically depend upon $\Gamma/\mu b$, being, e.g. 2.3 times for Ag than for Al.

Figure 12.13 shows the corresponding plot for Al, and also superiority of Eq. 12.30 over the alternative empirical form given by Eq. 12.17 which has been used to fit low-temperature deformation data. Therefore at least for single phase FCC materials, Eq. 12.29 with Eq. 12.30 can be considered to describe the PLB behaviour, in terms of an unspecified mechanism with stress-dependent activation energy. These results point to the need of examining the possibility of PLB regime with stress-dependent activation energy in a wide range of materials.

As discussed in the Appendix to this chapter, it is necessary to distinguish between the extrapolated steady-state stress designated σ_v (for Voce stress) in the Appendix, and the steady-state stress σ_s where θ actually drops to zero. σ_s and σ_v attain similar values for $T/T_M \sim 0.5$ and higher, but $\sigma_s - \sigma_v$ increases as T/T_M is lowered because of stage IV of work hardening. This difference of σ_s and σ_v is ignored in the preceding analysis, because the conclusions drawn here pertain to relatively high T/T_M levels. However, the scaled obstacle strength h determined above Eq. 12.30 pertains to τ_v rather than to τ_s.

12.7 Kinetics of Evolution of Deformation Substructure

Kinetics of evolution of deformation substructure, specifically its slowest transient, has been extensively characterized in terms of thermomechanical parameters in a wide variety of materials. For uniaxial non-cyclic deformation, this involves work hardening and creep studies. Ideally, work hardening analysis concerns evolution of σ with ε for the homogeneous deformation regime in an isothermal constant $\dot{\varepsilon}$ tension/compression test. The maximum strain levels that can be studied in these tests are limited by the onset of necking in tension tests and onset of barrelling in compression tests. Isothermal torsion tests with constant applied shear strain rate have also been employed for studying work hardening behaviour. These tests allow considerably larger strains without any relevant changes of sample dimensions, without necking, barrelling or friction effects, and without creating new surface area (though the increase of interfaces is as substantial as in any other testing mode). Also experimentally, it is easier to ensure that the applied shear strain rate in a torsion test remains constant, than to maintain constant strain rate in a tension/compression test. For constant σ isothermal creep tests, primary and secondary regimes are of interest. Usually creep tests are carried out at intermediate and high T/T_{M} levels and low stresses, corresponding to low strain rates, while tests for work hardening analysis are carried out at low and intermediate T/T_{M} levels but covering higher stress and strain rates. Therefore the regimes of deformation covered by these two types of tests are generally complementary.

Many empirical work hardening and primary creep equations have been proposed in the literature. These equations are not catalogued in this chapter. Even though some of these equations may not be consistent with the state variable approach, these may prove useful in applications where physical rationalization and extrapolative ability are not necessary. One such application is isothermal FE analysis. A viscoplastic formulation, discussed in the following Sect. 12.7.1, illustrates the usefulness of such an empirical formulation.

12.7.1 Formulation for Viscoplastic Deformation

A common engineering practice for isothermal viscoplastic deformation is to express the ε- and $\dot{\varepsilon}$- dependence of σ in a variable separable form:

$$\sigma = f_1(\varepsilon) \cdot f_2(\dot{\varepsilon})$$

The function f_2 may be designed so that $f_2 \to 1$ in the limit of (relatively) rate-insensitive behaviour. Such formulations have proved quite useful and popular in FE deformation modelling of complex structures. Consider a simple formulation for isothermal deformation:

$$\sigma = A_1 \cdot (\varepsilon)^{p_1} \cdot (\dot{\varepsilon})^{p_2} \tag{12.31}$$

with A_1, p_1, p_2 constants; for rate-insensitive deformation $p_2 = 0$. Equation 12.31 combines two dependencies in a multiplicative form:

(a) Hollomon equation for tensile work hardening: $\sigma \propto (\varepsilon)^{p_1}$ for constant $\dot{\varepsilon}$. This equation is an empirical generalization of Taylor work hardening model (Eq. 12.3b) where $p_1 = 0.5$. Hollomon equation has proved quite effective in correlating uniaxial deformation data for a variety of materials over large ranges of temperature and strain rate. Hollomon equation is not consistent with the state variable approach.
(b) Empirical power law dependence of σ on $\dot{\varepsilon}$ for constant ε (recall the Cottrell–Stokes law for constant deformation substructure, Example 12.4).

Example 12.5 ***Condition for Superplastic Flow***: The condition for superplastic flow can be understood by considering the condition for onset of plastic instability during tensile deformation (see Deiter 1988 and references cited therein).

Consider tensile deformation of a rod with length L and cross-sectional area A. During deformation, A continually decreases. When an incipient neck forms at a geometrical or microstructural inhomogeneity, it tends to shrink faster than the rest of the cross section of the specimen because of enhanced local stress on the section, and this eventually limits the ductility of the material. Necking can be delayed (i.e. ductility can be enhanced) by ensuring that incipient neck is not stable; i.e. it will not shrink faster than the rest of the specimen. That is, the condition for stable deformation is: $\mathrm{d}\dot{\mathsf{A}}/dA \geq 0$.

Since volume remains constant in plastic deformation, from the definition of strain, $\mathrm{d}\varepsilon \equiv \mathrm{dL/L} = -\mathrm{dA/A}$, ignoring elastic strain which is negligible compared to the plastic strain levels of interest in this problem. Therefore,

$$\mathrm{d}\varepsilon/\mathrm{dL} = -(1/\mathrm{A}) \cdot (\mathrm{dA/dL})$$

$$\mathrm{d}\dot{\varepsilon}/\mathrm{dL} = -(1/\mathrm{A}) \cdot \left(\mathrm{d}\dot{\mathrm{A}}/\mathrm{dL}\right) + \left(\dot{\mathrm{A}}/\mathrm{A}^2\right) \cdot (\mathrm{dA/dL})$$

If P is the instantaneous load, then $\sigma \equiv P/\mathsf{A}$, and since P does not vary with L,

$$\begin{aligned}\mathrm{d}P/\mathrm{dL} &= \sigma \cdot (\mathrm{dA/dL}) + \mathrm{A} \cdot (\mathrm{d}\sigma/\mathrm{dL}) \\ &= \sigma \cdot (\mathrm{dA/dL}) + \mathrm{A} \cdot \{(\partial\sigma/\partial\varepsilon) \cdot (\mathrm{d}\varepsilon/\mathrm{dL}) + (\partial\sigma/\partial\dot{\varepsilon}) \cdot (\mathrm{d}\dot{\varepsilon}/\mathrm{dL})\} \\ &= 0\end{aligned}$$

Substitution and rearrangement gives

$$\frac{\left(1/\dot{\mathrm{A}}\right) \cdot \left(\mathrm{d}\dot{\mathrm{A}}/\mathrm{dL}\right)}{(1/\mathrm{A}) \cdot (\mathrm{dA/dL})} = \frac{\left(\mathrm{d}\ln \dot{\mathrm{A}}/dL\right)}{(\mathrm{d}\ln \mathrm{A}/\mathrm{dL})} = \frac{(\partial \ln \sigma/\partial \ln \dot{\varepsilon}) + (1/\sigma) \cdot (\partial\sigma/\partial\varepsilon) - 1}{(\partial \ln \sigma/\partial \ln \dot{\varepsilon})}$$

The necessary condition to ensure that $\mathrm{d}\dot{\mathrm{A}}/\mathrm{dA} \geq 0$,

$$(\partial \ln \sigma/\partial \ln \dot{\varepsilon}) + \theta/\sigma \geq 1 \tag{12.32}$$

In a tension test, as deformation progresses, the work hardening rate θ gradually decreases and σ gradually increases. If rate sensitivity $\partial \ln \sigma/\partial \ln \dot{\varepsilon} \approx 0$ (i.e. relatively low temperatures), a stable neck forms when $\theta = \sigma$; this is the Considère criterion. This is the case where necking is opposed by strain hardening. At high temperatures on the other hand, work hardening is relatively small, and can be ignored. Under these conditions, very large tensile ductility can be ensured, i.e. neck formation can be suppressed, by ensuring $\partial \ln \sigma/\partial \ln \dot{\varepsilon} \geq 1$. In this case, necking is opposed by strain rate hardening. Deformation characterized by an equation of the form $\sigma \propto \dot{\varepsilon}$ is known as Newtonian viscous deformation; one example is hot glass which can be drawn from melt to fine fibres without the fibres necking down. The systematic increase in tensile ductility with increasing $\partial \ln \sigma/\partial \ln \dot{\varepsilon}$ has been verified in many materials.

12.7.2 Work Hardening at Constant T and $\dot{\varepsilon}$

In interpreting data from tension or compression tests, it is necessary to recognize that these tests are frequently carried out not at constant $\dot{\varepsilon}$, which requires sophisticated instrumentation, but at constant actuator speed. In such a test, $\dot{\varepsilon}$ gradually decreases as deformation progresses. This may be ignored when strain rate sensitivity of flow stress is low, e.g. FCC materials at lower T/T_{M} levels. Otherwise, it may be necessary to assess the effect of this deviation from the ideal condition.

The discussion in this section is in terms of tension/compression tests. The arguments however apply to torsion tests of polycrystalline materials, where the relevant variables are torque, angle of twist, and the twist rate. Quantitative conversion from tensile to torsional work hardening analysis requires using the appropriate values for $\bar{M}$ for these tests. The same observation applies for creep studies in tension and torsion dealt with in the next section.

From Eq. 12.4b, in general the work hardening relation should involve only state variables of deformation, and also should be consistent with a steady state. For uninterrupted uniaxial tensile/compressive deformation at constant T and $\dot{\varepsilon}$, a work hardening relation $\sigma = f(\varepsilon)$ (for the slowest transient of the substructure) implies that for a given stress–strain curve, a single parameter description should be adequate, even though ε is not a state variable of deformation (recall Hart's hardness parameter σ_{H}, Sect. 12.5.3). From Fig. 12.3 and Eq. 12.6, it follows that for a work hardening relation consistent with the state variable approach:

- θ at constant $\dot{\varepsilon}$ and T can be expressed as a function of σ alone.
- $\theta = 0$ as $\sigma = \sigma_s$, the steady-state (saturation) stress. To be consistent with steady-state creep, σ_s must be finite.

The first criterion is met by any work hardening relation of the functional form $\sigma = f(\varepsilon)$, for example, the Hollomon work hardening relation (Eq. 12.31 at constant $\dot{\varepsilon}$). For Hollomon relation, however, $\theta \to 0$ only for $\sigma \to \infty$, as $n_1 < 1$; this is not consistent with the second criterion above.

The following generic empirical work hardening relation (Armstrong et al. 1982) meets both the criteria indicated above:

$$\sigma = \sigma_s - A_s(\varepsilon) \cdot (\sigma_s - \sigma_0) \tag{12.33}$$

In this equation $A_s(\varepsilon)$ is a function of strain, and σ_0 is the (back-extrapolated) flow stress for $\varepsilon = 0$ corresponding to the slowest transient (see Fig. 12.1). $A_s = 1$ for $\varepsilon = 0$, $\sigma = \sigma_0$, and $A_s = 0$ for $\sigma = \sigma_s$. An arbitrary number of work hardening relations can be formulated, by choosing the function $A_s(\varepsilon)$ satisfying these boundary conditions. Note that Eq. 12.33 can be applied to both single and multiphase materials. Let $\varepsilon_{\theta=0}$ be the strain for the saturation state with $\sigma = \sigma_s$. $A_s(\varepsilon)$ can be chosen depending upon whether the $\varepsilon_{\theta=0}$ is envisaged to finite, or infinitely large (saturation state is approached asymptotically). For compressive deformation of iron, copper and an aluminium 1100 alloy to very large strains at ambient temperature, the following functional form for A_s, with p a positive constant, proved satisfactory (Armstrong et al. 1982):

$$A_s = 1 - (\varepsilon/\varepsilon_{\theta=0})^p$$

For example, for the aluminium alloy $p = 0.48$. Here $\varepsilon_{\theta=0}$ is finite. On the other hand, $\varepsilon_{\theta=0} \to \infty$ for the function

$$A_s = \exp\left[-N\varepsilon^p\right]$$

with p and N positive constants. Setting $p = 1$ in the last equation results in the well-known empirical Voce work hardening equation, with θ decreasing linearly with increasing σ. Voce formulation, and deviations from it at high strains are discussed in Appendix at the end of this chapter.

That a work hardening plot for an uninterrupted tension/compression test can be considered to reflect the evolution of a single structural parameter does not necessarily mean that single parameter description of structure would prove adequate when considering a set of work hardening curves, generated with different strain rate and temperature levels. Samuel et al. (2006) successfully applied the Hollomon equation to correlate the tensile work hardening data, generated at various strain rate and temperature levels, for a number of engineering materials with different initial thermomechanical treatments, nuclear irradiation damage, etc. They concluded that for many cases of practical interest, it should be adequate to use two structure parameters in phenomenological models for isotropic work hardening analysis including variations in initial microstructures.

12.7.3 Primary Creep at Constant T and σ

Creep studies are frequently carried out in the "non-ideal" condition of constant load, i.e. constant nominal stress. Then in tensile loading, true stress continually increases during creep because of decreasing specimen cross-sectional area. Consequently, $\dot{\varepsilon}_s$ also increases during creep. Even with materials of modest creep ductility ($\sim$5 to 10%) this can result in considerable distortion of the creep curves compared to the ideal condition of constant stress. The distortion is accentuated when primary creep strains are significant, i.e. at high stresses. At low stresses, the distortion is considerably smaller. This factor must be kept in view when interpreting data from constant load creep tests.

Many empirical equations for primary creep have been proposed in the literature. Such equations, of the functional form $\varepsilon(t)$ or $\dot{\varepsilon}(\varepsilon)$ for constant σ, T, pertain to the slowest transient of deformation substructure. This implies that for a given creep curve (and not necessarily for a set of creep curves generated at different combinations of constant stress and temperature), the evolution of deformation substructure can be effectively described by a single structural parameter. Then, from the state variable approach with single structural parameter, the admissible equations for describing primary (+secondary) creep for a given creep curve must be consistent with the functional form (McCartney and McLean 1976)

$$\mathrm{d}\dot{\varepsilon}/\mathrm{d}t = f(\dot{\varepsilon}, \dot{\varepsilon}_s) \quad \text{with} f = 0 \quad \text{for } \dot{\varepsilon} = \dot{\varepsilon}_s \tag{12.34}$$

Many such equations could be empirically constructed. One equation proposed is (McCartney and McLean 1976):

$$\mathrm{d}\dot{\varepsilon}/\mathrm{d}t = -\beta \cdot \dot{\varepsilon} \cdot (\dot{\varepsilon} - \dot{\varepsilon}_s) \tag{12.35}$$

β is a constant for constant σ, T.

Another creep formulation proposed by Amin et al. (1970) based on first-order kinetics that meets the criterion of Eq. 12.34 is:

$$\frac{\mathrm{d}}{\mathrm{d}t}(\dot{\varepsilon} - \dot{\varepsilon}_s) = -\beta \cdot \dot{\varepsilon}_s \cdot (\dot{\varepsilon} - \dot{\varepsilon}_s) \tag{12.36a}$$

The parameter β here is independent of stress and strain. These authors ascribed primary creep to climb-controlled dispersal of dislocations; justification for the assumed form Eq. 12.36a was provided a posteriori in terms of experimental verification. Integrating Eq. 12.36a with boundary conditions $\dot{\varepsilon}_0 = \dot{\varepsilon}(t=0)$, $\varepsilon(t=0)=0$,

$$\varepsilon = \dot{\varepsilon}_s t + \frac{(\dot{\varepsilon}_0 - \dot{\varepsilon}_s)}{\beta \dot{\varepsilon}_s} \cdot [1 - \exp(-\beta \dot{\varepsilon}_s t)] \tag{12.36b}$$

Consider the well-known empirical McVetty–Garofalo equation for primary + secondary creep expressed by the following equation:

$$\varepsilon = \dot{\varepsilon}_s t + \varepsilon_p \cdot [1 - \exp(-t/t_R)] \tag{12.36c}$$

where ε_p is the extent of primary creep and t_R, a characteristic time, that defines the rate constant for primary creep. It can be seen that Eq. 12.36b defines the parameters ε_p and t_R in Eq. 12.36c in terms of the reaction parameters. The proportionality between $\dot{\varepsilon}_s$ and $1/t_R$ (compare Eqs. 12.36b and 12.36c) has been observed for creep in some materials.

12.7.4 Micromechanical Modelling for Work Hardening and Creep

Micromechanical modelling gives insight into the expected work hardening and creep relations, and the dependence of the various parameters in these relations of thermomechanical variables. These in turn prove useful in phenomenological formulation of efficient MEOS. A combination of athermal work hardening and static recovery following Friedel's network coarsening model (Eq. 12.26e, Example 12.4) leads to the following work hardening equation:

$$\theta/\mu = \theta_0/\mu - A_f \cdot \frac{D_v}{b^2} \cdot \left(\frac{\mu b^3}{k_B T}\right) \cdot \left(\frac{\sigma}{\mu}\right)^3 \cdot \frac{1}{\dot{\varepsilon}} \tag{12.37}$$

This can be generalized to the form:

$$\theta/\mu = \theta_0/\mu - R_d(T) \cdot \left(\frac{\sigma}{\mu}\right)^{n_{PL}} \cdot \frac{1}{\dot{\varepsilon}} \tag{12.37a}$$

In this case, the temperature dependent parameter $R_d(T)$ is given by

$$R_d(T) = \dot{\varepsilon}_0 \cdot \exp(-Q_{SD}/RT), \dot{\varepsilon}_0 = \text{constant}$$

Equation 12.37a should apply at relatively high T/T_M levels, where static recovery dominates. For lower and intermediate T/T_M levels where dynamic recovery dominates, Eq. 12.37a proves inadequate as expected. For single phase FCC materials, an alternative formulation has been proposed in the literature:

$$\theta/\mu = \theta_0/\mu - R_d(T) \cdot \left(\frac{\sigma}{\mu}\right) \cdot \frac{1}{\dot{\varepsilon}^{1/n(T)}} \tag{12.37b}$$

In Eq. 12.37b, as explicitly shown, $R_d(T)$ and $n(T)$ depend upon temperature. The linear dependence of θ on σ in Eq. 12.37b is consistent with Voce formulation (Sect. 12.7.2).

For a homogeneous dislocation network, where dislocation tangles provide the obstacles to dislocation motion, Estrin and Mecking (1984) used the following kinetic model for work hardening including dynamic recovery (a_1, a_2 parameters):

$$\dot{\rho} = \dot{\varepsilon} \cdot \left(a_1 \cdot \sqrt{\rho} - a_2 \cdot \rho\right) \tag{12.38}$$

The first term corresponds to athermal work hardening due to dislocation storage, with mean free path of dislocations inversely proportional to $\sqrt{\rho}$ corresponding to similitude (see Eq. 12.3d). The second term corresponds to loss of dislocations at the network because of dynamic recovery. Estrin and Mecking showed that with Eq. 12.38, along with Eq. 12.1a for the relation between σ and ρ (incorporating Taylor factor), at constant $\dot{\varepsilon}$ and T the relation between θ and σ is linear, as in the Voce formulation. Estrin and Mecking also showed that for a homogeneous distribution of dislocations where transgranular precipitates provide obstacles to dislocation motion, a plot of $\theta\sigma$ (ordinate) against σ^2 (abscissa) from a constant $\dot{\varepsilon}, T$ test would be linear with ordinate intercept proportional to d^{-2}, where d is the mean inter-particle spacing. It is assumed that d remains constant during deformation. This model has been successfully used in correlating work hardening behaviour in tension and low cycle fatigue in precipitation-strengthened systems.

Both the above models consider homogeneous dislocation distribution, characterized by a single substructural parameter ρ. Many micromechanical models for work hardening and creep consider inhomogeneity of dislocation distribution. For example, the model of Ostrom and Lagneborg (1980) for single phase materials considers a statistical distribution of dislocation segment lengths. Under an applied stress σ, dislocations with link lengths larger than a length $\bar{l} \propto \mu b/\sigma$ become mobile, and expand until arrested by the dislocation network in the proximity. The model includes considerations on the mechanism for overcoming the local obstacles, the interaction of mobile and immobile dislocations, and the thermal recovery of network dislocations. Both work hardening and creep equations have been derived for many of these models, including those of Estrin and Mecking (1984) and Ostrom and Lagneborg (1980). Models have also been proposed for other microstructures. These models are not discussed in this chapter.

12.8 Dynamic Strain Ageing (DSA)

Static and dynamic strain ageing are important features of deformation in most structural alloys. The origin of strain ageing is well understood. Solute atoms can reduce the energy by locating themselves around edge dislocations, the larger solutes going to regions of tensile stress, and smaller solutes to regions of compressive stress. Interstitial solute–vacancy pairs, being non-hydrostatic defects, can interact with screw dislocations. Thus, solute atoms tend to segregate to the dislocation core creating an atmosphere, called the Cottrell atmosphere. At low temperatures, mobility of solute atoms is small. Therefore, once the dislocation breaks away form the cloud, it can move at much lower stresses. This is reflected in static strain ageing and yield point phenomena. The solute cloud condensed on the dislocation line constitute a linear barrier: it has been shown that only a short length of dislocation line needs to escape from the atmosphere for a catastrophic break away to set in. At sufficiently high temperatures on the other hand, solutes are too mobile, and as a result, even though solute cloud is large in size, strengthening due to solutes becomes small. Dynamic strain ageing (DSA) is linked to relative mobilities of the solute atmosphere and dislocations: the drag offered by the solutes is maximum when solute mobility and dislocation speed match. DSA effect is therefore strongest at some intermediate temperature for constant strain rate tests. Because of DSA, strain rate ceases to be a single valued function; this may lead to deformation instabilities. This can result in serrated tensile flow curves. Several other parameters, e.g. strength, ductility, work hardening parameters, fracture toughness, etc., show anomalous variations in the DSA regime. DSA has been extensively studied in the literature. Rodriguez (1984) presented a systematic exposition of the various manifestations of DSA.

12.8.1 Characterizing DSA

Onset of serrations in tensile flow curves at constant $\dot{\varepsilon}, T$ is found to be associated with a critical strain ε_c; in such cases, the $\dot{\varepsilon}$- and T-dependence of ε_c proves useful in elucidating the DSA mechanism. Absence of such serrations, however, does not mean that DSA is absent. Figure 12.14 schematically shows the transients in flow stress, on affecting $\dot{\varepsilon}$ jumps in the DSA regime when the tensile flow curve is free from serrations. The various parameters used to characterize the transients are also indicated in this figure.

Ray et al. (1992b) cited the relevant literature and summarized the method for quantitative interpretation of these transients. Ageing occurs during the period t_w the mobile dislocations are held up at obstacles awaiting thermal activation (see Sect. 12.5.1). Let t_{w_1} and t_{w_2} be the waiting times for strain rates $\dot{\varepsilon}_1$ and $\dot{\varepsilon}_2$ respectively. On affecting a jump in $\dot{\varepsilon}$ from $\dot{\varepsilon}_1$ to $\dot{\varepsilon}_2$, the quasi-steady-state ageing

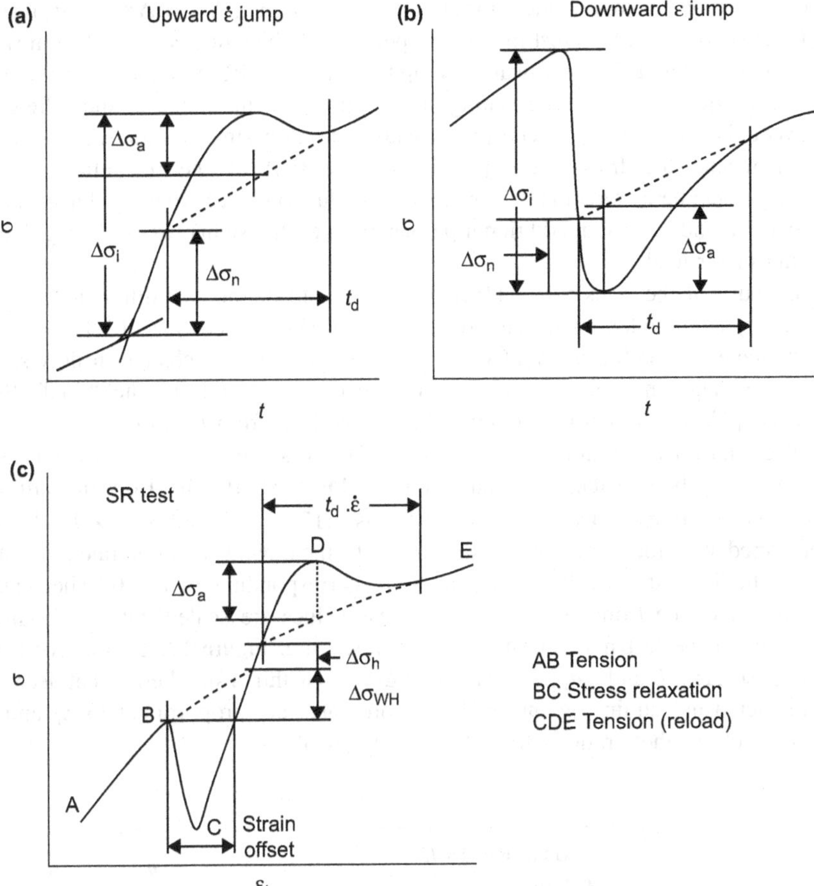

Fig. 12.14 Schematic diagram (not to scale) illustrating graphical analysis of yield transients from **a** upward and, **b** downward $\dot{\varepsilon}$-jump tests, and **c** tensile reloading after stress relaxation (called SR tests here), which involves an upward $\dot{\varepsilon}$ jump from $\dot{\varepsilon}$ level very close 0. Solid lines indicate test data and dashed lines the forward and backward extrapolations. Constructions to determine the various parameters for analysis are also shown. (Ray et al. 1992b)

time t_a changes from t_{w_1} to t_{w_2}. Assuming a first-order kinetics for the evolution of t_a,

$$t_a = t_{w_2} - \left(t_{w_2} - t_{w_1}\right) \cdot \exp(-t/t_R) \tag{12.39}$$

The characteristic time constant for the process $t_R \sim t_{w_2}$; also the duration of the transient t_d should be proportional to t_{w_2}. The duration of the yield transient t_d is taken as the time duration for the segment over which back-extrapolation is made. The "strength" of ageing $\Delta\sigma_a$ (see Sect. 12.8.2) may be obtained as the stress difference between the peak experimental stress, and the corresponding back-extrapolated stress at the same strain/time. Assuming $\Delta\sigma_i \approx \Delta\sigma_n + \Delta\sigma_a$ results in (slight) under- and overestimations respectively (which decreases with strain) for the (a) upward and, (b) downward $\dot{\varepsilon}$-jump tests. For the SR tests, the stress increase as a consequence of the intervening stress relaxation, measured as the difference between the back- extrapolated (quasi-steady state) flow stress for the second tensile transient and the flow stress just before the start of the relaxation, equals $\Delta\sigma_{WH} + \Delta\sigma_h$; here, $\Delta\sigma_{WH}$ is the stress increase due to work hardening during stress relaxation, and $\Delta\sigma_h$ is an additional permanent (non-transient) strengthening due to ageing under load.

Because of the transient yield, it is necessary to define two different (thermodynamic) strain rate sensitivities of flow stress (SRS) (see Sect. 12.5.2): (i) the "instantaneous" SRS $S_i \equiv \Delta\sigma_i/\Delta \ln \dot{\varepsilon}$, which corresponds to change in flow stress due to change in $\dot{\varepsilon}$ alone, with t_a unchanged at t_{w_1}; and (ii) the "final" SRS $S_f \equiv \Delta\sigma_n/\Delta \ln \dot{\varepsilon}$, which incorporates the change in t_a from t_{w_1} to t_{w_2}.

The data for a solution annealed type AISI 316 stainless steel at ambient temperature may be considered for an example (Ray et al. 1992b). Both upward and downward $\dot{\varepsilon}$ jumps between two fixed levels 3.17×10^{-5} and 3.17×10^{-4} were performed at various strain levels. In some tests, the downward strain rate jump step was actually a stress relaxation (SR) step, corresponding to $\dot{\varepsilon} = 0$. The tensile curves were free from serrations. The stress values are scaled with E, Young's modulus of the test material at the test temperature. Figure 12.15 plots for the $\dot{\varepsilon}$ jump tests $\Delta\sigma_n/E$ and $\Delta\sigma_i/E \approx \Delta\sigma_a/E + \Delta\sigma_n/E$ (within a small error that decrease with increasing strain) against σ/E; the ordinates are proportional to S_f and S_i respectively as the $\dot{\varepsilon}$ ratio is fixed for both types of tests.

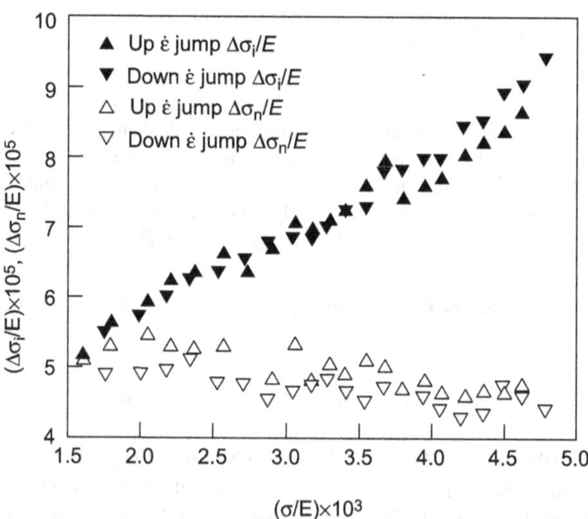

Fig. 12.15 Plots of $\Delta\sigma_n/E$ and $\Delta\sigma_i/E(\approx \Delta\sigma_a/E + \Delta\sigma_n/E)$ against σ/E for the $\dot{\varepsilon}$ jump tests (Ray et al. 1992b)

For analyzing these transients, it is convenient to consider only two contributions to the total stress σ , namely σ_d from dislocation interactions and σ_{sol} because of solute strengthening (see Sect. 12.5.2). The general form is then (see Eqs. 12.21b–12.21c):

$$S = \frac{\sigma_d}{m_d} + \frac{\sigma_{sol}}{m_{sol}} = \frac{\sigma}{m_d} + \left(\frac{1}{m_{sol}} - \frac{1}{m_d}\right) \cdot \sigma_{sol}$$

The slope of $S_i - \sigma$ plot equals $1/m_d$ for the base strain rate free from the dynamics of ageing effect. The $\Delta\sigma_i/E - \Delta\sigma/E$ data in Fig. 12.15 are therefore consistent with Cottrell–Stokes law; the slight curvature upwards at high stress levels, attributable to dynamic recovery. From linear fit of these data, $m_d \approx 205$. This is realistic for an FCC material at low T/T_M values. As strength of the DSA effect ($\Delta\sigma_a$) increases, so does $S_i - S_f$. It is recognized that the necessary, but not sufficient, condition for serrated flow is: $S_f \leq 0$. In Fig. 12.15, positive values of effect ($\Delta\sigma_a$) increases, so does $S_i - S_f$. It is recognized that the necessary, but not sufficient, condition for serrated flow is: $S_f \leq 0$. In Fig. 12.15, positive values of S_f explain why the tensile flow curves were free from serrations in this instance. With serrated tensile flow curves, only the parameter S_f can be determined in an average sense.

12.8.2 *Models for DSA*

Ray et al. (1992b) summarized certain basic aspects of the two important trends in modelling for DSA and also cited the relevant references. Ling and McCormick (1990) developed a model for the case when DSA effect is via the σ_{sol} term. Mobile dislocations acquire solutes by bulk diffusion, which is augmented by vacancies produced by deformation. The mobile dislocation density ρ_m and concentration of vacancies C_v are assumed to vary with ε as $\rho_m \propto \varepsilon^{\beta_1}$ and $C_v \propto \varepsilon^{\beta_2}$ respectively, with β_1, β_2 constants. When DSA is caused by interstitial solutes, vacancy concentration has no effect on diffusion and $\beta_2 = 0$. Then, the average elementary incremental strain per successful thermal activation (defined in Eq. 12.11a, Sect. 12.5.1) $\varepsilon_0 \propto \varepsilon^{\beta_1 + \beta_2}$, and

$$\dot{\varepsilon} \propto \varepsilon^{\beta_1 + \beta_2} \cdot \exp(-Q/RT) \tag{12.40a}$$

Here Q is the activation energy for diffusion of the solute species responsible for DSA. This model assumes $t^{2/3}$-kinetics for diffusion, proposed by Cottrell and Bilby for substitutional solutes, with a saturation effect at long ageing times. Therefore, stress increase with ageing time t_a can be expressed by a function of the form:

$$\Delta\sigma_a = \Delta\sigma_i - \Delta\sigma_n = B \cdot (A' t_d)^{2/3} \cdot \exp\left[-(A' t_d)^{2/3}\right] \tag{12.40b}$$

The constant A' in the above equation includes lattice diffusivity, strength of solute–dislocation locking, and the saturation and average concentrations of solute. The constant B is a measure of the maximum strengthening possible. For this model the two parameters that characterize the transient are t_d and $\Delta\sigma_a$. If the strain dependence stipulated above is invoked, then ignoring saturation behaviour,

$$\Delta\sigma_a \propto \varepsilon^{2(\beta_1 + \beta_2)/3} . \tag{12.40c}$$

Geometrical reasoning shows that the ε-dependence of ε_0 may actually be more complex than the power law relation assumed in this model.

Kocks et al. (1985) considered an alternative model where DSA arises via the σ_d term, due to pipe diffusion of solutes to the mobile-forest dislocation junctions, which alters the dislocation activation kinetics. It is argued that $\varepsilon_0 \propto (\sigma_d)^{p'}$, with p' a constant. For this model also, a saturation effect at long ageing times can be envisaged, and for a material obeying Cottrell–Stokes law, the following expression is obtained (Ray et al. 1992b)

$$\Delta\sigma_a/\sigma_d = B \cdot (A' t_d)^k \cdot \exp\left[-(A' t_d)^k\right] \tag{12.41}$$

The constant A' now includes dislocation core diffusivity, strength of solute–dislocation locking, and the saturation and average concentrations of solute. The exponent $k = 2/3$ in the model proposed by Kocks et al. assuming Cottrell–Bilby kinetics, but for pipe diffusion, $k = 1/3$ corresponding to a $t^{1/3}$-kinetics is more appropriate. In this model, the two parameters that characterize the yield transient are t_d and $\Delta\sigma_a/\sigma_d$.

Example 12.6 Samuel et al. (1988) studied serrated yielding in tension for an AISI type 316 stainless steel in solution annealed condition. Isothermal tension tests were carried out at various constant nominal strain rate in the temperature range 300–1123 K. A critical strain ε_c was found to be associated with onset of serrations. From Eq. 12.40a,

$$\varepsilon_c^{\beta_1 + \beta_2} \propto \dot{\varepsilon} \cdot \exp(Q/RT) \tag{12.42}$$

There were two distinct regimes for onset of serrations, one between ~523 to 623 K and the other between ~673 and 923 K. Figure 12.16a plots $\ln \varepsilon_c$ against reciprocal of absolute temperature for $\dot{\varepsilon} = 3 \times 10^{-4}\ \text{s}^{-1}$ and Fig. 12.16b plots $\ln \dot{\varepsilon}$ against $\ln \varepsilon_c$ for two constant temperature levels belonging to the two regimes. Using the value for $\beta_1 + \beta_2$ determined for slopes of the two lines in Fig. 12.16b, Q was determined from Eq. 12.42 to be 138 kJ/mol for ~523 to 623 K and 277 kJ/mol for ~673 to 923 K, as indicated in Fig. 12.16a. The Q values for low-temperature DSA is close to that for lattice diffusion of C or

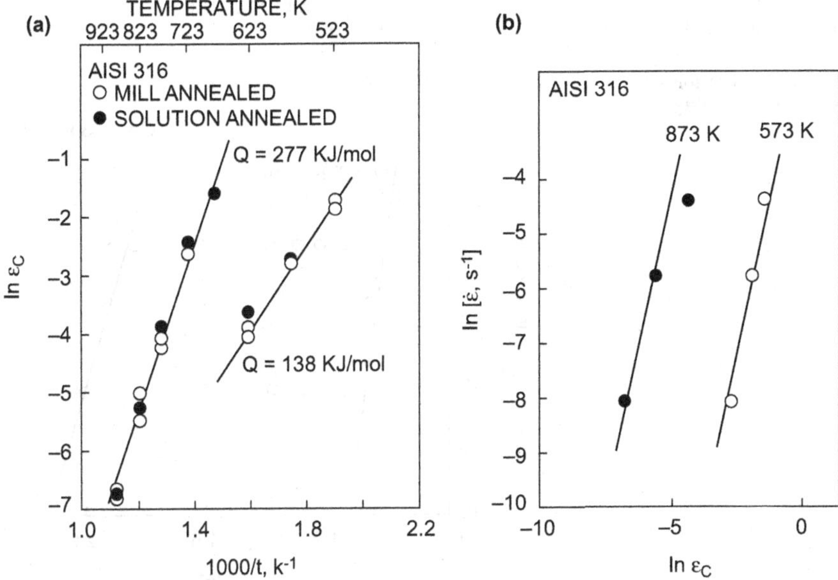

Fig. 12.16 **a** $\ln \varepsilon_c$ plotted against reciprocal temperature shows two distinct temperature regimes for serrated plastic flow; the corresponding activation energies are also indicated. **b** Slopes of the plots of $\ln \dot{\varepsilon}$ against $\ln \varepsilon_c$ at constant temperature levels of 573 and 873 K yield the values of $\beta_1 + \beta_2 \approx 2.3$, for both the temperatures [from Samuel et al. (8)]

N, and therefore the lower temperature DSA was attributed to interstitials. In phenomenological modelling, it may be necessary to consider DSA effects arising from each solute species (a specific example, for AISI type 316 alloy, is cited in Sect. 12.9.4; see Eq. 12.50). From the present example, it appears that a less rigorous but viable option may be to consider only two DSA contributions, one from the substitutional solutes taken together, and another from interstitial solutes taken together.

Figure 12.17 identifies the boundaries between stress–strain plots with and without serrations in a plot of $\ln \dot{\varepsilon}$ versus $1/T$. For the two regimes, Q determined from this plot ignoring variation of critical strain and invoking Arrhenius relation, are also indicated in this figure; the values are very close to the corresponding values indicated in Fig. 12.16a. The Q value for high-temperature regime is very close to that for lattice diffusion of Cr (255 kJ/mol) in this type of alloy. These authors therefore attributed the high-temperature DSA to substitutional solutes, (probably Cr).

Example 12.7 With solution annealed AISI SS 316 at 300 K, Ray et al. (1992b) carried out both upward and downward $\dot{\varepsilon}$-jump tests with base strain rates of 3.17×10^{-5} and 3.17×10^{-4} per s, and a tensile test at nominal strain rate of 3.17×10^{-4} per s interrupted at various strain levels by stress relaxations of

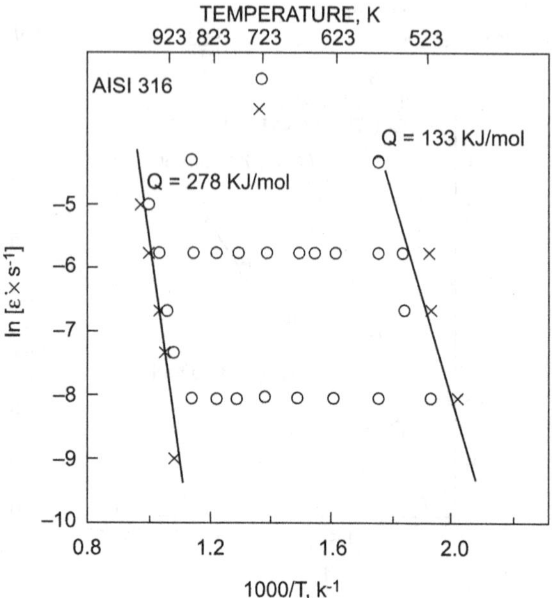

Fig. 12.17 Boundaries between stress and strain plots with and without serrations are identified in this plot (Samuel et al. 1988)

duration 1800 s (designated SR tests here). The tensile reloading after relaxation essentially involves an upward $\dot{\varepsilon}$-jump from a vey low base $\dot{\varepsilon}$ level. The tensile flow curves were free from serrations, but the stress humps following $\dot{\varepsilon}$-jumps indicated the presence of DSA. The S_i and S_f for the $\dot{\varepsilon}$-jump tests were shown in Fig. 12.15.

From the consideration of solute diffusivity, there is little doubt that this DSA in SS 316 at ambient temperature should be attributed to interstitial solutes. The stress values were scaled with Young's modulus at test temperature E. Composite stress–strain plot using the stress–strain data for points where $\dot{\varepsilon}$ jumps are affected (for the $\dot{\varepsilon}$-jump tests) or stress relaxation is commenced showed perceptible non-transient age hardening during stress relaxations ($\sigma_h > 0$). This is attributable probably to an increase in forest dislocation density due to strain ageing during relaxations. Figure 12.18 presents the $t_d\dot{\varepsilon}_2$ data (this should be proportional to ε_0) for the various tests.

As the figure shows, a factor of 10 and more for $\dot{\varepsilon}$-ratio was reduced for $t_d\dot{\varepsilon}_2$ to a factor of ~ 1 at low strains and ~ 1.5 at high strain levels. These results basically establish the underlying process as an ageing phenomenon, and are generally consistent with Eq. 12.39; also the data for high strain for the downward $\dot{\varepsilon}$-jump tests suggest a saturation effect for high ageing times, as postulated in Eqs. 12.40b and 12.41. $\Delta\sigma_a$ values were somewhat higher for the downward $\dot{\varepsilon}$-jump tests than for the higher $\dot{\varepsilon}$-jumps tests, reflecting higher t_w for the former type of test, Fig. 12.19a–b. From Fig. 12.19b, the values of β_1 in Eq. 12.40c (recall that $\beta_2 = 0$ for DSA due to interstitial solutes) was determined tobe 1.59 and 1.1 respectively for upward and downward $\dot{\varepsilon}$-jumps, and 1.68 for the transients following stress relaxations. Now $\beta_1 = 1$ reflects the case when average slip distance of mobile

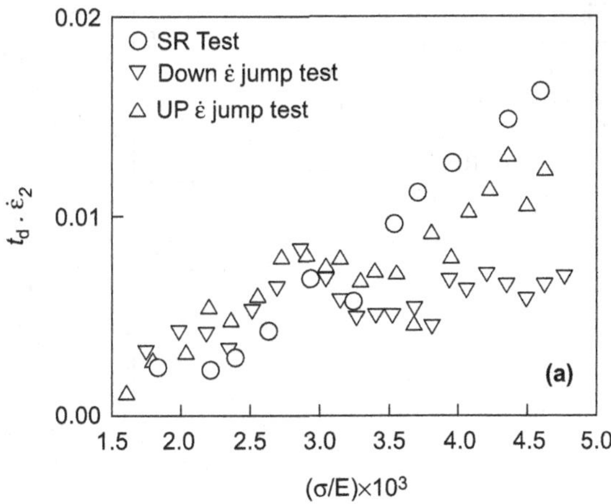

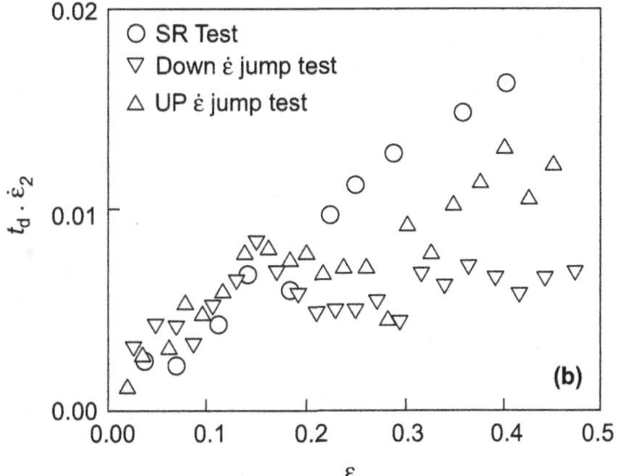

Fig. 12.18 Plots of $t_d\dot{\varepsilon}_2$ against **a** σ/E **b** ε (Ray et al. 1992b)

dislocations is independent of strain; therefore it is doubtful whether values of $\beta_1 > 1$ can be theoretically justified. From Fig. 12.19a–b, $\Delta\sigma_a = 0$ for an average stress $\sigma/E \approx 1.5 \times 10^{-3}$; this stress corresponds to $\varepsilon \approx 0$. For both the models considered in Sect. 12.8.2, strength of DSA equals zero for $\varepsilon = 0$ when $\sigma = \sigma_{sol}$ (Eqs. 12.40c and 12.41 assuming $\sigma_d = 0$ for $\varepsilon = 0$). Therefore, for either of these models, $\sigma_{sol}/E \approx 1.5 \times 10^{-3}$ for the base strain rate. (The same estimate could be obtained from Fig. 12.15 as the stress for $S_i = S_f$.) Using this value for σ_{sol}, an average initial solute concentration on the dislocation line has been estimated as one solute atom every $7b$ length of dislocation line, which may not be unrealistic.

Figure 12.20a, b show, with identical resolutions for the axe for easy visual comparison, the variation with $\ln t_d$ of the (a) $\ln(\Delta\sigma_a/\sigma_d)$ from strain rate jump and SR tests, and also $\ln((\Delta\sigma_a + \Delta\sigma_h)/\sigma_d)$ from SR tests (appropriate for the model of Kocks and coworkers, Eq. 12.41), and (b) $\ln(\Delta\sigma_a/E)$ from strain rate jump and SR tests, also $\ln((\Delta\sigma_a + \Delta\sigma_h)/E)$ from SR tests (appropriate for the model of

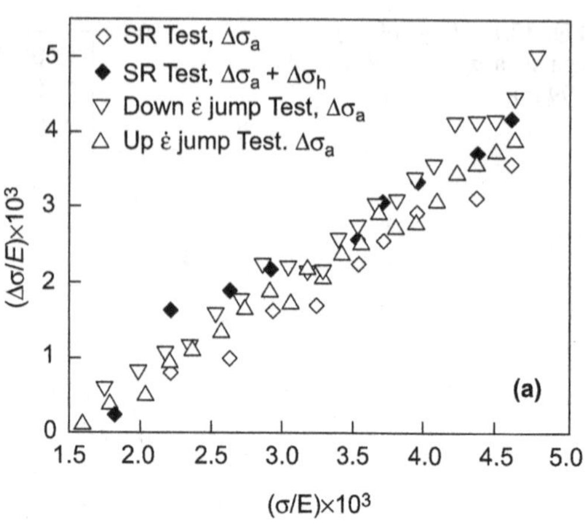

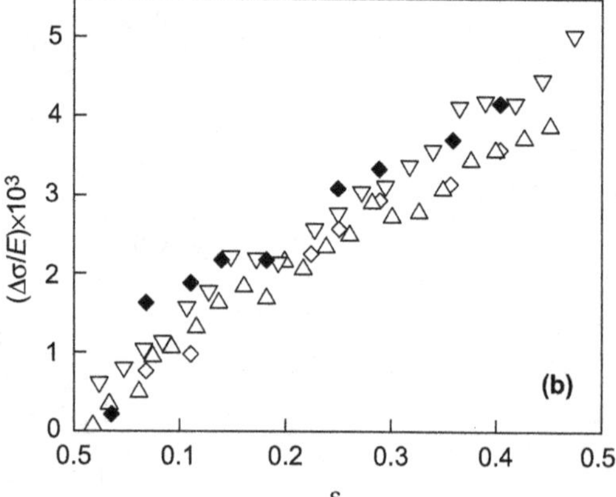

Fig. 12.19 Plots of $\Delta\sigma_a/E$ from the various tests, and also of $(\Delta\sigma_a + \Delta\sigma_h)/E$ from SR tests against **a** σ/E, and **b** ε. Symbols are explained in Fig. 12.19a (Ray et al. 1992b)

McCormick and coworkers, Eq. 12.40b). σ_d for Fig. 12.20a were calculated as $(\sigma - \sigma_{sol})$ using the estimate for σ_{sol} obtained above.

It is clear from these two figures that:

1. Over a large range, irrespective of the scaling used, $\Delta\sigma_a$ component from upward $\dot{\varepsilon}$-jumps and SR test show similar kinetics.
2. Both the plots reflect saturation behaviour for long ageing times as postulated in the two models.
3. In Fig. 12.20b, considerable differences (that can be clearly distinguished from data scatter) are noted between data for high and low t_d values.
4. The model of Kocks et al. (Fig. 12.20a) describe the data best over the entire t_d range.

Least square fitting of data (excluding the SR test $(\Delta\sigma_a + \Delta\sigma_h)/E$ data) in Fig. 12.20b to the equation

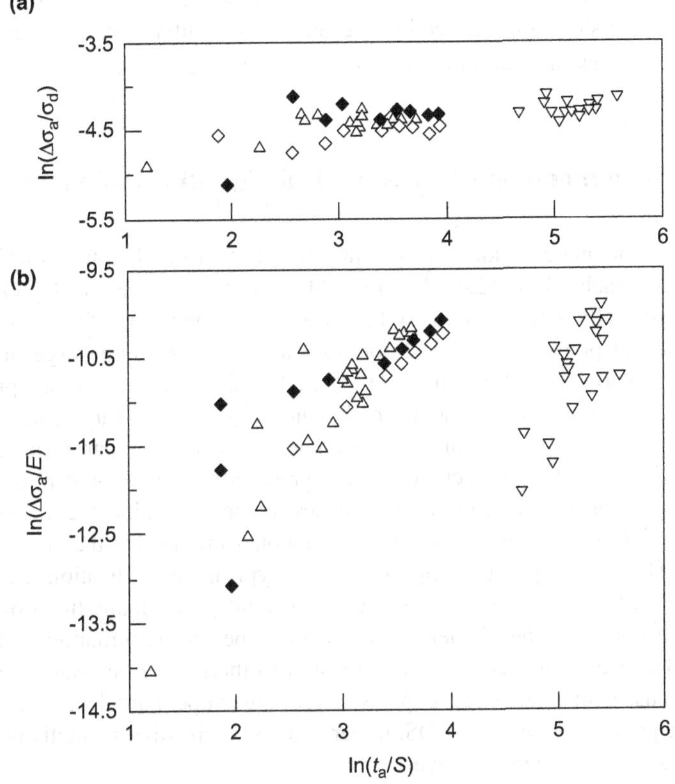

Fig. 12.20 Plots against $\ln t_d$ of **a** $\ln(\Delta\sigma_a/\sigma_d)$ from strain rate jump and SR tests, and also $\ln((\Delta\sigma_a + \Delta\sigma_h)/\sigma_d)$ from SR tests (Eq. 12.41), and **b** $\ln(\Delta\sigma_a/E)$ and $\ln((\Delta\sigma_a + \Delta\sigma_h)/E)$ from SR tests (appropriate for the model of McCormick and coworkers, Eq. 12.40b) (after Ray et al. 1992b)

$$\Delta\sigma_a = \Delta\sigma_i - \Delta\sigma_n = B \cdot (A' t_d)^k \cdot \exp\left[-(A' t_d)^k\right]$$

(compare with Eq. 12.40b and also Eq. 12.41) where the exponent k is also optimized yielded $k = 1.9$, i.e. the early kinetics appears to follow a $t^{1.9}$ law. This value for the exponent is much higher than the maximum value of $2/3$ for substitutional diffusion in the model of McCormick and coworkers. On the other hand, least square fitting of all data in Fig. 12.20a to Eq. 12.41 showed that the early kinetics seemed to follow a $t^{0.27}$ law. This value for the exponent is sufficiently close to the theoretically predicted value of $1/3$ for interstitials. It was thus concluded that DSA in solution annealed SS 316 at ambient temperature is due to ageing of mobile-forest dislocation intersections because of interstitials, with an early kinetics following $t^{1/3}$ law and saturation effect for long ageing times.

12.9 Phenomenological Mechanical Equations of State

Invariably (and understandably) all efficient MEOS formulations adopt the state variable approach (Sect. 12.4). For technological applications, an MEOS formulation is typically embedded in a FE programme, and used for predicting the deformation response of structural components which are often large in size and complex in geometry. Referring to Eqs. 12.4a–12.4b, it is therefore prudent to minimize the number of elements in Ψ, and select simple functional forms for Eqs. 12.4a–12.4b with the minimum number of adjustable parameters, without compromising the desired accuracy of the predictions. This would also minimize the burden of prior "calibration" tests necessary to determine the values for the adjustable parameters, and also the computational burden for the subsequent FE analysis. The second point is important—in a typical FE application, explicit formulations for Eqs. 12.4a–12.4b would have to be integrated many times over all the integration points of the elements used to describe the deformation of the component. Because of the same reasons, for non-isothermal FE analysis, it is advantageous to use simple analytical expressions for the temperature dependencies of the adjustable parameters of the MEOS, rather than use their experimentally determined values at several temperature levels.

Clearly, mechanistic insight is necessary for formulating an efficient MEOS. TASRA (Sect. 12.5) provides the logical foundation for the equation for plastic flow kinetics and its temperature dependence at constant structure, Eq. 12.4a. Also, equations for structure evolution (Eq. 12.4b), should be consistent with athermal hardening (corresponding to geometrical-statistical storage of dislocations), dynamic recovery and static recovery, steady state of deformation, and also otherwise the observed trends discussed in Sects. 12.6 and 12.7. Again, for typical engineering alloys, it is essential to appropriately integrate DSA with the formulation that is consistent with the basic features discussed in Sect. 12.8.

Several phenomenological MEOS of various degrees of complexities and for applications with various materials and deformation conditions, have been developed. Hart's rheological model (1976) for grain matrix deformation (Eqs. 12.23a–12.23b and Fig. 12.11, Sect. 12.5.3) is an example of an MEOS formulation that uses a single structure parameter σ_H. Several authors have used this model for predicting deformation response of structures.

A detailed review of the various MEOS formulations in literature is beyond the scope of the present chapter. The present section aims at introducing how the basic principles enunciated in the preceding paragraphs are implemented, with some examples. The version of the programme MATMOD (abbreviation for Material

Modelling, first proposed in 1974, and revised and adapted for specific applications since then) cited in this section as illustration, has been discussed by Miller (1987). References to isothermal versions of the other MEOS formulations cited may be found in Abdel-Kader et al. (1991) who compared their performance for a specific problem. As in the preceding sections, only uniaxial deformation is considered in this section, and the appropriate thermomechanical variables are $\dot{\varepsilon}$, σ and T (Sect. 12.5).

Scaling of stress and all strength terms with elastic modulus (Young's modulus E in MATMOD) is important for any MEOS that is meant application over a wide range of temperatures. This scaling is implied, but for simplicity, not explicitly indicated in the equations in this section.

12.9.1 The Isothermal Plastic Flow Equation

For general uniaxial deformation, it is necessary to use at least two structural parameters (Sect. 12.4). For isotropic hardening, a single parameter description for the component of structure that evolves with deformation may be viable for technologically important regimes as discussed in the previous sections. While specifically not discussed in this chapter, the same may be expected from a single parameter description for the kinematic hardening component. Many MEOS formulations for technological applications thus use only two strength terms, namely "drag stress" σ_D to account for isotropic hardening, and the "rest stress" σ_R to account for kinematic hardening. From Examples 12.1 and 12.3, a plausible equation for isothermal plastic flow kinetics is as follows:

$$\dot{\varepsilon} = \dot{\varepsilon}_0 \cdot \left(\frac{|\sigma - \sigma_R|}{\sigma_D}\right)^m \cdot \mathrm{Sgn}(\sigma - \sigma_R) \tag{12.43}$$

Sgn is the signum function. The manner in which σ_R appears in the equation can be understood by noting that $\sigma - \sigma_R$ is the driving force for plastic straining, and straining takes place in the direction in which $\sigma - \sigma_R$ is positive. For Eq. 12.43, there is no threshold stress for onset of plastic flow. The same is true for the following $\dot{\varepsilon}$ equations for isothermal conditions:

Walker model (Abdel-Kader et al. 1991) (β fitting parameter):

$$\dot{\varepsilon} = \frac{1}{\beta} \cdot \left[\exp\left(\frac{|\sigma - \sigma_R|}{\sigma_D}\right) - 1\right] \cdot \mathrm{Sgn}(\sigma - \sigma_R) \tag{12.44a}$$

Bodner model (Abdel-Kader et al. 1991) (D and n fitting parameters):

$$\dot{\varepsilon} = \left(\frac{2D}{\sqrt{3}}\right) \cdot \exp\left[-\frac{1}{2}\left(\frac{\sigma_{\mathrm{D}} + \sigma_{\mathrm{R}} \cdot \mathrm{Sgn}(\sigma)}{|\sigma|}\right)^{2n}\right] \cdot \mathrm{Sgn}(\sigma) \tag{12.44b}$$

MATMOD (Miller 1987) (B, p, n, and θ' fitting parameters):

$$\dot{\varepsilon} = B \cdot \theta' \cdot \left[\sinh\left(\left(\frac{|\sigma - \sigma_{\mathrm{R}}|}{\sqrt{\sigma_{\mathrm{D}}}}\right)^{p}\right)\right]^{n} \cdot \mathrm{Sgn}(\sigma - \sigma_{\mathrm{R}}) \tag{12.45}$$

With Eq. 12.45, for high purity Al for example, the experimentally determined values were: $p = 2.0$,$n = 5.0$. The "temperature correction factor" in MATMOD θ' is discussed in Sect. 12.9.2, and see Miller (1987) for the rationale for the functional form of Eq. 12.45.

In contrast to Eqs. 12.43–12.45, in the Chaboche model (Abdel-Kader et al. 1991), σ_{D} is interpreted as current yield stress that sets the threshold for viscoplastic behaviour. This is reflected in the corresponding isothermal plastic flow equation (n, K fitting parameters):

$$\begin{aligned} \dot{\varepsilon} &= \left[\frac{(|\sigma - \sigma_{\mathrm{R}}| - \sigma_{\mathrm{D}})}{K}\right]^{n} \cdot Sgn(\sigma - \sigma_{\mathrm{R}}) \quad &&\text{for } |\sigma - \sigma_{\mathrm{R}}| > \sigma_{\mathrm{D}} \\ \dot{\varepsilon} &= 0 &&\text{for } |\sigma - \sigma_{\mathrm{R}}| \le \sigma_{\mathrm{D}} \end{aligned} \tag{12.46}$$

12.9.2 Temperature Dependence of Plastic Flow

Consider deformation at high enough temperatures corresponding to the PL creep regime. Then comparing with the models of power law creep, in Eq. 12.43, $\dot{\varepsilon}_0 \propto \exp(-Q_{\mathrm{SD}}/RT)$, and m is expected to be a constant. This suggests that for deformation in this regime, the T-dependence of $\dot{\varepsilon}$ can be incorporated by replacing $\dot{\varepsilon}$ in all isothermal equations for plastic flow and structure evolution with $\dot{\varepsilon}/\exp(-Q_{\mathrm{SD}}/RT)$. If on the other hand, Eq. 12.43 is applied to regime of deformation where Cottrell–Stokes law is valid (i.e. relatively low temperatures), then Example 12.3 suggests that activation energy is stress-dependent and in Eq. 12.43, $m \propto 1/T$. Also, as mentioned in Sect. 12.5.1, $\dot{\varepsilon}_0$ could be considered to be (effectively) constant, or at most vary with a power of stress.

The situation becomes considerably complex if the model needs to encompass both high and low temperatures (corresponding respectively to constant and stress-dependent activation energy), mediated by the PLB regime discussed in Sect. 12.6.2. An expeditious practical approach is adopted in MATMOD (Miller 1987) for realistic representation of experimental data without complicating the resultant functional form too much. Basically, with some support from literature data, activation energy is set equal to that for alloy diffusion for $T \ge T_{\mathrm{t}}$, and linearly varying with T for $T < T_t$. T_t ($\sim 0.6T_M$) is a material dependent transition temperature; for high purity Al for example, $T_t = 461$ K. Accordingly, the parameter θ' in Eq. 12.45 is given by the following equations:

$$\begin{aligned} \theta' &= \exp(-Q_{SD}/RT) && \text{for } T \geq T_t \\ \theta' &= \exp\{[-Q_{SD}/RT_t] \cdot [\ln(T_t/T) + 1]\} && \text{for } T \leq T_t \end{aligned} \tag{12.47}$$

In Eq. 12.45, the entire T-dependence of $\dot{\varepsilon}$ is considered to be contained in the θ' term, and the material parameters B, p and n are therefore independent of temperature.

12.9.3 DSA in MEOS Formulations

Dynamic Strain Ageing (DSA) is explicitly considered in all MEOS formulations for typical engineering alloys. For this purpose, for isothermal condition, the strength contribution due to DSA may be described by an expression of the form:

$$F_C = F_1 \cdot \exp\left[-F_2 \cdot \left|\log\frac{|\dot{\varepsilon}|}{\dot{p}_0}\right|^{F_3}\right] \tag{12.48}$$

F_1, F_2, F_3 and $\dot{p}_0$ are adjustable material parameters. F_1 is the peak value of strengthening due to DSA, that occurs at $|\dot{\varepsilon}| = \dot{p}_0$, and the parameter F_2 sets the width of the peak. Figure 12.21 shows an example of the contribution from DSA for a specific model.

MATMOD prescription for incorporating the temperature dependence of DSA in the above equation is as follows. DSA is a strong function of dislocation speed which may be taken to be approximately proportional to $\dot{\varepsilon}$, and the effective solute diffusivity D_{eff} which for simplicity may be replaced by the temperature correction factor θ' defined by Eq. 12.47. Temperature dependence of DSA can thus be accounted for by adjusting the strain rate for peak DSA effect, e.g. by replacing $|\dot{\varepsilon}|$ in Eq. 12.49 by $|\dot{\varepsilon}|/\theta'$. MATMOD-4V (Miller 1987) describes a more elaborate and theoretically better justified strategy for DSA in typical engineering alloys (see Sect. 12.9.4).

12.9.4 σ_D and σ_R in MATMOD (Miller 1987)

The strength of MATMOD lies in its detailed quantitative consideration of the microstructural contributions to σ_D and σ_R, and their evolution with deformation, including synergistic effects. This allows great flexibility in choosing the specific details for economic "tuning" of MATMOD to the specific material and

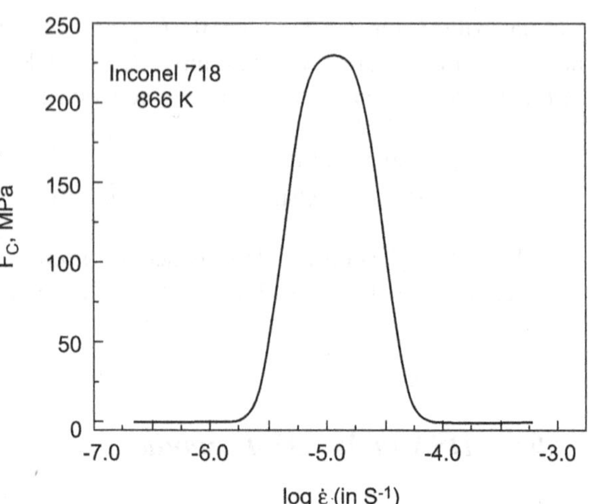

Fig. 12.21 Estimated contribution from DSA for Inconel 718 at 866 K computed using data reported by in Abdel-Kader et al. (1991) for the Chaboche model. $\dot{\varepsilon}$ in s^{-1}

deformation regime under consideration taking into account mechanistic aspects, without sacrificing accuracy of prediction. Some of the important features of the formulation are highlighted below. The equations are too numerous to be reproduced in this introductory chapter.

In general, MATMOD sets $\sigma_R = \sigma_{R_a} + \sigma_{R_b}$. σ_{R_a} represents short-range back stresses (probably from dislocations blocked at weaker obstacles such as other individual dislocations), while the σ_{R_b} component represents long range back stresses (probably associated with blockage of dislocations at strong obstacles like subgrain walls). This separation is consistent with the experimental observation that in strain cycling, the variation of internal stress seems to indicate both a rapidly varying component and a slowly varying component.

For isotropic hardening, let F_{def} represent the sum of all contributions which vary with deformation level, and F_{sol} the sum of contributions which do not. Then $\sigma_D = F_{\text{def}} + F_{\text{sol}}$. F_{def} may be considered to be the sum of two contributions:

- $F_{\text{def},\rho}$, the contribution from homogenously distributed dislocations. Considering only strengthening from forest dislocations with density ρ and invoking Taylor work hardening relationship, this contribution is of the form $\sigma|_T = f(\dot{\varepsilon}) \cdot \sqrt{\rho}$.
- $F_{\text{def},\lambda}$ which represents the effects of heterogeneous dislocation substructure. If only contribution of subgrains of size λ to flow stress is considered, then the strength contribution is of the form (see Sect. 12.2) $\sigma|_T = f(\dot{\varepsilon}) \cdot \lambda^{-1}$.

Similar structure for strain rate relation obtains for strain-independent isotropic strengthening contributions summed in F_{sol}:

- grain boundary strengthening obeying the Hall–Petch relation expressed in the form $\sigma|_T = f(\dot{\varepsilon}) \cdot \left(k_1 + k_2 \cdot d^{-1/2}\right)$ where f is the appropriate function, d the grain size, and k_1, k_2 constants;

- strengthening due to precipitates or dispersoids; for mean inter-particle spacing l, $\sigma|_T = f(\dot{\varepsilon}) \cdot \left(1/\sqrt{l}\right)$ from Orowan equation.
- solute effects, responsible for DSA, which can be strong in appropriate deformation regime in typical engineering alloys.

$f(\dot{\varepsilon})$ in the above formulations for the contributions to F_{def} and F_{sol} are the appropriate functions of strain rate.

The MATMOD prescription for accounting for the strongly temperature dependent DSA has been indicated in Sect. 12.9.3. MATMOD sets $F_3 = 2$ in Eq. 12.48. For an engineering alloy with multiple solutes, total DSA contribution will be obtained as the sum of the contributions from each diffusing species. When solution and strain hardening act in a synergistic manner, intrinsic strengthening can be considered to comprise two terms $F_{\mathrm{sol,1}}$ and $F_{\mathrm{sol,2}}$, and

$$\sigma_{\mathrm{D}} = F_{\mathrm{def}}\left(1 + F_{\mathrm{sol,2}}\right) + F_{\mathrm{sol,1}}$$

This is the case with austenitic AISI SS 316 steel for example. For this material, $F_{\mathrm{sol,1}}$ is the sum of contributions from three substitutional solute species of similar diffusivities, and expressed as

$$F_{\mathrm{sol,1}} = F_{\mathrm{sol,10}} + \sum_{\mathrm{i=1}}^{3} M_{\mathrm{i}} \left\{ \exp\left[-\left(\frac{\log\left(|\dot{\varepsilon}| / \theta'_{\mathrm{sol,1}} \right) - J_{\mathrm{i}}}{B_{\mathrm{i}}} \right)^{2} \right] \right\} \tag{12.49}$$

($F_{\mathrm{sol,10}}$, M_{i}, B_{i}, J_{i} fitting constants); for more precise description, instead of θ', $\dot{\varepsilon}$ is here scaled with $\theta'_{\mathrm{sol,1}} = \exp\left(-Q_{\mathrm{sol,1}}/RT\right)$, with $Q_{\mathrm{sol,1}}$ determined as ~ 280 kJ/mole. Likewise, two interstitial solute species are considered to contribute to $F_{\mathrm{sol,2}}$, and in the relevant expression written following Eq. 12.49, strain rate is scaled with $\theta'_{sol,2} = \exp\left(-Q_{\mathrm{sol,2}}/RT\right)$, with $Q_{\mathrm{sol,2}}$ determined as ~ 160 kJ/mole. For a well-annealed material, F_{def} is small at low strain, and therefore the engineering yield strength at $\varepsilon = 0.002$ is dominated by $F_{\mathrm{sol,1}}$, while strength at high strains is dominated by $F_{\mathrm{sol,2}}$. With these refinements, the strain rate equation for SS 316 can be expressed as:

$$\dot{\varepsilon} = B_{\mathrm{M}} \cdot \theta' \cdot \left[\sinh\left(\frac{\left| \frac{\sigma}{E} - \sigma_{\mathrm{R}} \right|}{\sqrt{F_{\mathrm{sol,1}} + F_{\mathrm{def}}\left(1 + F_{\mathrm{sol,2}}\right)}} \right)^{1.5} \right]^{n} \cdot \mathrm{Sgn}\left[\frac{\sigma}{E} - \sigma_{\mathrm{R}} \right] \tag{12.50}$$

$$\theta' = \exp(-Q_1/RT) + 4 \cdot 10^{-7} \cdot \exp(-Q_2/RT)$$

In this equation, n is the stress exponent of steady-state creep, and Q_1 and Q_2 have been determined to be respectively 280 and 168 kJ/mole.

12.9.5 $\dot{\sigma}_{\mathrm{D}}$ and $\dot{\sigma}_{\mathrm{R}}$ for Isothermal Deformation

In principle, strain hardening, dynamic recovery, and static recovery need to be considered in writing phenomenological equations for $\dot{\sigma}_{\mathrm{D}}$ and $\dot{\sigma}_{\mathrm{R}}$ (Sect. 12.3). Often extent of dynamic recovery is taken to be proportional to the strain increment. Then, the functional forms for a simple formulation could be as follows [cf. Miller (1987)]:

$$\begin{aligned} \delta\sigma_{\mathrm{D}} &= H \cdot |\delta\varepsilon| - D(\sigma_{\mathrm{D}}) \cdot |\delta\varepsilon| - R(\sigma_{\mathrm{D}}, T) \cdot \delta t \\ \delta\sigma_{\mathrm{R}} &= H \cdot \delta\varepsilon - D(|\sigma_R|) \cdot |\delta\varepsilon| \cdot \mathrm{Sgn}(\sigma_{\mathrm{R}}) - R(|\sigma_{\mathrm{R}}|, T) \cdot \mathrm{Sgn}(\sigma_{\mathrm{R}}) \cdot \delta t \end{aligned} \tag{12.51}$$

The strain hardening coefficient $H = \theta_0$ if strain hardening is considered to be athermal, as in MATMOD. $D(\cdot)$ and $R(\cdot)$ are respectively dynamic and static recovery functions, which appear with negative signs, because these processes lead to decrease in strength. Isotropic strengthening is driven by magnitude of strain increment, while kinematic strain hardening represents increase in strength in the forward direction of straining. This is reflected in the manner strain increment appears in the first terms on the right-hand side of these two equations. Recovery is generally found to depend on the strength level. This is depicted by identifying appropriate arguments of the functions $D(\cdot)$ and $R(\cdot)$. The equations also recognize that for the kinematic strength term, recovery acts exactly opposite to the applied stress. Note that Eq. 12.51 does not envisage interaction between σ_{D} and σ_{R} so far as their evolution with deformation is concerned.

Equation 12.51 provides a convenient perspective for the isothermal versions of a few promising MEOS formulations (from Abdel-Kader et al. 1991) cited in this Section as illustrations. The expression for $\dot{\sigma}_{\mathrm{D}}$ in the Chaboche model (Abdel-Kader et al. 1991) (Eq. 12.46 for $\dot{\varepsilon}$-equation) and Walker model (Abdel-Kader et al. 1991) (Eq. 12.44a for $\dot{\varepsilon}$-equation) is identical (b_{C}, σ_{D_1} fitting parameters, of course with different values for the two models):

$$\dot{\sigma}_{\mathrm{D}} = b_{\mathrm{C}} \cdot (\sigma_{\mathrm{D}_1} - \sigma_{\mathrm{D}}) \cdot |\dot{\varepsilon}| \tag{12.52}$$

even though σ_{D} has very different interpretations in these two models. Equation 12.52 includes strain hardening and dynamic recovery, but excludes static recovery. Evolution of σ_{D} is assumed to follow a first-order kinetics; $\mathrm{Sgn}(\sigma_{\mathrm{D}_1} - \sigma_{\mathrm{D}})$ determines whether material hardens or softens with progressive deformation, and $\sigma_{\mathrm{D}_1} = \sigma_{\mathrm{D}}$ in steady state. In Chaboche model the DSA contribution F_{C} (Eq. 12.48) is added to σ_{D} because DSA is considered to be isotropic. Integrating $\dot{\sigma}_{\mathrm{D}}$ for the isothermal Chaboche model:

$$\sigma_{\mathrm{D}} = \sigma_{\mathrm{D}_1} + (\sigma_{\mathrm{D}_0} + \sigma_{\mathrm{D}_1} + F_{\mathrm{C}}) \cdot \exp\left(-b_{\mathrm{C}} \int_0^t |\dot{\varepsilon}| \, \mathrm{d}\tau\right)$$

which introduces a material parameter $\sigma_{D_0} = (\sigma_D)_{\varepsilon=0}$.

The expression for $\dot{\sigma}_D$ in Bodner model (Abdel-Kader et al. 1991) (Eq. 12.44b for $\dot{\varepsilon}$-equation) is:

$$\dot{\sigma}_D = m_{B_1} \cdot (\sigma_{D_1} - \sigma_D) \cdot |\sigma\dot{\varepsilon}| - A_{B_1} \cdot \sigma_{D_1} \cdot \left[\frac{\sigma_D - \sigma_{D_2}}{\sigma_{D_1}}\right]^{r_{B_1}} \tag{12.53}$$

(m_{B_1}, A_{B_1}, r_{B_1}, σ_{D_1} and σ_{D_2} fitting parameters). The functional form for the combined contribution from work hardening and dynamic recovery (the first term on the right-hand side of Eq. 12.53) is identical to that for Eq. 12.52 except for one crucial difference: unlike Eqs. 12.51 and 12.52, this contribution is driven by incremental plastic work in Eq. 12.53. Also, Eq. 12.53 includes the static recovery term (the second term on the right-hand side). It basically adopts the Friedel network coarsening model (Example 12.4) but introduces a material parameter σ_{D_2} as a threshold value of σ_D for static recovery to set in. In this model too, DSA is considered to be isotropic, but F_3 in Eq. 12.48 set equal to 2.

For all these three models, the expressions for $\dot{\sigma}_R$ include terms for hardening, and dynamic and static recovery. The static recovery terms follow Friedel network coarsening model (Example 12.4) without any threshold strength. For the Chaboche model, the expression is (c_C, α_C, γ_C, m_C fitting parameters):

$$\dot{\sigma}_R = c_C \cdot (\alpha_C\dot{\varepsilon} - \sigma_R\,|\dot{\varepsilon}|) - \gamma_C \cdot |\sigma_R|^{m_C} \cdot Sgn(\sigma_R) \tag{12.54a}$$

For the Walker model, the expression is (n_{W_1},n_{W_2},n_{W_3}, m_W fitting parameters):

$$\dot{\sigma}_R = n_{W_1} \cdot \dot{\varepsilon} - (n_{W_2} - F_C) \cdot |\dot{\varepsilon}| \cdot \sigma_R - n_{W_3} \cdot |\sigma_R|^{m_W} \cdot \mathrm{Sgn}(\sigma_R) \tag{12.54b}$$

Equations 12.54a and 12.54b are very similar in form except for the strengthening from DSA F_C. In Walker model, DSA is considered to have a directional character, and the F_C term (with F_3 in Eq. 12.48 set equal to 1) is added to the σ_R term. The expression for $\dot{\sigma}_R$ in Bodner formulation is more complicated, and is not reproduced here.

Unlike the other formulations considered above, In MATMOD, $\dot{\sigma}_D$ and $\dot{\sigma}_R$ are defined in terms of the deformation dependence of the various contributions to F_{def} and σ_R. These are described adopting the approach of Eq. 12.51. The temperature dependence of static recovery is described by multiplying with the factor θ'. The dynamic recovery term for each of these variables is designed such that, as discussed with Fig. 12.3, if the current value of the structure variable is less than its steady-state value computed for current $\dot{\varepsilon}$ and T, then straining increases its value; otherwise, strain softening reduces its value. An important difference of MATMOD with the other three models considered in this chapter is that in MATMOD, structure variable interactions, for example effects of σ_{R_b} on $\dot{F}_{def,\lambda}$, $F_{def,\rho}$ upon $\dot{\sigma}_{R_a}$, $F_{def,\lambda}$ on $\dot{\sigma}_{R_b}$, and $F_{def,\lambda}$ on $\dot{F}_{def,\lambda}$, are incorporated in the evolutionary equations. The relevant kinetic equations are too numerous to be listed here, and may be found in Miller (1987).

Detailed consideration of all the microstructural features and their interactions in MATMOD means that it is potentially more accurate than the other models lacking this microstructural nexus. This improvement is achieved at the cost of increased number of adjustable parameters. However, as mentioned earlier, considering only those microstructural features, which are known to be important for the specific problem under consideration, can reduce this number.

12.10 Review Questions

1. Schematically show the variation of $\sigma \cdot \theta$ with σ for a single phase polycrystalline FCC material at low and intermediate homologous temperatures, and explain the variation in terms of dislocation processes.
2. For single parameter (ψ) description of deformation substructure, the following identity holds $\left(\frac{\partial \psi}{\partial \ln \sigma}\right)_{\dot{\varepsilon}} \cdot \left(\frac{\partial \ln \sigma}{\partial \ln \dot{\varepsilon}}\right)_{\psi} \cdot \left(\frac{\partial \ln \dot{\varepsilon}}{\partial \psi}\right)_{\sigma} = -1.$
   If Eq. 12.5 (Example 12.1) holds, then use this identity to show that $\theta = (\partial\sigma/\partial\varepsilon)_{\dot{\varepsilon},T}$ and $C \equiv (\partial(1/\dot{\varepsilon})/\partial t)_{\sigma,T}$ evaluated at identical combinations of $(\sigma, \dot{\varepsilon}, T)$, are related by the equation
   $$\theta = (\sigma/m) \cdot C$$
   Note that the parameter C pertains to primary creep. Is it possible to experimentally verify this predicted relation between θ and C?
3. Consider thermally activated motion of a dislocation on a glide plane. Using schematic diagram(s), qualitatively explain the concepts of reaction path and activation area A. Can A increase with increasing stress τ? Schematically show the $\Delta G - \tau$ profiles, and write down the expressions for strain rate $\dot{\gamma}$ (assuming that the pre-exponential factor $\dot{\gamma}_0$ varies as a power of τ) when for the rate-controlling mechanism,
   (i) A is independent of τ,
   (ii) A decreases linearly with increasing τ.
4. Equations 12.14a–12.14b give the expressions for $\mathcal{A}$ and Q in the total stress formalism. Write down the equations for the corresponding parameters $\mathcal{A}^*$ and Q^* in the effective stress formalism.
5. Verify that for Cottrell–Stokes law expressed by Eq. 12.18d,
   $$m/m^* = Q/k_B T.$$
6. How will dynamic recovery at high deformation levels (i.e. high stress levels) affect the Haasen plot for linear superposition rule shown schematically in Fig. 12.7?

7. Schematically show the Haasen plot when rule of superposition of squares of stresses (i.e. Equations 12.20a–12.20b) applies (Note: see Kocks et al. 1975; Diak et al. 1998).
8. Why is it important to maintain stable temperature during a long duration stress relaxation test? How will you estimate the required degree of stability?
9. Discuss the usefulness and limitations of the stress relaxation testing for determining the dependence of activation area A on τ. Include in your considerations (i) dynamic recovery, (ii) static recovery, and (iii) dynamic strain ageing.
10. Suppose U-type stress relaxation is due to gradual softening of dislocation structure due to dynamic recovery. With the help of Fig. 12.3, qualitatively argue that if single parameter description of deformation substructure applies, then $\mathrm{d}\ln(-\dot{\sigma})/\mathrm{d}\ln\sigma \to n$ as duration of relaxation $t \to \infty$, where n is the stress exponent for steady-state creep at the temperature of stress relaxation test (i.e. slope of the line CC in Fig. 12.3).
11. Show that Friedel's network coarsening model (Example 12.4, Eq. 12.26a) can be expressed in the form

$$\dot{\rho} \propto -D_{\mathrm{v}} \cdot \left(\frac{\mu b^3}{k_{\mathrm{B}}T}\right) \cdot \rho^2$$

12. Consider the functional form for viscoplastic uniaxial non-reversed deformation: $\sigma = f_1(\varepsilon) \cdot f_2(\dot{\varepsilon})$, with $f_2(\dot{\varepsilon}) \to 1$ in the limit of rate-insensitive deformation (Sect. 12.7.1). Is this separation of variables theoretically justified? Two empirical functional forms for $f_2(\dot{\varepsilon})$ proposed in the literature are:

$$f_2(\dot{\varepsilon}) = (1 + \dot{\varepsilon}/\zeta)^{\vartheta} \text{ and } f_2(\dot{\varepsilon}) = 1 + (\dot{\varepsilon}/\zeta)^{\vartheta}$$

ζ and ϑ are T-dependent viscoplastic material parameters, and $\zeta \to \infty$ for rate-insensitive deformation. Plan an efficient experimental campaign to determine these viscoplastic constants for a given material for a given temperature.
13. Consider first-order kinetics for evolution of σ for a tension test in the form

$$\dot{\sigma} = k(\sigma_{\mathrm{v}} - \sigma)$$

where σ_{v} is the Voce stress (i.e. the extrapolated steady-state stress; see Appendix), and the parameter k is constant for a given tension test, i.e. constant $\dot{\varepsilon}, T$. Show that this formulation is consistent with the Voce work hardening relation, Eq. 12.55 in Appendix. Hence write down the $\dot{\varepsilon}, T$-dependence of the parameter k in the above equation.
14. Creep test is carried out in a constant load creep-testing machine at a constant nominal stress (i.e. initial stress) σ_0. From the $\varepsilon - t$ plot, the minimum creep rate is determined to be $\dot{\varepsilon}_{\mathrm{m}}$, and the corresponding strain is ε_{m}. Estimate the minimum creep rate that would be obtained in a constant true stress

creep-testing machine for the stress σ_0 at the same temperature. Justify any assumptions you make.

15. Show that McCartney–McLean equation for creep (1976) (Eq. 12.35):

$$d\dot{\varepsilon}/dt = -\beta \cdot \dot{\varepsilon} \cdot (\dot{\varepsilon} - \dot{\varepsilon}_s), \beta \text{ constant for given} \sigma, T$$

can be integrated, with $\varepsilon(t = 0) = 0$ and $\dot{\varepsilon}_0 = \dot{\varepsilon}(t = 0)$ to derive:

$$\varepsilon = \frac{1}{\beta} \cdot \ln[1 + (\dot{\varepsilon}_0/\dot{\varepsilon}_s) \cdot \{\exp(\beta\dot{\varepsilon}_s t) - 1\}]$$

How will you determine the value for the constant β from strain–time data for the primary + secondary regimes of a creep test?

16. Identify the important features of the Amin–Mukherjee–Dorn creep equation (1970) (Eqs. 12.36a–12.36b) that can be experimentally verified.
17. Consider the Estrin–Mecking model (1984) for work hardening in a single phase material where dislocation tangles provide obstacles to dislocation motion (Eq. 12.38). Comment on the $\dot{\varepsilon}$- and T-dependence of the parameters a_1, a_2 in this equation.
18. Suggest an extension of Eq. 12.38 to include static recovery. How will you proceed to experimentally examine the validity of your proposed model?
19. Starting from Eq. 12.38 for the Estrin–Mecking model (1984), derive the equation for primary creep at constant σ and T (Note: refer to Estrin and Mecking 1984).
20. Consider a micromechanical model for evolution of mobile dislocation density ρ_m during creep (a_1, a_2 parameters, constants for given σ, T):

$$\dot{\rho}_m = a_1 \cdot \rho_m - a_2 \cdot \rho_m^2$$

J.C.M. Li considered this model in 1963. The first term right-hand side of corresponds to dislocation multiplication, and the second term corresponds to loss of mobile dislocations by mutual interactions. Use Orowan equation, $\dot{\varepsilon} \propto \rho_m$ for constant σ, T, to show that this model leads to the McCartney–McLean equation (1976) for primary creep (Eq. 12.35). Hence, comment on the expected σ- and T-dependence of β in Eq. 12.35.

21. The work hardening analysis discussed in Sect. 12.7.2 tacitly assumed that the tensile flow curves are free from serrations. However, in engineering alloys, the tensile flow curves are often serrated, because of DSA. Is it meaningful to carry out work hardening analysis for such serrated tensile flow curves? [Note: see Rodriguez (1984) for a classification of the types of serrations].

Appendix

Voce Work Hardening Relation, Stages IV and V of Deformation (Kocks and Mecking 2003)

Figure 12.22 schematically illustrates the features of work hardening in FCC polycrystals. The sharp decrease of θ at the low σ end corresponds to a very small segment of the stress–strain curve at the beginning of yielding up to $\sim 0.2\%$ strain. This stage may be ascribed to effects of grain boundaries on dislocation storage. Beyond this small amount of strain, the θ-σ plot merges into an almost linearly decreasing branch, consistent with the Voce relation. The intercept of this line with the axes is determined by extrapolation (the dashed line in Fig. 12.22). The intercept with the ordinate yields an athermal $\theta_0 \sim \mu/50$. The intercept with the abscissa is the Voce stress σ_v, corresponding to the extrapolated steady state of deformation, which carries the entire $\dot{\varepsilon}$- and T-dependence of work hardening.[4]

An athermal θ_0/μ is analogous to the stage II work hardening in FCC single crystals. The regime of (near-) linear variation of θ with σ is analogous to stage III deformation in FCC single crystals, corresponding to dynamic recovery. The Voce work hardening relation can be stated in the following form of the equation of a master curve:

$$\frac{\theta}{\mu} = \frac{\theta_0}{\mu} \cdot \left(1 - \frac{\sigma/\mu}{\sigma_v(\dot{\varepsilon}, T)/\mu}\right) \tag{12.55}$$

Work hardening data for FCC metals can be reasonably well described this master plot for intermediate range of σ/σ_v. For Cu polycrystals tested in tension (T range: ambient to 673 K; $\dot{\varepsilon}$: 10^{-4} and $10^{-1}\ \mathrm{s}^{-1}$), the master plotted with a $\sigma/\sigma_v(\dot{\varepsilon}, T)$ and θ/μ scaled with $\theta_0/\mu \approx 0.05$, resulted in the data for the scaled stress σ/σ_v in the range of ~ 0.2 to 0.8 being effectively merged (within experimental scatter) onto a single linear "master" plot, with tails with upward curvature at both high and low stress ends, that did not merge so well (Kocks and Mecking 2003). This linear variation however may be only approximate, and in some cases, stage III may be better described by a slightly curvilinear relation between σ/σ_v and θ/θ_0.

The extrapolated steady state defined by σ_v is not established for T/T_M levels up to ~ 0.5: a new stage IV intervenes at large strains when θ has dropped to a

[4]For convenience, the discussion in Appendix 12.1 is in terms of data from tensile tests. Deformation data for large strains are generated by torsion testing. Conversion from torsion to tension data requires using appropriate values for $\bar{M}$ for these two tests. Note, however, that in Fig. 12.22 both axes have been scaled by the intercepts on the corresponding axes. If deformation dependence of $\bar{M}$ can be ignored, then this scaled plot is independent of the testing method—tension or torsion (σ/σ_v equals the ratio of torque to the extrapolated torque for the Voce relation in a torsion test, and likewise for the scaled work hardening axis).

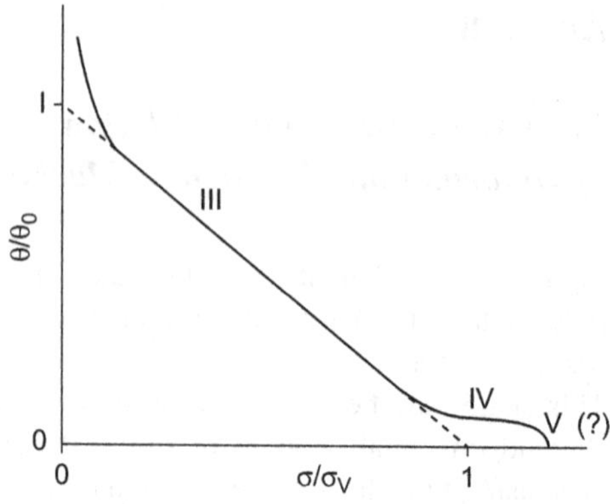

Fig. 12.22 Schematic diagram illustrating the stages of work hardening in a polycrystal in a scaled θ versus σ plot. Single crystals oriented for polyslip also show similar behaviour. As mentioned in the text, Stage III may show a slight curvature

particular level much below θ_0. In tension or compression tests, deformation instability intervenes before this small work hardening level is reached. Therefore, torsion tests are used to study this phenomenon. As shown in Fig. 12.22, the condition $\theta = 0$ may be reached eventually at a stress $\sigma_s > \sigma_v$. This has been described by a continuously decreasing hardening rate in stage IV, and also by a separate stage V. The hardening level in stage IV and its length increase with decreasing temperature. Thus, σ_s (where θ actually drops to zero) and σ_v (marking end of stage III) attain similar values at high temperatures ($T/T_M \sim 0.5$ and higher), but their difference increases as temperature is lowered. Stage IV has been found in many pure metals and alloys in polycrystals under many deformation modes, in wire drawing as well as in rolling and in torsion. Stage IV appears to have a number of variants, depending on material and temperature. The onset of stage IV may be defined by the appearance of a more or less sharp kink in a plot like Fig. 12.22 (Kocks and Mecking 2003). For many materials in different testing modes, the ratio of (work hardening/Voce stress) for the onset of stage IV is found to be constant. For polycrystals, this ratio depends on $\bar{M}$. For FCC polycrystals with randomly oriented grains, for uniaxial deformation using normal stress–longitudinal strain, this ratio (i.e. θ/σ_v) is found to be in the range 0.15–0.3. When this proportionality holds, it means that the T- and strain rate dependence of stage IV is basically the same as that of stage III: no new thermal activation mechanism is called for. Also any dependence on SFE is dominated by stage III processes since the slope in stage IV stays in a fixed relationship to the scaling stress for stage III. Several mechanisms for stage IV and also stage V (when it is considered to be a separate stage) of deformation have been considered in literature. Discussion on these mechanisms is beyond the scope of the present chapter. Suffice it to note here that the interdependencies between the work hardening parameters in the stages III and IV led Kocks and Mecking (2003) to conclude that at low and medium temperature, hardening kinetics in Stages III and IV is governed by stress driven slip processes, which control storage, mutual annihilation and rearrangement of dislocations; there is no necessity for introducing diffusion as a controlling mechanism.

As already mentioned, the discussion on stress-dependent activation energy for the PLB regime in Sect. 12.6.2 pertains really σ_v to rather than to σ_s; the difference can be ignored for relatively high T/T_M levels. As shown there, the $\dot{\varepsilon}$- and T-dependence of σ_v can be expressed in a form consistent with thermally activated dislocation process:

$$\frac{\Delta G}{\mu b^3} \equiv \frac{k_B T}{\mu b^3} \cdot \ln \frac{\dot{\varepsilon}_0}{\dot{\varepsilon}} = g\left(\frac{\sigma_v/\mu}{\sigma_{v_0}/\mu_0}\right) \tag{12.56}$$

or, equivalently

$$\frac{\sigma_v}{\mu} = \frac{\sigma_{v_0}}{\mu_0} \cdot f\left(\frac{1}{g^0} \cdot \frac{k_B T}{\mu b^3} \cdot \ln \frac{\dot{\varepsilon}_0}{\dot{\varepsilon}}\right) \tag{12.57}$$

A specific functional form derived by Kocks and Mecking (2003) using data for Ag, Cu, Ni and Al is:

$$\frac{\sigma_v}{\mu} = \frac{\sigma_{v_0}}{\mu_0} \cdot \left\{1 - \left(\frac{1}{g^0} \cdot \frac{k_B T}{\mu b^3} \cdot \ln \frac{\dot{\varepsilon}_0}{\dot{\varepsilon}}\right)^{1/2}\right\}^2 \tag{12.58}$$

with $\dot{\varepsilon}_0 = 10^7\ \text{s}^{-1}$ (compare Eq. 12.58 with the stress-dependent $\dot{\varepsilon}_0$ in Eq. 12.29, and the functional form for g). As with Eq. 12.29, for Eq. 12.58 also the results were consistent with both g^0 and σ_{v0}/μ_0 being determined by $\Gamma/\mu b$. g^0 varied from 0.95 for Ag ($\Gamma/\mu b = 0.0018$) to 0.65 for Al ($\Gamma/\mu b = 0.023$); the limiting values for g^0 suggested by the data trend are $g^0 = 1.01$ for $\Gamma/\mu b = 0.0$, and $g^0 \to 0.41$ as $\Gamma/\mu b \to \infty$. As Kocks and Mecking pointed out, σ_{v_0}/μ_0 determined from experimental data using Eq. 12.58 could be sensitive to deformation, because of change in Taylor factor $\bar{M}$ as deformation texture slowly develops. However, g^0 determined should qualify as a material property; since g^0 relates to local activation process, it is expected to depend on physical properties like $\Gamma/\mu b$, but not on microstructure, texture, etc.

References

Abdel-Kader, M.S., El-Hefnawy, N.N., Eleiche, A.M.: Nucl. Eng. Des. **128**, 369 (1991)

Alden, T.H.: Mechanical testing for deformation model development. Rohde, R.W., Swearengen, J.C. (eds.) ASTM STP 765, p 29 (1982)

Amin, K.E., Mukherjee, A.K., Dorn, J.E.: J. Mech. Phys. Solids **18**, 413 (1970)

Armstrong, P.E., Hockett, J.E., Sherby, O.D.: J. Mech. Phys. Solids **30**, 37 (1982)

Bendarsky, L., Rosen, A. Mukherjee, A.K.: Int. Met. Rev. **30**, 1 (1985)

Cadek, J.: Creep in Metallic Materials. Elsevier, Amsterdam (1988)

Diak, B.J., Upadhyaya, K.R., Saimoto, S.: Progr. Mater. Sci. **43**, 223 (1998)

Dieter, G.E.: Mechanical Metallurgy, SI Metric Edition. McGraw Hill, London, Section 8–10, p 307 (1988)
Estrin, Y., Mecking, H.: Acta Metall. **32**, 57 (1984)
Friedel, J.: Dislocations. Addison Wesley (1964)
Hart, E.W.: J. Eng. Mater. Technol. **98**, 193 (1976)
Kocks, U.F., Argon, A.S., Ashby, M.F.: Progr. Mater. Sci. **19**, 1 (1975)
Kocks, U.F., Cook, R.E., Mulford, R.A.: Acta Metall. **33**, 623 (1985)
Kocks, U.F., Mecking, H.: Progr. Mater. Sci. **48**, 171 (2003)
Kocks, U.F.: Trans. ASME Ser. H. J. Eng. Mater. Tech. **98**, 76 (1976)
Ling, C.P., McCormick, P.G.: Acta Mater., Vol **38** (1990), p 2631
McCartney, L.M., McLean, D.: J. Mech. Eng. Sci. **18**, 39 (1976)
Mecking, H., Nicklas, B., Zaubova, N., Kocks, U.F.: Acta Metall. **34**, 527 (1986)
Miller, A.K.: Unified Constitutive Equations for Creep and Plasticity. Elsevier Applied Science, England, p 139 (1987)
Nabarro, F.R.N.: Acta Metal. Mater. **38**, 164 (1991)
Narutami, T., Takamura, J.: Acta Metal. **39**, 2037 (1981)
Ostrom, P., Lagneborg, R.: Strength of metals and alloys. In: Haasen, P., Gerold, V., Kostrz, G. (eds.) Proceedings ICSMA 5. Pergamon Press, New York, p 277 (1980)
Poirier, J.-P.: Creep of Crystals: High Temperature Deformation Processes in Metals, Ceramics and Minerals. Cambridge: Cambridge University Press (1985)
Ray, S.K., Rodriguez, P.: Mater. Trans. JIM **33**, 910 (1992)
Ray, S.K., Samuel, K.G., Rodriguez, P.: Scripta Metall. **27**, 271 (1992b)
Ray, S.K., Sasikala, G., Rodriguez, P., Sastry, D.H.: Mater. Sci. Tech. **9**, 1079 (1993)
Ray, S.K., Rodriguez, P., Sastry, D.H.: Trans. IIM **45**, 383 (1992a)
Rodriguez, P.: Bull. Mater. Sci. **6**, 653 (1984)
Rodriguez, P., Ray, S.K.: Bull. Mater. Sci. **10**, 133 (1988)
Rohde, R.W., Jones, W.B., Swaearengen, J.C.: Acta Metal. **29**, 41 (1981)
Samuel, K.G., Mannan, S.L., Rodriguez, P.: Acta Metall. **36**, 2323 (1988)
Samuel, K.G., Ray, S.K., Rodriguez, P.: J. Mech. Behav. Mater. **17**, 401 (2006)
Taylor, G.: Progr. Mater. Sci. **36**, 29 (1992)

Chapter 13
Kinetics of Creep Fracture

13.1 Introduction

It was mentioned in the previous chapter that plastic deformation kinetics can be studied from different perspectives. Chapter 12 dealt with the phenomenological modelling of plastic deformation kinetics. The present chapter focuses on phenomenological modelling of the kinetics of creep fracture under uniaxial loading (i.e. constant stress, or more commonly, constant load). Purely empirical methods, such as artificial neural network modelling, are not discussed.

Mechanisms of creep fracture

At relatively high stresses and strain rates typical of tensile test conditions, transgranular ductile fracture takes place by nucleation of microvoids in the matrix, followed by their growth and eventual coalescence. This creates easy path for crack propagation that leads to final fracture. At high enough temperatures, however, rupture by necking dominates because of dynamic recrystallization. Time to fracture in a typical tension test is seldom more than 1–2 h. Making realistic predictions about service lives ($\sim$20 years or more) of high-temperature structural components on the other hand is an important technical challenge, because it involves extrapolations from laboratory test data of considerably smaller durations. In engineering literature, often the term creep is used to describe the conditions of low stress and strain rates, and also high temperatures resulting in long rupture lives. This usage is adopted in the present chapter also, even though this is not fully consistent with the state variable approach (see Sect. 12.3).

Thermal activation plays a crucial role in determining the kinetics of creep fracture, analogous to the strong temperature dependence of deformation under these conditions (see Chap. 12). For engineering applications, the strong temperature dependence of creep deformation is often expressed by the well-known Norton equation that relates steady-state creep rate $\dot{\varepsilon}_s$ to the stress σ and temperature T.

H. S. Ray and S. Ray, *Kinetics of Metallurgical Processes*,
Indian Institute of Metals Series, https://doi.org/10.1007/978-981-13-0686-0_13

In terms of the reference strain rate $\dot{\varepsilon}_{ref}$ at a reference stress σ_{ref} (arbitrarily chosen) at the same temperature, the Norton equation can be written as:

$$\dot{\varepsilon}_s = \dot{\varepsilon}_{ref} \cdot (\sigma/\sigma_{ref})^n, \quad \dot{\varepsilon}_{ref} \propto \exp -\frac{Q_c}{RT} \tag{13.1}$$

For high-temperature creep, the activation energy for creep Q_c and the stress exponent n are usually assumed to be constant (see Sect. 12.6.1). For engineering alloys, often the minimum creep rate $\dot{\varepsilon}_m$ is used because most often, clearly discernible ("true") steady-state creep does not obtain.[1]

This section is devoted to a brief account of the mechanisms of creep cavity nucleation and growth, as necessary perspective for modelling for of kinetics of creep fracture, with rupture lives t_r ranging from a few 10^2 to a few 10^5 h.

Considerable progress has been made in understanding the process of ductile fracture, see, e.g., Cadek (1988), Cocks and Ashby (1982), Riedel (1987) and Kassner and Hayes (2003). In a typical long-duration creep test, the damage process dominates the tertiary creep regime, analogous to the post-necking regime in a constant strain rate tension test. The tertiary creep could result from several sources. For particle-strengthened creep resistant alloys, coarsening of the particles during the prolonged creep exposure leads to gradual reduction in strength and thus accelerates tertiary creep rate. For other alloys, tertiary creep results from cavitations: (i) inhomogeneous nucleation of cavities within the matrix (typical of higher stresses and lower temperatures) or on grain boundaries (typical of higher temperatures and with lower stresses), (ii) their growth, followed by (iii) their coalescence leading to easy paths for rapid crack propagation and final fracture. Other mechanisms that can give rise to the accelerating creep rate in tertiary creep are deformation instability in the form of initiation and propagation of a neck, or loss of load-bearing section because of, e.g., surface oxidation during exposure to hostile environments. Corrosive environments also play crucial role in some circumstances.

13.1.1 Cavity Nucleation

Cavity nucleation at relatively low temperatures has been attributed to the stress concentration at the head of a dislocation pile up against hard particles (on grain boundaries or grain interiors) reaching a critical value that causes particle fracture, or decohesion at the matrix/particle interface (Gandhi 1985). At these temperatures, a critical strain (e.g. estimated as $\sim$0.35 in mild steel at ambient temperature) is necessary for reaching this critical stress by work hardening. In contrast, creep

[1]Note that to be consistent with the current literature on deformation and creep fracture, the symbol Q (with appropriate suffixes) is used to indicate the various activation energies.

cavitation has been detected at even <0.2% strains. Experimental evidence indicates that in engineering alloys, creep cavities nucleate continuously, with rates strongly dependent on impurity content and varying over a few decades; cavities continue to nucleate up to a sizeable fraction of t_r and even close to fracture. Many authors have suggested empirical relationships for the strain dependence of creep cavity nucleation; for several steels, the relationship is found to be linear.

Observation of stable cavities at very small strains under creep conditions plus the fact that work hardening effects are considerably weaker in creep indicate that under creep conditions, new cavity nucleation mechanisms become operative:

(i) vacancy condensations at stress concentrations like grain boundary particles and
(ii) grain boundary sliding, resulting in void formation at grain boundary triple points or grain boundary tensile ledges.

For either of these mechanisms, a small threshold strain for stable cavity nucleation can be envisaged. With γ_s is the surface energy, for applied stress σ, the critical size for stable cavity nuclei is given by $2\gamma_s/\sigma$. Therefore, for the first mechanism, plastic deformation must create the necessary critical concentration of excess vacancies. For the second mechanism, a critical amount of grain boundary shear is necessary. In unstable alloys, the density of sites for cavity nucleation may itself evolve during creep exposure.

13.1.2 Cavity Growth

Matrix deformation controls growth of intragranular cavities (e.g. in tension tests) at lower temperatures, whereas the mechanisms of intergranular cavity growth depend upon stress and temperature. For typical creep conditions, the fraction of t_r spent in cavity coalescence can generally be ignored; this is because once cavities coalesce to assume a crack-like morphology, these grow relatively rapidly leading to complete fracture. If continuous nucleation of cavities is ignored, and stable cavities are assumed to be present from the very start of the test, then t_r would equal the time for growth of these stable nuclei to sizes approximately equal to their mean spacing. The relevant expressions have been derived in the literature [see particularly References Cocks and Ashby (1982), Riedel (1987)]. Table 13.1 from Ray et al. (1995) summarizes the salient features, viz. the $(\partial \ln t_r/\partial \ln \sigma)_T$ values and diffusivities expected for the different rate-controlling mechanisms for growth of intergranular cavities. The operative mechanism of intergranular cavity growth at high temperature may also depend upon the instantaneous cavity size. Riedel (1987) derived expressions for various cavity growth mechanisms including continuous nucleation of cavities. Comparison with limited experimental results available shows that these expressions are able to predict t_r within a factor of $\sim$2. It is, however, interesting to note that for constrained cavity growth with diffusive

Table 13.1 Slope $(\partial \ln t_r/\partial \ln \sigma)_T$ for the different mechanisms of intergranular cavity growth and the corresponding diffusivities (Ray et al. 1995)

σ	T	Slope	Cavity growth mechanism	Relevant diffusivity
Low	High	$-\infty$	Cavities sinter	Surface
↓	↑	–1	Constrained; diffusive accommodation	Boundary (coble creep)
		$-n$	Constrained or inhibited; creep accommodation	Boundary
		–5	Fluctuations near sintering stress	Boundary
		–1	Equilibrium diffusive growth	Boundary
		–3	Crack-like diffusive	Surface
		–3/2	Crack-like diffusive	Surface and boundary
		$-n$	Continuum hole growth	Lattice
High	Low			

Note n is the Norton stress exponent (Eq. 13.1)

accommodation, incorporating continuous cavity nucleation did not seem to significantly influence the σ dependence of t_r. Table 13.1, however, shows that in an expression for t_r derived assuming cavity growth mechanisms, the dominant temperature dependence would manifest via activation energy terms.

13.2 Some General Observations on Modelling for Creep Damage Evolution

Technologically, it is necessary to predict the creep behaviour of load-bearing components in service, using data from short duration accelerated laboratory test as the basis. Ideally, this could be achieved by a reasonably accurate modelling for the (tertiary) creep deformation and fracture. For a typical modern creep resistant alloy, t_r can be influenced by

(i) alloy chemistry including tramp elements,
(ii) nature, size and distribution of inclusions (sulphides, oxides, etc.),
(iii) heat treatment, and of course
(iv) the creep testing conditions, which in turn can modulate the phase compositions, and the density cavity nucleating phases initially present and also appearing during creep.

These factors can modulate the phase compositions, and the density of cavity nucleating phases initially present, and also being nucleated during creep exposure. In additions, some alloys are resistant to intergranular cavitation because of dynamic recrystallization during creep. Propensity to dynamic recrystallization is linked to grain boundary mobility and thus on stress, temperature, as well as segregations and precipitations on grain boundaries.

Clearly, in general, ab initio physical modelling for creep damage kinetics for the purpose at hand may prove to be a formidable task and would in addition require extensive body of material data. It is therefore common is to adopt a semi-empirical or an empirical approach. Since the ultimate objective involves extrapolation, it is necessary to ensure that

(i) the creep equation and the extrapolation relations adopted closely reflect the underlying physical processes and
(ii) the dominant creep damage mechanism does not change in from laboratory tests to service conditions.

More realistic the mechanistic basis of a creep relation, higher is its reliability for extrapolation, allowing economy in generating the necessary base data. The last feature is important, because the cost of a creep test is roughly proportional to its duration. Again, extrapolation of the entire creep curves typically requires a large pool of precise experimental data. The common practice therefore is to seek simplifications and generalizations. Frequently, the contribution of primary creep particularly for service conditions can be ignored. Since in practice a few key properties may be used in design (e.g. time to 1% strain, time to onset of tertiary creep, rupture life), these values, rather than the entire creep curves, need to be extrapolated.

13.3 Stress Rupture Parameters

Historically, the search for suitable extrapolation methods particularly for stress rupture lives t_r has been essentially empirical. Starting with the Larson–Miller approach proposed in 1952 (see e.g., Endo et al. 1999), more than thirty such approaches have been proposed; the original references are too numerous to cite here. In this section, a comprehensive summary of the stress rupture parametric approach is presented with examples.

13.3.1 General Kinetic Formulation

It is interesting to note that many of the popular extrapolation methods had been proposed before generalization based upon reaction rate theory was considered by Grounes (1969) in 1969. Such an empirical generalization, however, follows naturally from the differential approach discussed in Sect. 2.2. Suppose a scalar parameter ω measures creep damage in some sense and varies between two fixed limits ω_{i} and ω_{f} for the undamaged and fully damaged conditions, respectively. For example, if ω equals the life fraction (the ratio of creep exposure period to rupture life t_r for the creep conditions), then $\omega_{\mathrm{i}} = 0$, $\omega_{\mathrm{f}} = 1$. It is not necessary to specifically define ω for the following derivation. We assume that the damage kinetics at a constant stress and temperature can be described by an equation of the form

$$\dot{\omega} = f(\omega, \sigma, T) \cdot \exp\left(-\frac{Q(\sigma, T)}{\mathrm{RT}}\right) \tag{13.2}$$

This is a very general statement, without any specific indications as to the mechanism (s) operative. Note, however, the separation of variables assumed in this equation, and also while Q (an activation energy term) is assumed to be independent of ω, the pre-exponential term (function f) is allowed to depend on all the three variables ω, σ and T. Then for isothermal creep at constant stress, time for a given level of damage ω is obtained in the form

$$t_\omega = \exp\left(-\frac{Q(\sigma, T)}{\mathrm{RT}}\right) \cdot \int_{\omega_\mathrm{i}}^{\omega} \frac{\mathrm{d}\omega}{f(\omega, \sigma, T)} = \exp\left(-\frac{Q(\sigma, T)}{\mathrm{RT}}\right) \cdot \Psi(\omega_\mathrm{i}, \omega, \sigma, T)$$

For rupture then, the general expression becomes:

$$t_r = \Psi(\omega_\mathrm{i}, \omega = \omega_\mathrm{f}, \sigma, T) \cdot \exp\left(-\frac{Q(\sigma, T)}{\mathrm{RT}}\right)$$

For fixed $\omega_\mathrm{i}, \omega_\mathrm{f}$, this correlation can be expressed as

$$\ln t_r = \ln \Psi(\sigma, T) + \frac{Q(\sigma, T)}{\mathrm{RT}} \tag{13.3}$$

or, equivalently,

$$T \cdot [\ln t_r - \ln \Psi(\sigma, T)] = \frac{Q(\sigma, T)}{\mathrm{RT}} \tag{13.4}$$

For this general expression, the apparent activation energy for rupture Q_r is obtained as

$$Q_\mathrm{r} = \frac{\partial \ln t_r}{\partial(1/\mathrm{RT})} = \frac{\partial \ln \Psi(\sigma, T)}{\partial(1/\mathrm{RT})} + Q(\sigma, T) \tag{13.5}$$

This shows that, in general, Q_r is stress and temperature dependent, even if Q is not. The formulation above generalizes as many as thirty-one creep rupture parameters proposed in the literature, apart from opening up the possibilities for other possible formulations not yet considered.

Robinson's life fraction rule

Consider a special case of Eq. 13.2, when creep damage kinetics can be expressed in a variable separable form

$$\dot{\omega} = H(\sigma) \cdot \Theta(T) \cdot \Omega(\omega) \tag{13.6}$$

With this form, for creep at constant σ and T, the rupture life is

$$t_r = H(\sigma) \cdot \Theta(T) \cdot \int_0^1 \frac{d\omega}{\Omega(\omega(t))}$$

For creep under varying σ and/or T conditions, usually Robinson's life fraction rule (Robinson 1952) is used to estimate creep life. According to this rule, creep failure in non-steady creep conditions occurs when

$$\sum_i \left(\frac{t}{t_r}\right)_i = 1 \tag{13.7}$$

i.e. the sum of ratios of exposure period t to the corresponding t_r for each of the ith (σ, T) combinations of exposure and reaches a value of unity. This implies that for each steady creep exposure segment, the ratio t/t_r gives the corresponding accumulated creep damage, and for failure corresponds to total creep damage reaching a value of 1. Cocks and Ashby (1982) have shown that this rule naturally follows by substituting Eq. 13.7 in Eq 13.6:

$$\int_0^1 \frac{dt}{t_r} = \int_0^1 \frac{dt}{H(\sigma) \cdot (\Theta T) \cdot \int_0^1 \frac{d\omega}{\Omega(\omega)}} = \int_0^1 \frac{d\omega}{\Omega(\omega) \int_0^1 \frac{d\omega}{\Omega(\omega)}} = 1 \tag{13.8}$$

Thus, considering Table 13.1 and Eq. 13.6, the life fraction rule should be applicable when a single mechanism controls the creep damage kinetics over the entire life under consideration. However, if creep damage mechanism changes because of change of σ and/or T, or if creep damage kinetics is controlled by a coupled mechanism, then depending upon the sequence of mechanisms, the life fraction rule may be unsafe in prediction.

The practical limitations of life fraction rule have led to several alternative creep life summation rules being proposed in the literature (Woodford 1979; Roberts et al. 1979). Some examples are given in the following table; some of these expressions include both life fraction and "strain fraction" $(\varepsilon/\varepsilon_r)_i$, defined in a manner analogous to the life fraction (Table 13.2).

Table 13.2 Creep life summation rules alternative to Robimson's summation of life fraction rule (Woodford 1979; Roberts et al. 1979)

Lieberaman	$\sum_i (\varepsilon/\varepsilon_r)_i = 1$
Freeman and Voortes	$\sum_i [(t/t_r)_i \cdot (\varepsilon/\varepsilon_r)_i]^{1/2} = 1$
El Ata and Finnie	$K \cdot \sum_i (t/t_r)_i + (1-K) \cdot \sum_i (\varepsilon/\varepsilon_r)_i = 1, \quad K < 1$
Oding and Burudsky	$\sum_i [(t/t_r)_i]^m = 1$

Similar to Robinson's life fraction rule, an empirical mixed damage rule for creep-low cycle fatigue has proved very popular in engineering design

$$\sum_i (t/t_r)_i + \sum_j (N/N_f)_j = 1$$

Here N is the number of fatigue cycles for a given set of operating conditions and N_f the corresponding fatigue life (cycles to failure), and thus N/N_f is the "cycle fraction"; i and j refer respectively to incidences of creep and fatigue.

Parametric formulation for stress rupture

Equations 13.3 and 13.4 show that it should be possible to extrapolate stress rupture life (or any other significant creep property that can be uniquely associated with the damage parameter ω), using suitable *parametric formulations*, i.e. defining the functions Ψ and Q. The approach is best appreciated with the help of a specific example.

Example 13.1 White and LeMay (1978) presented the $\dot{\varepsilon}_m$ and t_r data for an SS 316/316L weldment covering wide ranges of stress and temperatures. The t_r values ranged from 1.65 to 1030 h. Figure 13.1a shows that at constant stress levels, variation of $\ln t_r$ with $1/T$ could be considered to be linear; for the best-fit lines shown in Fig. 13.1a, the σ dependences of slopes and intercepts are shown in Fig. 13.1b–c. The results taken together indicate a correlation of the form

$$\ln t_r = (a + b \cdot \sigma) + (c + d \ln \sigma)/T \tag{13.9}$$

Values for the four adjustable constants a, b, c, d can be determined from Fig. 13.1b–c. Comparing Eq. 13.9 with Eq. 13.4, both Ψ and Q depend on σ but not on T.

Figures 13.2a, however, show that for the same data, at constant stress levels, variation of $\ln t_r$ with T could be considered to be linear; for the best-fit lines shown in Fig. 13.2a, the σ dependences of slopes and intercepts are shown in Fig. 13.2b–c. The results taken together indicate a correlation of the form

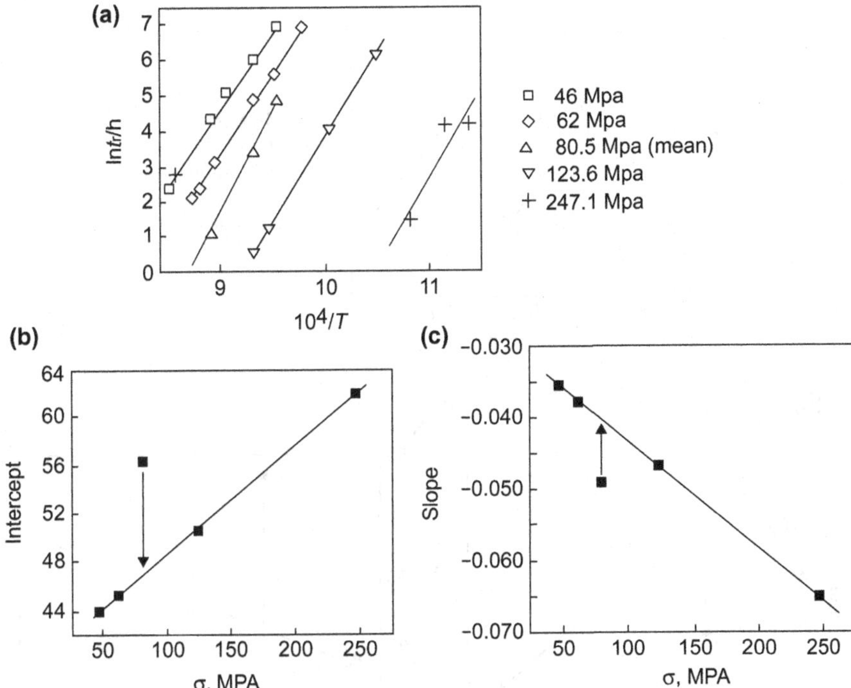

Fig. 13.1 **a** Stress rupture data for SS 316/316L weldment (White and LeMay 1978) suggest that, ln t_r varies linearly with $1/T$ at constant σ [compare with Fig. 16.1a of White and LeMay (1978)], **b–c** for the best-fit lines shown in Fig. 13.1a, the stress dependence of the intercepts and slopes is shown in (**b**) and (**c**) respectively

$$\ln t_r = (a + b \cdot \sigma) + (c + d \cdot \sigma) \cdot T \tag{13.10}$$

Values for the four fitting constants a, b, c, d can be determined from Fig. 13.2b–c. For Eq. 13.10, Ψ now depends only σ but Q depends on both σ and T.

Operationally, both Eqs. 13.9 and 13.10 have four adjustable constants. Note that in Figs. 13.1a and 13.2a, the lines have been obtained by least square optimization, without imposing any other conditions.

In this example, simplest analyses led to two alternative stress rupture parameters for the same data set. With alternative graphical or numerical approaches, many more parametric formulations could be attempted. Note also that both Eqs. 13.9 and 13.10 are consistent with the general formulation, Eq. 13.4. One reason as to why the same data set could lead to two possibilities with very different functional forms is not enough body of data particularly at high rupture lives. This actually is typical of all structural alloys, even though in many instances, rupture data are now available for much higher fractions of service lives than in this example.

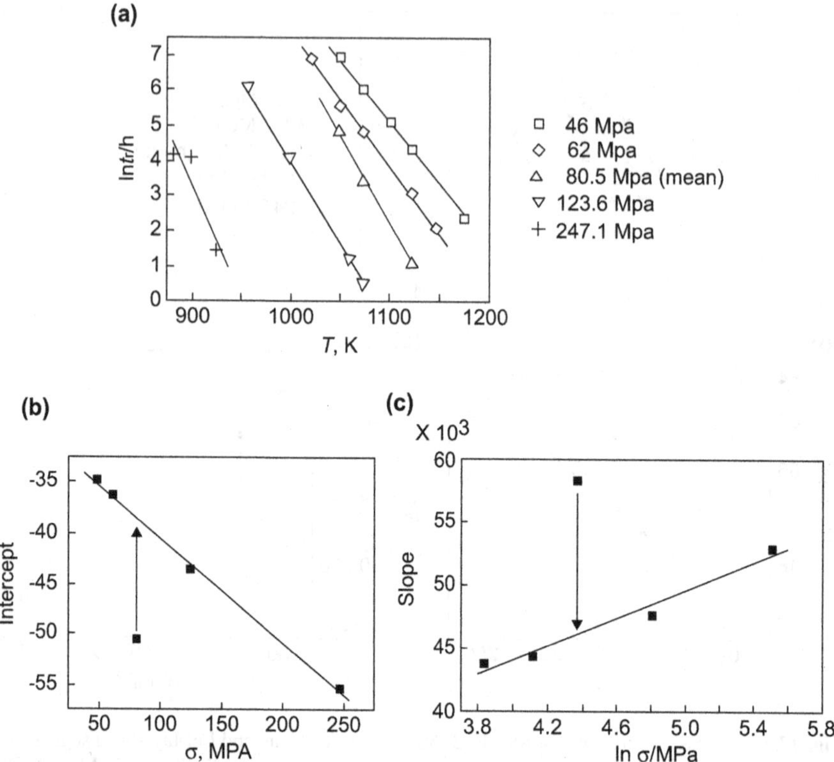

Fig. 13.2 **a** Replot of the data in Fig. 13.1a suggests that at constant σ, $\ln t_r$ varies linearly with T [compare with Fig. 11 of White and LeMay (1978)], **b–c** for the best-fit lines shown in Fig. 13.2a, the stress dependence of the intercepts and slopes are shown in (**b**) and (**c**) respectively

13.3.2 Popular Stress Rupture Parameters

Many of the stress rupture parameters proposed in the literature are based essentially upon empirical analyses of the kind described in Example 13.1. The commonly used functions can be expressed in a parametric form

$$P(\ln t_r, \sigma, T) = G(\sigma) \tag{13.11}$$

Some of the more popular parameters are depicted graphically in Fig. 13.3. The first stress rupture parameter proposed was by Larson and Miller, and probably because of its simplicity, it continues to be the most popular choice to this day. For this parameter, Ψ is constant, while Q depends on stress. In contrast, Orr–Sherby–Dorn parameter is obtained when Q is constant and Ψ depends only on stress.

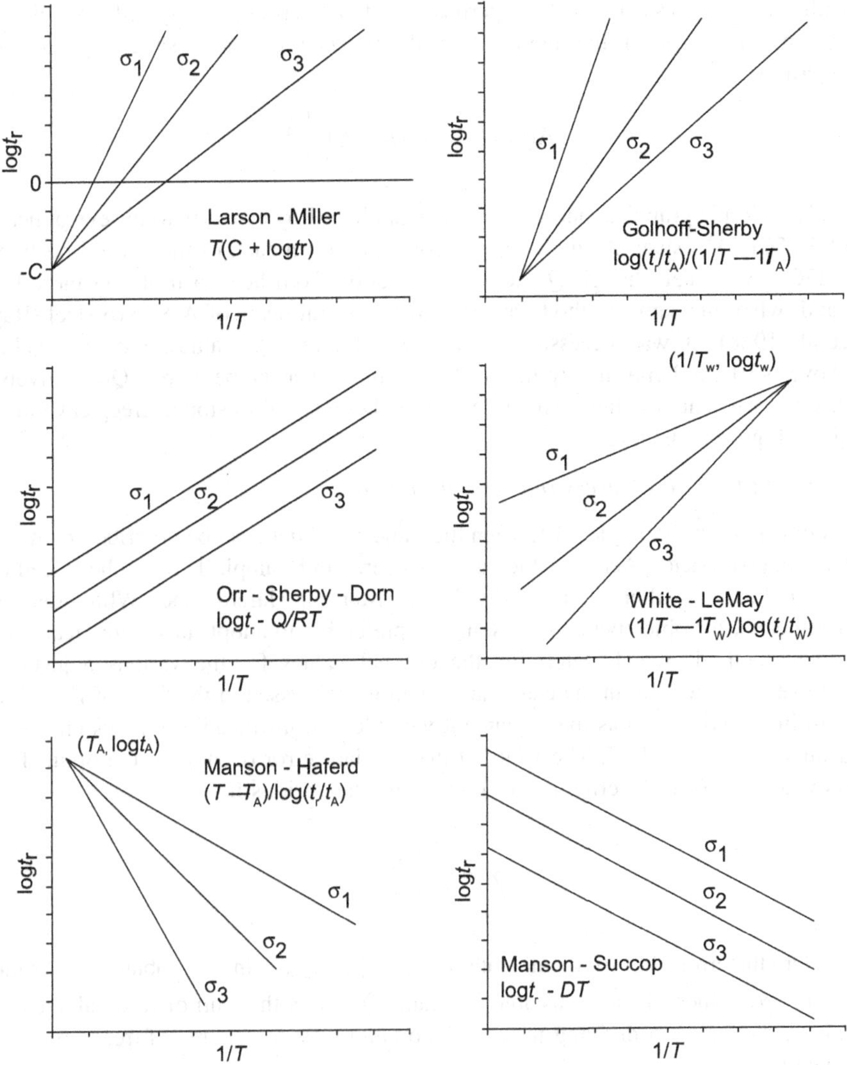

Fig. 13.3 Schematic representation of some time temperature parameters. Parameter is shown in each plot ($\sigma_3 > \sigma_2 > \sigma_1$)

These authors equated Q with the activation energy for self (or alloy) diffusion, i.e. the same as for high-temperature creep, but did not specify the creep damage mechanism. In practice, Q would be chosen so as to minimize data spread. For this parameter, and in general if Ψ depends weakly on temperature, the isothermal plots with $(\ln \sigma, \ln t_r)$ as axes can be superposed on to a "master plot" by shifts parallel to the $\ln t_r$ axis. From the temperature dependence of these shifts, Q can be determined:

$$\Delta(\ln t_r)_\sigma = (Q/R) \cdot \Delta\left(\frac{1}{T}\right)$$

This simple graphical analysis has been applied (Ray et al. 1988) to several heats of 2.25Cr-1Mo steel. A single Q value obtained from each of the several heats of 2.25Cr-1Mo steel, though Q showed a variation from heat to heat. On the other hand, when the same method was applied to several heats of AISI 316 steel (Ray et al. 1988), it was necessary to use two different Q values, with the value above ~910 K being nearly half of that at the lower temperatures. Qualitatively, this is consistent with increasing role of grain boundary diffusion in creep cavitation at the higher temperatures.

Choosing the "best" stress rupture correlation

As illustrated in Example 13.1, often the same set of data can be described by more than one parameter. Actually, the data considered in Example 13.1 can be forced to fit to Larson–Miller, Orr–Sherby–Dorn, Manson–Haferd and White–LeMay parameters, for objective comparison, it is preferable to adopt fully numerical least square methods which determine the optimal values for the various adjustable parameters appearing in the candidate equation, expressed in the form of Eq. 13.3, with $\ln t_r$ (and *not* t_r) as the dependent variable; the justification for this choice is dealt with in Sect. 13.7. The value of the standard error of fit χ for the optimal fit provides an objective criterion for comparing the various fits:

$$\chi = \sqrt{\frac{1}{n_{\mathrm{f}}} \sum_{\mathrm{i}} e_{\mathrm{i}}^2}$$

Here the error of fit for the ith data set, $e_{\mathrm{i}} = \left(\ln t_{\mathrm{r,data}} - \ln t_{\mathrm{r,fit}}\right)_{\mathrm{i}}$ obtained with the optimized values of the adjustable constants, $\sum_{\mathrm{i}} e_{\mathrm{i}}^2$ is the sum of residual squares for the data used in deriving the correlation and n_{f} is the degree of freedom of fit, given by

$$n_{\mathrm{f}} = \mathrm{i} - \text{number of adjustable constants optimized} - 1$$

Least square fitting of data to an equation in the form of Eq. 13.3 implies that the scaled error of fit $\kappa \equiv \left(\ln t_{\mathrm{r,data}} - \ln t_{\mathrm{r,fit}}\right)/\chi$ has a Gaussian distribution, with zero mean and unit standard deviation. It is important to verify that distribution of κ does meet these requirements. This can be verified by rigorous statistical methods. Simpler is to graphically verify that variations of κ with the input parameters do not show any systematic trends. Finally, it is also necessary to ensure that the correlation does not result in untoward curvatures or unacceptable intersections of isothermals, etc., particularly in extrapolation. The choice for the function $G(\sigma)$ is important in this regard.

Choice for the function $G(\sigma)$

For numerical data fitting to Eq. 13.11, it is first necessary to define the function $G(\sigma)$. Most often a polynomial in σ or $\log\sigma$ has been used, the degree of the polynomial also being optimized in the course of the computation. Both experience and understanding of the mechanisms involved indicate that the stress rupture isotherms should be monotonic with σ and $\left(\partial^2 \ln t_r / \partial(\ln\sigma)^2\right) \leq 0$.[2] Such monotonic behaviour cannot be assured a priori in a polynomial of degree higher than two, while a quadratic polynomial may not be precise enough for data correlation. The implications of these requirements have been considered in the literature. The following alternative forms for $G(\sigma)$ meeting these requirements (a, b, c, d... constants) have been considered in the literature

$$G = a - b \cdot \log\sigma - c \cdot \sigma - d \cdot \sigma^2, \quad a, b, c, d \geq 0$$
$$G(\sigma) = a + b \log \sinh(c\sigma)$$

Yet another possibility is to use a spline function; limited experience with such spline functions suggests that the benefits may not be commensurate with the effort necessary to write the complicated computer program.

Example 13.2 This example shows a typical stress rupture parametric analysis to illustrate some of the points mentioned above. NRIM creep data sheet No 47 (National Research Institute for Metals (NRIM) 1999) documents the various creep properties for one heat of an Fe-21Cr-32Ni-Ti–Al (alloy 800H) superalloy covering stress range ~60–400 MPa, and temperature range 823–1073 K, with stress rupture lives beyond ~57,500 h. This data sheet also contains results of analysis of the stress rupture data using the Manson–Haferd parameter. The optimal correlation developed was (T in K, σ in MPa, t_r in hours):

$$\begin{aligned} &\log t_r = 9.314026 + (T - 660) \cdot P, \text{ with} \\ &P = \sum_{k=0}^{4} b_k \cdot (\log\sigma)^k \\ &b_0 = -1.345394,\ b_1 = 2.526676,\ b_2 = -1.772799, \\ &b_3 = 5.475845 \times 10^{-1},\ b_4 = -6.350638 \times 10^{-2} \\ &\chi = 0.117 \end{aligned} \tag{13.12}$$

[2]Exceptions are when cavity sizes are very small, close to the critical size for stability, or in precipitation strengthened steels, if the intrinsic creep strength of the matrix is approached. These processes are typical of very low σ levels. These are not considered in this chapter.

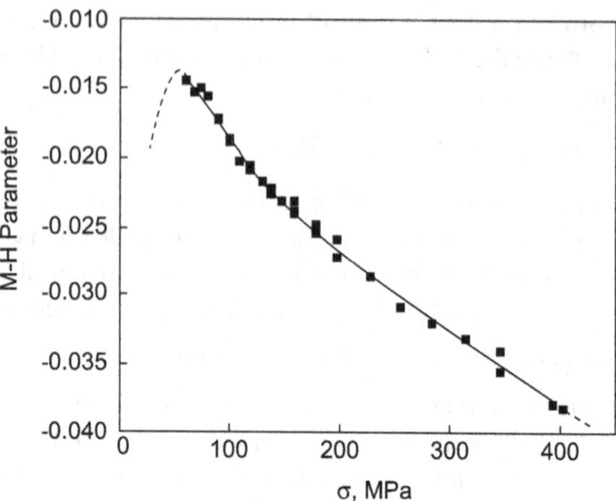

Fig. 13.4 Manson–Haferd parametric correlation (Eq. 13.12) for a heat of Alloy 800H. The symbols represent the data, the firm line the polynomial derived for the range covered by the data, and the dashed line the extension of the polynomial beyond the data range

Figure 13.4 plots the parametric correlation. Clearly, extrapolations from this parametric correlation for stresses lower than the minimum stress in the data would be erroneous because of the non-monotonic variations. The solution in this instance is to restrict extrapolations for t_r to the acceptable segment of the actual range of parameter values, i.e. firm line in Fig. 13.4. Figure 13.5 suggests that no significant error is anticipated in considering κ as a unit normal deviate.

For this correlation, the apparent activation energy for creep rupture at stress σ and temperature T is obtained as

$$Q_{\mathrm{r}} = \left(\frac{\partial \ln t_r}{\partial\left(\frac{1}{RT}\right)}\right)_{\sigma} = 2.3026 \cdot R \cdot P(\sigma) \cdot \frac{\partial(T - 660)}{\partial\left(\frac{1}{T}\right)} = -2.3026 \cdot R \cdot P(\sigma) \cdot T^2$$

For the range of the data covered, Q_{r} computed varies from 493 kJ/mole for $\sigma = 400$ MPa, $T = 823$ K, to 311 kJ/mole for $\sigma = 60$ MPa, $T = 1073$ K. The isoQ_{r} contours for three levels are shown in Fig. 13.6. Q_{r} increases with increasing σ, and also T. The variation of Q_{r} with σ seems to agree, and with T seems to disagree, with the expectations from mechanistic considerations summarized in Table 13.1. Also, magnitudes of Q_{r} seem to be rather high for lattice diffusion to be the operative mechanism with highest activation energy.

The following table presents for some σ-T pairs the predicted mean t_r values, and the upper and lower bounds using $\pm 2\chi$ (66.5% probability) criterion. The last column, taken from the data sheet gives information about the actual experimental situations—the time at which test was reported as continuing, or test was terminated. The predictions are generally satisfactory. Note that these extrapolations are within the acceptable range (firm line in Fig. 13.4), and as such are free from potential problems with polynomial representation of the $G(\sigma)$ function (Table 13.3).

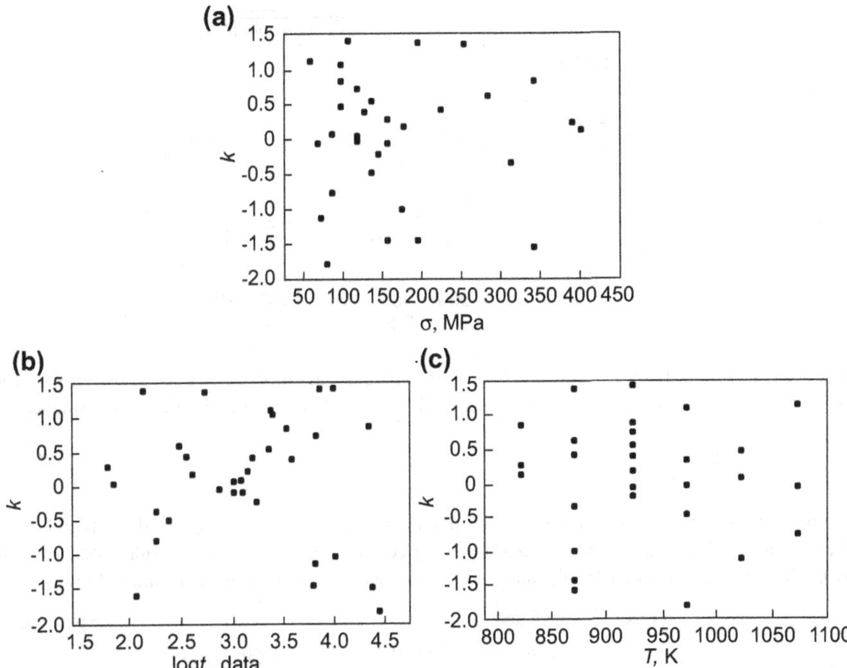

Fig. 13.5 For the fit shown in Fig. 13.4, κ values plotted against **a** σ, **b** experimentally determined $\log t_r$ or **c** T are generally consistent with κ being distributed with zero mean and unit standard deviation

13.3.3 Generalizations

The very large number of possible parametric correlations, plus the fact that the same body of data can often be forced-fit to several alternative forms, has led to attempts at developing generalized analysis schemes (Goldhoff 1979; Manson and Ensign 1979).

Method of Booker and Rummel

The generalized regression scheme adopted by Booker and Rummel involves fitting the data to equations of the form

$$\log t_r = \alpha_0 + \sum_{\mathrm{i}} \alpha_{\mathrm{i}} \cdot g_{\mathrm{i}}(\sigma, T) \tag{13.13}$$

The set of functions g_{i} contain various plausible combinations of σ and T but no adjustable constants; only the set $\alpha_0, \alpha_{\mathrm{i}}, \quad i = 1, 2, \ldots$ appearing linearly in the

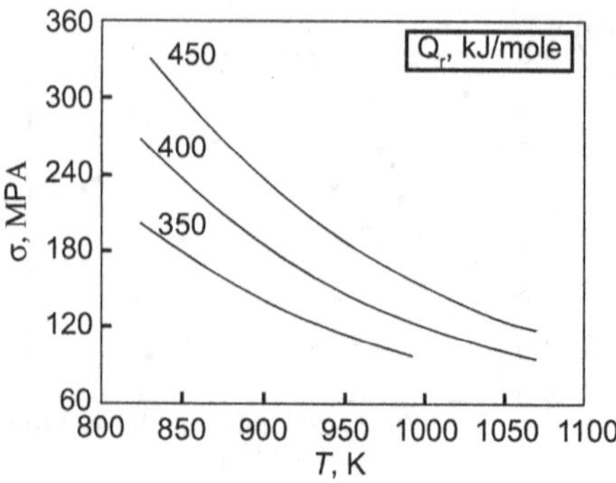

Fig. 13.6 Contours for three constant levels of apparent activation energy for stress rupture for the stress rupture parametric correlation derived for Alloy 800H (Eq. 13.12) and displayed in Fig. 13.4

Table 13.3 For Alloy 800H superalloy creep ruptures data from NRIM creep data sheet No 47 (National Research Institute for Metals (NRIM) 1999), comparison of the predicted results (with statistics) by the Manson-Haferd parametric fit carried out in Example 13.2, with actual experimental results

No.	σ MPa	T K	Predicted t_r, h			Exptl. t_r h
			Mean	Lower bound	Upper Bound	
1	167	873	11408.4	6656.2	19553.5	>13368
2	147	873	24122.8	14074.3	41345.5	>13368
3	137	873	36361.4	21214.9	62321.8	>57481
4	118	873	86041.9	50200.7	147472.1	>57481
5	98	873	243274.7	141937.5	416962.5	>57481
6	98	923	29110.4	16984.3	49894.0	>57312
7	69	923	230033.2	134211.7	394267	>12888

equation are the unknowns constants to be optimized. Therefore, for each functional form chosen, only linear least square fitting of data is necessary which considerably reduces the programming and the computational burden. The method consists in selecting by trial and error from among a very large number of "candidate" $g_i(\sigma, T)$ functions the particular set that best describe the data; the final selection is made based upon standard error of fit, subject to the predicted extrapolated isothermals and their extrapolations being free of undesirable curvatures or illogical intersections, etc. This type of analysis requires considerable experience and judgment.

Manson's General Parameter

Some of the popular stress rupture parameters can be expressed in a common form (called the general parameter) suggested by Manson:

$$P = \sigma^{-q}(\log t_r - \log t_A)/(T - T_A)^r$$
$$\text{i.e,} \quad \log t_r = \log t_A + \sigma^q \cdot (T - T_A)^r \cdot G(\sigma) \tag{13.14}$$

where q, r, t_A, T_A are adjustable constants:

Manson–Brown parameter:$q = 0$.
Manson–Haferd parameter: $q = 0,\ r = 1$
Larson–Miller parameter: $q = 0,\ r = -1,\ T_A = 0$

Typically, a single numerical optimization program, with provision for prior selection of alternative special forms, is used to analyse the data. An alternative will be to fit the data to the generalized form to optimize the values for all the four adjustable constants, plus those appearing in the G function, and examine post-optimization, if values for one or more of q, r or T_A are close to their special values.

Manson's Universal Parameter

Constants for the Manson–Haferd parameter derived in Example 13.2 are very close to the "universal" values in the parameter $(\log t_r - 10)/(T - 600)$ that seems to be reasonably satisfactory for many materials. Similarly, a "universal" Larson–Miller parameter $(T + 460) \cdot (20 + \log t_r)$ has also been proposed. Recognizing that the predicted stress for extrapolation to long rupture lives were often optimistic with the "universal" L–M parameter, and conservative with the "universal" M–H parameter, Manson developed a compromise parameter for steels and superalloys, intended to yield values midway between these two predictions:

$$P \equiv \log t_r + (\log t_r)^2/40 - 40000/(T + 460) = G(\sigma) \tag{13.15}$$

Example 13.3 For the data for Alloy 800H dealt with in the previous example, Fig. 13.7a presents the correlation by this "universal" parameter. The high scatter compared to Fig. 13.4 is obvious. Interestingly, least square fit of the data to the parameter

$$P \equiv \log t_r + (\log t_r)^2/A_1 - A_2/(T + A_3)$$
$$P = b_0 + b_1 \cdot \log\sigma + b_2 \cdot (\log\sigma)^2 + b_3 \cdot (\log\sigma)^3 \tag{13.16}$$

so as to optimize all the parameters appearing in the above yields the optimal solution shown in the following table. The corresponding fit is shown in Fig. 13.7b. In effect, the parameter in this instance reduces to

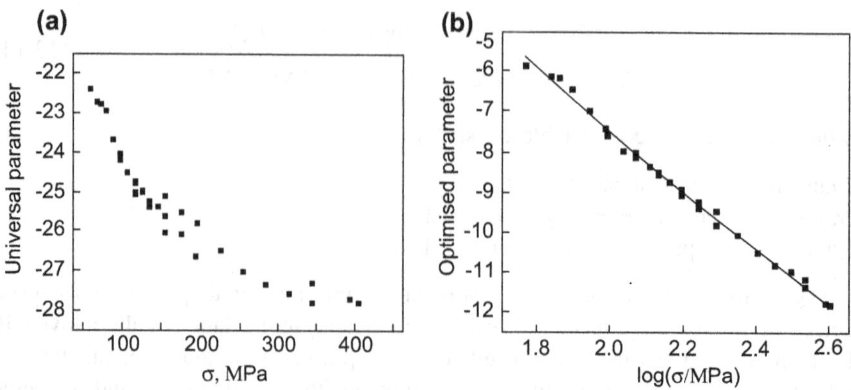

Fig. 13.7 **a** Manson's Universal Parameter correlation (Eq. 13.15) for the data analysed in Figs. 13.4, 13.5 and 13.6. Symbols represent the test data. **b** Improved correlation by optimizing the constants in Eq. 13.16 compared to Fig. 13.7a. Symbols represent the test data, and the line the optimal fitted polynomial

$$P = \log t_r - A(2)/(T + A(3))$$

With $A_1 = -9.552670779553266e + 015$

$A_2 = 6.105547295224357e + 003,\quad A_3 = -4.142762331287060e + 002,$

$b_0 = 1.921018176736574e + 001,\quad b_1 = -2.112080583289197e + 001$

$b_2 = 4.998534439854191e + 000,\quad b_3 = -5.611239561623699e - 001$

Minimum Commitment Method

In this method, a general stress rupture equation, of the form

$$\log t_r + A \cdot P(T) \cdot \log t_r + P(T) = G(\sigma) \tag{13.17}$$

is assumed to be applicable to all materials. Here, P is a function of T, and G a function of stress. A is a characteristic material constant. The analytic expression

$$P = R_1 \cdot (T - T_{\mathrm{m}}) + R_2 \cdot (1/T - 1/T_{\mathrm{m}})$$

with R_1 and R_2 constants and T_{m} mid-range temperature is reported to be adequate for many materials. The possible functional forms for G have been discussed earlier. Determining A has proved to be a difficult problem. A reflects the microstructural stability of the material: materials known to be highly unstable require large negative values of A ($\sim$–0.15), and stable materials are best correlated with $A = 0$.

Since all isothermals would intersect at a value of $\log t_r = G = -1/A$ (hence the name of this version of MCM is *focal point convergence method*), a possible way to determine A is to determine the point of convergence of the isothermals, by extrapolation if necessary. This way, A can be determined independently from P and G. Thus for the Larson–Miller parameter with the universal value of 20 (with $\log t_r$) for the Larson–Miller constant, $A = 0.05$; for Orr–Sherby–Dorn and Manson–Succop parameters, $A = 0$; and for Manson–Haferd parameter with $\log t_a > 0$, A is negative. For the focal point convergence method, the limiting value of $(\partial \ln t_r/\partial \ln \sigma)_T$ as $\sigma \to 0$ varies with temperature. This is in contrast with the predictions from the cavity growth mechanisms (Table 13.1): excluding cavity sintering at very low stresses and high temperature, in general, $-1 \leq (\delta \ln t_r/\delta \ln \sigma)_T \leq n$, and $\left(\delta^2 \ln t_r/(\delta \ln \sigma)^2\right)_T \leq 0$, with a smooth variation in curvature between the limits. In the context of comparing two multiple heat stress rupture correlations, Sasikala et al. (1999) carried out isothermal fitting of $\ln t_r$-$\ln \sigma$ data at several temperature levels, each for several heats of 0.5Mo–0.15C steel, a 9Cr–1Mo steel, two grades of 2.25Cr–1Mo steels, and two grades of AISI type 316 stainless steels. These authors strongly recommended imposing the above restrictions for general applications. In fact, Manson and Ensign (1979) gave two examples where results were more consistent with the alternative *oblique translation convergence* method (illustrated in Fig. 13.8 and discussed in the next section): for a 25Cr-1Mo steel, they ascribed this to precipitation processes, and for a 17-22-AS, to transition from transcrystalline to intercrystalline failure.

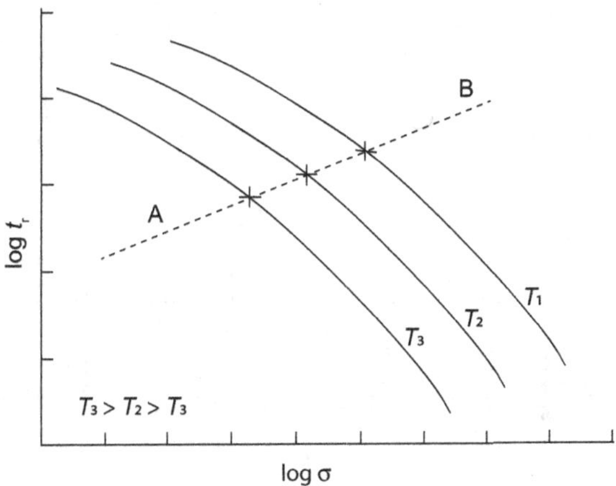

Fig. 13.8 Schematic illustration of oblique translation convergence. Any isothermal can be can be brought into superposition by shifting parallel to the axes, such that if $\Delta \log \sigma$, $\Delta \log t_r$ are the extents of shifts, the ratio $M = \Delta \log t_r/\Delta \log \sigma$ remains constant. Graphically, the dashed line AB has slope M, but is otherwise arbitrarily located. Its intersections with the isothermal are marked and would coincide after oblique translation

13.4 The Monkman–Grant Relation

Monkman and Grant (1956) proposed that the product of steady-state creep rate and rupture life $(\dot{\varepsilon}_s \cdot t_r)$ is a constant for a given material. It has often been observed that a modified form of Monkman–Grant relation, $\dot{\varepsilon}_s^{\alpha} \cdot t_r =$ constant, where α is a material dependent constant that is close to unity is more appropriate. Dobes and Milicka (1976) proposed a modified Monkman–Grant relationship: $\dot{\varepsilon}_s \cdot t_r / \varepsilon_r = C_{MMG}$ where C_{MMG} is a material dependent constant and ε_r is the true strain at fracture. The damage tolerance parameter λ of Ashby and Dyson (1984), which assesses the ability to accumulate strain without fracture and thus creep ductility (see Fig. 13.9 for a schematic illustration) is the reciprocal of C_{MMG}.

The modified Monkman–Grant relationship has been found to apply for many materials. Figure 13.10 presents a typical modified Monkman–Grant-type plot for the weld metal of a nitrogen-bearing variant of an austenitic stainless steel. This figure also illustrates a common practice: for engineering alloys, most often the creep curve does not show an extended steady state and therefore in this type of analysis, the minimum creep rate $\dot{\varepsilon}_m$ is used in place of $\dot{\varepsilon}_s$. The Monkman–Grant relation (or its modifications) establishes a link between creep deformation and creep fracture. The practical implications of such a link will be clear from the simple Example 13.4.

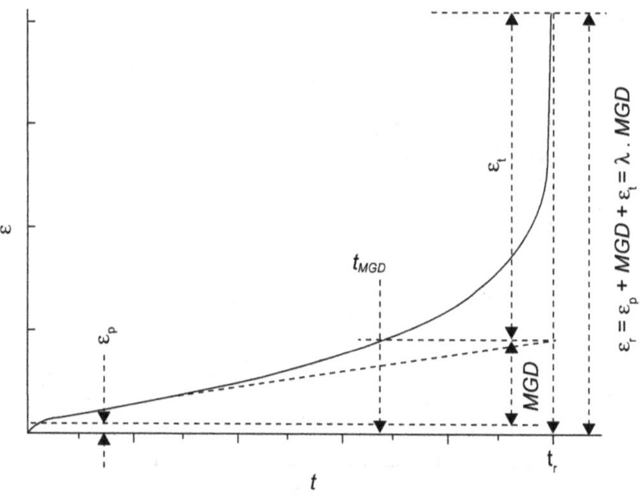

Fig. 13.9 Schematic creep curve for negligible primary creep strain ε_p. The Monkman–Grant ductility, MGD $= \dot{\varepsilon}_s \cdot t_r$, and ε_t is the tertiary creep strain, The rupture strain $\varepsilon_r = \varepsilon_p + MGD + \varepsilon_t = \lambda \cdot MGD$, where λ is the damage tolerance parameter. t_{MGD} is the time to creep strain equal to MGD. [After Phaniraj et al. (2003)]

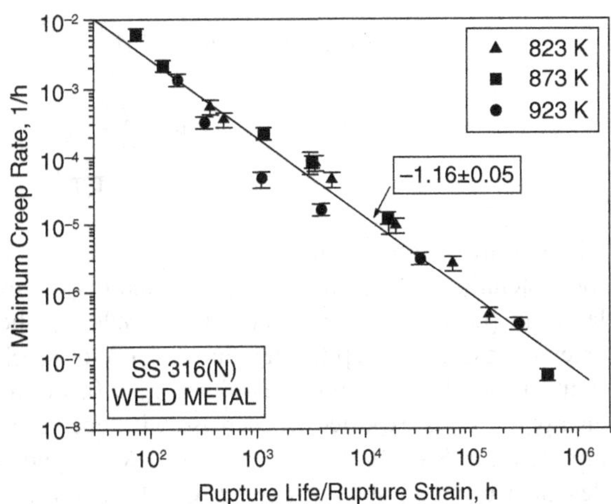

Fig. 13.10 For a SS 316 (N) weld metal, double logarithmic plot of $\dot{\varepsilon}_m$ against t_r/ε_r displays the Monkman–Grant relation modified by Dobes and Milicka. Sasikala (2001), Sasikala et al. (2004)

Example 13.4 For a given creep temperature T and stress σ_1, aluminium specimens fail in 1000 h in laboratory tests, and it is required to estimate the stress σ_2 for a service life of 200,000 h at the same temperature. If power law creep is assumed to obtain extending over laboratory and plant conditions, then a value of $n = 4$, the "universal " stress exponent for PL creep in fcc materials, should be a reasonably good estimate for the Norton stress exponent. Invoking Monkman–Grant relation,

$$(\sigma_1/\sigma_2)^4 = 200000/1000 = 200, \text{ and therefore } \sigma_2 = 0.266\sigma_1.$$

Attempts have been made in the literature to examine the validity of Monkman–Grant relation for various creep damage mechanisms. Considering Table 13.1 such a link would be expected when matrix deformation substantially controls the creep damage kinetics. With the modified Monkman–Grant relationship $\dot{\varepsilon}_s^\alpha \cdot t_r = C$ (C and α material dependent constants), and invoking the equation for power law creep, the stress rupture equation is of the form

$$\ln t_r = A - n\alpha \cdot \ln(\sigma/\mu) + \alpha Q_{SD}/RT \tag{13.18}$$

where, as before, n is the Norton exponent. The apparent activation energy for stress rupture is thus $Q_r = \alpha Q_{SD}$. Note that this equation is consistent with the Orr–Sherby–Dorn parameter, and $Q_r \approx Q_{SD}$ since α in general is close to unity (see also Fig. 13.14), From Table 13.1, such a relation can be expected for intra- or intergranular creep cavitation controlled by matrix creep ($\alpha = 1$). For constrained intergranular cavitation, the stress exponent would be expected to be the same, but the activation energy would equal that for boundary diffusion. If instead, power law creep modulated by a constant back stress σ_b is applicable, then the equation for stress rupture life is of the form

$$\ln t_r = A - n_{\text{PL}}\,\alpha \cdot \ln\left(\frac{\sigma - \sigma_{\text{b}}}{\mu}\right) + \alpha\frac{Q_{\text{SD}}}{\text{RT}}$$
$$Q_{\text{r}} = n_{\text{PL}}\,\alpha \cdot \frac{\mu}{\sigma - \sigma_b}\frac{\partial\left(\frac{\sigma_{\text{b}}}{\mu}\right)}{\partial\left(\frac{1}{\text{RT}}\right)} + \alpha Q_{\text{SD}} \tag{13.19}$$

with the stress exponent for power law creep $n_{\text{PL}} \approx 3$ for ferritic steels and $n_{\text{PL}} \approx 4$ for austenitic steels. σ_{b} in precipitation and dispersion hardened alloys can be very large. In precipitation strengthened materials, σ_{b} should decrease with increasing creep exposure or creep temperature because of precipitation coarsening and in the limit of very high rupture life values, $\sigma_{\text{b}} \approx 0$. In other words, apparent activation energy for $\dot{\varepsilon}_{\text{m}}$ is expected to decrease with decreasing stress levels. This has been reported for SS 316 by White and LeMay, and their parameter was actually designed to incorporate this trend. On the basis of Eq. 13.19, White and LeMay observed that

(i) for Orr–Sherby–Dorn parameter to be applicable, the back stresses should be minimal as is the case for relatively pure metals and single phase alloys,
(ii) for White–LeMay and Manson–Haferd parameters, back stresses are significant and Q_{r} decreases with decreasing σ and
(iii) for Larson–Miller and Goldhoff–Sherby parameters to be applicable when the applied stress aids in thermal activation and therefore Q_{r} decreases with increasing σ.

Applicability of Monkman–Grant-type relations indicates that in a stress rupture correlation, stress should be scaled with shear modulus μ. This subtle refinement may become significant when dealing with data from a wide range of temperature and variation in μ with T is too large to be ignored. For nitrogen-bearing SS 316 weld material with substantially larger t_r ranges than those in White–LeMay, it has been shown (Sasikala 2001) that the apparent activation energy for $\dot{\varepsilon}_{\text{m}}$ is higher than that for t_r, even though modified Monkman–Grant relationship was found to be applicable, Fig. 13.10. This difference in the activation energies has been attributed to precipitates coarsening during tertiary creep regime. One may thus conclude that for such complex engineering alloys tested to large t_r levels, the apparent validity of Monkman–Grant or modified Monkman–Grant relations really reflects the insensitivity of these models to this variation in activation energy. For data fitting purposes, it is more convenient to ignore σ_{b}. Then Norton stress exponent n varies continuously with stress and temperature, with a value of 4 at low-stress/high-temperate (high rupture life) limits for fcc materials and three for other alloys, which continues to increase with increasing stress and/or decreasing temperature to values $\sim$10–40 or even higher depending on material.

The above discussion implies, as does Eq. 13.4, that a single activation energy term is adequate to account for the strong temperature effect on creep rupture life. This is expected when one of the mechanisms in Table 13.1 is (primarily) rate controlling. Cocks and Ashby (1982) have shown that with realistic simplifications, void growth controlled by any one or two of the three important mechanisms, namely matrix deformation, boundary diffusion and surface diffusion, the stress rupture correlation would be expected to be of the general form (Cocks and Ashby 1982)

$$\begin{aligned} P_2 &= G(P_1) \\ P_1 &= \log\sigma - H\cdot\left(\frac{1}{T}-\frac{1}{T_{\mathrm{m}}}\right) - \log\left(\frac{\mu(T)}{\mu(T_{\mathrm{m}})}\right) \\ P_2 &= \log t_r - J\cdot\left(\frac{1}{T}-\frac{1}{T_{\mathrm{m}}}\right) \end{aligned} \tag{13.20}$$

with T_{m} a convenient reference (e.g. mid-range) temperature and J and H are constants. Note the scaling of σ with temperature dependent shear modulus μ. For example for cavity growth by coupled matrix deformation—boundary diffusion,

$$H = \frac{Q_{\mathrm{c}} - Q_{\mathrm{b}}}{2.303R(n-1)}, \quad J = \frac{nQ_{\mathrm{b}} - Q_{\mathrm{c}}}{2.303R(n-1)}$$

Q_{c} and Q_{b} are respectively (constant) creep and boundary diffusion activation energies, and n, the Norton stress exponent, is also assumed to be constant. Cocks and Ashby (1982) have shown that Eq. 13.20 should apply for creep rupture resulting from precipitate coarsening, and even for simple instances of nucleation dominated fracture. It can be readily verified that Eq. 13.20 corresponds to oblique translation convergence method, with isothermal stress rupture data plotted with $\log\sigma$, $\log t_r$ as axes. Cocks and Ashby used a graphical construction based on this feature to show the applicability of Eq. 13.20 in a few instances. However, application of this "temperature correction" to problems of creep life inter- and extrapolation using numerical methods, optimizing J and H in the process, seems to be very limited (Ray et al. 1995).

13.5 Continuum Damage Mechanics Approach

The continuum damage mechanics approach, pioneered by Kachanov (1990), Rabotnov (1969), can be considered as an extension of the approach in Sect. 13.3.1. Specifically, the evolution of strain rate $\dot{\varepsilon}$ is included to obtain the coupled differential equations:

$$\begin{aligned} \dot{\varepsilon} &= \dot{\varepsilon}(\sigma,\omega,T) \\ \dot{\omega} &= \dot{\omega}(\sigma,\omega,T) \end{aligned} \tag{13.21}$$

The entire creep history leading to rupture would be obtained by integrating this equation pair. Much of the attraction of such a continuum damage mechanics-based formulation is lost unless the damage parameter is physically identified and can be independently measured. Specifically, for "creep brittle" materials, Kachanov identified ω with the (effective) fractional loss of original load-bearing area A_0 due to creep cavity formation with limiting values 0 in the undamaged (virgin) condition, and 1 for a fully damaged condition, i.e. complete rupture. With this definition for ω, for a constant load creep test at load P, the effective stress becomes

$$\frac{P}{A_0(1-\omega(t))}$$

He also proposed that effective elastic modulus can be used as an independent measure of ω; higher the value of ω, lower is the effective elastic modulus.

Some intuitive simplification of Eq. 13.21 is possible, and practically necessary. Had there been no damage accumulation during tertiary creep, the creep rate would have remained constant at its value for the undamaged condition, $\dot{\varepsilon}_{t=0} = \dot{\varepsilon}_0$. Therefore, irrespective of the details of the micromechanism or kinetics of creep damage, the parameter $\dot{\varepsilon}/\dot{\varepsilon}_0$, the ratio of instantaneous to initial creep rates, is a measure of the current damage level ω. The following constitutive equation for uniaxial creep deformation is consistent with this intuitive reasoning. With current applied stress $\sigma = P/A$ and σ_0 its initial value,

$$\begin{aligned}\dot{\varepsilon} &= \dot{\varepsilon}_0 \cdot \left(\frac{\sigma}{\sigma_0}\right)^n = \dot{\varepsilon}_0 \cdot \left(\frac{1}{1-\omega}\right)^n \\ \dot{\omega} &= \dot{\omega}_0 \cdot \left(\frac{\sigma}{\sigma_0}\right)^m = \dot{\omega}_0 \cdot \left(\frac{1}{1-\omega}\right)^m\end{aligned} \tag{13.22}$$

The power law form for $\dot{\varepsilon}$ is a straightforward extension of the Norton equation; n is the Norton stress exponent for the true steady-state creep in the *undamaged* material, $\dot{\varepsilon}_0$. The power law form for $\dot{\omega}$ is empirical, requiring experimental validation. Ideally, with this model, fracture occurs when $\omega = 1$. Actual measurements on crept specimens, however, show that the fraction of cross-sectional cavitation is $\sim 0.1 - 0.2$, much lower than the ideal value 1. This indicates that fracture is preceded by a regime of fast cavity coalescence. In this regime, Eq. 13.22 may cease to be valid; however, since this regime occupies a small fraction of total life, this factor alone should not significantly influence the accuracy of t_r estimation by this approach.

With constant m, n and $m+1 > n$, for a constant stress, integration of Eq. 13.22 gives (Riedel 1987),

$$\frac{\varepsilon}{\varepsilon_r} = 1 - \left(1 - \frac{t}{t_r}\right)^{1/\lambda}, \quad \lambda = \frac{m+1}{m+1-n} \tag{13.23}$$

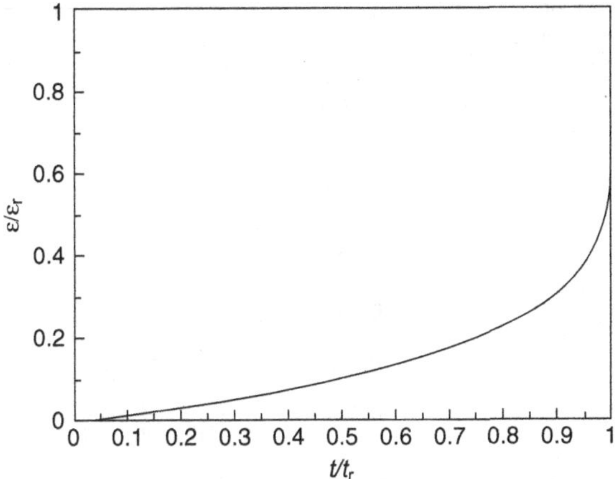

Fig. 13.11 Plots of reduced creep curves predicted by Eq. 13.23 for $\lambda = 6$

Here creep strain $\varepsilon = 0$ at $t = 0$, and $\varepsilon = \varepsilon_r$, the rupture strain at $t = t_r$, the rupture life. λ can be identified as the damage tolerance parameter of Ashby and Dyson.

Suppose $m = n = 5$; then $\lambda = 6$ and Eq. 13.23 can be used to compute the $\varepsilon/\varepsilon_r - t/t_r$ relation as shown in Fig, 13.11. As long as n and m (and therefore, λ) remain constant, each unique value of life fraction corresponds to a unique value of strain fraction, and either of these is a measure of current creep damage level. Indeed, as Eq. 13.21 represents a special case of the general variable separable form for damage kinetics, Eq. 13.6, for constant n and m (and therefore, for constant λ), it is consistent with the life fraction rule for creep under non-steady conditions, Eq. 13.7.

It has been shown (Ashby et al. 1984) that each individual damage micromechanism results in a characteristic shape of creep curve, with a corresponding value for $\lambda : \lambda = 2$ to 5 for necking and boundary diffusion-controlled void growth; $\lambda = 1$ for surface diffusion-controlled void growth giving rise to crack-like damage resulting in low ductility and also for void growth controlled by power law creep accompanied by large ductility; and $\lambda > 10$ for damage in the form of degradation of microstructure by thermal coarsening of particles during creep in particle-strengthened materials. From Eq. 13.23, the time t_{MGD} where creep strain ε equals $\dot{\varepsilon}_0 \cdot t_r = \varepsilon_r/\lambda$ is given by the expression (Phaniraj et al. 2003)

$$t_{\mathrm{MGD}}/t_r = 1 - [(\lambda - 1)/\lambda]^{\lambda}$$

Figure 13.12 shows the λ dependence of t_{MGD}/t_r. For this equation, the ratio t_{MGD}/t_r approaches $1 - 1/e = 0.63$ (e is the base of natural logarithm) (Phaniraj et al. 2003). The trend is schematically shown in Fig. 13.12. Experimental data for SS 304, and also 9Cr-1Mo steel at low and high stresses covering λ in the range 2.1–10 matched the predictions (Phaniraj et al. 2003).

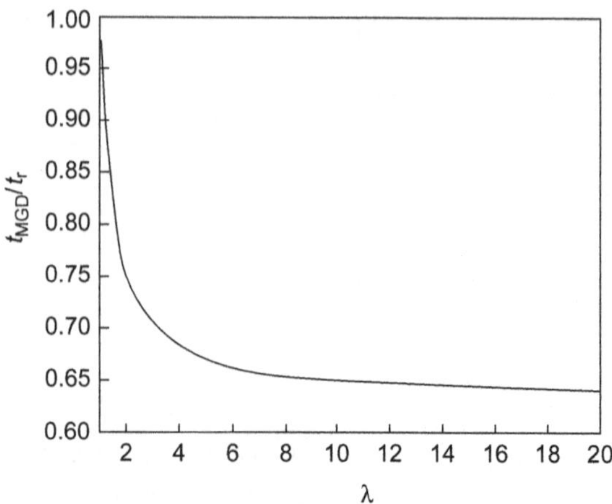

Fig. 13.12 Variation of t_{MGD}/t_r with λ for Eq. 13.23

Since the ratio becomes essentially constant for $\lambda > 4$ (this would cover practically all structural steels in high-temperature service) independent of change of damage mechanism or variations in values of n and m, it was suggested (Phaniraj et al. 2003) that t_{MGD} rather than t_r should be used in conservative structural design.

Cocks and Ashby (1982) pointed out a conceptual limitation of Eq. 13.22: $\dot{\omega} > 0$ for $\omega = 0$, i.e. non-existing damage grows at a finite rate. This problem can be circumvented by separately considering

(i) a nucleation stage of duration t_n during which damage nucleates, resulting in an initial damage ω_i and
(ii) a growth stage of duration t_g during which damage grows, following Eq. 13.23 from the initial value ω_i to the final value corresponding to rupture, ω_f.

The rupture life is then obtained as $t_r = t_n + t_g$. An empirical simplification leading to a conservative prediction of t_r is to set $t_n = 0$ and ω_i a small value, say 0.005. This would, however, lead to undue conservatism where the duration of primary creep is a significant fraction of rupture life. Empirical alternatives to rigorous modelling for t_n based on void nucleation kinetics have been considered in the literature.

Cocks and Ashby (1982) compared Eq. 13.22 with their mechanistic models for void growth kinetics under uniaxial loading controlled by (i) matrix creep, (ii) surface diffusion or (iii) boundary diffusion. In their model for a random array of voids growing by power law creep, $m = n$, $\dot{\omega}_0 = \dot{\varepsilon}_0$, and damage growth rate is given by the following expression

$$\dot{\omega} = \dot{\omega}_0 \cdot \left(\frac{\sigma}{\sigma_0}\right)^n \cdot \left\{\left(\frac{1}{1-\omega}\right)^n - (1-\omega)\right\}$$

With this model, $\dot{\omega} = 0$ for $\omega = 0$. For large ω approaching unity, the above form reduces essentially to Eq. 13.22 (with $m = n$), whereas for small ω values, it reduces to

$$\dot{\omega} \approx \dot{\omega}_0 \cdot \left(\frac{\sigma}{\sigma_0}\right)^n \cdot \left\{\left(\frac{1}{1-\omega}\right)^n - 1\right\}$$

with damage growth much smaller than that given by Eq. 13.22. Therefore, when the initial damage ω_{i} is small, t_r predicted by the mechanistic model is much larger than that by Eq. 13.22. The same is true for their mechanistic models for boundary or surface diffusion-controlled void growth. In these mechanistic models, with ω increasing, $\dot{\omega}$ first decreases and then increases, in contrast to the monotonically increasing $\dot{\omega}$ for Eq. 13.22.

Equation 13.22 is the foundations of several refinements of varying degrees of complexities, including more complicated functional forms for $\dot{\omega}$, extension to multiaxial loading where, for example, σ in the $\dot{\varepsilon}$ expression may be replaced by von Mises equivalent stress σ_{e} and $\dot{\omega}$ expression may be modified to reflect that transgranular cavity growth controlled by matrix deformation increases exponentially with triaxiality factor (the ratio of hydrostatic stress to σ_{e}), etc. The resultant formulations are often material and condition specific. These issues are not addressed in the present chapter.

13.6 Tertiary Creep Damage in $\varepsilon - \dot{\varepsilon}$ Framework

13.6.1 The Concept

A phenomenological approach to describe the tertiary creep relates $\dot{\varepsilon}/\dot{\varepsilon}_0$, a measure of creep damage, to strain (and not time), in the form

$$\dot{\varepsilon}/\dot{\varepsilon}_0 = f(\varepsilon) \qquad (13.24)$$

Relating creep damage evolution with strain is consistent with the Monkman–Grant relation (or its modifications), and the damage tolerance parameter concept of Ashby and Dyson. It should be justified for all realistic conditions for engineering alloys, where matrix creep controls the two important cavity growth mechanisms: matrix cavitation, and constrained grain boundary cavity growth (Ashby et al. 1984). Deviations may be expected when diffusive cavity growth mechanisms become important, i.e. high T and low σ. It may thus be anticipated that in general, of the various life summation rules indicated in Sect. 13.3.1, the strain fraction rule is likely to be more appropriate where cavity growth is controlled essentially by matrix deformation, whereas with increasing contributions from diffusive cavity growth, the time fraction should become increasingly important.

Endo and coauthors (1999, 1995) used the following tertiary creep equation for describing creep curves

$$\frac{\dot{\varepsilon}}{\dot{\varepsilon}_0} = \exp(\Omega\varepsilon) \tag{13.25}$$

In this equation, $\dot{\varepsilon}_0$ is the initial strain rate and Ω the strain rate accelerating factor. Ω depends on σ and T. With $\varepsilon = 0$ for $t = 0$, the integral of Eq. 13.25 is given by the following equation

$$\varepsilon = -\frac{1}{\Omega}\ln(1 - \Omega\dot{\varepsilon}_0 t)$$

If primary creep is ignored, then t_r can be estimated as

$$t_r = \frac{1}{\Omega\,\dot{\varepsilon}_0}$$

corresponding to rupture strain $\varepsilon_r = \infty$. Endo and coauthors used this method for predicting creep lives for ferritic and martensitic steels power plant applications

For 9–12% Cr-Mo-V steels, an alternative form for the tertiary creep has been proposed (Polcik et al 1998)

$$\ln\dot{\varepsilon} = \ln\dot{\varepsilon}_0 + C\cdot(\varepsilon/\varepsilon_R)^k \tag{13.26}$$

In this equation m, the parameters C, ε_R and k are constants at constant σ and T. It may be noted that neither of these two formulations include a threshold strain for creep damage. This is quite understandable, because the dominant creep damage mechanism in ferritic and martensitic steels is coarsening of precipitates which starts from the very beginning of the creep test ($\varepsilon = 0$); primary creep may, however, mask its manifestation.

13.6.2 Tertiary Creep Damage Evolution in a N-Bearing SS 316 Weldment

A specific application of the concept embodied in Eq. 13.24 is illustrated here, with the example of the creep damage kinetics in N-bearing SS 316-type weld metal, reported by Sasikala et al. (Sasikala et al. 2004) The initial microstructure of the weld metal contained δ ferrite ($\sim$7 Ferrite Number). The morphology of δ ferrite was vermicular or skeletal, and its distribution was fairly uniform in the austenite matrix. Constant nominal stress ($\sim$120–375 MPa) creep tests were carried out at three temperatures: 823, 873 or 923 K. The maximum stress rupture life from these tests exceeded 20,000 h. It was established that cavities nucleate at the transformation products of δ ferrite in the matrix. Creep damage was found to be primarily due to the growth and eventual coalescence of these matrix cavities. The extents of δ-ferrite transformation were measured in all creep tested specimens. It exceeded 90% for all tests at 923 K and was less than 50% for all tests at 823 K. For 873 K, the extent of transformation increased from $\sim$40 to $\sim$90% with t_r increasing to $\sim$500 h and exceeded 90% for $t_r > 1000$ h.

Figure 13.10 showed the modified Monkman–Grant plot for these test data. This plot justifies adopting $\dot{\varepsilon} - \varepsilon$ framework for describing damage evolution kinetics in the present instance. Figure 13.13a–c shows the evolution of $\dot{\varepsilon}$ with ε during the creep tests for the three test temperatures. Each plot is characterized by an initial regime where $\dot{\varepsilon}$ decreases with increasing ε. This can be identified with the primary creep regime (designated I). This transforms to the regime where $\dot{\varepsilon}$ increases with increasing ε. Clearly, this is the tertiary creep regime (designated III). The transition between the two regimes fail to show a true "steady state" (stage II) where $\dot{\varepsilon}$ remains constant over some measurable range of ε, Therefore, true steady state does not obtain in the test material and conditions. Therefore, steady-state creep rate $\dot{\varepsilon}_0$ (corresponding to zero tertiary creep damage) cannot be measured directly from the test data; only minimum creep rate $\dot{\varepsilon}_m$ can be measured. It is also clear from the plots that $\dot{\varepsilon}_m > \dot{\varepsilon}_0$.

As Fig. 13.13a for 823 K clearly shows, regime III actually comprises two stages. In the first stage (designated as IIIA), the $\ln \dot{\varepsilon}$—ε plot is curved downward, whereas in the next stage (designated as IIIB), the plot is curved upward. The regime IIIA occupies a large fraction of t_r, even though the strain accumulated in this regime is relatively small. This regime can be attributed to mostly growth of isolated cavities. The regime IIIB on the other hand occupies a small fraction of t_r, even though the strain accumulated in this regime can be a sizeable fraction of total strain to rupture. This regime may be attributed to linkage of cavities and also possibly specimen necking. As Fig. 13.13a–c shows for lower stresses (and also possibly depending on creep test data resolution), the second regime may not be discernible in some $\ln \dot{\varepsilon}$—ε plots.

For stage IIIA of creep data at constant nominal stress and temperature, Sasikala et al. (2004) chose the following functional description:

$$\ln \dot{\varepsilon} = \ln \dot{\varepsilon}_0 + k\,(\varepsilon - \varepsilon_0)^{v},\ \ \varepsilon \geq \varepsilon_0 \tag{13.27}$$

In this equation, $\varepsilon_0 > 0$ is threshold strain for tertiary creep damage, $\dot{\varepsilon}_0$ is the true steady-state creep rate which must be determined from data fitting, and k and υ are the other two fitting constant. Since a critical concentration of vacancies is required for cavity nucleation at the embrittling transformation products of δ-ferrite (σ and Laves phases), the necessity of a finite threshold strain in Eq. 13.27 is justified.

The method of least square fitting the test data to determine the adjustable constants is quite involved (Sasikala et al. 2004), and only salient points are highlighted here. It has been shown (Sasikala et al. 2004) that the stress dependence of $\dot{\varepsilon}_0$ term in Eq. 13.27 can be assumed to be given by $\dot{\varepsilon}_0 \propto \sigma^n$, where n is a constant stress exponent (Norton equation). Then Eq. 13.27 can be rewritten as:

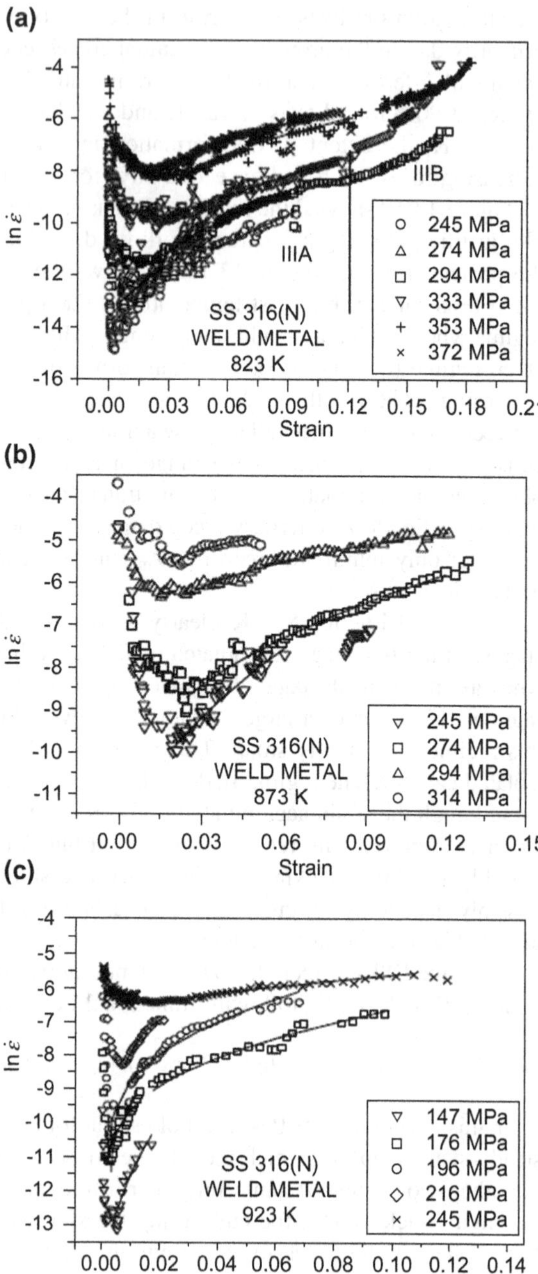

Fig. 13.13 **a** Plots of ln $\dot{\varepsilon}$–ε for SS 316(N) weld metal for 823 K for different stress levels. The solid lines show the fits to Eq. 13.27a. The vertical line gives an example of transition between regimes IIIA and IIIB. Sasikala et al. (2004), **b** plots of ln $\dot{\varepsilon}$–ε for SS 316(N) weld metal for 873 K for different stress levels. The solid lines show the fits to Eq. 13.27a. Sasikala et al. (2004), **c** plots of ln $\dot{\varepsilon}$–ε for SS 316(N) weld metal for 923 K for different stress levels. The solid lines show the fits to Eq. 13.27a. Sasikala et al. (2004)

$$\ln \dot{\varepsilon} = \ln \dot{\varepsilon}_{0,\mathrm{ref}} + n \ln\left(\frac{\sigma}{\sigma_{\mathrm{ref}}}\right) + k(\varepsilon - \varepsilon_0)^{\nu},\ \varepsilon \geq \varepsilon_0 \qquad (13.27\mathrm{a})$$

In this equation, σ_{ref} is a "reference" stress and $\dot{\varepsilon}_{0,\mathrm{ref}}$ the corresponding value of $\dot{\varepsilon}_0$. This correlation clearly obtained for 823 and 923 K, with $n \sim 8.5$, and apparent activation energy $Q_{\mathrm{a}} \sim 300\ \mathrm{kJ\ mol^{-1}}$. Q_{a} is thus similar to the activation energy for lattice diffusion $\left(\approx 280\ \mathrm{kJ\ mol^{-1}}\right)$ in this kind of steel, but significantly lower than those reported (Sasikala 2001) from Arrhenius fits for $\dot{\varepsilon}_{\mathrm{m}}$ or t_r in this material. Also, for this material, reported (Sasikala 2001) values for the Norton stress exponents calculated for $\dot{\varepsilon}_{\mathrm{m}}$ varied with stress and temperature, but were consistently and significantly higher than 8.5. These results justify the use of $\dot{\varepsilon}_0$ (determined from fitting the data) rather than $\dot{\varepsilon}_{\mathrm{m}}$ determined from the $\varepsilon - t$ plots in Eq. 13.27 and thus validate Eq. 13.27a. For 873 K, this power law dependence clearly obtained separately for the low stress (245 and 274 MPa) and the high stress (294 and 314 MPa) data, but not across these two groups. Based upon post-creep fracture microscopic evidence, the authors argued that the difference in behaviour for the intermediate temperature arises because in creep damage accumulation, in situ transformation of δ ferrite would play virtually no role at 923 K, a small role at 823 K, and crucial role at 873 K depending upon the stress level.

In Eqs. 13.27, and 13.27a, it may be anticipated that k, ε_0 and υ would vary with T, but for a given T should be relatively insensitive to σ. Therefore, least square fitting of test data to Eq. 13.27a to determine the optimized values for the fitting parameters was carried separately for (i) all the test data for 823 K using $\sigma_{\mathrm{ref}} =$ 245 MPa and (ii) all the test data for 923 K using $\sigma_{\mathrm{ref}} = 147$ MPa. Both these analyses used $n = 8.5$. The results showed that with increase in test temperature from 823 to 923 K, ε_0 decreased drastically from ~ 0.0031 to ~ 0.0014, υ decreased from ~ 0.42 to ~ 0.29. and k decreased from ~ 14.5 to ~ 13.3. That is, with increasing test temperature and the consequent increase in δ-ferrite transformation level, cavity nucleation becomes much easier and isolated cavity growth faster, as expected. The same analysis was carried out for the intermediate temperature of 873 K taking data for only the two lowest stress levels (245 and 274 MPa), with $n = 8.5$ as for the other two temperatures, and $\sigma_{\mathrm{ref}} = 245$ MPa. However, the optimized values of the parameters k, ε_0 and υ were not consistent with the results for the other two temperatures in terms of their variations with T, possibly because of significant levels of δ-ferrite transformation levels during the creep tests.

Sasikala et al. (2004) also showed how the above analysis can prove useful in solving actual design problems. They considered determining the design stress for rupture life $t_r = 10^5$ h at 823 K. For this purpose, they first determined a revised Monkman–Grant product. Since stage IIIA constitute the major fraction of t_r, they determine a revised Monkman–Grant product $\dot{\varepsilon}_0 \cdot t_r$ by integrating Eq. 13.27a for constant k, υ for various ε_{r} values:

$$\dot{\varepsilon}_0 \cdot t_r \approx \int_{\varepsilon_0}^{\varepsilon_r} \exp(-k(\varepsilon - \varepsilon_0)^{\nu}) \cdot d\varepsilon = \Gamma(\varepsilon_0, \varepsilon_r, \kappa, \nu) \tag{13.28}$$

Figure 13.14 shows their results for 823 and 923 K in the form of plots of the computed values for the integral Γ for the reference conditions against ε_r values. As the figure shows, with increasing ε_r, $\dot{\varepsilon}_0 \cdot t_r$ initially rises, but then tends to saturate beyond a small ε_r value of ~0.05. The saturation values of $\dot{\varepsilon}_0 \cdot t_r$ were 0.00481 for 823 K and 0.00183 for 923 K. The authors attributed this variation (by a factor ~2.6) to large difference in extents of δ ferrite transformation, and thus evolution of embrittling phases at these two temperatures. Now, for typical design situations, ε_r is expected to be above 0.05 (at 823 K, for $t_r \sim 2 \times 10^4$ h, the ε_r was about 0.15), and also any stress dependence of the other arguments of the integral Γ should be unimportant. Therefore, the saturation values for Γ can be used in design.

With this $\dot{\varepsilon}_0 \cdot t_r = 0.00481$ for 823 K, for the design life $t_r = 10^5$ h, $\dot{\varepsilon}_0$ is obtained as 4.81×10^{-8} h^{-1}. To determine the corresponding stress, the model of steady-state creep modulated by back stress σ_b from precipitates (see Sect. 12.6.2; Fig. 12.12) naturally comes into consideration, albeit using $\dot{\varepsilon}_0$ rather than $\dot{\varepsilon}_m$, with better justification (see Sect. 13.6.1, and also discussion following Eq. 13.27a):

$$\dot{\varepsilon}_0 = A \cdot (\sigma - \sigma_b)^4 \cdot \exp\left(-\frac{Q_{SD}}{RT}\right).$$

Here A is a constant. Now the constants of fits to Eq. 13.27a are used to compute $Z_0 \equiv [\dot{\varepsilon}_0 \cdot \exp(Q_{SD}/RT)]^{1/4}$ for the data for 823 and 923 K, which are plotted against σ in Fig. 13.15; Q_{SD} is taken as 280 kJ.mol^{-1} (the results are not sensitive to say ~10% variation in Q_{SD} value). Figure 13.15 shows the results. The error bars in this diagram correspond to those for $\dot{\varepsilon}_0$ determined during data fitting to

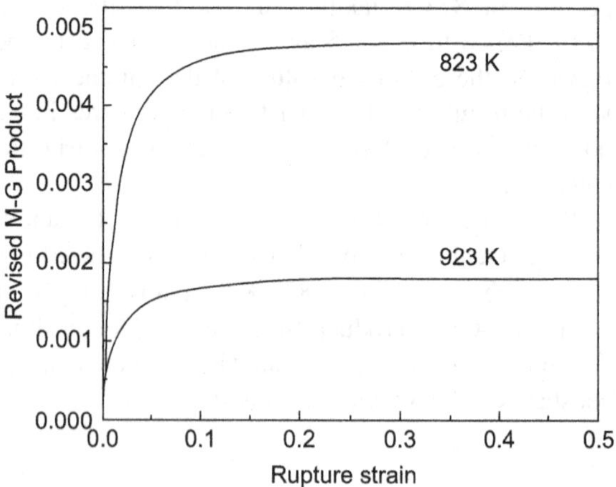

Fig. 13.14 ε_r dependence of the revised Monkman–Grant product $\Gamma \approx \dot{\varepsilon}_0 \cdot t_r$ for the reference conditions. Sasikala et al. (2004)

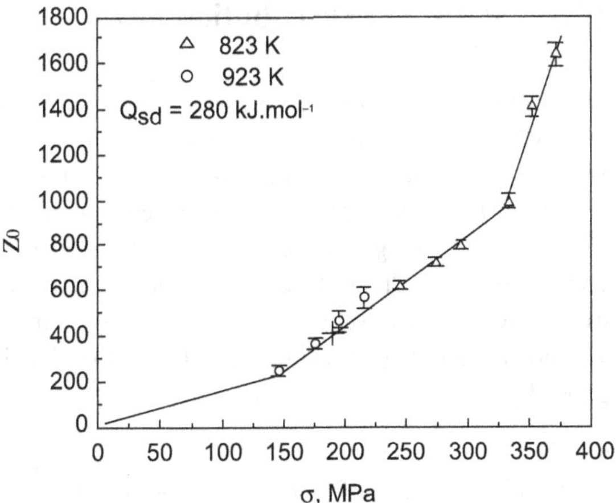

Fig. 13.15 Variation of Z_0 with applied stress for 823 and 923 K. The point marked by "+" corresponds to the "design" point. Sasikala et al. (2004)

Eq. 13.27a. The disposition of the data is consistent with a constant back stress for the highest stress levels, (its magnitude equals the σ-offset for the straight line for the highest stress levels), but which gradually decreases as the creep exposure increases. The datum point for the lowest stress level has been (arbitrarily) joined with the origin. This is because theoretically a very small back stress ~1 MPa is expected, which can be ignored for the stress range in this plot.

For a relatively simple precipitation-strengthened system, a bi-linear behaviour described by these two lines would have been expected. The design condition for $\dot{\varepsilon}_0 = 4.81 \times 10^{-8}\,\mathrm{h}^{-1}$ is marked by + in the diagram, and the design stress can now be read off as ~190 MPa.

Several points are pertinent for the analysis scheme described above:

(i) The body of stress rupture data available for this weld metal was grossly inadequate for adopting the conventional stress rupture parametric approach (Sect. 13.3) for determining the design stress at any temperature.
(ii) Test data at the higher temperature 923 K proved crucial for obtaining the design stress at the lower temperature 823 K by interpolation.
(iii) In ignoring primary creep, the analysis scheme underestimates t_r by the time required for threshold strain ε_0. This is acceptable for the material under study, but may not apply for other materials.
(iv) In strain limited design, maximum permissible strain is usually low and may be set as low as 1%. In such cases, ignoring primary creep may not be justified.

13.7 Statistical Distribution of t_r

It was mentioned in Sect. 13.3.2 that least square fitting of data to an equation in the form of Eq. 13.4 implies that t_r obeys a log-normal distribution, which must be verified after optimized fitting. Raj et al. 2004 have shown that log-normal distribution for t_r can be expected for quite realistic conditions. Assume that in Eq. 13.3, the activation energy is constant, and the statistical scatter in Ψ can be ignored compared to that due to temperature via the activation energy term. It is reasonable to assume that temperature obeys a Gaussian distribution: that is, if its peak value is T_p and the standard deviation S_T, then its probability distribution function pdf is given by

$$\mathrm{pdf}\,(T) = \frac{1}{\sqrt{2\pi}\cdot S_\mathrm{T}}\cdot \exp\left[-\frac{1}{2}\left(\frac{T-T_\mathrm{p}}{S_\mathrm{T}}\right)^2\right] \tag{13.29}$$

Then, if temperature variations are small, linear approximation gives

$$\frac{Q}{\mathrm{RT}} = \frac{Q}{R}\cdot\left[\frac{1}{T_\mathrm{p}} - \frac{1}{T_\mathrm{p}^2}\cdot(T-T_\mathrm{p})\right]$$

Substituting this in Eq. 13.3

$$\ln t_r = \ln\Psi + \frac{2Q}{\mathrm{RT_p}} - \frac{Q}{\mathrm{RT_p}}\cdot\frac{T}{T_\mathrm{p}}$$

$$\text{i.e.,}\quad T = \frac{\mathrm{RT}_p^2}{Q}\cdot\left[\ln\Psi + \frac{2Q}{\mathrm{RT}_p} - \ln t_r\right]$$

Now, from statistical principles:

$$\mathrm{pdf}(\ln t_r) = \mathrm{pdf}(T)\cdot\left|\frac{\mathrm{d}T}{\mathrm{d}\ \ln t_r}\right| = \mathrm{pdf}(T)\cdot\frac{\mathrm{RT_p^2}}{Q}$$

This yields

$$\mathrm{pdf}(\ln t_r) = \frac{1}{\sqrt{2\pi}\left(S_\mathrm{T}Q/\mathrm{RT_p^2}\right)}\cdot\exp\left[-\frac{1}{2}\left(\frac{\ln\Psi + Q/\mathrm{RT_p} - \ln t_r}{S_\mathrm{T}Q/\mathrm{RT_p^2}}\right)^2\right] \tag{13.30}$$

This shows that for a given σ, the distribution of $\ln t_r$ is Gaussian, with mean value of $\ln\Psi + Q/\mathrm{RT_p}$, and standard deviation $S_\mathrm{T}Q/\mathrm{RT_p^2}$. This justifies least square fitting of data to a parametric correlation of the form given by Eq. 13.3, and also the necessity for post-fit verification for the assumed distribution. The implication of this analysis, as pointed out by Raj et al, is: even small variations in temperature can lead to large scatter in lives, which is accentuated as activation energy increases.

13.8 Review Questions

1 How does the value of the Larson–Miller constant C change with the units (K vs. Rankine for temperature and MPa vs. ksi for stress)?

2 Write down the expressions for the apparent activation energy for creep rupture for the (i) Manson–Haferd and (ii) White–LeMay parameters.

3 Manson and Ensign (1979) suggested the following spline form for the isothermal variation of $\ln t_r$ with σ:

$$\ln t_r = a_1 - b_1 \cdot \ln \sigma - c_1 \cdot \sigma^{d_1} \quad \text{for } \sigma \leq \sigma_{\text{k}}, \quad \text{and}$$
$$\ln t_r = a_2 - b_2 \cdot \ln \sigma - c_2 \cdot \sigma^{-d_2} \quad \text{for } \sigma \geq \sigma_{\text{k}}.$$

The constants appearing in the above equations in general would depend upon temperature. This function has nine adjustable constants including the stress for the knot σ_{k}; however, three of these can be eliminated by imposing the conditions of continuity of value, slope and curvature at the knot, leaving six constants that need be optimized. Show that if $d_1, d_2 > 0$, then

$$(\partial \ln t_r/\partial \ln \sigma)_{\sigma \to 0} = -b_1 \quad \text{and} \quad (\partial \ln t_r/\partial \ln \sigma)_{\sigma \to \infty} = -b_2.$$

Discuss the constraints which should be imposed on the various constants in the above formulation, taking into account the mechanistic features summarized in Table 13.1.

4. Consider the form

$$G(\sigma) = A + B \log \sinh D\sigma, \quad A, B, D \text{ constants,}$$

introduced in the text. Discuss restraints (if any) that need to be imposed upon the constants for consistency with Table 13.1.

5. Examine if the Orr–Sherby–Dorn parameter can be considered as a special case of the general parameter (Eq. 13.14).

6. Write down the stress rupture parameter that would obtained by combining Garofalo's sinh equation for steady-state creep in the power law and power law breakdown regimes (see Sect. 12.6.2; Eq. 12.27) and modified Monkman–Grant relationship. Is it possible to comment on the ranges of stress and temperature where this parameter would be expected to be valid?

7. Verify that Eq. 13.20 corresponds to oblique translation method.
8. Explore a possible damage kinetics formulation for the strain fraction rule following the derivation of the life fraction rule. Is it possible to identify the damage mechanism(s) that may lead to the strain fraction rule?
9. Using Table 13.1 and References Cocks and Ashby (1982) and (1987) as basis, comment on the possible limitations of $\varepsilon - \dot{\varepsilon}$ formulations for damage kinetics modelling.

References

Ashby, M.F., Dyson, B.F.: In: Valluri, S.R., Owen, D.R.J. (eds.) Advances in Fracture Research, vol. 1, p. 3. Pergamon Press (1984)
Cadek, J.: Creep in Metallic Materials. Elsevier, Amsterdam (1988)
Cocks, A.C.F., Ashby, M.F.: Prog. Mater Sci. **27**, 189 (1982)
Dobes, F., Milichka, K.: Metal. Sci. B. **10**, 382 (1976)
Endo, T., Shi, J.: In: Bicego, V., Nitta, A., Viswanathan, R. (eds.) Materials Ageing and Component Life Extension, vol. 1, p. 429. CISE (1995)
Endo, T., Park, K.S., Masuyama, F.: In: Bicego, V., Nitta, A., Price, J.W.H., Viswanathan, R. (eds.) proceedings of an International Symposium on Case Histories on Integrity and Failures in Industry, Sept. 28–Oct. 1 1999, Milan, Italy, p. 831. Engineering Materials Advisory Services Ltd., EMAS, Warley, U.K (1999)
Gandhi, C.: Fracture mechanism maps for metals and alloys. In: Rishi, R (ed.) Flow and Fracture at Elevated Temperatures, p. 88. ASM, Metals Park, Ohio (1985)
Goldhoff, R.M., (ed.): Development of a Standard Methodology for the Correlation and Extrapolation of Elevated Temperature Creep and Rupture Data. Electric Power Research Institute, Research Project 638–1, Final Report (1979, April)
Grounes, M.: ASME J. Basic Eng. (1969)
Kachanov, L.M.: Introduction to Continuum Damage Mechanics. Martinus Nijhoff, Dordricht, Netherlands (1990)
Kassner, M.E., Hayes, T.A.: Int. J. Plast **19**, 1715 (2003)
Manson, S.S., Ensign, C.R.: J. Eng. Mater. Technol. **101**, 317 (1979)
Monkman, F.C., Grant, N.J.: Proc. ASTM **56**, 593 (1956)
National Research Inst fro Metals (NRIM): NRIM Creep data sheet 47: Data Sheets on the Elevated Temperature Properties of Iron-Based 21Cr-32Ni-Ti-Al Alloy for Corrosion Resisting and Heat Resisting Superalloy Bar (NCF 800H-B) (1999)
Phaniraj, C., Choudhary, B.K., Rao, K.B.S., Raj, B.: Scripta. Mater. **48**, 1313 (2003)
Polcik, P., Straub, S., Henes, D., Blum, W.: In: Strang, A., Cawley, J., Greenwood, G.W. (eds.) Microstructural Stability of Creep Resistant Alloys for High Temperature Plant Applications, p. 405. Institute of Materials, London (1998)
Rabotnov, Y.N.: Creep Problems in Structural Members. North Holland, Amstredam (1969)
Raj, R., Kong, J.S., Frangopol, D.M., Raj, L.E.: Metall. Mater. Trans. A **35A**, 1471 (2004)
Ray, S.K., Mathew, M.D., Rodriguez, P.: Res. Mech. **25**, 41 (1988)
Ray, S.K., Sasikala, G., Rodriguez, P.: Trans. Ind. Inst. Metals **48**, 453 (1995)
Riedel, Hermann: Fracture at High Temperatures. Springer-Verlag, Berlin (1987)
Roberts, B.W., Ellis, F.V., Bynum, J.E.: J. Eng. Mater Technol. **101**, 331 (1979)
Robinson, E.L.: Trans. Am. Inst. Min. Eng. **74**, 777 (1952)

Sasikala, G.: Creep deformation and fracture behaviour of type 316L(N) stainless steel and its weld metal. Ph.D. thesis, University of Madras (2001)
Sasikala, G., Ray, S.K., Rodriguez, P.: Mater. Sci. Eng. **A260**, 284 (1999)
Sasikala, G., Ray, S.K., Mannan, S.L.: Acta. Mater. **52**, 5677 (2004)
White, W.E., LeMay, I.: J. Eng. Mater. Technol. **100**, 319 (1978)
Woodford, D.A.: J. Eng. Mater Technol. **101**, 311 (1979)

Index

[Chapter and Section/Subsection/Subsubsection Numbers are Indicated.]

H. S. Ray and S. Ray, *Kinetics of Metallurgical Processes*,
Indian Institute of Metals Series, https://doi.org/10.1007/978-981-13-0686-0

Printed in the United States
By Bookmasters

Printed in the United States
By Bookmasters